THIRD EDITION

PRACTICAL NONPARAMETRIC STATISTICS

W. J. CONOVER

Texas Tech University

JOHN WILEY & SONS, INC.

New York • Chichester • Weinheim • Brisbane • Singapore • Toronto

ACQUISITIONS EDITOR   Brad Wiley II

ASSISTANT EDITOR   Mary O'Sullivan

FREELANCE PRODUCTION MANAGER   Jeanine Furino

DESIGNER   Karin Kincheloe

ILLUSTRATION EDITOR   Jaime Perea

FREELANCE PRODUCTION SERVICES   Elm Street Publishing Services, Inc.

This book was set in Palatino by Bi-Comp Inc.

This book is printed on acid-free paper. ∞

The paper in this book was manufactured by a mill whose forest management programs include sustained yield harvesting of its timberlands. Sustained yield harvesting principles ensure that the number of trees cut each year does not exceed the amount of new growth.

To order books or for customer service please, call 1(800)-CALL-WILEY (225-5945).

Library of Congress Cataloging in Publication Data:
Conover, W. J.
    Practical nonparametric statistics / W. J. Conover. — 3rd ed.
        p.   cm.
    Includes bibliographical references and index.
    ISBN 0-471-16068-7 (alk. paper)
    1. Nonparametric statistics.   I. Title.
QA278.8.C65   1999
519.5'3—dc21   98-8521
                                                              CIP

10 9 8 7

SHIPPING LABEL

FROM:
USED BOOKSELLER
59 MARKET STREET
NEWARK, NJ 07102

Amanda Stent
380 MAIN ST APT 18

CHATHAM   07928-2113
NJ
UNITED STATES

PACKING SLIP:

AMAZON MARKETPLACE ITEM: PRACTICAL NONPARAMETRIC STATISTICS
TRANSACTION ID: 103-1599410-6224244
PURCHASED 2009-06-21 20:01:39 PST
SHIPPED BY: USEDBOOKSELLER@YAHOO.COM

PREFACE

When I decided to write a book on nonparametric statistics over thirty years ago, I wanted a book that could be used as a textbook in a one-semester course on nonparametric (or distribution-free) statistics. But I also wanted a book that could be used as a quick reference to the most useful nonparametric methods, so research workers could easily find what they were looking for. The result was the first edition of this book, published by John Wiley & Sons in 1971. The editor at that time was Joe Frank, my former roommate in college.

Neither Joe nor I could have foreseen the tremendous success of our project. None of the six other publishing houses I sent my manuscript to was interested in this book. But Joe Frank, and John Wiley & Sons, gave me a chance for which I am eternally grateful.

After two editions and dozens of printings, this book has gained a reputation for being a solid textbook that can be relied upon to be sound in theory and clear in exposition. At the same time it is used by researchers throughout the world as a quick reference book where they can find clear, concise instructions on how and when to use the most popular nonparametric procedures.

Now it is time for a third edition. The second edition was published almost twenty years ago, and some updating is required. I have included in the third edition some procedures that have withstood the test of time and are now used by many practitioners, such as the *Fisher Exact Test* for two-by-two contingency tables, the *Mantel-Haenszel Test* for combining several contingency tables, the *Kaplan-Meier estimate of the survival curve*, the *Jonckheere-Terpstra Test* and the *Page Test* for ordered alternatives, and a discussion of the bootstrap method. This third edition adds many new exercises and problems, some new worked-out examples, and many updated references to related material. It omits some procedures that are no longer considered practical, primarily because they can be used only on independent samples of equal size, and most data sets do not have that luxury. Also some proofs formerly found in Chapters 1 and 2 have been reduced in size at the request of many readers.

Considerable thought was given to integrating one or more computer packages into this third edition. On the one hand, users of this book would find computer instructions very useful, but on the other hand, computer packages change more rapidly than this book does, and computer tips become outdated quickly. A compromise position was decided upon. Some helpful computer tips are included where they are expected to remain relevant for several years. However, comprehensive information on individual computer packages will have to remain in the domain of periodical reviews found in journals and magazines.

Several computer software packages with extensive nonparametric programs are mentioned in this book. Many more packages are available than are listed here. Interested readers may want to browse the Internet to obtain more information on some software packages designed for nonparametric statistical analyses. In many cases demonstration packages can be downloaded from the website. At the risk of quickly becoming out of date, here are several statistics packages, popular at the time of this writing, and the websites where more information can be obtained.

Minitab—http://www.minitab.com

PASS or *NCSS*—http://www.ncss.com

Resampling Stats—http://www.statistics.com

SAS—http://www.sas.com/rnd

SPSS—http://www.spss.com

STATA—http://www.stata.com

STATISTICA—http://www.statsoft.com

StatMost—http://www.dataxiom.com

StatXact—http://www.cytel.com

SYSTAT—http://www.spss.com/software/science/systat

S-Plus—http://www.mathsoft.com/splus

The *STATA* website contains links to these and other software providers.

As with the previous editions, the only absolute prerequisite to reading and understanding this book is a good working knowledge of college algebra, and a modicum of mathematical ability. The purpose of Chapters 1 and 2 is to bring such a student up to the level of knowledge required to understand the theory and methods in the rest of the book. In practice, however, most readers of this book have taken one or more previous statistics courses, and they find they can omit some parts of the first two chapters because it is simply review material for them.

This book has been used successfully as a textbook both at the graduate and undergraduate levels. At the undergraduate level most instructors find the Problems and Theory sections too challenging for their students, and they simply omit this material from their course with no difficulty. I have taught this course at the graduate level countless times, and only once did I omit Chapters 1 and 2, with disastrous results I might add! Other instructors have told me they omitted Chapters 1 and 2 without a problem, but their students have sometimes told me they had to go back and read those chapters on their own before they could grasp the later material.

When I teach this course I give an exam over Chapters 1 and 2, a second exam over Chapters 3 and 4, and third exam over Chapter 5. The final exam covers the entire book, including Chapter 6. I have successfully used the Review Problems as the basis for take-home exams in some cases. I don't use the computer when I teach this course, but about half the instructors I talk to do use the computer to varying degrees.

Many individuals deserve my thanks for their part in creating this book. I can't list them all. The ones who immediately come to mind are my wife, Susan, for her support and encouragement, the many editors of Wiley whom I have worked with through the years, especially Joe Frank, Andy Ford, and Brad Wiley II, and the countless individuals whose comments have shaped this and previous editions, including Barney Bissinger, Ben Duran, Ron Iman, Mark Johnson, Hossien Mansouri, and countless reviewers, students, instructors, and practitioners. My sincere thanks go to all of these, and to God for guiding my life, often against my will, through the maze of paths that led me here, where I am very happy to be.

W. J. Conover

CONTENTS

INTRODUCTION

One of the dictionary definitions of the word "science" is given as "truth ascertained by observation, experiment, and induction." A vast amount of time, money, and energy is being spent by society today in the pursuit of science. This pursuit is quite often frustrating because, as any scientist knows, the processes of observation, experiment, and induction do not always lay bare the "truth." One experiment, with one set of observations, may lead two scientists to two different conclusions.

For example, a scientist places a rat into a pen with two doors, both closed. One door is painted red and the other blue. The rat is then subjected to 20 minutes of music of the type popular with today's teenagers. After this experience, both doors are opened and the rat runs out of the pen. The scientist notes which color door the rat chose. This experiment is repeated 10 times, each time using a different rat.

At the end of the composite experiment, the experimenter notes that the rats chose the red door 7 out of 10 times and concludes the "truth" as being that the treatment used causes rats to prefer the red door to the blue door. However, a colleague overhears this conclusion and jokingly tests the scientist: "If I tossed a coin 10 times getting 7 heads and, before each toss, I whistled 'Yankee Doodle,' would you conclude that my whistling caused the coin to prefer heads?" Seeing the analogy between a rat choosing one of two doors and a coin landing on one of its two sides, the scientist realizes the error and decides that the outcome of the experiment could easily have been the result of chance.

Later the scientist conducts a second experiment. He injects a certain drug into the bloodstream of each of 10 rats. Five minutes later he examines the rats and finds that 7 are dead, and the other 3 are apparently healthy. However, since only 7 are dead, he recalls the previous experiment and concludes that such a result could easily have occurred by chance and therefore there is no proof that the drug injections are dangerous.

His colleague again interrupts, saying, "With your first experiment each rat had a 50-50 chance of choosing the red door, without the music, and therefore we can compare that experiment to tossing a coin. In this experiment, the chances of a rat dying within five minutes are quite slim indeed, if the drug has no effect. Since your experiment resulted in 7 of these rare events out of only 10 possibilities, it seems safe to conclude that the drug injections caused the deaths."

And so goes research. It soon becomes apparent to most scientists that the ideal way of expressing results of experiments, such as the preceding, is to be able to say something like, "Without the treatment I administered, experimental results as extreme as the ones I obtained would occur only about 3 times in 1000. Therefore I conclude that my treatment has a definite effect." In this way every scientist who reads of this experiment knows just how much subjectivity, or opinion, entered into the stated conclusion.

The purpose of that field of science known as "statistics" is to provide the means for measuring the amount of subjectivity that goes into the scientists' conclusions and thus to separate "science" from "opinion." This is accomplished by setting up a theoretical "model" for the experiment, such as the model called "tossing a coin," which was set up for the first experiment discussed. Laws of probability are applied to this model in order to determine what the "chances" (probabilities) are for the various possible outcomes of the experiment under the assumption that chance alone, and not music or drug injections, determines the outcome of the experiment. Then the experimenter has an objective basis for deciding whether the results were a result of the treatments that were applied, or whether the same results could have easily occurred by chance alone with no treatment.

Although it is sometimes difficult to describe an appropriate theoretical model for the experiment, the real difficulty often comes after the model has been defined, in the form of finding the probabilities associated with the model. Many reasonable models have been invented for which no probability solutions have ever been found. Thus statisticians have often changed the model slightly in order to be able to solve for the desired probabilities in the hope that the change in the model was slight enough so that the changed model was still fairly realistic. Then they were able to obtain exact solutions for these "approximate problems." This body of statistics is sometimes called "parametric statistics," and embodies such well-known tests as the "t test," the "F test," and others.

In the late 1930s a different approach to the problem of finding probabilities began to gather some momentum. This approach involved making few, if any, changes in the model and using simple and unsophisticated methods to find the desired probabilities, or at least a good approximation to those probabilities. Thus approximate solutions to exact problems were found, as opposed to the exact solution to approximate problems furnished by parametric statistics. This new package of statistical procedures became known as "nonparametric statistics."

Nonparametric methods have become essential tools in the workshop of the applied scientist who needs to do statistical analyses. When the price for making a wrong decision is high, applied scientists are very concerned that the statistical methods they are using are not based on assumptions that appear to be invalid, or are impossible to verify.

Besides the advantage of using a simpler model, nonparametric statistical methods often involve less computational work and therefore sometimes are easier and quicker to apply than other statistical methods. A third advantage of nonparametric statistical techniques is that much of the theory behind the nonparametric methods may be developed rigorously, using no mathematics beyond high school algebra. A scientist who understands the theory behind the statistical method is less apt to use that method in a situation where such usage would be incorrect and is better able to develop his or her own statistical methods if the model is one that has not yet been considered by statisticians.

The parts of nonparametric statistics that require the use of more advanced mathematics will be presented without deriving them but, whenever convenient, there will be a reference to a source where the proof may be found.

The fourth, and most important, reason for preferring nonparametric methods is that they are often more powerful than the parametric methods if the assumptions behind the parametric model are not true. Power is the probability that the experimenter will be able to prove what he or she is trying to prove in the experiment. Nonparametric methods of analysis often make more efficient use of the data than parametric methods, when the parametric methods are inappropriately applied.

This brings us to the subject of why this book contains none of the many nonparametric tests based on *runs*. A *run* is a sequence of similar observations. For example, if daily measurements on the Dow-Jones Industrial Average are recorded, a *run up* consists of consecutive measurements of the DJIA where each day's DJIA is higher than the previous day's DJIA. Similarly, a *run down* can be defined. The *total number of runs* counts both types of runs, and can be used as a measure of independence of measurements from one day to the next.

Runs tests can be used to test independence of observations, to test whether two samples came from the same population (by arranging the two samples from smallest to largest and counting runs of observations from the same sample), and in other situations. Tests based on the number of runs are easy to perform (it is usually easy to count the number of runs) and the probability distribution of the number of runs is often known exactly. However, runs tests have very little power, and can be replaced in every case by another nonparametric test with much greater power. Therefore tests based on the number of runs are not practical, and for that reason are not included in this book.

This formulation of parametric statistics versus nonparametric statistics is merely an attempt to give a rough idea concerning the subject of this book. A more precise distinction between the two branches of statistics will be given in Chapter 2, where the philosophy of scientific experimentation is discussed in greater detail. In order to present examples and illustrations in Chapter 2, a preliminary knowledge of some elementary aspects of probability is needed. This is the concern of Chapter 1.

From Chapter 3 onward there is a heavy reliance on the concepts introduced in Chapters 1 and 2. These later chapters present various nonparametric procedures, organized according to the type of model that is being analyzed rather than to the type of experiment being conducted. When most people think of nonparametric statistics, they think only of the methods based on ranks presented in Chapters 5 and 6. This is because methods for analyzing qualitative (nominal) data, given in Chapters 3 and 4, are necessarily nonparametric. Some of the problems at the ends of the sections show little-known but useful connections between parametric statistics and nonparametric statistical methods for analyzing both qualitative and quantitative data.

For convenience to the experimenter who wants to examine the body of techniques that may be used in analysis, a cross-referencing table is presented inside the front cover, listing the techniques given in the book according to the type of problem they are intended to solve.

A word about the numbering of examples, equations, and figures is appropriate at this time. Example 4.2.3 refers to Example 3 in Section 4.2. When referring to an example within the same section, only the last number is used. For instance, within Section 4.2, Example 4.2.3 is referred to simply as Example 3. The same is true for equations, figures, and problems. No such economy is used with regard to section numbers, so that Section 4.2 is always called Section 4.2, even within Chapter 4.

For those who wish to obtain more information about nonparametric procedures, many references are included at the end of each appropriate section. The bibliography by Savage (1962) is quite useful for obtaining earlier references on each topic.

CHAPTER 1

PROBABILITY THEORY

PRELIMINARY REMARKS

One of the attractive qualities of nonparametric statistical methods is that it is not necessary to be an expert in probability theory to understand the theory behind the methods. With a few easily learned, elementary concepts, the basic fundamentals underlying most nonparametric statistical methods become quite accessible. This chapter introduces those basic concepts. All that is required is patience, confidence, and a good understanding of high school algebra.

This book is arranged so that readers can go directly to the statistical procedure they want to use and follow the step-by-step instructions from beginning to end. However, they will not necessarily understand what they are doing or why they are doing it. Such lack of understanding often leads to mishandled data and misstated conclusions. By spending a little time in Chapters 1 and 2, readers should understand thoroughly the nonparametric procedure being used and may even be able to adapt it slightly so it will apply better to the particular set of data being analyzed.

The recommended procedure for studying each section is to read the text, pencil through the examples, and then work the exercises and problems at the end of the section. This will prepare readers for the next section and will develop the patience and confidence first mentioned.

1.1 COUNTING

The process of computing probabilities often depends on being able to count, in the usual sense of counting, "1, 2, 3," and so on. The usual way of counting becomes quite tedious in some complicated situations, so some sophisticated methods of counting are developed in this section to handle those complicated situations.

When we speak of tossing a coin, we will consider only two possible outcomes: either a head (H) appears, or a tail (T) appears. If a coin is tossed once there are two possible outcomes: H or T. If a coin is tossed twice there are $2^2 = 4$ possible outcomes: HH, HT, TH, TT, where HT means a head occurs on the first toss and a tail on the second. Each time we consider one additional toss of the coin, the number of possible outcomes is doubled, since the last toss may result in either of two outcomes. Thus, if a coin is tossed n times there are 2^n possible outcomes.

Experiment

Generalizing this discussion somewhat, we may refer to the tossing of a coin as one example of an experiment. Whether the coin is tossed once, twice or, in general, n times, the procedure may be considered to be an experiment. Since tossing a coin three times may be considered to be an experiment and is a composite of three separate experiments where the coin is tossed only once each time, we may refer to the shorter experiments as trials and the collection of trials as "the experiment." In general, an experiment is the process of following a well-defined set of rules, where the result of following those rules is not known prior to the experiment.

Model

Few scientists seriously consider coin tossing as an experiment worthy of merit by itself. The value of coin tossing is that it serves as a prototype for many different models in many different situations. If an unbiased coin is being considered, one in which each face is equally likely to result, the experiment is not unlike experiments involving rats that have two choices of doors, consumers choosing between two products, educators determining which of two teaching methods is more effective, market analysts deciding whether the market tends to be higher or lower on Mondays, and many other situations.

If we allow the coin to be biased, where one face is more likely to turn up than the other, a much broader class of experiments is included under the same model. Examples include experiments where a drug is injected into the bloodstream of rats to see if the drug is lethal, a new cure is tested on sick patients, a consumer is given several choices of a product and asked to choose one where only one of the products is manufactured by Company X, and other situations. In each case there are two outcomes of interest, such as "life" versus "death," "cure" versus "no cure," "our brand" versus "other brands," and the two outcomes might not be equally likely to occur.

Throughout this chapter and the next, models involving tossing coins, rolling dice, drawing chips from a jar, placing balls into boxes, and so on, will be discussed as if they were experiments worthy of merit, while actually the value of these models lies mainly in the fact that they serve as useful and simple prototypes of many more complicated models arising from experimentation in diverse areas such as electron physics, psychology, sociology, education, biology, economics, chemistry, etc. An excellent study of the diversity of such models is given by Feller (1968). Some justification for the study of these models will be presented in this chapter, but for the most part the justification will be deferred until later chapters where the various nonparametric procedures are introduced.

Event

Thus we may refer to coin tossing as an experiment and each individual toss of the coin as a trial. The possible outcomes of one trial, several trials, or the entire experiment will be called *events*. The coin tossing experiment just described consists of n trials, where each trial may result in either the event H or the event T. A combination of events may itself be an event. Therefore it is permissible to consider each of the 2^n possible outcomes of the experiment as an event. Examples of other events would include the event "at least one head," the event "a tail on the fourth toss," and the event "at least twice as many heads as tails."

Further generalization leads to the following rule:

Rule 1 If an experiment consists of n trials where each trial may result in one of k possible outcomes, there are k^n possible outcomes of the entire experiment.

EXAMPLE 1

Suppose an experiment is composed of seven trials, where each trial consists of throwing a ball into one of three boxes. The first throw may result in one of three different outcomes. There are $3^2 = 9$ outcomes associated with the first two trials combined. This reasoning extends to the seven throws comprising the experiment, resulting in $3^7 = 2187$ different outcomes of the experiment. ■

Now consider a box containing n plastic chips numbered 1 to n. One chip is selected from the box and placed on the table so the number is showing. This chip could be any of the n chips that were in the box, so we say there are n ways of selecting the first chip. A second chip is then selected from the chips remaining in the box and placed next to the first chip, so that its number is showing also. Because there were $n - 1$ chips remaining in the box, the second chip could be selected in any one of $n - 1$ different ways. Because each of the n ways of drawing the first chip has associated with it $n - 1$ ways of drawing a second chip, there are all together $n(n - 1)$ ways of drawing first one chip and then a second chip.

A third chip can be drawn in $n - 2$ different ways and placed on the table next to the second chip. Now there are $n(n - 1)(n - 2)$ ways of drawing three chips in sequence. If the process is continued until the last chip is drawn (there is only one way of drawing the last chip, since only one chip is left in the box) we can see that there are

$$n(n - 1)(n - 2) \cdots (3)(2)(1) = n! \tag{1}$$

(read "n factorial") ways of drawing n numbered chips out of a box, or $n!$ ways of arranging any n distinguishable objects into a row. Note that for convenience we will define $0! = 1$, in accordance with conventional usage.

   Rule 2   There are $n!$ ways of arranging n distinguishable objects into a row.

EXAMPLE 2

Consider the number of ways of arranging the letters A, B, and C in a row. The first letter can be any of the three letters, the second letter can be chosen two different ways once the first letter is selected, and the remaining letter becomes the final letter selected, for a total of $(3)(2)(1) = 6$ different arrangements. The six possible arrangements are ABC, ACB, BAC, BCA, CAB, and CBA. ∎

EXAMPLE 3

Suppose that in a horse race there are eight horses. If you correctly predict which horse will win the race and which horse will come in second and wager to that effect, you are said to "win the exacta." Suppose you want to be sure to win the exacta. That means you need to purchase $(8)(7) = 56$ betting tickets, one for each of the 56 possible ways the first and second places might result. The complete race results, for all eight positions at the finish line, could occur in any one of $8! = 40,320$ different ways. ∎

   If the n objects are distinguishable one from another, then each of the $n!$ arrangements is unique. But suppose two of the objects are identical. Then for each arrangement of the n objects, there is a second arrangement that is indistinguishable from the first—the arrangement in which $n - 2$ of the objects are in the same position as in the first arrangement, but the two identical objects are interchanged. Each of the $n!$ arrangements may be paired in this manner with another identical arrangement. The number of different arrangements is thus $n!/2$, or $n!/2!$.
   Suppose three of the objects are identical, and $n - 3$ are distinguishable from each other. If we divide the $n!$ arrangements into groups of identical arrangements, we find there are $3!$ arrangements in each group. This is because the three identical objects may be placed $3!$ different but indistinguishable ways into their

three positions, using Rule 2. Then the number of different arrangements, equal to the number of groups of identical arrangements, is $n!/3!$. If exactly n_1 objects are identical, the $n!$ arrangements may be divided into groups of identical arrangements, each group being of size $n_1!$. If there are n_1 identical objects of type 1, and n_2 identical objects of a different type 2, then for each arrangement of the objects of type 1 there are $n_2!$ identical arrangements of type 2. So there are, in all, $n_1! \, n_2!$ arrangements in each group of identical arrangements. Therefore, the number of groups is $n!/(n_1! \, n_2!)$. This leads to another counting rule:

Rule 3   If a group of n objects is composed of n_1 identical objects of type 1, n_2 identical objects of type 2, . . . , n_r identical objects of type r, the number of distinguishable arrangements into a row, denoted by $\begin{bmatrix} n \\ n_i \end{bmatrix}$, is

$$\begin{bmatrix} n \\ n_i \end{bmatrix} = \frac{n!}{n_1! n_2! \ldots n_r!} \tag{2}$$

In particular, if a group of n objects is composed of k identical objects of one kind and the remaining $(n - k)$ objects are identical objects of a second kind, the number of distinguishable arrangements of the n objects into a row, denoted by $\binom{n}{k}$, is given by

$$\binom{n}{k} = \frac{n!}{k!(n - k)!} \tag{3}$$

Throughout this book we will use the convention that $\binom{n}{k} = 0$ if k is greater than n. This is natural because there is no way of considering arrangements of n objects where more than n of them are alike.

To justify the use of Rule 3, let us divide the $n!$ arrangements into groups of identical arrangements. Each group then has $n_1! \, n_2! \ldots n_r!$ arrangements in it. Because no arrangement may appear in two different groups, the number of groups is $n!/(n_1! \, n_2! \ldots n_r!)$. We may assume without loss of generality that $n_1 + n_2 + \cdots + n_r = n$, because some of the n_i may equal 1, representing objects that are similar only to themselves. Because $1! = 1$, and because dividing Equation 2 by 1 does not affect the numerical value, Rule 3 remains unaffected by the preceding assumption. It is also apparent now that Rule 2 is a special case of Rule 3, where all of the $n_i = 1$.

EXAMPLE 4

In Example 2 we listed the six ways of arranging the letters A, B, and C in a row. Suppose now that the letters A and B are identical. We will denote them by the letter X. Then the arrangements ABC and BAC become indistinguishable, denoted by XXC. Also ACB and BCA become XCX. The original $3! = 6$ arrangements are reduced to

$$\binom{3}{2} = \frac{3!}{2!1!} = \frac{(3)(2)(1)}{(2)(1)(1)} = 3$$

distinguishable arrangements, that is, XXC, XCX, and CXX.   ∎

EXAMPLE 5

In a coin tossing experiment where a coin is tossed five times, the result is two heads and three tails. The number of different sequences of two heads and three tails equals the number of distinguishable arrangements of two objects of one kind and three objects of another, which is $\binom{5}{2} = 10$. Note that the 10 arrangements are as follows, where H = "head" and T = "tail."

HHTTT	*THHTT*	*TTHHT*
HTHTT	*THTHT*	*TTHTH*
HTTHT	*THTTH*	*TTTHH*
HTTTH		

∎

How many different groups of k objects may be formed from n objects? We can use Rule 3 to answer this question. Suppose that the n objects are lined up in a row, and we have k identical tags to place on k of the n objects. It is easy to see that the number of ways of placing the k tags on k of the n objects which, in turn, equals the number of distinguishable arrangements of k tagged positions and $n - k$ untagged positions, is $\binom{n}{k}$, as given by Rule 3. In this situation $\binom{n}{k}$ is often read "the number of ways of taking n things k at a time."

EXAMPLE 6

Consider again the three letters A, B, and C. The number of ways of selecting two of these letters is $\binom{3}{2} = 3$, that is, AB, AC, and BC. To see how this relates to the previous discussion, we will "tag" two of the three letters with an asterisk (*) denoting the tag.

A^*B^*C gives AB
A^*BC^* gives AC
and AB^*C^* gives BC ∎

Note the similarity between this example and Example 4.

Binomial Coefficient

For still another way of using the term $\binom{n}{k}$, consider the expression $(x + y)^n = (x + y)(x + y) \cdots (x + y)$. The term x^n occurs only when the x term from the first factor is multiplied by the x term from the second factor, and so on for all n factors. The term $x^{n-1}y$ results from multiplying the x term from $n - 1$ of the factors times the y term from one factor. Since the y term may be selected from any one of the n factors, expansion of $(x + y)^n$ results in n terms involving $x^{n-1}y$. Similarly, for each value of k, the term $x^k y^{n-k}$ results from the selection of k xs from k of the factors, and then $n - k$ ys from the remaining $n - k$ factors. There are $\binom{n}{k}$ ways of selecting k factors for the xs, with the remaining factors contributing ys. Therefore the term $x^k y^{n-k}$ appears $\binom{n}{k}$ times in the expansion of $(x + y)^n$. Since all terms in the expansion are added together, we may write

$$(x + y)^n = x^n + \binom{n}{n-1} x^{n-1} y^1 + \binom{n}{n-2} x^{n-2} y^2 + \cdots$$
$$+ \binom{n}{2} x^2 y^{n-2} + \binom{n}{1} x^1 y^{n-1} + y^n \tag{4}$$

Recall that $0! = 1$, so $\binom{n}{0} = 1$ and $\binom{n}{n} = 1$. If we use the notation

$$\sum_{i=a}^{b} C_i = C_a + C_{a+1} + C_{a+2} + \cdots + C_{b-1} + C_b$$

which is read as "the sum of the terms C_i as i goes from a to b," we may write

$$(x + y)^n = \sum_{i=0}^{n} \binom{n}{i} x^i y^{n-i} \tag{5}$$

which is known as the "binomial expansion" and is found in most high school algebra textbooks. This illustrates why the term "binomial coefficient" is often used to describe the symbol $\binom{n}{k}$. Similarly, it may be noted that the coefficient of $x_1^{n_1} x_2^{n_2} \cdots x_r^{n_r}$ in the expansion of $(x_1 + x_2 + \cdots + x_r)^n$ is given by the "multinomial coefficient" $\begin{bmatrix} n \\ n_i \end{bmatrix}$.

EXAMPLE 7

We will use the binomial expansion to evaluate $(2 + 3)^4$. Of course, we know the answer is $5^4 = 625$. From the binomial expansion in Equation 5 we have

$$(2 + 3)^4 = \sum_{i=0}^{4} \binom{4}{i} 2^i 3^{4-i}$$

$$= \binom{4}{0} 2^0 3^4 + \binom{4}{1} 2^1 3^3 + \binom{4}{2} 2^2 3^2 + \binom{4}{3} 2^3 3^1 + \binom{4}{4} 2^4 3^0$$

$$= (1)(1)(81) + (4)(2)(27) + (6)(4)(9) + (4)(8)(3) + (1)(16)(1)$$

$$= 81 + 216 + 216 + 96 + 16 = 625$$ ■

EXERCISES

1. How many four-digit numbers (from 0000 to 9999) may be formed using the 10 digits 0 through 9, where each digit may be repeated any number of times?

2. How many different four-letter arrangements are there, using the 26 letters in the alphabet, where each letter may be used repeatedly?

3. How many ways are there of arranging the letters L, O, V, E into four-letter "words," where each letter is used only once?

4. How many ways are there of seating five people in a row?

5. In how many ways may a committee of three be chosen from a club with 12 members?

6. What is the coefficient of $x^3 y^3$ in the expansion of $(x + y)^6$?

7. What is the coefficient of $x^2 y^4 z$ in the expansion of $(x + y + z)^7$?

8. What is the coefficient of $x^2 y^5$ in the expansion of $(w + x + y + z)^7$? (*Hint.* $x^2 y^5 = w^0 x^2 y^5 z^0$)

9. Evaluate $\sum_{i=1}^{3} \binom{4}{i}$.

10. Evaluate $\sum_{i=0}^{3} \binom{4}{i} \left(\frac{1}{2}\right)^2$.

11. Evaluate $\sum_{i=3}^{5} \binom{6}{i} \left(\frac{1}{3}\right)^i \left(\frac{2}{3}\right)^{6-i}$.

12. Evaluate $\sum_{i=1}^{4} 5$.

PROBLEMS

1. How many ways are there of choosing n_1 objects of the first kind, n_2 objects of the second kind, and so forth, to n, objects of the kth kind, where there are altogether N_1 objects of the first kind, N_2 objects of the second kind, and so on? How many ways are there if n_i is greater than N_i for some i?

2. Show that $\sum_{i=0}^{n} \binom{n}{i} p^i(1-p)^{n-i} = 1$.

1.2 PROBABILITY

Now we are ready to apply the three counting rules in Section 1.1 to find some interesting and useful probabilities. First we must introduce some standard terminology used in statistics. Correct understanding of the terms defined in this section and elsewhere will make communication of statistical concepts much easier.

Sample Space

We will define the important terms *sample space* and *points in the sample space* in connection with an experiment.

Definition 1 The *sample space* is the collection of all possible different outcomes of an experiment.

Definition 2 A *point in the sample space* is a possible outcome of an experiment.

Each experiment has its own sample space, which consists essentially of a list of the different outcomes of the experiment that are possible. It is tacitly assumed that the sample space is subdivided as finely as reasonably possible with each subdivision being called a point. Also, it is tacitly assumed that each possible outcome is represented by one and only one point.

EXAMPLE 1

If an experiment consists of tossing a coin twice, the sample space consists of the four points *HH, HT, TH,* and *TT.* ∎

EXAMPLE 2

An examination consisting of 10 "true or false" questions is administered to one student as an experiment. There are $2^{10} = 1024$ points in the sample space, where each point consists of the sequence of possible answers to the 10 successive questions, such as "*TTFTFFTTTT.*" ∎

Event

It is now possible to define *event*, in terms of the points in the sample space.

Definition 3   An *event* is any set of points in the sample space.

In Example 1 we may speak of the event "two heads," which consists of the single point *HH*; the event "one head," which consists of the two points *HT* and *TH*; the event "at least one tail," which consists of the points *TH, HT,* and *TT*; as well as the event "four heads," which has no points in it. A set with no points in it is sometimes called *the empty set*. The event consisting of all points in the sample space is sometimes called *the sure event* because it is certain to occur every time the experiment is performed.

Two different events may have points common to both. The events "at least one tail" and "at least one head" have two points *TH* and *HT* in common. If two events have no points in common, they are called *mutually exclusive events* because the occurrence of one event automatically excludes the possibility of the other event occurring at the same time.

If all of the points in one event are also contained in a second event, we say that the first event *is contained in* the second event, or that the second event *contains* the first event. The event "at least one head" contains the event "two heads." Each event therefore contains itself.

Probability

To each point in the sample space there corresponds a number called *the probability of the point* or *the probability of the outcome*. These probabilities may be any number from 0 to 1. If we can conceive of a long series of repetitions of the experiment under fairly uniform conditions, the relative frequency of the occurrence of the point or event in mind represents an approximation to the probability of that point or event.

Definition 4   If *A* is an event associated with an experiment, and if n_A represents the number of times *A* occurs in *n* independent repetitions of the experiment, the *probability of the event A*, denoted by $P(A)$, is given by

$$P(A) = \lim_{n \to \infty} \frac{n_A}{n} \tag{1}$$

which is read "the limit of the ratio of the number of times *A* occurs to the number of times the experiment is repeated, as the number of repetitions approaches infinity."

A formal definition of *independent* is deferred until later. For now we may think of experiments as independent if the outcome of any one experiment does not influence the outcome of the other experiments.

The definition of the probability of an event includes the definition of the probability of an outcome as a special case, since an event may be considered as consisting of a single outcome. It is apparent from the definition that the probability of an event equals the sum of the probabilities of all outcomes comprising the event, since the number of times the event occurs equals the sum of the numbers of times the mutually exclusive outcomes comprising the event occur.

Probability Function

In practice, the set of probabilities associated with a particular sample space is seldom known, but the probabilities are assigned according to the experimenter's preconceived notions. That is, the experimenter formulates a model as an idealized version of the experiment. Then the sample space of the model experiment is examined, and the probabilities are assigned to the various points of the sample space in some manner which the experimenter feels can be justified.

EXAMPLE 3

In an experiment consisting of the single toss of an unbiased coin, it is reasonable to assume that the outcome H will occur about half the time. Thus we may assign the probability $1/2$ to the outcome H, and the same to the outcome T. We write this as $P(H) = 1/2$, $P(T) = 1/2$. ∎

EXAMPLE 4

In an experiment consisting of three tosses of an unbiased coin, it is reasonable to assume that each of the $2^3 = 8$ outcomes $HHH, HHT, HTH, HTT, THH, THT, TTH, TTT$ is equally likely. Thus the probability of each outcome is $1/8$. Also $P(3$ tails$) = 1/8$, $P($at least one head$) = 7/8$, and $P($more heads than tails$) = P($at least 2 heads$) = 4/8 = 1/2$. ∎

We have been working with *probability functions* in the previous two examples.

> **Definition 5** A *probability function* is a function that assigns probabilities to the various events in the sample space.

In Example 3 the probability function was given by $P(H) = 1/2$, $P(T) = 1/2$. It is necessary that the probability function assign a probability to each point in the sample space. Then the probabilities of all events in the sample space are automatically specified by the probabilities of the sample points contained in the events.

Several properties of probability functions become apparent. Let S be a sample space and let A be any event in S. Then, if P is a probability function, $P(S) = 1$, because

$$P(S) = \lim_{n \to \infty} \frac{n}{n} = 1$$

$P(A) \geq 0$, because $n_A \geq 0$, and therefore

$$\lim_{n \to \infty} \frac{n_A}{n} \geq 0$$

and $P(\overline{A}) = 1 - P(A)$, where $\overline{A}$ is the event "the event A does not occur," because $n_{\overline{A}} = n - n_A$, and

$$\lim_{n \to \infty} \frac{n_{\overline{A}}}{n} = \lim_{n \to \infty} \frac{n - n_A}{n} = \lim_{n \to \infty} \left(1 - \frac{n_A}{n}\right) = 1 - \lim_{n \to \infty} \frac{n_A}{n} = 1 - P(A)$$

Conditional Probability

We mentioned earlier that while the various outcomes of an experiment are mutually exclusive, the various events associated with an experiment do not necessarily have that property. In our experiment of tossing a coin three times the events "three heads" and "at least two heads" may both occur at the same time. Now consider the probability of the event "three heads" if we are given that the event "at least two heads" has occurred. If at least two heads have occurred, we know that several points in the sample space, that is, TTT, TTH, THT, and HTT, may be eliminated. The possible outcomes of the experiment are reduced to four equally likely points. Therefore the probability of each point is now $1/4$, and hence the probability of the event "three heads," or HHH, is $1/4$, if we are given the fact that at least two heads have occurred. The additional information that we are given has the effect of eliminating some of the outcomes from consideration and thus artifically reducing the sample space.

In another experiment, consider rolling a die. Let S be the sample space, let A be the event "a 4, 5, or 6 occurs," and let B be the event "an even number (2, 4, or 6) occurs," as depicted by Figure 1. The probability that the event A has

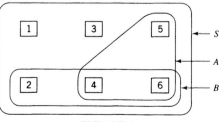

FIGURE 1

occurred, given that B has occurred, is written $P(A|B)$ and is usually read "the probability of A given B." Since we know that B has occurred, we may not only eliminate the points that are in neither A nor B, that is, the points 1 and 3, but we may even eliminate the point in A that is not in B, or 5. Thus all points not in B are eliminated, and the sample space is just the set of points in B. The only points in B that can result in the event A are the points in both A and B, or 4 and 6. These points represent the event "both A and B occur."

> **Definition 6**   If A and B are two events in a sample space S, the event "both A and B occur," representing those points in the sample space that are in both A and B at the same time, is called *the joint event A and B* and is represented by AB. The probability of the joint event is represented by $P(AB)$.

Then the probability of "A given B" is given by the probability of "AB" relative to the reduced sample space "B". Or, symbolically,

$$P(A|B) = \frac{P(AB)}{P(B)} \tag{2}$$

Looking at it another way, suppose that the preceding experiment is repeated n times. However, only those outcomes resulting in the event B are recorded and the outcomes not resulting in B are ignored. Let n_B represent the number of times B occurs, and let n_{AB} represent the number of times A occurs when B occurs. Then

$$P(A|B) = \lim_{n \to \infty} \frac{n_{AB}}{n_B} = \lim_{n \to \infty} \frac{n_{AB}/n}{n_B/n} = \frac{P(AB)}{P(B)} \tag{3}$$

We have intuitively justified the following definition.

> **Definition 7**   The conditional probability of A given B is the probability that A occurred given that B occurred and is given by

$$P(A|B) = \frac{P(AB)}{P(B)} \tag{4}$$

where $P(B) > 0$. If $P(B) = 0$, $P(A|B)$ is not defined.

EXAMPLE 5

Consider the rolling of a fair die, so that each of the six possible outcomes has probability $1/6$ of occurring. As before, let A be the event "a 4, 5, or 6 occurs" and let B be the event "an even number occurs." Then $P(AB) = P(4 \text{ or } 6) =$

$2/6 = 1/3$. Also, $P(B) = 3/6 = 1/2$. Then the conditional probability $P(A|B)$ is given by

$$P(A|B) = \frac{P(AB)}{P(B)} = \frac{1/3}{1/2} = \frac{2}{3}$$ ∎

We should note the reasonableness of this answer, since we are given that an even number (i.e., event B) has occurred and the outcome of the experiment is either a 2, 4, or 6. We now want to know the probability that a number greater than 3 (i.e., event A) also occurred. Since two of the three even numbers are greater than 3, our answer is 2/3.

Independent Events

The idea of conditional probability leads quite naturally into the idea of independent events. If the probability of A, given that B occurs, is the same as the probability of A without any information on the occurrence or nonoccurrence of B, we feel that the occurrence or nonoccurrence of A is independent of whether or not B occurs. That is, we feel that A is independent of B if $P(A|B) = P(A)$. In fact, this may be used as the definition of independence, but it is not clear from this form of the definition whether B then is also independent of A. So it is better to substitute $P(A)$ for $P(A|B)$ in Equation 4, the definition of conditional probability. This leads to the following.

Definition 8   Two events A and B are *independent* if

$$P(AB) = P(A)P(B) \tag{5}$$

Because of the symmetry of Equation 5 it is readily apparent that if A is independent of B, B is also independent of A, and so it is better to say "A and B are independent," where it is meant that they are independent of each other.

EXAMPLE 6

In an experiment consisting of two tosses of a balanced coin, the four points in the sample space are assumed to have equal probabilities. Let A be the event "a head occurs on the first toss" and let B be the event "a head occurs on the second toss." Then A has the points HH and HT, B has the points HH and TH, and AB has the point HH. Also $P(A) = 2/4$, $P(B) = 2/4$, and $P(AB) = 1/4$. Therefore Equation 5 is satisfied and A and B are independent. ∎

The following example illustrates that the independence of two events is not always intuitively obvious and should always be determined directly from the definition and Equation 5.

EXAMPLE 7

Consider again the experiment consisting of one roll of a balanced die, where the sample space consists of the six equally likely points 1, 2, 3, 4, 5, and 6. Let A be the event "an even number occurs," including the points 2, 4, and 6. Let B be the event "at least a 4 occurs," including the points 4, 5, and 6. Finally, let C be the event "at least a 5 occurs," including the points 5 and 6. Then A and B are not independent, because $P(A)P(B) = (1/2)(1/2) = 1/4$ while $P(AB) = 1/3$. However, A and C are independent, because $P(A)P(C) = (1/2)(1/3) = 1/6$, the same as $P(AC)$.   ■

Sometimes the notions of *independent events* and *mutually exclusive events* are confused with each other, because both notions give the impression that "the two events do not have anything to do with each other." The property of independence depends not only on the two events being considered but also on the particular probability function defined on the sample space. It is possible for $P(AB)$ and $P(A)P(B)$ to be equal to each other with one set of probabilities and to be unequal with another set of probabilities. But "mutually exclusive" simply means the two events have no points in common, and no matter what probability function is defined on the sample space, AB is empty, so $P(AB) = 0$. If A and B are mutually exclusive, they will be independent only if either $P(A)$ or $P(B)$ equals zero, since Equation 5 must be satisfied.

Independent Experiments

Now we will define the concept of *independent experiments*.

> **Definition 9** Two experiments are independent if for every event A associated with one experiment and every event B associated with the second experiment,
>
> $$P(AB) = P(A)P(B)$$

It is equivalent to define two experiments as independent if every event associated with one experiment is independent of every event associated with the other experiment.

It is quite tedious to examine every pair of events associated with two experiments to see if they satisfy Definition 9. However, it is sufficient to verify the definition only for those events consisting of a single point each. Then the definition is automatically verified for all other events.

In practice, the model is usually set up assuming independence, and the assumption of independence is then used to find $P(AB)$ using $P(A)$ and $P(B)$ in Definition 9. This is the main value of the definition of independence. Thus it is reasonable to extend the definition of independent experiments to cover the eventuality of more than two experiments being involved.

> **Definition 10**   *n experiments are mutually independent* if for every set of *n* events, formed by considering one event from each of the *n* experiments, the following equation is true.

$$P(A_1 A_2 \cdots A_n) = P(A_1)P(A_2) \cdots P(A_n) \tag{6}$$

> where A_i represents an outcome of the *i*th experiment, for $i = 1, 2, \ldots, n$.

The word "mutually" may be omitted in the preceding definition if no confusion results.

EXAMPLE 8

Let an experiment consist of one toss of a biased coin, where the event H has probability p and the event T has probability $q = 1 - p$. Consider three independent repetitions of the experiment, where a subscript will be used to denote the experiment with which the outcome is associated. Thus $H_1 T_2 H_3$ means the first experiment resulted in H, the second in T, and the third in H. Because of our assumption of independence,

$$P(H_1 T_2 H_3) = P(H_1)P(T_2)P(H_3) = pqp$$

If we consider the event "exactly two heads" associated with the combined experiments, this may occur $\binom{3}{2} = 3$ ways, and hence

$$P(\text{exactly two heads}) = 3p^2 q$$ ∎

Obviously the preceding might just as well have been described as one experiment with three independent trials. The extension to considering an experiment consisting of *n* independent tosses may be made. The probability of obtaining "exactly *k* heads" then equals the term $p^k q^{n-k}$ times the number of times that term can appear. Therefore, in *n* independent tosses of a coin,

$$P(\text{exactly } k \text{ heads}) = \binom{n}{k} p^k q^{n-k} \tag{7}$$

where $p = P(H)$ on any one toss.

The three preceding definitions, as is true for all definitions, work both ways. Example 6 presents a situation where the satisfaction of Equation 5 implies that two events are independent. Example 8 presents a situation where the assumption of independence implies that Equation 6 is satisfied. It follows then that if Equation 6 is not satisfied, the experiments are not independent, and conversely if the experiments are not independent, Equation 6 is not satisfied for at least one set of events $A_1 A_2 \cdots A_n$.

EXERCISES

1. In an experiment consisting of three tosses of a coin, where the order of the tosses (first to third) is important, list the points in the sample space.

2. Referring to Exercise 1, give:
   (a) Two mutually exclusive events.
   (b) Two events that are not mutually exclusive.

3. If the probability of rain is 0.15, what is the probability of no rain?

4. If the probability of arriving at an intersection while the traffic light is green is 0.35 and the probability of the light being yellow is 0.10, what is the probability that the light is red?

5. If a football team has an equal probability of winning or losing each game (assuming no ties occur and independent outcomes from game to game), what is the probability of the team losing at least seven games in an eight-game season?

6. If a football team has probability 0.4 of winning each game, independent of the other games, what is the probability that they win only 1 game or less in a 10-game season?

7. If the probability of getting a torn dollar bill is 0.05, what is the probability that two out of the three dollar bills obtained are torn (assume independence)?

8. If 60% of all stolen cars are recovered and 2% of all cars are stolen each year, what is the probability of a person having a car stolen and never recovered?

9. The probability of a customer buying a certain brand of cleaner is 0.15. Forty percent of the customers that buy that cleaner also buy a dispenser. What is the probability that a customer buys both?

10. In three independent tosses of an unbiased coin, what is the probability of obtaining at least one tail?

11. In three independent tosses of an unbiased coin, what is the probability of obtaining three heads if we know that at least one head has occurred?

12. In four independent tosses of an unbiased coin, what is the probability of getting at least three heads, if we are given that at least two heads have occurred?

13. In three independent tosses of an unbiased coin, what is the probability of obtaining three heads if we know that the first toss resulted in a head? (*Note.* Exercises 11 and 13 have different answers.)

14. If 75% of all student accounts at a bank are closed within one year, and 20% of the bank's accounts are student accounts, what is the probability of an account at the bank being a student account and remaining open more than one year?

15. In four independent tosses of an unbiased coin, what is the probability of obtaining at least three heads if we know that at least one tail has occurred?

16. A lottery game consists of drawing three digits from 0 to 9 at random, independent of each other, such as 212 or 935.

(a) What is the probability of successfully predicting, in one attempt, the three numbers in the correct order?

(b) Is this probability the same for the prediction 555 as it is for the prediction 212 or the prediction 935?

17. A lottery game consists of drawing three digits from 0 to 9 at random, independent of each other, such as 212 or 935.

(a) What is the probability of successfully predicting, in one attempt, the three numbers in any order? That is, the prediction 215 is correct even if the numbers were drawn in the order 512 or 152.

(b) Is the probability the same for the prediction 555 as it is for the prediction 212 or the prediction 935?

PROBLEMS

1. Show that in a sample space with n points, there are exactly $2^n - 1$ events containing at least one point.

2. This problem shows that pairwise independence does not imply three-way independence. Consider drawing a number from a hat, where the numbers 1 to 9 are in the hat and each number is equally likely to be drawn. If the number is 1, 2, or 3, event A occurs. A 1, 4, or 5 drawn produces event B, and a 2, 4, or 6 produces event C.

(a) Show events A and B are independent. Show events A and C are independent. Show events B and C are independent.

(b) Are events A, B, and C mutually independent? Why not?

(c) Find the probability that none of the events A, B, or C occurs.

1.3   RANDOM VARIABLES

Random Variable

Outcomes associated with an experiment may be numerical in nature, such as the score on an examination, or nonnumerical, such as the choice "red door" by a rat escaping from a pen. In order to analyze the results of an experiment, it is often useful to assign numbers to the points in the sample space. Any rule for assigning such numbers is called a *random variable*.

Definition 1 A *random variable* is a function that assigns real numbers to the points in a sample space.

We will usually denote random variables by the capital letters W, X, Y, or Z, with or without subscripts. The real numbers attained by the random variables will be denoted by lowercase letters.

EXAMPLE 1

In an experiment where a consumer is given a choice of three products, soap, detergent, or Brand A, the sample space consists of the three points representing the three possible choices. Let the random variable X assign the number 1 to the choice "Brand A" and the number 0 to the other two possible outcomes. Then $P(X = 1)$ equals the probability that the consumer chooses Brand A. ∎

At times it is convenient to define more than one random variable for a single sample space, such as in the following example.

EXAMPLE 2

Six girls and eight boys are each asked whether they communicate more easily with their mother or their father. Let X be the number of girls who feel they communicate more easily with their mothers and let Y be the total number of children who feel they communicate more easily with their mothers. If $X = 3$, we know the event "3 girls feel they communicate more easily with their mothers" has occurred. If, at the same time, $Y = 7$, we know that the event "3 girls and $7 - 3 = 4$ boys feel they communicate more easily with their mothers" has occurred. ∎

If X is a random variable, "$X = x$" is a shortcut notation that we use to correspond to some event in the sample space, specifically the event consisting of the set of all points to which the random variable X has assigned the value "x."

EXAMPLE 3

In an experiment consisting of two tosses of a coin, let X be the number of heads. Then "$X = 1$" corresponds to the event containing only the points HT and TH. ∎

Thus "$X = x$" is sometimes referred to as "the event $X = x$," when the intended meaning is "the event consisting of all outcomes assigned the number x by the random variable X."

Because of this close correspondence between random variables and events, the definitions of *conditional probability* and *independence* apply equally well to random variables.

Definition 2   The *conditional probability of X given Y*, written $P(X = x | Y = y)$, is the probability that the random variable X assumes the value x, given that the random variable Y has assumed the value y.

The equation for determining conditional probabilities may be obtained from Definition 1.2.7 as

$$P(X = x | Y = y) = \frac{P(X = x, Y = y)}{P(Y = y)} \quad \text{if } P(Y = y) > 0 \tag{1}$$

EXAMPLE 4

Let X be the number of girls that communicate more easily with their mothers out of six girls, as in Example 2, and let Y be the total number of children who communicate more easily with their mothers. For convenience let $Z = Y - X$, so Z equals the number of boys, out of eight boys, who communicate more easily with their mothers. Assume that the answers given by the children are independent of each other and that each child has the same probability p (unknown) of saying he or she communicates more easily with his or her mother. We will find the conditional probability $P(X = 3 | Y = 7)$.

First, by the preceding assumptions, $X = 3$ and $Z = 4$ are independent events. Because the event $(X = 3, Y = 7)$ is the same as the event $(X = 3, Z = 4)$ we have the joint probability

$$P(X = 3, Y = 7) = P(X = 3, Z = 4)$$
$$= P(X = 3)P(Z = 4) \text{ by independence,}$$
$$= \binom{6}{3} p^3 (1 - p)^3 \binom{8}{4} p^4 (1 - p)^4 \tag{2}$$

because of Example 1.2.8. By the same example, we conclude that

$$P(Y = 7) = \binom{14}{7} p^7 (1 - p)^7 \tag{3}$$

so the conditional probability $P(X = 3 | Y = 7)$ is

$$P(X = 3 | Y = 7) = \frac{P(X = 3, Y = 7)}{P(Y = 7)} = \frac{\binom{6}{3}\binom{8}{4}}{\binom{14}{7}} = .408 \tag{4}$$

because all of the factors involving the unknown p cancel each other.   ∎

Probability Function

Just as the points in a sample space are mutually exclusive, the values that a random variable may assume are mutually exclusive. That is, for a single outcome of an experiment, the random variable defined for that experiment furnishes us with only one number. Thus the entire set of values that a random variable may assume has many of the same properties as a sample space. The individual values assumed by the random variable correspond to the points in a sample space, a set of values corresponds to an event, and the probability of the random variable assuming any value within a set of values equals the sum of the probabilities associated with values within the set. For example,

$$P(a < X < b) = \sum_{a<x<b} P(X = x)$$

where the summation extends over all values of x between, but not including, the numbers a and b, and

$$P(X = \text{even number}) = \sum_{x \text{ even}} P(X = x)$$

where the summation applies to all values of x that are even numbers. Because of this similarity between the set of possible values of X and a sample space, the description of the set of probabilities associated with the various values X may assume is often called the *probability function of the random variable X*, just as a sample space has a probability function. However, the probability function of a random variable is not an arbitrary assignment of probabilities, as is the probability function for a sample space, because once the probabilities are assigned to the points in a sample space and once a random variable X is defined on the sample space, the probabilities associated with the various values of X are known and the probability function of X is thus already determined.

> **Definition 3**   The *probability function of the random variable X*, usually denoted by $f(x)$, is the function that gives the probability of X taking the value x, for any real number x. In other words,

$$f(x) = P(X = x) \tag{5}$$

The probability function always equals 0 at values of x that X cannot assume.

Sometimes it is convenient to represent the probability function as a bar graph, with the values of the random variable as the abscissa (along the horizontal axis) and the probabilities as the ordinate (the height of the bar). For instance, if $P(X = 1)$ equals 0.3, $P(X = 2)$ equals 0.4, and $P(X = 4)$ equals 0.3, the bar graph of the probability function looks like this. The heights of the bars represent the various probabilities associated with the random variable X.

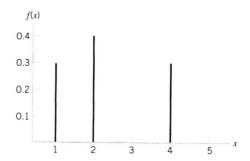

It is not always convenient to use $f(x)$ to denote the probability function of a random variable. Other expressions that may be used include $f_0(x)$, $f_1(x)$, $f_2(x)$, $g(x)$, $h(x)$, and so on. However, the meaning of the various expressions used will always be clear from the context.

Distribution Function

We have seen that the distribution of probabilities associated with a random variable may be described by a probability function. Another way of accomplishing the same thing is by means of a *distribution function,* which describes the accumulated probabilities.

> **Definition 4**   The *distribution function of a random variable X,* usually denoted by $F(x)$, is the function that gives the probability of X being less than or equal to any real number x. In other words,
>
> $$F(x) = P(X \leq x) = \sum_{t \leq x} f(t) \tag{6}$$

where the summation extends over all values of t that do not exceed x. Distribution functions are often called cumulative distribution functions (c.d.f. for short) to emphasize their property of presenting cumulative probabilities.

Distribution functions also may be represented graphically, with x as the abscissa and $F(x)$ as the ordinate. As an illustration, suppose, as before, that

$P(X = 1) = 0.3$, $P(X = 2) = 0.4$, and $P(X = 4) = 0.3$. Then the graph of $F(x)$ looks like this.

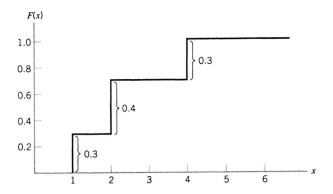

The graph actually consists only of the horizontal lines; the vertical lines are drawn in merely to give the graph a somewhat "connected" appearance, and to assist in the finding of *quantiles* as explained in the next section. The lengths of the vertical lines are the same as the lengths of the bars in the graph of the probability function.

Binomial Distribution

Some probability distributions are well known and consequently have been given names.

Definition 5 Let X be a random variable. The *binomial distribution* is the probability distribution represented by the probability function

$$f(x) = P(X = x) = \binom{n}{x} p^x q^{n-x} \qquad x = 0, 1, \ldots, n \qquad (7)$$

where n is a positive integer, $0 \le p \le 1$, and $q = 1 - p$. Note that we are using the usual convention that $0! = 1$.

The distribution function is then

$$F(x) = P(X \le x) = \sum_{i \le x} \binom{n}{i} p^i q^{n-i} \qquad (8)$$

where the summation extends over all possible values of i less than or equal to x. Table A3 (see appendix) gives the values of $F(x)$ for some selected values of the parameters n and p.

EXAMPLE 5

An experiment consists of n independent trials where each trial may result in one of two outcomes, "success" or "failure," with probabilities p and q, respectively, such as with the tossing of a coin. Let X equal the total number of "successes" in the n trials. Then, as was shown by Equation 1.2.7,

$$P(X = x) = \binom{n}{x} p^x q^{n-x}$$

for integer x from 0 to n. Thus X has the binomial distribution.   ■

Discrete Uniform Distribution

Another useful probability distribution is the *discrete uniform distribution*.

> **Definition 6**   Let X be a random variable. The *discrete uniform distribution* is the probability distribution represented by the probability function
>
> $$f(x) = \frac{1}{N} \qquad x = 1, 2, \ldots, N \tag{9}$$

Thus X may assume any integer value from 1 to N with equal probability, if X has the discrete uniform probability function.

EXAMPLE 6

A jar has N plastic chips, numbered 1 to N. An experiment consists of drawing one chip from the jar, where each chip is equally likely to be drawn. The sample space has N points, representing the N chips that may be drawn. Let X equal the number on the drawn chip. Then X has the discrete uniform distribution.   ■

Joint Distributions

When several random variables are defined on the same sample space or when several experiments, each with one or more random variables defined for them, are considered as a combined experiment, it becomes useful to consider joint distributions, described by *joint probability functions* and *joint distribution functions*.

> **Definition 7**   The *joint probability function* $f(x_1, x_2, \ldots, x_n)$ of the random variables $X_1, X_2, \ldots, X_n$, is the probability of the joint occurrence of $X_1 = x_1, X_2 = x_2, \ldots,$ and $X_n = x_n$. Stated differently,
>
> $$f(x_1, x_2, \ldots, x_n) = P(X_1 = x_1, X_2 = x_2, \ldots, X_n = x_n) \tag{10}$$

Definition 8 The *joint distribution function* $F(x_1, x_2, \ldots, x_n)$ of the random variable $X_1, X_2, \ldots, X_n$ is the probability of the joint occurrence of $X_1 \leq x_1$, $X_2 \leq x_2, \ldots,$ and $X_n \leq x_n$. Stated differently,

$$F(x_1, x_2, \ldots, x_n) = P(X_1 \leq x_1, X_2 \leq x_2, \ldots, X_n \leq x_n) \qquad (11)$$

EXAMPLE 7

Consider the random variables X and Y as defined in Example 2. Let $f(x, y)$ and $F(x, y)$ be the joint probability function and the joint distribution function, respectively. Then, from Example 4,

$$f(3, 7) = P(X = 3, Y = 7) = \binom{6}{3}\binom{8}{4} p^7 (1 - p)^7 \qquad (12)$$

and

$$F(3, 7) = P(X \leq 3, Y \leq 7) = \sum_{\substack{0 \leq x \leq 3 \\ x \leq y \leq 7}} f(x, y) \qquad (13)$$

where

$$f(x, y) = \binom{6}{x} p^x (1 - p)^{6-x} \binom{8}{y - x} p^{y-x} (1 - p)^{8-(y-x)}$$

and where the summation in Equation 13 extends over all values of x and y such that $x \leq 3$ and $y \leq 7$, with the usual restriction that x and $y - x$ be nonnegative integers. Note that Equations 12 and 13 cannot be evaluated without knowing the value of p. ∎

Definition 9 The *conditional probability function* of X given Y, $f(x|y)$, is

$$f(x|y) = P(X = x | Y = y) \qquad (14)$$

From Equation 1 we see that

$$f(x|y) = P(X = x | Y = y) = \frac{P(X = x, Y = y)}{P(Y = y)}$$

$$= \frac{f(x, y)}{f(y)} \qquad (15)$$

where $f(x, y)$ is the joint probability function of X and Y and $f(y)$ is the probability function of Y itself.

EXAMPLE 8

As a continuation of Example 7, let $f(x|y)$ denote the conditional probability function of X given $Y = y$. Then

$$f(3|7) = P(X = 3|Y = 7) = 0.408$$

from Equation 4. To find a formula for $f(x|y)$ in general (i.e., for any values of x and y we may choose), first let $f(x, y)$ denote the joint probability function of X and Y. This is given in Example 7 as

$$f(x, y) = \binom{6}{x} p^x (1 - p)^{6-x} \binom{8}{y - x} p^{y-x} (1 - p)^{8-(y-x)}$$

which originally was a general form for Equation 2. Also, let $f(y)$ be the probability function of Y. From Example 4 again we can generalize to get

$$f(y) = P(Y = y) = \binom{14}{y} p^y (1 - p)^{14-y}$$

By Definition 9 we can now write the conditional probability function of X given $Y = y$.

$$f(x|y) = \frac{f(x, y)}{f(y)} = \frac{\binom{6}{x} \binom{8}{y - x}}{\binom{14}{y}} \quad \begin{array}{l} 0 \le x \le 6 \\ 0 \le y - x \le 8 \end{array} \tag{16}$$

where all of the terms involving the unknown parameter p conveniently cancel out.

∎

Hypergeometric Distribution

In the previous examples we worked with a probability distribution known as the *hypergeometric distribution*. In its more general form we usually refer to having A objects of one kind and B objects of a second kind (the total numbers of girls and boys in the examples). Then the probability of selecting x of the A objects, given that altogether k of the $A + B$ objects total are selected, under the assumption that each object has the same chance of being selected, is given by the hypergeometric probability function.

Definition 10 Let X be a random variable. The *hypergeometric distribution* is the probability distribution represented by the probability function

$$f(x) = P(X = x) = \frac{\binom{A}{x} \binom{B}{k - x}}{\binom{A + B}{k}} \quad \begin{array}{l} 0 \le x \le A \\ 0 \le k - x \le B \end{array} \tag{17}$$

where A, B, and k are nonnegative integers and $k \le A + B$.

Mutually independent random variables may be defined in a manner similar to Definitions 1.2.9 and 1.2.10 of independent experiments.

Definition 11 Let $X_1, X_2, \ldots, X_n$ be random variables with the respective probability functions $f_1(x_1), f_2(x_2), \ldots, f_n(x_n)$ and with the joint probability function $f(x_1, x_2, \ldots, x_n)$. Then $X_1, X_2, \ldots, X_n$ are *mutually independent* if

$$f(x_1, x_2, \ldots, x_n) = f_1(x_1)f_2(x_2) \cdots f_n(x_n) \tag{18}$$

for all combinations of values of $x_1, x_2, \ldots, x_n$.

EXAMPLE 9

Consider the experiment described in Example 8. Then the probability function of X, the number of girls who feel they communicate more easily with their mothers, out of 6 girls, is given by

$$f_1(x) = P(X = x) = \binom{6}{x} p^x (1 - p)^{6-x} \tag{19}$$

and the probability function of Y, the total number of children who feel they communicate more easily with their mothers, out of 14 children, is given by

$$f_2(y) = P(Y = y) = \binom{14}{y} p^y (1 - p)^{14-y} \tag{20}$$

Since

$$f(x, y) = P(X = x, Y = y) = P(X = x | Y = y)P(Y = y)$$

the use of Equations 16 and 20 results in the joint probability function of X and Y being given by

$$f(x, y) = \frac{\binom{6}{x}\binom{8}{y-x}}{\binom{14}{y}} \binom{14}{y} p^y (1 - p)^{14-y}$$

$$= \binom{6}{x}\binom{8}{y-x} p^y (1 - p)^{14-y}$$

But, since

$$f_1(x)f_2(y) = \binom{6}{x}\binom{14}{y} p^{x+y}(1 - p)^{20-x-y}$$

we see that

$$f(x, y) \neq f_1(x)f_2(y)$$

and, therefore, X and Y are not independent. ∎

EXERCISES

1. If $f(x)$ is the binomial probability function with $n = 6$ and $p = 1/3$, find
   (a) $f(6)$.   (b) $f(0)$.
   (c) $f(2.5)$.   (d) $F(2.5)$.
   (e) $F(-3)$.   (f) $F(7)$.
   (g) Draw a bar graph of the probability function.
   (h) Draw a graph of the distribution function.

2. Suppose $f(x)$ is the discrete uniform probability function with N equal to 12. Find
   (a) $f(2)$.   (b) $f(12)$.
   (c) $f(0)$.   (d) $f(1.5)$.
   (e) $F(0)$.   (f) $F(3.1)$.
   (g) $F(1000)$.   (h) $F(-1000)$.
   (i) Draw a bar graph of the probability function.
   (j) Draw a graph of the distribution function.

3. Let X and Y be independent, binomially distributed random variables, with parameters $n = 3$, $p = 1/2$ for X, and $n = 4$, $p = 1/2$ for Y. Let $f(x, y)$ denote the joint probability function of X and Y. Find
   (a) $f(0, 0)$.   (b) $f(0, 1)$.
   (c) $f(1, 0)$.   (d) $f(3, 4)$.
   (e) $f(4, 4)$.   (f) $F(0, 0)$.
   (g) $f(1, 1)$.   (h) $F(3, 4)$.

4. Let $f(x)$ be the hypergeometric probability function, where $A = 3$ and $B = 4$. Find
   (a) $f(0)$ given $k = 0$.   (b) $f(1)$ given $k = 1$.
   (c) $f(2)$ given $k = 1$.   (d) $f(1)$ given $k = 5$.
   (e) $f(1)$ given $k = 6$.

5. A diner selects one sandwich at random out of six possible sandwich varieties.
   (a) What is the sample space?
   (b) What is the probability function on the sample space?
   (c) Define a random variable on the sample space such that the random variable has a discrete uniform distribution.

6. Seven boys and 10 girls take an examination, and each student has probability 0.2 of failing the examination.
   (a) What is the sample space for this experiment?
   (b) Given that three students failed the examination, what is the probability that all three are boys?
   (c) What is the name of the probability distribution you are using?
   (d) If the probability of each failure is 0.8 instead of 0.2, what is the answer to part b?

PROBLEMS

1. Which of the following functions are possible probability functions? Justify your answer.

   (a) $f(x) = 1/6$ for $x = 1, 2, 3, 4,$
          $= 0$ elsewhere

   (b) $f(x) = (1/4)^x$ for $x = 1, 2, 3, 4, \ldots,$
          $= 0$ elsewhere

   (c) $f(x) = (1 - p)p^x$ for $x = 0, 1, 2, \ldots,$
          $= 0$ elsewhere, where p is a constant between 0 and 1

2. Assume that every patient with a particular type of disease has probability 0.1 of being cured within a week, if the patient is given no treatment for the disease. Ten patients with that type of disease are given a new type of drug. After one week 9 out of the 10 patients are cured.

   (a) What is the probability of at least nine patients being cured if the drug is assumed to have no curative effects?

   (b) In your opinion, would you consider this drug to be beneficial?

   (c) What sample space did you use in this analysis?

   (d) What probability function did you define on the sample space?

   (e) What random variable did you define on the sample space?

   (f) What is the name of the probability distribution of your random variable?

1.4   SOME PROPERTIES OF RANDOM VARIABLES

We have already discussed some of the properties associated with random variables, such as their probability functions and their distribution functions. The distribution function describes all of the properties of a random variable that are of interest, because the distribution function reveals the possible values the random variable may assume and the probability associated with each value. At times, however, it is inconvenient or confusing to present the entire distribution function to describe a random variable, and some sort of a "summary description" of the random variable is needed. We will now introduce some other properties of random variables that may be used to present a brief, but incomplete, description of the distribution of the random variable.

Quantiles

The most common method used in this book for summarizing the distribution of a random variable is by giving some selected *quantiles* of the random variable. The term "quantile" is not as well known as the terms "median," "quartile," "decile," and "percentile," yet these latter terms are popular names given to particular quantiles. The median of a random variable, for example, is some number that the random variable will exceed with probability one-half or less and will be smaller than with probability one-half or less. This definition may be extended as follows.

Definition 1 The number x_p, for a given value of p between 0 and 1, is called the *pth quantile of the random variable X*, if $P(X < x_p) \leq p$ and $P(X > x_p) \leq 1 - p$.

If more than one number satisfies the definition of the *p*th quantile, we will avoid confusion by adopting the convention that x_p equals the average of the largest and the smallest numbers that satisfy Definition 1.

That is, X is less than x_p with probability p or less, and X exceeds x_p with probability $1 - p$ or less. The *median* is the 0.5 quantile, the third *decile* is the 0.3 quantile, the *upper and lower quartiles* are the 0.75 and 0.25 quantiles, respectively, and the sixty-third *percentile* is the 0.63 quantile.

Perhaps the easiest method of finding the *p*th quantile involves using the graph of the distribution function of the random variable. The *p*th quantile is the abscissa of the point on the graph that has the ordinate value of p, as illustrated in the following example.

EXAMPLE I

Let X be a random variable with the following probability distribution.

$$P(X = 0) = 1/4$$
$$P(X = 1) = 1/4$$
$$P(X = 2) = 1/3$$
$$P(X = 3) = 1/6$$

Then the distribution function of X may be represented by the graph in Figure 2. The 0.75 quantile $x_{0.75}$, called the upper quartile, may be found by drawing a

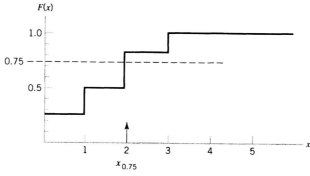

FIGURE 2

horizontal line through 0.75 on the vertical axis, as indicated by the dotted line in Figure 2. The value of x where the dotted line intersects the graph is the upper quartile, which equals 2 in this example. Therefore we may say that $x_{0.75} = 2$, which may be verified directly from the definition, since

$$P(X < 2) = 1/2$$

which is less than 0.75, and since

$$P(X > 2) = 1/6$$

which is less than $1 - 0.75$.

Similarly, the median is found by drawing a line through 0.5 on the vertical scale. The median is any value from 1 to 2 inclusive, and it is easy to see that any of these values satisfies the definition of the median. By our convention we select 1.5 as the median. ∎

Certain random variables called "test statistics" play an important role in most statistical procedures. These test statistics are useless unless their distribution functions are at least partially known. Most of the tables in the appendix give information concerning the distribution functions of various test statistics used in nonparametric statistics. This information is condensed with the aid of quantiles; otherwise the tables would be inconveniently bulky.

Often we will define a random variable and, instead of working with that random variable, we will work with a function of the random variable. A real valued function of a random variable X is a rule for assigning new real numbers to the sample space instead of the usual numbers assumed by X. For example, if $Y = X + 4$, Y is a real valued function of X; if $X = x$, then $Y = x + 4$. If $X = 3$, then $Y = 7$. This is usually written as $Y = u(X)$, where $u(X)$ in this case is $X + 4$. Other functions might include $u(X) = X^2$, $u(X) = X$, and $u(X) = (X - a)^2$ for some constant a. Since Y also assigns real numbers to points in the sample space, even though Y uses X in the process, we see that Y is a random variable. It is true in general that a real valued function of a random variable is also a random variable.

Expected Value

Another very useful property of a random variable is its *expected value*. First we will present a general definition of expected value; it will be followed by some particular cases.

Definition 2 Let X be a random variable with the probability function $f(x)$ and let $u(X)$ be a real valued function of X. The expected value of $u(X)$, written $E[u(X)]$, is

$$E[u(X)] = \sum_x u(x)f(x) \tag{1}$$

where the summation extends over all possible values of X. If the sum on the right side of Equation 1 is infinite, or does not exist, we say that the expected value of $u(X)$ does not exist.

Our interest is confined mainly to two special expected values, the mean and the variance of X.

Definition 3   Let X be a random variable with the probability function $f(x)$. The mean of X, usually denoted by μ, is

$$\mu = E(X) \qquad (2)$$

From Equation 1 we have

$$\mu = E(X) = \sum_x xf(x) \qquad (3)$$

which shows our mean to be the same as the "centroid" in physics. The mean, as does the centroid, marks a central point, a point of balance. If weights, in proportion to the various probabilities, were placed on a yardstick at the appropriate values of X, the yardstick would balance right at the mean. Because of this tendency to "locate" the center of the distribution, the mean is sometimes called a "measure of location." The mean and the median, discussed previously, are the two most commonly used measures of location.

EXAMPLE 2

Consider a simple experiment that results in "success" with probability p or "failure" with probability q equal to $1 - p$. Let X equal 1 if a "success" occurs and 0 if a "failure" occurs. Thus X has the binomial distribution with n equal to 1. From Equation 3 the expected value of X is determined as follows.

$$E(X) = 1(p) + 0(1 - p) = p \qquad (4)$$

The mean of X is equal to p. If the outcomes have equal probability, p equals 1/2 and the mean of X equals 1/2. ∎

EXAMPLE 3

Consider a businessman who always eats lunch at the same restaurant, which has lunches priced at $4.00, $4.50, $5.00,and $5.50. The businessman knows from past experience that on any given day he will select the $4.00 lunch with probability 0.25, the $4.50 lunch with probability 0.35, and the remaining two lunches with probability 0.20 each. Let X be the price of the lunch, in dollars. The probability function of X is

$$P(X = 4) = 0.25$$
$$P(X = 4.5) = 0.35$$
$$P(X = 5) = 0.20$$
$$P(X = 5.5) = 0.20$$

The mean of X is found using Equation 3.

$$E(X) = (4)(0.25) + (4.5)(0.35) + (5)(0.20) + (5.5)(0.20) = 4.675$$

Over a long period of time the businessman may expect the average luncheon expense to be somewhere near $4.67\frac{1}{2}$, even though no single lunch will cost that amount. ∎

Scale

Just as the mean and the median are called measures of location, the properties of the random variable that measure the amount of spread, or variability, of the random variable are called "measures of scale." One measure of scale based on quantiles is the *interquartile range*, the number obtained by subtracting $x_{0.25}$ from $x_{0.75}$. Another measure of scale, based more directly on the probability function, is the range, which equals the largest possible value of the random variable minus its smallest possible value. The most common measure of scale is the *standard deviation*, which equals the square root of the variance, defined as follows.

Definition 4   Let X be a random variable with mean μ and the probability function $f(x)$. The variance of X, usually denoted by σ^2 or by Var (X), is

$$\sigma^2 = E[(X - \mu)^2] \tag{5}$$

Using Equation 1 the variance of X may be written as

$$\sigma^2 = \sum_x (x - \mu)^2 f(x)$$
$$= \sum_x (x^2 - 2\mu x + \mu^2) f(x)$$
$$= \sum_x x^2 f(x) - 2\mu \sum_x x f(x) + \mu^2 \sum_x f(x) \tag{6}$$

Because $\sum_x f(x)$ equals 1, and because of Equation 3, Equation 6 becomes

$$\sigma^2 = E(X^2) - 2\mu^2 + \mu^2 = E(X^2) - \mu^2 \tag{7}$$

which is often a more useful form of the variance for computing purposes.

The positive square root of the variance is called the *standard deviation* of X and is usually denoted by σ.

EXAMPLE 4

If X has the binomial distribution with n equal to 1, then $P(X = 1) = p$ and $P(X = 0) = 1 - p$. In Example 2 the mean of X was found to equal p. Therefore, from Equation 6,

$$\sigma^2 = (1 - p)^2(p) + (0 - p)^2(1 - p)$$
$$= p(1 - p)$$
$$= pq \tag{8}$$

Alternatively, we might use Equation 7 to compute σ^2. Then we would first compute $E(X^2)$ using Equation 1.

$$E(X^2) = (1)^2(p) + (0)^2(1 - p)$$
$$= p$$

The variance of X is then found to be

$$\sigma^2 = E(X^2) - \mu^2$$
$$= p - p^2$$
$$= p(1 - p)$$

as before. The standard deviation of X is $\sqrt{p(1 - p)}$.   ■

EXAMPLE 5

There are six identical chips numbered 1 to 6. A monkey selects one chip and gives it to its trainer. The sample space is the chip selected by the monkey. Let X be the number on the chip. If each chip has probability $1/6$ of being selected, then X has the discrete uniform distribution. For the mean of X we have

$$E(X) = 1(\tfrac{1}{6}) + 2(\tfrac{1}{6}) + 3(\tfrac{1}{6}) + 4(\tfrac{1}{6}) + 5(\tfrac{1}{6}) + 6(\tfrac{1}{6})$$
$$= 3\tfrac{1}{2}$$

The expected value of X^2 is given by

$$E(X^2) = 1(\tfrac{1}{6}) + 4(\tfrac{1}{6}) + 9(\tfrac{1}{6}) + 16(\tfrac{1}{6}) + 25(\tfrac{1}{6}) + 36(\tfrac{1}{6})$$
$$= 15\tfrac{1}{6}$$

The variance of X is computed using Equation 7.

$$\text{Var}(X) = E(X^2) - \mu^2$$
$$= 15\tfrac{1}{6} - (3\tfrac{1}{2})^2$$
$$= 2\tfrac{11}{12}$$

The standard deviation is the square root of the variance and, for this example, equals 1.71.   ■

Definition 2 defined the expected value of a function of a single random variable. An extension of the definition may be made to include functions of several random variables considered jointly. This extended definition leads us into consideration of the *covariance* of two random variables and enables us to find the mean and variance of the sum of several random variables.

Definition 5   Let $X_1, X_2, \ldots, X_n$ be random variables with the joint probability function $f(x_1, x_2, \ldots, x_n)$, and let $u(X_1, X_2, \ldots, X_n)$ be a real valued function of $X_1, X_2, \ldots, X_n$. Then the *expected value of* $u(X_1, X_2, \ldots, X_n)$ is

$$E[u(X_1, X_2 \ldots, X_n)] = \sum u(x_1, x_2, \ldots, x_n) f(x_1, x_2, \ldots, x_n) \tag{9}$$

where the summation extends over all possible combinations of values of $x_1, x_2, \ldots, x_n$.

One of the simpler functions of $X_1, X_2, \ldots, X_n$ is

$$Y = X_1 + X_2 + \cdots + X_n \tag{10}$$

That is, each value of the random variable Y associated with the combined experiment involving the X_is is obtained simply by adding the values achieved by all the X_is. Then

$$E(Y) = \sum (x_1 + \cdots + x_n) f(x_1, \ldots, x_n)$$
$$= \sum x_1 f(x_1, \ldots, x_n) + \cdots + \sum x_n f(x_1, \ldots, x_n) \tag{11}$$

where each summation extends over all possible combinations of the values of $x_1, \ldots, x_n$. Using Definition 5, Equation 11 immediately becomes

$$E(Y) = E(X_1) + \cdots + E(X_n) \tag{12}$$

The result of these calculations may be stated as a theorem.

Theorem 1   Let $X_1, X_2, \ldots, X_n$ be random variables and let

$$Y = X_1 + X_2 + \cdots + X_n$$

Then $E(Y) = E(X_1) + E(X_2) + \cdots + E(X_n)$.

The statement in Theorem 1 holds true in all cases, whether the random variables are independent or not. Often the apparently difficult problem of finding the mean of the sum of several random variables reduces to a trivial exercise with the use of this theorem.

The results of the next two examples will be used in later chapters.

EXAMPLE 6

Let Y be the total number of "successes" in n independent trials, where each trial results in either "success" or "failure" with probability p and $q = 1 - p$, respectively. Then Y has the binomial distribution with parameters n and p. However, Y may be regarded as the sum of n independent random variables $X_1, X_2, \ldots, X_n$, where $X_i = 1$ if the ith trial results in "success" and $X_i = 0$ if the ith trial results in failure, for each i from 1 to n. Then

$$Y = X_1 + X_2 + \cdots + X_n$$

and, from Theorem 1,

$$E(Y) = E(X_1) + E(X_2) + \cdots + E(X_n)$$

In Example 2 the mean of X_i was found to equal p. Therefore

$$E(Y) = np \tag{13}$$

gives the mean for the binomial distribution.   ∎

Note that in the binomial distribution the trials are assumed to be independent and, therefore, the X_i are independent. This assumption is not needed in order to find the mean.

The following lemma is needed in Example 7. This lemma presents a convenient equation for expressing the sum of consecutive integers.

Lemma 1

$$\sum_{i=a}^{N} i = \frac{(N + a)(N - a + 1)}{2} \quad \text{and} \quad \sum_{i=1}^{N} i = \frac{(N + 1)N}{2}$$

Proof   The desired sum may be written two ways. Let $S = \sum_{i=a}^{N} i$. Then

$$S = a + (a + 1) + (a + 2) + \cdots + (N - 1) + N$$
$$S = N + (N - 1) + (N - 2) + \cdots + (a + 1) + a$$

Adding the two equations together gives

$$2S = (N + a) + (N + a) + (N + a) + \cdots + (N + a) + (N + a)$$
$$= (N + a)(N - a + 1)$$

Therefore

$$S = \sum_{i=a}^{N} i = \frac{(N + a)(N - a + 1)}{2}$$

For $a = 1$, this becomes

$$\sum_{i=1}^{N} i = \frac{(N+1)N}{2}$$

completing the proof.

EXAMPLE 7

There are N chips in a jar, numbered from 1 to N. One by one, n of those chips, where n is less than N, are drawn from the jar, the number noted, and they are put aside. Let Y be the sum of the numbers on the n drawn chips. Assume the drawings are random; that is, each chip is equally likely to be selected.

The mean of Y would be difficult to find without using Theorem 1. The successive drawings are not independent, because once a number is recorded, no other chip can have that same number. However, we may regard Y as the sum of the random variables $X_1, X_2, \ldots, X_n$, where each X_i is the number on the ith chip drawn, with the probability function

$$P(X_i = k) = \frac{1}{N}, \quad \text{for} \quad k = 1, 2, 3, \ldots, N$$

Therefore, with the assistance of Lemma 1, we have the following.

$$E(X_i) = \sum_{k=1}^{N} k \left(\frac{1}{N}\right)$$
$$= \frac{(N+1)}{2} \tag{14}$$

Equation 14 furnishes us with the mean of a discrete uniform random variable. Since Y equals $X_1 + X_2 + \cdots + X_n$, we have

$$E(Y) = E(X_1) + E(X_2) + \cdots + E(X_n) = n\frac{N+1}{2} \tag{15}$$

∎

Covariance

A particularly useful function of two random variables is

$$[X_1 - E(X_1)][X_2 - E(X_2)]$$

whose expected value is called the *covariance of X_1 and X_2*. In particular, a comparison of Definition 4 with the following reveals that the variance of X_1 may be considered as the covariance of X_1 with itself.

Definition 6   Let X_1 and X_2 be two random variables with means μ_1 and μ_2, probability functions $f_1(x_1)$ and $f_2(x_2)$, respectively, and joint probability function $f(x_1, x_2)$. The *covariance of X_1 and X_2* is

$$\text{Cov}(X_1, X_2) = E[(X_1 - \mu_1)(X_2 - \mu_2)] \tag{16}$$

The definition of expected value, Definition 5, may be used to give

$$
\begin{aligned}
\text{Cov}(X_1, X_2) &= E[(X_1 - \mu_1)(X_2 - \mu_2)] \\
&= \sum (x_1 - \mu_1)(x_2 - \mu_2) f(x_1, x_2)
\end{aligned}
\tag{17}
$$

where the summation extends over all values of x_1 and x_2. This expands as

$$
\begin{aligned}
\text{Cov}(X_1, X_2) &= \sum (x_1 x_2 - \mu_1 x_2 - \mu_2 x_1 + \mu_1 \mu_2) f(x_1, x_2) \\
&= E(X_1 X_2) - \mu_1 \mu_2 - \mu_2 \mu_1 + \mu_1 \mu_2 \\
&= E(X_1 X_2) - \mu_1 \mu_2
\end{aligned}
\tag{18}
$$

Equation 18 is often easier to use than Equation 17 when calculating a covariance.

EXAMPLE 8

An insurance company has noticed that the probability of any particular person having an automobile accident within a given year is about 0.1. However, this probability becomes 0.3 if it is known that the person had an automobile accident the previous year.

     Let X_1 equal 0 or 1, dependent on whether a particular person has no accident or at least one accident, respectively, during the first year of his or her insurance period. Let X_2 be similarly defined for the second year of that same person's insurance period. The probability function of X_1, and therefore X_2 also, is

$$P(X_1 = 0) = 0.9$$
$$P(X_1 = 1) = 0.1$$

From Example 2 we obtain

$$E(X_1) = 0.1$$
$$E(X_2) = 0.1$$

The joint probability function of X_1 and X_2 at $X_1 = 1$ and $X_2 = 1$ may be found as follows.

$$
\begin{aligned}
f(1, 1) &= P(X_1 = 1, X_2 = 1) \\
&= P(X_2 = 1 | X_1 = 1) P(X_1 = 1) \\
&= (0.3)(0.1) \\
&= 0.03
\end{aligned}
$$

The computation of $E(X_1X_2)$ follows directly from Definition 5.

$$E(X_1X_2) = (1)(1)f(1, 1) \text{ plus "zero" terms}$$
$$= 0.03$$

The covariance of X_1 and X_2 is then obtained using Equation 18.

$$\text{Cov}(X_1, X_2) = E(X_1X_2) - E(X_1)E(X_2)$$
$$= 0.03 - (0.1)(0.1)$$
$$= 0.02 \qquad \blacksquare$$

Correlation Coefficient

We will now define the *correlation coefficient*, which is used as a measure of linear dependence between two random variables. Although we will not prove it here, the correlation coefficient is always between -1 and $+1$. It equals zero when the two random variables are independent, although it may equal zero in other cases also.

> **Definition 7** The *correlation coefficient* between two random variables is their covariance divided by the product of their standard deviations. That is, the correlation coefficient, usually denoted by ρ, between two random variables X_1 and X_2 is given by
>
> $$\rho = \frac{\text{Cov}(X_1, X_2)}{\sqrt{\text{Var}(X_1)\text{Var}(X_2)}} \qquad (19)$$

A lemma that will be used in the next example is now presented. This lemma furnishes us with a convenient formula for expressing the sum of the squares of the first N consecutive integers.

Lemma 2

$$\sum_{i=1}^{N} i^2 = \frac{N(N+1)(2N+1)}{6}$$

Proof Let $S = \sum_{i=1}^{N} i^2$. Then

$$
\begin{aligned}
S &= 1^2 + 2^2 + 3^2 + 4^2 + \cdots + N^2 \\
&= 1 + 2 + 3 + 4 + \cdots + N \\
&\quad + 2 + 3 + 4 + \cdots + N \\
&\quad\quad + 3 + 4 + \cdots + N \\
&\quad\quad\quad + 4 + \cdots + N \\
&\quad\quad\quad\quad \cdots \quad \cdots \\
&\quad\quad\quad\quad\quad + N
\end{aligned}
$$

where the sum of the numbers in the ith column is i^2. However, instead of adding down the columns, we will add across the rows. The sum of the numbers in the jth row from the top is found by Lemma 1 to be

$$j + (j + 1) + (j + 2) + \cdots + N = \frac{(N + j)(N - j + 1)}{2}$$
$$= \tfrac{1}{2}(N^2 + N + j - j^2)$$

Adding these row sums together gives

$$S = \sum_{j=1}^{N} \tfrac{1}{2}(N^2 + N + j - j^2)$$
$$= \tfrac{1}{2}\sum_{j=1}^{N} N^2 + \tfrac{1}{2}\sum_{j=1}^{N} N + \tfrac{1}{2}\sum_{j=1}^{N} j - \tfrac{1}{2}\sum_{j=1}^{N} j^2$$
$$= \tfrac{1}{2}(N \cdot N^2) + \tfrac{1}{2}(N \cdot N) + \tfrac{1}{2} \cdot \frac{(N + 1)N}{2} - \tfrac{1}{2}S$$

since the last sum in the middle equation is denoted by S in this proof. Rearranging gives

$$\tfrac{3}{2}S = \tfrac{1}{4}(2N^3 + 3N^2 + N) = \tfrac{1}{4}N(N + 1)(2N + 1)$$

so that

$$S = \sum_{i=1}^{N} i^2 = \frac{N(N + 1)(2N + 1)}{6}$$

completing the proof.

EXAMPLE 9

A jar contains N plastic chips numbered 1 to N, as in Example 7. An experiment consists of drawing n of these chips from the jar, where $n \leq N$. We assume that each chip is equally likely to be selected and that the drawing is without replacement. Let $X_1, X_2, \ldots, X_n$ be random variables where X_i equals the number on the ith chip drawn from the jar, for $i = 1, 2, \ldots, n$. In this example we will find the covariance of X_i and X_j. From Example 7, we have the mean of X_i

$$E(X_i) = \frac{N + 1}{2}$$

Also, from Equation 7 and Lemma 2 we have the variance of X_i

$$\text{Var}(X_i) = E(X_i^2) - [E(X_i)]^2 = \sum_{k=1}^{N} k^2 \frac{1}{N} - \left(\frac{N+1}{2}\right)^2$$

$$= \frac{1}{N} \frac{N(N+1)(2N+1)}{6} - \left(\frac{N+1}{2}\right)^2$$

$$= \frac{(N+1)(N-1)}{12} \tag{20}$$

Now consider the covariance of two random variables X_i and X_j, where $i \neq j$. Their joint probability function is

$$f(x_i, x_i) = P(X_i = x_i, X_j = x_j) = P(X_i = x_i | X_j = x_j) \cdot P(X_j = x_j)$$

$$= \frac{1}{N-1} \cdot \frac{1}{N} \quad \text{for } x_i, x_j = 1, 2, \ldots, N; \ x_i \neq x_j \tag{21}$$

The covariance of X_i and X_j, using Definition 6, is

$$\text{Cov}(X_i, X_j) = E\{[X_i - E(X_i)][X_j - E(X_j)]\}$$

$$= \sum_{\substack{k=1 \\ k \neq s}}^{N} \sum_{s=1}^{N} \left(k - \frac{N+1}{2}\right)\left(s - \frac{N+1}{2}\right) \frac{1}{(N-1)N}$$

where the summation extends over all k and s from 1 to N, except that k does not equal s because X_i and X_j cannot equal the same number at the same time. If we, at the same time, add and subtract those terms for $k = s$ the covariance becomes

$$\text{Cov}(X_i, X_j) = \frac{1}{(N-1)N} \sum_{k=1}^{N} \left(k - \frac{N+1}{2}\right) \sum_{s=1}^{N} \left(s - \frac{N+1}{2}\right) - \frac{1}{N-1} \sum_{k=1}^{N} \left(k - \frac{N+1}{2}\right)^2 \frac{1}{N} \tag{22}$$

To simplify Equation 22 we note that

$$\sum_{i=1}^{N} \left(i - \frac{N+1}{2}\right) = 0 \tag{23}$$

and from the definition of variance and from Equation 20 we have

$$\text{Var}(X_i) = \sum_{k=1}^{N} \left(k - \frac{N+1}{2}\right)^2 \frac{1}{N} = \frac{(N+1)(N-1)}{12} \tag{24}$$

Substitution of Equations 23 and 24 into Equation 22 yields

$$\text{Cov}(X_i, X_j) = -\frac{N+1}{12} \tag{25}$$

∎

The fundamental importance of the covariance is based on what happens to the covariance in the case of two independent random variables. Let X_1 and

X_2 be two independent random variables with probability functions $f_1(x_1)$ and $f_2(x_2)$, and means μ_1 and μ_2, respectively. Then the covariance of X_1 and X_2 is

$$\text{Cov}(X_1, X_2) = \sum_{x_1 x_2} x_1 x_2 f_1(x_1) f_2(x_2) - \mu_1 \mu_2$$

$$= \left[\sum_{x_1} x_1 f_1(x_1)\right]\left[\sum_{x_2} x f_2(x_2)\right] - \mu_1 \mu_2$$

$$= \mu_1 \mu_2 - \mu_1 \mu_2 = 0$$

Therefore independence of two random variables implies that their covariance is zero, which in turn implies that their correlation coefficient equals zero.

Theorem 2   If X_1 and X_2 are independent random variables, the covariance of X_1 and X_2 is zero.

The converse of Theorem 2 is not necessarily true as the following example illustrates. That is, zero covariance does not necessarily imply that the random variables are independent, even though the implication is an error made often in practice.

EXAMPLE 10

Define the joint probability function of two random variables as follows.

$$P(X = 0, Y = 0) = 1/2$$
$$P(X = 1, Y = 1) = 1/4$$
$$P(X = -1, Y = 1) = 1/4$$

The probability function of X is then

$$P(X = 0) = 1/2$$
$$P(X = 1) = 1/4$$
$$P(X = -1) = 1/4$$

and the probability function of Y is

$$P(Y = 0) = 1/2$$
$$P(Y = 1) = 1/2$$

The expected values of X and Y are

$$E(X) = 0$$
$$E(Y) = 1/2$$

The covariance of X and Y is

$$\text{Cov}(X, Y) = E(XY) - E(X)E(Y)$$
$$= (1)\,(\tfrac{1}{4}) + (-1)(\tfrac{1}{4}) - (0)(\tfrac{1}{2})$$
$$= 0$$

However, X and Y are not independent, because

$$P(X = 0, Y = 0) = 1/2$$

which is not equal to

$$P(X = 0)P(Y = 0) = (\tfrac{1}{2})(\tfrac{1}{2})$$
$$= 1/4$$

Therefore X and Y have zero covariance, even though they are not independent. ∎

We are now equipped to find the variance of the sum of several random variables. Let Y equal $X_1 + X_2 + \cdots + X_n$, where the X_is may or may not be independent. We wish to find the variance of Y.

$$\text{Var}(Y) = E\{[Y - E(Y)]^2\}$$
$$= E\{[X_1 + X_2 + \cdots + X_n - E(X_1) - E(X_2) - \cdots - E(X_n)]^2\}$$
$$= E\left\{\sum_{i=1}^{N} [X_i - E(X_i)]^2 + \sum_{i=1}^{N} \sum_{\substack{j=1 \\ i \neq j}}^{N} [X_i - E(X_i)][X_j - E(X_j)]\right\}$$

But since the expected value of a sum of random variables equals the sum of the expected values of the random variables,

$$\text{Var}(Y) = \sum_{i=1}^{n} E\{[X_i - E(X_i)]^2\} + \sum_{i=1}^{n} \sum_{\substack{j=1 \\ i \neq j}}^{n} E\{[X_i - E(X_i)][X_j - E(X_j)]\}$$
$$= \sum_{i=1}^{n} \text{Var}(X_i) + \sum_{i=1}^{n} \sum_{\substack{j=1 \\ i \neq j}}^{n} \text{Cov}(X_i, X_j) \tag{26}$$

If $X_1, \ldots, X_n$ are mutually independent then, from Theorem 2, we have $\text{Cov}(X_i, X_j) = 0$, and

$$\text{Var}(Y) = \sum_{i=1}^{N} \text{Var}(X_i) \tag{27}$$

We may summarize this as a theorem.

Theorem 3 Let $X_1, X_2, \ldots, X_n$, be random variables and let

$$Y = X_1 + X_2 + \cdots + X_n$$

Then

$$\text{Var}(Y) = \sum_{i=1}^{n} \text{Var}(X_i) + \sum_{i=1}^{n} \sum_{\substack{j=1 \\ i \neq j}}^{n} \text{Cov}(X_i, X_j)$$

Furthermore, if $X_1, X_2, \ldots, X_n$ are mutually independent,

$$\text{Var}(Y) = \sum_{i=1}^{n} \text{Var}(X_i)$$

EXAMPLE 11

In continuation of Example 9, let X_i equal the number on the ith chip drawn, as before, and let Y equal the sum of the X_is as in Example 7. Then Theorem 3 gives us

$$\text{Var}(Y) = \sum_{i=1}^{n} \text{Var}(X_i) + \sum_{i=1}^{n} \sum_{\substack{j=1 \\ i \neq j}}^{n} \text{Cov}(X_i, Y_j)$$

The various terms in this equation are given by Equations 20 and 25. The variance term appears n times, and the covariance term appears $n(n-1)$ times.

$$\text{Var}(Y) = n \frac{(N+1)(N-1)}{12} + n(n-1)\left(-\frac{N+1}{12}\right)$$

$$= \frac{n(N+1)(N-n)}{12} \tag{28}$$

Note that $\text{Var}(X_i)$ is only a special case of $\text{Var}(Y)$ for $n = 1$. ∎

The variance of a random variable that has the binomial distribution is found in Example 12.

EXAMPLE 12

Consider n independent trials, where each trial may result in "success" with probability p, or "failure" with probability q, where $p + q$ equals 1. As in Examples 4 and 6, let X_i equal 0 or 1, depending on whether the ith trial results in "failure" or "success," respectively, and let Y equal the total number of "successes" in the n trials. Since the X_i are mutually independent, Theorem 3 states that

$$\text{Var}(Y) = \sum_{i=1}^{n} \text{Var}(X_i)$$

From Example 4, Var (X_i) equals pq, so

$$\text{Var}(Y) = npq$$

furnishes us with the variance of Y, which has the binomial distribution. ∎

The results of some of the preceding examples will be used later in this book, and so they are stated separately as theorems, for convenience.

Theorem 4 Let X be a random variable with the binomial distribution

$$P(X = k) = \binom{n}{k} p^k q^{n-k}$$

Then the mean and variance of X are given by

$$E(X) = np$$
$$\text{Var}(X) = npq$$

Theorem 5 Let X be the sum of n integers selected at random, without replacement, from the first N integers 1 to N. Then the mean and variance of X are given by

$$E(X) = \frac{n(N + 1)}{2}$$
$$\text{Var}(X) = \frac{n(N + 1)(N - n)}{12}$$

EXAMPLE 13

An advertising agency drew 12 sample magazine ads for one of their customers and ranked the ads from 1 to 12 on the basis of the agency's opinion of which ads would be the most effective in selling the product. The "most effective" ad was given the rank 1, and so on. The customer, the manufacturer of the product, selected 4 ads for purchase. They were ranked 4, 6, 7, and 11 by the agency.

Assuming that the customer's choice and the agency's rankings were independent, the sum of the ranks on the selected ads should be distributed the same as the sum of the numbers on 4 chips selected at random out of 12 chips numbered 1 to 12. Let X equal the sum of the ranks of 4 ads if they are selected independently of the ranks. Then Theorem 5 states that the mean of X is

$$E(X) = \frac{(4)(12 + 1)}{2} = 26$$

and the variance of X is

$$\text{Var}(X) = \frac{(4)(12 + 1)(12 - 4)}{12} = 34\tfrac{2}{3}$$

The standard deviation of X is

$$\sigma = \sqrt{\text{Var}(X)} = 5.9$$

The observed value of X is

$$X = 4 + 6 + 7 + 11 = 28$$

which is close to the mean of X under the preceding assumptions. ∎

EXERCISES

1. If $P(X = 0) = 1/3$, $P(X = 1) = 1/3$, $P(X = 2) = 1/6$, and $P(X = 3) = 1/6$, find

   (a) $E(X)$.   (b) Var (X).
   (c) $E(X^2 + 2X)$.   (d) The median.
   (e) $x_{1/3}$.   (f) The fourth decile.

2. If $P(X = 0) = 0$, $P(X = 1) = 1/2$, $P(X = 2) = 1/4$, and $P(X = 4) = 1/4$, find

   (a) $E(X)$.   (b) Var (X).
   (c) $E(-X)$.   (d) The median.
   (e) The upper quartile. (f) The thirty-seventh percentile.

3. If $P(X = 0, Y = 0) = 1/4$, $P(X = 0, Y = 1) = 1/4$, $P(X = 1, Y = 0) = 1/4$, and $P(X = 1, Y = 1) = 1/4$, find

   (a) $E(X)$.   (b) $E(Y)$.
   (c) $E(XY)$.   (d) $E(X + Y)$.
   (e) Cov (X, Y).   (f) $P(X = 0)$.
   (g) $P(X = 1)$.   (h) Are X and Y independent?

4. If $P(X = 0, Y = 0) = 1/8$, $P(X = 0, Y = 1) = 3/8$, $P(X = 1, Y = 0) = 3/8$, and $P(X = 1, Y = 1) = 1/8$, find

   (a) $E(X)$.   (b) $E(Y)$.
   (c) $E(XY)$.   (d) $E(X^2Y)$.
   (e) Cov (X, Y).   (f) $P(X = x)$ for all x.
   (g) Are X and Y independent?

5. What is the sum of the 66 integers from 1 to 66?

6. What is the sum of the 30 integers from 70 to 99?

7. If X equals the number of spots showing on one roll of a balanced die, find

   (a) $E(X)$.
   (b) Var (X).
   (c) $E(X^2 + X)$.

8. If 30 tickets are numbered consecutively from 1 to 30, and 2 tickets are selected at random without replacement, find

   (a) E(sum of the numbers on the two tickets).
   (b) Var (sum of the numbers on the two tickets).

9. If 10 customers are numbered consecutively from 1 to 10, and 2 of these customers are selected randomly for an interview, find the mean, the variance, and the range of the sum of the numbers of the customers selected for review.

10. If 100 customers are numbered consecutively from 1 to 100, and 12 of these customers are selected randomly for an interview, find the mean, the variance, and the range of the sum of the numbers of the customers selected for interview.

11. Let $P(X = 0, Y = 0) = 1/3$, $P(X = 0, Y = 1) = 1/3$, and $P(X = 1, Y = 0) = 1/3$.

    (a) Find the marginal probability distribution of X.
    (b) Find $E(X^2Y)$.
    (c) Find Cov (X, Y).
    (d) Find the correlation coefficient between X and Y.

12. The bivariate random variable (X, Y) takes the values $(1, 1)$ with probability 0.25, $(2, 1)$ with probability 0.25, and $(1, 2)$ with probability 0.50. Find the median of X, the variance of Y, and the covariance of X and Y. Let $F(x, y)$ be the distribution function of (X, Y). Find $F(1, 2)$. Draw the distribution function of X and the probability function of Y.

13. Eight horses are in the big race. Three horses are from Lubbock. Post positions (1 through 8) are assigned at random, without replacement, to the horses. Let X be the sum of the post positions of the three horses from Lubbock. Find the mean and the variance of X.

14. Let X and Y be independent binomial random variables, with $n = 3$, $p = 1/3$ for X, and $n = 4$, $p = 1/4$ for Y. Find $F(1, 1)$. Also find $E(XY)$.

PROBLEMS

1. Prove that a vertical line drawn at the mean of a random variable X divides the distribution function of X in such a way that the area under the distribution function, to the left of the mean, and above 0, equals the area above the distribution function, to the right of the mean, and below 1. (*Hint.* Draw a picture of the distribution function, and the areas in question.)

2. In the same way that Lemma 2 was obtained using Lemma 1, prove Lemma 3,

$$\sum_{i=1}^{N} i^3 = \frac{N^2(N + 1)^2}{4}$$

using Lemma 2. [A general extension is given by Iman (1970).]

1.5 CONTINUOUS RANDOM VARIABLES

All of the random variables that we have introduced so far in this chapter have one property in common: their possible values can be listed. The list of possible values assumed by the binomial variables is 0, 1, 2, 3, 4, . . . , $n - 1$, n. No other values may be assumed by the binomial random variable. The list of values that may be assumed by the discrete uniform random variable could be written as 1, 2, 3, . . . , N. Similar lists could be made for each random variable introduced in the previous definitions and examples.

These lists may be infinite in length, such as in an experiment where the random variable X equals the number of times a monkey pushes the "wrong" button before finally pushing the "right" button and getting a reward. Then X may equal 0 if the right button is pushed the first time, or X may equal 1000 if the monkey has difficulty finding the right button. Theoretically there is no limit to the number of times the monkey may choose the wrong button before pushing the correct one. The possible values of X may be listed, even though the list may be infinitely long. The infinitely long list of possible values is a characteristic of the model and not of the actual experiment in this example and in most situations, because real factors such as the eventual death of the monkey, the absence of research funds, or the waning enthusiasm of the experimenter will prevent the actual experiment from being prolonged to the point of absurdity. Nevertheless, the model may be reasonable, and the random variable in the model may have an infinity of possible values.

Discrete Random Variable

A more precise way of stating that the possible values of a random variable may be listed is to say that there exists a *one-to-one correspondence* between the possible values of the random variable and some or all of the positive integers. This means that to each possible value there corresponds one and only one positive integer, and that positive integer does not correspond to more than one possible value of the random variable. Random variables with this property are called *discrete*. All of the random variables we have considered so far are discrete random variables. However, the theorems we have proven hold for all random variables, even though we proved them only for discrete random variables.

> **Definition 1**   A random variable X is *discrete* if there exists a one-to-one correspondence between the possible values of X and some or all of the positive integers.

The distribution function of a discrete random variable is always a *step function*, that is, the graph looks like a series of stair steps, although the steps may not be very uniform in appearance, and there may be an infinite number of steps. If any portion of the graph rises gradually instead of rising only in clear-cut steps, the associated random variable is not discrete.

Continuous Random Variable

If the graph of a distribution function has no steps but rises only gradually where it rises, the distribution function is called *continuous,* and the random variable with that distribution function is called a *continuous random variable.* Figure 3 is an example of the graph of a continuous distribution function. Saying that a

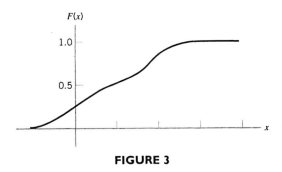

FIGURE 3

distribution function has no steps is the same as saying that no two horizontal lines will intersect the graph at the same value as measured along the horizontal axis. That is, if there is a step in the graph of a distribution function, at least two horizontal lines may be drawn, say at heights p_1 and p_2, closely enough to each other so that they intersect the graph at the same value, as measured along the horizontal axis. Since this describes the graphical method of finding quantiles, we may say that if there is a step in the distribution function, there are at least two quantiles x_{p_1} and x_{p_2} that are equal to each other. Conversely, if there are no two quantiles exactly equal to each other, there are no steps in the distribution function, and the function is continuous. This leads to a method of defining *continuous random variable.*

> **Definition 2** A random variable X is *continuous* if no two quantiles x_{p_1} and x_{p_2} of X are equal to each other, where p_1 is not equal to p_2. Equivalently, a random variable X is continuous if $P(X \leq x)$ equals $P(X < x)$ for all numbers x.

EXAMPLE 1

The distribution function graphed in Figure 4 is a continuous distribution function, and any random variable with the distribution function $F(x)$ is a continuous random variable. Typical continuous random variables involve measuring time, weight, distance, volume, and so forth. ∎

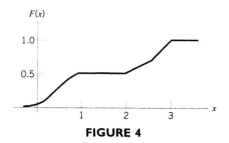

$F(x)$

1.0

0.5

1   2   3 x

FIGURE 4

In practice, no actual random variable is continuous, because the observed values of actual random variables are always the result of measurements of some sort, and measurements are made with tools that have only a finite capacity for discriminating between two values. Continuous random variables exist only in theory, such as in a model of an actual experiment. At times it is preferable to assume a random variable to be continuous, even though it is known to be discrete, as in Example 2.

EXAMPLE 2

The time it takes a racehorse to run a mile race is a continuous quantity, because time is generally a continuous quantity. In practice, however, the time is usually measured to the nearest $\frac{1}{5}$ second. It is not unusual for a horse to run two races in identical lengths of time (i.e., measured time). The actual lengths of time will be exactly equal with probability zero; therefore it is reasonable to assume that the time of a race, measured exactly, is a continuous random variable that is approximately equal to the measured time of the race, a discrete random variable. If two horses run in the same race and cross the finish line ahead of all other horses, with identical measured times, the winner of the race is then determined by examining a photograph taken at the finish line at the moment the horses crossed. Only rarely does this fail to determine the actual winner of the race. This is meant to illustrate that even though the random variable (measured time) is discrete, it is used as an approximation to the actual time because the order in which the horses finished the race still may be determined even though two or more measured times seem to be identical. ∎

Another reason for considering continuous random variables is that the distribution function of a discrete random variable sometimes may be approximated by a continuous distribution function, resulting in a convenient method for computing desired probabilities associated with the discrete random variable. Two continuous distribution functions commonly used for this purpose are the *normal distribution* and the *chi-squared distribution.*

Normal Distribution

The distribution function in the following definition might frighten those who are unfamiliar with elementary calculus. There is no cause for alarm, however, because the distribution function is well tabulated. Such tables may be found in most statistics texts, as well as in Table A1. Also most statistics programs on the computer furnish quantiles and probabilities when requested.

> **Definition 3** Let X be a random variable. Then X is said to have the *normal distribution* if the distribution function of X is given by
>
> $$F(x) = P(X \le x) = \int_{-\infty}^{x} \frac{1}{\sqrt{2\pi}\sigma} e^{-1/2[(y-\mu)/\sigma]^2} dy \tag{1}$$
>
> where it can be shown (using calculus) that the parameters μ and σ are the mean and standard deviation of X. The *standard normal distribution* is the normal distribution with μ equal to 0 and σ equal to 1.

The normal distribution function cannot be evaluated directly, and so Table A1 may be used to find approximate probabilities associated with normal random variables. Table A1 gives over one thousand quantiles of the standard normal random variable. Quantiles of normal random variables with mean μ and variance σ^2 may be found from Table A1, with the aid of the equations given without proof in the following theorem.

Theorem 1 For a given value of p, let x_p be the pth quantile of a normal random variable with mean μ and variance σ^2 and let z_p be the pth quantile of a standard normal random variable. The quantile x_p may be obtained from z_p by using the relationship

$$x_p = \mu + \sigma z_p \tag{2}$$

Similarly, z_p may be obtained from x_p with the aid of the relationship

$$z_p = \frac{x_p - \mu}{\sigma} \tag{3}$$

EXAMPLE 3

Let Z be a random variable with the standard normal distribution. To find the probability that Z will not exceed 1.42, Table A1 is used. From Table A1 we see that

$$P(Z \le 1.4187) = 0.922$$

and

$$P(Z \leq 1.4255) = 0.923$$

We simply use the quantile closest to 1.42 to obtain

$$P(Z \leq 1.42) \cong 0.922 \qquad \blacksquare$$

EXAMPLE 4

Let X be the IQ of a person selected at random from a large group of people. Assume that X has the normal distribution with mean 100 and standard deviation 15. Suppose we want to find the probability that X will exceed 125. We have

$$P(X > 125) = 1 - P(X \leq 125)$$

so it suffices to find $P(X \leq 125)$. The quantile z_p, of the standard normal random variable, corresponding to the quantile $x_p = 125$ is found from Equation 3.

$$z_p = \frac{x_p - \mu}{\sigma}$$

$$= \frac{125 - 100}{15}$$

$$= 1.6667$$

From Table A1 we see that if $z_p = 1.6667$, then $p = 0.952$. Therefore 125 is the 0.952 quantile of X.

$$P(X \leq 125) = 0.952$$
$$P(X > 125) = 0.048$$

The desired probability is 0.048.

To find the upper 1 percentile, called the 99th percentile, we want to find the number $x_{0.99}$, where

$$P(X \leq x_{0.99}) = 0.99$$

Since, from Table A1, $z_{0.99} = 2.3263$, $x_{0.99}$ may be found from Equation 2 to be

$$x_{0.99} = \mu + \sigma z_{0.99}$$
$$= 100 + 15(2.3263)$$
$$= 135$$

Therefore the probability of the randomly selected person having an IQ less than 135 is about 0.99. $\qquad \blacksquare$

EXAMPLE 5

A railroad company has observed over a period of time that the number X of people taking a certain train seems to follow a normal distribution with mean

540 and standard deviation 32. How many seats should the company provide on the train if it wants to be 95% certain that everyone will have a seat?

We wish to find the 95th percentile. From Equation 2 we have that

$$x_{0.95} = \mu + \sigma z_{0.95}$$
$$= 540 + 32(1.6449)$$
$$= 593$$

where $z_{0.95}$ is obtained from Table A1. The company needs 593 seats on the train so that they can be 95% certain that there will be enough seats for everyone on any one run of that train. ∎

In Example 5 the random variable X is actually a discrete random variable that assumes only the nonnegative integers as values. Therefore X cannot possibly have a normal distribution. The normal approximation to the distribution of X was used partly for convenience and partly out of necessity, because a realistic discrete distribution might be difficult to formulate. In other problems a discrete distribution function that agrees well with the data may be known, yet the normal approximation still might be used for ease in calculations. The validity of using the normal approximation usually depends on the central limit theorem.

Central Limit Theorem

The so-called central limit theorem appears in many different forms. All forms have in common the purpose of stating conditions under which the sum of several random variables may be approximated by a normal random variable. The theorem says that the distribution function of the sum of several random variables approaches the normal distribution function, as the number of random variables being added becomes large (i.e., goes to infinity), and when certain other general conditions are met. These "other general conditions" may be stated many different ways, giving rise to many forms for the central limit theorem. A thorough discussion of this theorem is well beyond the scope of this book.

Theorem 2 (Central Limit Theorem) Let Y_n be the sum of the n random variables $X_1, X_2, \ldots, X_n$, let μ_n be the mean of Y_n, and let σ_n^2 be the variance of Y_n. Under some general, easily met conditions, as n, the number of random variables, goes to infinity, the distribution function of the random variable

$$\frac{Y_n - \mu_n}{\sigma_n}$$

approaches the standard normal distribution function.

In practice, the number of random variables summed never goes to infinity. But the value of the central limit theorem is that in situations where the theorem holds, the normal approximation is usually considered to be "reasonably good" as long as n is "large." The terms "reasonably good" and "large" are subjective terms; therefore much latitude exists in the practice of using the normal approximation. Usually if n is greater than 30 the normal approximation is satisfactory. However, sometimes when n is as small as 5 or 10 the normal approximation can be quite good.

A useful illustration of the normal approximation to the binomial distribution is presented in Example 6.

EXAMPLE 6

Let Y_n be a random variable with the binomial distribution (Definition 1.3.5) with mean np and variance npq (Theorem 1.4.4). Then Y_n may be regarded as the sum of n independent random variables, each with the binomial distribution where n equals 1 (Example 1.4.12). For large n, the random variable

$$\frac{Y_n - np}{\sqrt{npq}}$$

is distributed approximately the same as a standard normal random variable. This is equivalent to saying that for large n, the distribution function of Y_n may be approximated by the normal distribution with mean np and variance npq (Theorem 1). ∎

The following example illustrates another application of the central limit theorem, which will be useful in Chapter 5.

EXAMPLE 7

Consider the sampling scheme where n integers are selected at random, without replacement, from the first N integers, 1 to N. Let X_i be the ith integer selected, and let

$$Y_n = X_1 + X_2 + \cdots + X_n$$

be the sum of the integers selected. For large n and large N, the distribution function of

$$\frac{Y_n - \dfrac{n(N+1)}{2}}{\left(\dfrac{n(N+1)(N-n)}{12}\right)^{\frac{1}{2}}}$$

(Theorem 1.4.5) may be approximated by the standard normal distribution function. In other words, the distribution function of Y_n may be approximated by the normal distribution function with mean $n(N+1)/2$ and variance $n(N+1)(N-n)/12$ (Theorem 1). ∎

Chi-squared Distribution

The widespread applicability of the central limit theorem makes it a very useful theorem. Since it justifies to some extent the use of the normal approximation, the normal distribution is a valuable distribution. Other distributions that are related to the normal distribution also become important, such as the *chi-squared distribution*.

In the following definition, the chi-squared distribution function is given using the "integral" notation of calculus and the "gamma function" $\Gamma(k/2)$. This notation needs no explanation or even understanding because the tabulated values of the chi-squared distribution function given in Table A2 will be used whenever values of the distribution function are needed. Most statistics programs on the computer also furnish quantiles and probabilities when requested.

> **Definition 4** A random variable X has the *chi-squared distribution with k degrees of freedom* if the distribution function of X is given by
>
> $$F(x) = P(X \le x) = \int_0^x \frac{y^{(k/2)-1}e^{-y/2}}{2^{k/2}\Gamma(k/2)}\,dy \qquad \text{if } x > 0$$
> $$= 0 \qquad \text{if } x \le 0 \qquad (4)$$

The distribution function Equation 4 shows that a chi-squared random variable may assume only nonnegative values, since $F(x) = 0$ for negative values of x. The degrees of freedom, k, is merely a parameter. The values of k are usually restricted to the integers 1, 2, 3, and so forth. For different values of the parameter k, the distribution functions are different also. Table A2 gives some selected quantiles of a chi-squared random variable, for $k = 1, 2, 3$, up to 30, and for some values of k greater than 30. For k greater than 100 the central limit theorem may be used to obtain approximate quantiles, which will be justified later in this section.

It is shown in most introductory books on mathematical statistics that if X is a random variable with the chi-squared distribution with k degrees of freedom, the mean and variance of X are given by

$$E(X) = k \qquad (5)$$
$$\text{Var}(X) = 2k \qquad (6)$$

The following theorem is proved in Freund (1962, p. 194).

Theorem 3 Let $X_1, X_2, \ldots, X_k$ be k independent and identically distributed standard normal random variables. Let Y be the sum of the squares of the X_i.

$$Y = X_1^2 + X_2^2 + \cdots + X_k^2 \qquad (7)$$

Then Y has the chi-squared distribution with k degrees of freedom.

EXAMPLE 8

A child psychologist asks each of 100 children to tell which of two trucks they would rather play with. The two trucks are identical in all respects, except that one is red and the other is green. The psychologist is interested in knowing whether children have a color preference.

Forty-two children selected the green truck, and the other 58 chose the red truck. In the model "no preference" is assumed, so the random variable X, equal to the number of children who selected the green truck, should have the binomial distribution with mean $np = 50$ and variance $npq = 25$. The normal approximation to the distribution function of X seems appropriate, so

$$\frac{X - 50}{5}$$

is considered to be approximately the same as a standard normal random variable. However, the psychologist is interested in determining differences in either direction; that is, she wants to know whether X is much smaller than 50 as well as whether X is much larger than 50, so she uses the square of the difference essentially, but actually examines the random variable

$$X^* = \left(\frac{X - 50}{5}\right)^2$$

because it may be compared with a chi-squared random variable with 1 degree of freedom. In this experiment $X^* = [(42 - 50)/5]^2$, or 2.56. The probability of getting a number smaller than 2.56, corresponding to a value of X closer to 50, is found by interpolation in Table A2, $k = 1$, to be about 0.88. Therefore the psychologist concludes that there is some indication of a color preference among the children. (More will be said concerning this method of drawing a conclusion in later chapters.) ∎

In Example 8 the distribution function of the random variable $(X - 50)/5$ was considered to be approximately equal to the standard normal distribution function. Therefore the chi-squared approximation, with 1 degree of freedom, was used for the distribution of X^*. The desired probability could have been found by using both tails of the normal distribution function. That is,

$$P(X^* \le (-1.6)^2) = P\left(-1.6 < \frac{X - 50}{5} < +1.6\right)$$

The probability on the left, found from Table A2, should equal the probability on the right, obtained from Table A1. The only difference between the two probabilities results from using interpolation in the two tables. If more than 1 degree of freedom is involved, then Table A1 may not be used as an alternative to Table A2.

EXAMPLE 9

In continuation of the experiment described in Example 8, the psychologist obtains two toy telephones, identical except that one is white and the other is blue. She asks each of 25 children to choose one to play with. Seventeen children chose the white telephone, and the other 8 preferred the blue telephone. Let Y be the random variable equal to the number of children selecting the white telephone. Since

$$\frac{Y - np}{\sqrt{npq}} = \frac{Y - (1/2)(25)}{5/2}$$

is approximately a standard normal random variable under the assumption of no color preference, the random variable

$$Y^* = \left(\frac{Y - (1/2)(25)}{5/2}\right)^2$$

may be compared with a chi-squared random variable with 1 degree of freedom. Since $Y = 17$, $Y^* = 3.24$. The probability of a chi-squared random variable with 1 degree of freedom being less than 3.24 is found from Table A2 to equal about 0.92 using interpolation. Therefore, if the assumption that each toy was equally likely to be chosen is in fact true, such a large deviation from the expected value of 12.5 would occur only about 8% of the time.

Since the experiment in Example 8 and this one were designed for the same purpose, it would seem desirable to be able to combine the results in some way. If X^* and Y^* may be considered as independent random variables, a reasonable consideration here, then Theorem 3 may be used to combine X^* and Y^* as

$$W = X^* + Y^*$$

and the distribution function of W may be approximated by the chi-squared distribution function with 2 degrees of freedom. Then

$$W = 2.56 + 3.24$$
$$= 5.80$$

The probability of a chi-squared random variable with two degrees of freedom being greater than 5.80 is only about 0.06, which was obtained by interpolation in Table A2.

In this example more information concerning the presence of color preference among children was obtained by combining the information gained in the two studies.

It should be noted that if Y had been defined as the number of children preferring the blue telephone, instead of the way Y was defined in this example, Y^* would still have the same value, because the deviation of Y from the mean was squared, eliminating the directional influence of the difference. ■

In Example 9, two approximate chi-squared random variables were added, and their sum was an approximate chi-squared random variable with 2 degrees of freedom. This method of combining independent chi-squared random variables is valid in general. More discussion of this method is given by Radhadkrishna (1965) and Nelson (1966).

The following theorem may be found in Freund (1962, p. 194).

Theorem 4 Let $X_1, X_2, \cdots, X_n$ be independent chi-squared random variables with $k_1, k_2, \ldots, k_n$ degrees of freedom, respectively. Let W equal the sum of the X_i. Then W is a chi-squared random variable with k degrees of freedom, where

$$k = k_1 + k_2 + \cdots + k_n$$

Theorem 4 will be used later in this book to approximate the distribution function of the sum of several random variables, where the random variables may be assumed to be independent and to be distributed approximately as chi-squared random variables.

Since a chi-squared random variable with k degrees of freedom may be considered to be the sum of k random variables, each having the chi-squared distribution with 1 degree of freedom, the conditions on the central limit theorem are met. The mean and variance of a chi-squared random variable with k degrees of freedom are given by Equations 5 and 6 to be k and $2k$, respectively. Therefore, if W is a chi-squared random variable with k degrees of freedom, the distribution function of

$$Z = \frac{W - k}{\sqrt{2k}} \tag{8}$$

may be approximated by the standard normal distribution function if k is large. From Theorem 1, if z_p is a quantile from Table A1, the quantile w_p for Table A2 may be approximated, for large k, by

$$w_p = k + \sqrt{2k}\, z_p \tag{9}$$

This is not as good as the approximations

$$w_p = \tfrac{1}{2}(z_p + \sqrt{2k - 1})^2 \tag{10}$$

or

$$w_p = k\left(1 - \frac{2}{9k} + z_p \sqrt{\frac{2}{9k}}\right)^3 \tag{11}$$

given at the bottom of Table A2.

EXERCISES

1. Let Z be a standard normal random variable. Find

    (a) $P(Z \leq 0)$.   **(b)** $P(Z \leq 1.96)$.
    (c) $P(Z > 1)$.   **(d)** $P(-1 < Z < 1)$.
    (e) $P(-4 < Z < 0)$.   **(f)** The upper quartile of Z.

2. Let X be a normal random variable with mean 0.5 and standard deviation 3. Find

    (a) $P(X \leq 0)$.   **(b)** $P(X \leq 1)$.
    (c) $P(X > -0.5)$.   **(d)** $P(-1 < X < 1)$.
    (e) The median of X.   **(f)** The upper quartile of X.

3. Suppose that X is the amount of time (in minutes) it takes a certain high school athlete to run 1 mile. Assume that X has a normal distribution with mean 4.30 and standard deviation 0.05. What is the probability that the athlete will break the school record of 4.15 minutes at the annual track meet?

4. Let X be the number of policyholders who make at least one claim to a large insurance company. Assume that there are 2000 policyholders and that each one has probability 0.2 of making at least one claim during the year. What is the probability that no more than 500 policyholders will make claims during any given year?

5. If the distribution of weights of a certain class of individuals is approximately normal, with mean 160 and variance 400, how high should a set of bathroom scales be calibrated so that about 99% of the people will be able to weigh themselves?

6. If Y is a binomial random variable with parameters $n = 60, p = 0.5$, estimate the probability that the random variable

    $$\frac{(Y - np)^2}{np(1 - p)}$$

    will exceed 5.

7. Let W be a chi-squared random variable with k degrees of freedom. Find

    (a) The 0.95 quantile of W, if $k = 4$.
    (b) The 0.95 quantile of W, if $k = 8$.
    (c) The 0.95 quantile of W, if $k = 200$.

8. If X, Y, and Z are independent chi-squared random variables with 3, 2, and 3 degrees of freedom, respectively, find the probability that W will exceed 15, where $W = X + Y + Z$.

9. Let X be the number of people who make a pledge during a drive for contributions to a school. Assume there are 500 people contacted in the drive, and each has probability 0.15 of making a pledge, independent of one another. Estimate $P(80 < X)$.

10. Let X be the number of consumers who visit the Dairy Queen in Plains, Texas, at least once during October. Assume there are 2000 people living in Plains, and each has probability 0.25 of visiting the Dairy Queen, independent of one another. Estimate $P(460 < X < 540)$.

11. Out of 100 attempts to score a goal, a basketball player scored on 43 of the attempts. Find the probability of scoring on 43 or fewer attempts if the real probability of the player scoring is 60%.

12. Ten companies are ranked according to profitability, from 1 (most profitable) to 10 (least profitable). Four companies are selected at random from the ten. Let X be the sum of the ranks of the four companies selected.

    (a) If $F(x)$ is the distribution function of X, find $F(14)$.

    (b) Use the normal approximation to find the approximate value of $F(14)$.

PROBLEMS

1. Let W be a chi-squared random variable with 100 degrees of freedom. Compare the approximations for the 0.95 quantile of W obtained using Equations 9, 10, and 11, with the exact value obtained from Table A2.

2. Let X be a binomial random variable with parameters $n = 100$, $p = 0.3$. Estimate

$$P(20 \le X \le 40)$$

using Tables A1 and A2.

1.6 REVIEW PROBLEMS FOR CHAPTER 1

1. If 75% of all graduate students who are admitted to Texas Tech actually enroll in graduate programs at Texas Tech, and 40% of all students who apply for admission are admitted, then what percentage of students who apply are admitted and do not enroll at Texas Tech?

2. A balanced coin is tossed eight times. What is the probability of getting exactly two heads if we are given the information that there were at least two heads in the eight tosses?

3. Consider five dice being rolled, and X equals the sum of the numbers (1 through 6) on the five dice. If the dice are balanced (i.e., equal probabilities of getting 1 through 6 for each die) and independent of each other, find the mean and the variance of X.

4. Consider a bivariate random variable (X, Y) that takes the value $(0, 0)$ with probability $1/4$, the value $(1, 1)$ with probability $1/2$, and the value $(2, 0)$ with probability $1/4$.
    (a) Find the covariance of X and Y.
    (b) Are X and Y independent? Explain.

5. In how many ways can a club with eight members select a President, Vice President, Secretary, and a Treasurer from its membership?

6. A new brand of snacks called Yummies is being field tested against Brand A and Brand B. Four parties are supplied equal amounts of all three snacks, and the amounts left after each party are compared. Let Y be the number of times Yummies is the most popular

snack. If there is no difference in preferences, assume Y has a binomial distribution with $n = 4$ and $p = 1/3$.

(a) Sketch the graph of the distribution of Y.

(b) Find the lower quartile of Y.

(c) Find the inter-quartile range of Y.

(d) Find the mean of Y.

(e) Find the standard deviation of Y.

(f) Find the exact probability that Y is 2 or less.

(g) Use the normal approximation to estimate the probability that Y is 2 or less. Compare your answer with the exact value found in part f.

7. A customer is equally likely to select each of six product brands, labeled "brand one," and so forth. Let X equal the brand number selected. Let Y equal 3 if one of the first three brands is selected and 6 if one of the last three is selected. Let Z equal 1 if the brand number is even and 2 if the brand number is odd.

(a) List the points in the sample space.

(b) Describe the probability function on the sample space.

(c) What is the name of the distribution of X?

(d) Find the interquartile range of Y.

(e) Find the variance of Z.

(f) Find the covariance of X and Z.

(g) Are Y and Z independent?

8. Twelve diamonds are ranked from 1 to 12 according to quality. Three diamonds are selected at random without replacement from the 12. Let X equal the sum of the ranks of the 3 diamonds.

(a) How many points are in the sample space?

(b) Describe any one point in the sample space.

(c) Find $P(X = 3)$.

(d) If $f(x)$ is the probability function of X, find $f(10)$.

(e) If $F(x)$ is the distribution function of X, find $F(10)$.

(f) Use the central limit theorem to approximate $F(10)$ and compare this approximation with the exact value found in part e.

9. The top 10 students in a large high school graduation class are ranked from 1 (best) to 10 (tenth best). Assume that each rank is equally likely to be assigned to a male student or a female student. Let X equal the sum of the ranks (from 1 to 10) that are assigned to female students, that is if all of the top 10 students are girls, $X = 1 + 2 + \cdots + 10 = 55$.

(a) How many points are in the sample space?

(b) Describe one point in the sample space.

(c) Describe the probability function on the sample space.

(d) Find $P(X = 0)$.

(e) Find $P(X = 1)$.

(f) If $f(x)$ is the probability function of X, find $f(3)$.

(g) If $F(x)$ is the distribution function of X, find $F(3)$.

10. About 49% of all human births are female and 51% are male. In a family with five children:

(a) What is the expected number of girls?

(b) What is the median number of girls?

(c) What is the probability that there are four boys and one girl?

(d) What is the most likely distribution of boys and girls?

(e) What additional assumptions did you need to make in order to answer these questions?

11. Consider two independent rolls of a balanced die. Let $\overline{X}$ be the average number of spots (i.e., $\overline{X} = (X_1 + X_2)/2$ where X_1 and X_2 are the number of spots on rolls 1 and 2).

 (a) Find the probability that $\overline{X} = 2$.

 (b) Draw a bar graph of the entire probability distribution of $\overline{X}$.

 (c) Draw a graph of the distribution function of $\overline{X}$.

 (d) Find the mean and variance of $\overline{X}$.

 (e) Find the exact value of $F(3)$.

 (f) Find an approximate value of $F(3)$ using the central limit theorem and compare it with part e.

12. Assume that about 10% of the people who make airline reservations on a particular flight do not show up for the flight. In order to accommodate as many people as possible, airlines customarily make more reservations than the airplane will hold because of the people who do not show. If the airplane holds 100 passengers, how many reservations can the airline make and still be 90% sure that everyone who shows up for the flight with reservations can be accommodated?

13. To play the lottery game Texas Lotto, a player selects six of the fifty numbers from 1 to 50, without replacement (i.e., without selecting the same number twice). The Lottery Commission then selects six of the numbers from 1 to 50 at random (i.e., every six-number combination is equally likely to be selected), also without replacement. The player wins if at least three of the player's numbers match numbers drawn by the Lottery Commission, in any order.

 Let X equal the number of "matched numbers" achieved by the player. The player wins if X equals 3, 4, 5, or 6.

 (a) How many combinations of six numbers out of fifty are there, disregarding the order in which they were drawn? What is the probability of each combination when the numbers are drawn at random?

 (b) What is the probability that all six of the player's numbers match the six drawn at random? That is, what is $P(X = 6)$?

 (c) Find the probability that X equals 5. $P(X = 4)$? $P(X = 3)$?

 (d) What is the name of the probability distribution of X? What are the values of the parameters in that probability distribution?

 (e) One drawing produced 241,024 tickets that had exactly three numbers correct. How many tickets do you think were purchased for that drawing?

 (f) The same drawing referred to in part e produced 12,422 tickets that had exactly four numbers correct. Is this consistent with the number of tickets that had exactly three numbers correct?

14. A game-show contestant on "Let's make a deal" has a chance to win a brand new car. All she has to do is to pick the correct garage door, the one hiding the car, out of three identical garage doors labeled A, B, and C. She picks garage door A.

 Before opening garage door A, the master of ceremonies, named Monte Hall, asks the contestant if she wants to change her mind. To make the game more interesting, Monte Hall (who knows where the car is) opens door B, so everyone can see the car is not there, and asks again if the contestant wants to change her mind and select door C instead of door A. At this point, should the contestant change her mind and select door C?

(a) Before Monte Hall opened door B, what was the probability of door A being the correct door, and the contestant winning the car?

(b) After Monte Hall opened door B, what was the probability of door A being the correct door? Door B? Door C?

CHAPTER 2

STATISTICAL INFERENCE

PRELIMINARY REMARKS

The concepts of probability theory introduced in the previous chapter do not cover the entire field of probability theory. But this small glimpse into the area of probability theory is all that is needed to understand the basic principles behind most of the nonparametric methods commonly used. We now bridge the gap between probability theory and its application to data analysis. In this chapter we introduce concepts of the basic science for data analysis called *statistics*.

The field of statistics owes many, if not most, of its significant ideas to people in the applied sciences who had difficult questions concerning their data. These people all had some mathematical ability, some training in mathematics, and a great deal of common sense. Their ideas gradually evolved into a few basic concepts that we present in this chapter.

2.1 POPULATIONS, SAMPLES, AND STATISTICS

Much of our knowledge concerning the world we live in is the result of samples. We eat at a restaurant once and we form an opinion concerning the quality of the food and service at that restaurant. We know 12 people from England and we feel we know the English people. Quite often the opinions we form from the sample are not accurate. However, samples that are obtained according to scientifically derived methods can give very accurate information about the entire population.

Experiment

The process of forming scientific opinions is often placed within the framework of an *experiment*. As we discussed in Chapter 1, an experiment is the process of following a well-defined procedure, where the outcome of following the procedure is not known prior to the experiment.

An experiment to determine the effect of a new medication consists of administering the medication according to a stated procedure for selecting and treating patients, and then observing the effect of the medication. An experiment to determine the quality of a manufactured product consists of obtaining and examining samples of the product according to a well-defined procedure, and then noting the results.

Population

The collection of all elements under investigation is called the *population*. A population may be a group of people, a group of animals, or even a group of inanimate objects such as items coming off an assembly line in a factory. Some populations are small, such as the population of past presidents of the United States, in which case the entire population may be examined. Other populations are large, such as the population of all Texans, or essentially infinite, such as the population of all human beings, in which case any study of the population requires examining only a sample from that population.

Sample

A *sample* is a collection of some elements of a population. There are several different categories of samples, depending on how the sample was obtained. A *convenience sample* is a collection of the elements that are easiest to obtain, such as "citizen on the street" interviews, or TV call-in surveys. It is not possible to obtain accurate estimates of population parameters from samples such as these. A *probability sample* on the other hand allows accurate statements to be made about the unknown population parameters. Probability samples require that every element in the population have a known, nonzero probability of being included in the sample. The probability samples considered in this book are *random samples*, which will be defined later in this section.

The Target Population and the Sampled Population

Suppose a psychologist wishes to study the effect of constantly interrupted sleep on the emotional balance of a person. He might consider the population to be all human beings of contemporary times. To conduct his experiment, he uses paid volunteers, obtained through an ad in a college newspaper. He can hardly consider his subjects to be representative of the population because they are all college students, at one university, in a rather restricted age group, and possessing an emotional makeup that prompts them to reply to an ad in a newspaper and volunteer for a somewhat personal study. And yet he is forced to use this type of sample for his experiment for practical reasons such as limited funds and limited time available for research, or to abandon his experiment entirely. Thus it is advisable to speak of two populations: the population targeted for investigation, and the population actually sampled.

The population about which information is wanted is called the *target population*. The population to be sampled is called the *sampled population*. Our example considered all contemporary human beings as the target population and all human beings who responded to the ad as the sampled population. All experimenters must necessarily work with the sampled population, and the validity of each experiment rests on the assumption that the sampled population is similar to the target population, at least with respect to the properties under investigation.

Random Sample

The statistical methods presented in this book usually assume that the sample is a *random sample,* so it is important to discuss the idea of a random sample.

There are two different ways of defining a random sample. The first definition refers to populations with a finite number of elements, N, where N may be large (all people in the world) or small (former presidents of the United States). Each element in the population is of equal importance and has equal probability of being selected in the sample. A *random sample of size n*, less than N, may be obtained by numbering all of the elements of the population from 1 to N and then drawing n numbers in a manner such that all groups of n numbers are equally likely to be drawn. The n numbers drawn correspond to the n elements in the population that are to be included in the sample. The drawing is usually *without replacement,* so that the same element can't appear more than once in the sample, but the definition applies even to those cases where the drawing is *with replacement,* so the same element may appear more than once.

> **Definition 1**   A sample of size n from a finite population is a *random sample* if each of the possible samples of that size was equally likely to be obtained.

The definition may seem a little strange in that the term "random" does not really refer to the sample itself but to the method by which the sample was obtained. In fact, we cannot look at a sample to see if it is a random sample or not. Instead, we look at the means by which the sample is obtained.

If the finite population has N elements total, then, as seen in Section 1.1, there are $\binom{N}{n}$ possible samples of size n if the sample is obtained without replacement. If the sampling is with replacement, there are N^n possible samples. If each of these possible samples is equally likely to be obtained, the method of sampling is considered to be random, and the resulting sample is a random sample.

The preceding definition of a random sample seems to be satisfactory for most situations where the population is finite. But suppose we are examining the number of dreams a certain individual has in one night. We think of a "random sample" in this case as the number of dreams she has in one night and the number she has another night, and so on for, say, seven nights. Even under ideal conditions the sampling method does not fit into the framework of Definition 1, with a concept of "equally likely." What is equally likely? Not the individual, because presumably we are studying only the individual, not a representative of some population (although this may be the ultimate objective in the back of our minds). Are we to select the nights for study in some equally likely fashion out of the remaining nights that the individual can expect to be alive? Clearly this is impossible. We may conclude that at least one more definition of random sample is needed.

The definition of random sample that is standard among mathematical statisticians is the following.

Definition 2 A *random sample of size n* is a sequence of n independent and identically distributed random variables $X_1, X_2, \ldots, X_n$.

Definitions 1 and 2 are identical only if the drawing in Definition 1 is *with replacement*, for then and only then are the observations independent. Drawing without replacement induces a dependence, because once an item is selected without replacement it cannot be selected again. However, if the population size N is very large, there is very little practical difference between drawing a sample with replacement and drawing a sample without replacement, so the slight dependence from one observation to the next is usually ignored. Formulas are derived in this book assuming independence among observations in the sample. Corrections to these formulas, which correct for the finite nature of the population, exist but are not considered in this book. The effect of the correction is negligible as long as the sample size n is less than about 10% of the size of the population.

Multivariate Random Variable

Each element selected for inclusion in the random sample in Definition 1, and each random variable X_i in Definition 2, may have several associated characteristics measured or observed by the experimenter. In that case the random variables representing the characteristics usually have two subscripts, such as Y_{ij}, where the first subscript identifies the individual element selected for the sample and the second subscript identifies which characteristic is being measured or observed.

That is, X_i may really represent the k-variate random variable $(Y_{i1}, Y_{i2}, \ldots, Y_{ik})$, where the X_is are still independent and identically distributed but where the individual Y_{ij} random variables within each X_i may or may not be independent and/or identically distributed.

As an example, consider the "dream" experiment just described. The random variable X_i could be the number of dreams counted during the ith night of observation. Then it may not be too unreasonable to assume that the X_is are independent, as defined by Definition 1.3.11, and identically distributed (meaning each X_i has the same distribution function). But suppose that each night the experimenter records not only the total number of dreams but also the total amount of sleep, which we call Y_{i1} and Y_{i2}, respectively. The number of dreams and the length of sleep during any one night may be related variables, so Y_{i1} and Y_{i2} are probably not independent. However, the sleep pattern on one night may be independent of the sleep pattern on another night. Mathematically, this means that the joint probability function of Y_{i1}, Y_{i2}, Y_{j1}, Y_{j2}, may be factored as follows

$$f(y_{i1}, y_{i2}, y_{j1}, y_{j2}) = f_1(y_{i1}, y_{i2}) f_2(y_{j1}, y_{j2}) \tag{1}$$

where f_1 and f_2 are the joint probability functions of (Y_{i1}, Y_{i2}) and (Y_{j1}, Y_{j2}), respectively. If the joint probability distribution of the sleep patterns does not change from one night to the next, f_1 is identical with f_2, and we say that (Y_{i1}, Y_{i2}) and (Y_{j1}, Y_{j2}) are identically distributed. A more convenient method of expressing the facts of "between" independence but not necessarily "within" independence, and "between" identical distributions but not necessarily "within" identical distributions is to let X_i represent Y_{i1} and Y_{i2} jointly. X_i is called a *bivariate random variable*, and a value of X_i actually consists of two numbers, one for Y_{i1} and one for Y_{i2}. Then all of the prior statements may be summarized by saying, "The X_is are independent and identically distributed."

Similarly, we may consider k measurements being taken each night, and the resulting k random variables Y_{i1}, Y_{i2}, . . . , Y_{ik} being represented by X_i, which is called a *k-variate random variable*, or also a *multivariate random variable*. Then independence of the X_is, in the sense of Definition 1.3.11, means that the joint probability distribution of all of the Y_is may be factored into the product of n joint probability functions, each being the joint probability function of Y_{i1}, Y_{i2}, . . . , Y_{ik} for some i. Identically distributed X_is mean that the joint probability functions just mentioned are identical functions.

We now have two definitions of random sample. The first definition applies only to samples from a finite population and may be directly related to the sample space. If each possible sample (of size n) is represented by one point in the sample space and if each point in the sample has equal probability of being selected as the sample, then the sampling method is random and the resulting sample is a random sample. The concepts of sample space and probability function are used in that definition, but there is no mention, explicit or implicit, of a random variable.

EXAMPLE I

A psychologist would like to obtain four subjects for individual training and examination. He advertises and 20 volunteers respond. He has several ways of selecting a sample of 4 from his sampled population of size 20.

He might select the first 4 to volunteer. Thus he may be biasing his selection toward those volunteers who tend to be more prompt or aggressive. This is probably not a random sample.

He might adhere strictly to Definition 1 and consider that there are $\binom{20}{4} = 4845$ ways in which a sample of size 4 may be selected. Then he obtains 4845 pieces of paper that are identical and writes 4 names on each piece of paper, a different combination each time, and puts them in a basket. One slip is randomly drawn, and those 4 people are used. This is a random sample, but the sampling method is not practical.

Another way of obtaining a random sample would be to write each of the names on a slip of paper, 20 slips in all, and one by one draw 4 slips in some random manner, such as from a hat. This method also satisfies the definition of a random sample. Computer programs are usually employed to simulate this procedure. ∎

The second definition of random sample is concerned directly with random variables and does not mention the sample space. However, since a random variable is a function defined on a sample space, a sample space is implicitly involved, although it remains in the background. Also, as mentioned in Section 1.3, the set of possible values of a random variable resembles a sample space. At times it will be necessary to list the points in this pseudo sample space in order to solve statistical problems that may arise. In fact, often no confusion will result if the possible measurements themselves (the values assumed by the random variables) are considered to be the points in the sample space. We usually think of these measurements as being numbers, but sometimes the numerical values of the measurements are obscure. So it would be well to discuss the various types of measurements.

Measurement Scales

The types of measurements are usually called *measurement scales* and are discussed at some length in various publications, including in an excellent paper by Stevens (1946). We will proceed from the "weakest" scale of measurement, the nominal scale, through the ordinal scale and the interval scale to the "strongest" scale, the ratio scale.

Nominal Scale

The *nominal scale* of measurement uses numbers merely as a means of separating the properties or elements into different classes or categories. The number assigned to the observation serves only as a "name" for the category to which the observation belongs, hence the title "nominal." We used the nominal scale of measurement when we defined a random variable that equaled 1 if a coin landed as a "head," and 0 if the coin landed as a "tail." We could, just as appropriately, have used the numbers 7.3 and 3.9 to represent head and tail, respectively. Our choice of 1 and 0 was primarily for convenience when we later desired to count the total number of heads in several tosses of the coin. When 12 subjects are arbitrarily numbered 1 to 12, a nominal scale of measurement is being used and the assignment of the numbers is a form of random variable. When classifying objects according to color, the categories may be labeled 1, 2, 3, or blue, yellow, red, or A, B, C. The numbers are merely category names. The numbers may be replaced by other unused numbers, as long as the categories remain intact.

Ordinal Scale

The *ordinal scale* of measurement refers to measurements where only the comparisons "greater," "less," or "equal" between measurements are relevant. The numeric value of the measurement is used only as a means of arranging the elements being measured in order, from the smallest to the largest. It is this ability to order the elements, on the basis of the relative size of their measurements, that gives the name of the *ordinal scale*. If some of the elements are equal to each other, we say ties exist. When a person is asked to assign the number 1 to the most preferred of three brands, the number 3 to the least preferred, and the number 2 to the remaining brand, she is using an ordinal scale of measurement and is using the numbers merely as a convenient way of representing her order of preference. Instead of the numbers 1, 2, 3 she could have used any three numbers, say 16, 20, 75, as long as the numbers are assigned to the brands in such a way that the relative order of the number represents the relative preference of the brand.

Interval Scale

The third scale, the *interval scale* of measurement, considers as pertinent information not only the relative order of the measurements as in the ordinal scale but also the size of the interval between measurements, that is, the size of the difference (in a subtraction sense) between two measurements. The interval scale involves the concept of a unit distance, and the distance between any two measurements may be expressed as some number of units. A good example is the scale by which we usually represent temperature. One unit (degree) increase in temperature is defined by a particular change in volume of mercury in a thermometer; consequently the difference between any two temperatures may be measured in units, or degrees. The actual numerical value of the temperature is merely a comparison with an arbitrary point called "zero degrees." The interval scale of measurement requires a zero point as well as a unit distance (it is not possible to have the latter without the former), but it is not important which measurement is declared to be zero or which distance is defined to be the unit distance. Temperature has been measured quite adequately for some time by both the Fahrenheit and Celsius scales, which have different zero temperatures and different definitions of 1 degree, or unit. The principle of interval measurements is not violated by a change in scale or location or both.

Ratio Scale

Finally, the *ratio scale* of measurement is used when not only the order and interval size are important, but also the ratio between two measurements is meaningful. If it is reasonable to speak of one quantity being "twice" another quantity, the ratio scale is appropriate for the measurement, such as when measuring crop yields, distances, weights, heights, income, and so on. Actually, the only distinction between the ratio scale and the interval scale is that the ratio scale has a natural measurement that is called zero, while the zero measurement is defined arbitrarily in the interval scale. As in the interval scale, the unit distance of the ratio scale is arbitrarily defined.

It is not possible to look at the measurements themselves in order to tell which scale of measurement is appropriate. Instead, one looks at the quantities being measured and the method of measurement and then determines the amount of meaning that may be attached to the numeric value of the measurement.

There is no universal agreement among scientists on these four scales of measurement. Some scientists prefer to use additional scales, and some measurements do not clearly fall into one of the four scales just defined. So this classification of measurements into four scales may be too simplistic, but it is sufficient for the purposes of this book.

Most of the usual parametric statistical methods require an interval (or stronger) scale of measurement. Most nonparametric methods assume either the nominal scale or the ordinal scale to be appropriate. Of course, each scale of measurement has all of the properties of the weaker measurement scales; therefore statistical methods requiring only a weaker scale may be used with the stronger scales also.

Statistics

Thus far in this section we have been concerned with populations, samples from populations, and measurement scales for measuring sample properties of interest. Measurement scales relate to random variables, because a system for measuring elements of the sample is in reality a random variable. Therefore measurement scales relate to statistics, because a statistic is a random variable. To a mathematical statistician the term "statistic" is interchangeable with the term "random variable." But popular usage of the word statistic indicates that it is more than just a random variable.

The word statistic originally referred to numbers published by the state, where the numbers were the result of a summarization of data collected by the government. Thus some people think of a statistic as a number that is based on several numbers, such as the average of several numbers in a sample, the proportion of a population that is in a particular category, and so on. In this sense a statistic is just a number. However, if we stop to consider that the numbers being averaged may vary from one sample to the next or that the population may change from one year to the next, we can justify extending our idea of a statistic from being only a number to being a rule for finding the number. Then "the average of the numbers in the sample" is the statistic, and the actual average obtained in one sample is a value of the statistic. As a rule for obtaining a number, a statistic meets the requirements of being a random variable, a function that assigns numbers to the points in the sample space (for an appropriately defined sample space). A statistic also conveys the idea of a summarization of data, so usually a statistic is considered to be a random variable that is a function of several other random variables. Then a value assumed by the statistic is implicitly assumed to be the result of some arithmetic operations performed on other numbers (the data) that, in turn, are the values assumed by several random variables. Since a random variable is a function defined on a sample space, a statistic may be defined as a function defined on a special sample space, a sample space whose points are the possible values of an *n*-variate random variable. A formal definition and an example may clarify the concept.

> **Definition 3**   A *statistic* is a function that assigns real numbers to the points of a sample space, where the points of the sample space are possible values of some multivariate random variable. In other words, a *statistic* is a function of several random variables.

Each sentence in Definition 3 would suffice as a definition of statistic. Both sentences are included for clarity.

EXAMPLE 2

Let $X_1, X_2, \ldots, X_n$ represent test scores of n students. Then each X_i is a random variable. Let W equal the average of the test scores.

$$W = \frac{X_1 + X_2 + \cdots + X_n}{n} = \frac{1}{n} \sum_{i=1}^{n} X_i \tag{2}$$

Then W is a statistic. If $X_1 = 76$, $X_2 = 84$, and $X_3 = 85$ represent the scores of three students, $W = (\frac{1}{3})(76 + 84 + 85) = 81\frac{2}{3}$. The statistic W satisfied the second sentence in Definition 3 by being a function of the random variables $X_1, X_2, \ldots, X_n$. The first sentence in Definition 3 is also satisfied because W assigns real numbers to the values of the multivariate random variable $(X_1, X_2, \ldots, X_n)$. In this case, if (X_1, X_2, X_3) assumes the multivariate value $(76, 84, 85)$, W assumes the value $81\frac{2}{3}$ as shown. This particular statistic is used often in statistics (the science). It is called the "sample mean" and will be discussed further in the next section.   ■

Order Statistic

We will often have occasion to use a particular class of statistics called *order statistics*, particularly when we are dealing with ordinal-type measurements. Suppose an observation $(x_1, x_2, \ldots, x_n)$ on a multivariate random variable $(X_1, X_2, \ldots, X_n)$ is "ordered"; that is, the elements are arranged from smallest to largest. We will denote the *ordered observation* by $x^{(1)} \leq x^{(2)} \leq \cdots \leq x^{(n)}$.

> **Definition 4** The *order statistic of rank k, $X^{(k)}$*, is the statistic that takes as its value the kth smallest element $x^{(k)}$ in each observation $(x_1, x_2, \ldots, x_n)$ of $(X_1, X_2, \ldots, X_n)$.

Therefore $X^{(1)}$, the order statistic of rank 1, always takes the smallest element in $(x_1, x_2, \ldots, x_n)$ as its value, and $X^{(n)}$ takes the largest. In Example 2, $X^{(1)} = 76$, $X^{(2)} = 84$, and $X^{(3)} = 85$. If another observation on (X_1, X_2, X_3) yields $(93, 73, 81)$, the values of the order statistics are $X^{(1)} = 73$, $X^{(2)} = 81$, and $X^{(3)} = 93$. If $(X_1, X_2, \ldots, X_n)$ is a *random sample*, sometimes $X^{(1)} \leq X^{(2)} \leq \cdots \leq X^{(n)}$ is called the *ordered random sample*.

Section 2.2 introduces many other useful statistics. There we will further discuss some uses of statistics in the analysis of experimental results.

EXERCISES

1. A congressional committee wishes to examine the effect of proposed legislation on the nation's high schools. It randomly selects five high schools from the Washington, D.C. area and conducts a study on those five schools.

    (a) What is the target population?
    (b) What is the sampled population?
    (c) If there are 100 high schools in the Washington, D.C. area, how many different samples are possible?
    (d) What is the probability of each sample in part c?

2. A Topeka television station asks the question "Should liquor by the drink be allowed in Kansas" and reports 372 phone calls, in which 164 persons said "no" and the remainder said "yes."

    (a) What was the target population?
    (b) What was the sampled population?
    (c) Was the sample a random sample? Explain.
    (d) Three statistics are indicated in this exercise. What are they, and what numerical values did they assume?
    (e) What measurement scale is used in the voting counts?
    (f) What measurement scale is used in registering each phone call as a "yes" or "no"?

3. A track meet awards a trophy to the team that accumulates the most points. A team receives 5, 3, or 1 point each time a member of that team finishes first, second, or third, respectively, in competition.

   (a) What measurement scale is used in awarding points?

   (b) Which statistic is (implicitly) mentioned, and what is it used for?

4. Football players on a team wear numbers on their uniforms. What measurement scale do these numbers represent?

5. What measurement scale is used in the following?

   (a) Postal zip codes.

   (b) Local telephone numbers.

   (c) Telephone area codes.

   (d) Social security numbers.

6. What measurement scale is used in the following?

   (a) Monthly salary.

   (b) Gallons measured on a gasoline pump.

   (c) The price of coffee per pound.

   (d) Intelligence as measured by IQ scores.

7. In order to select a law firm at random, a list of all lawyers of that city was obtained, and a lawyer was selected at random. The law firm to which that lawyer belonged was the selection. Was the law firm selected at random?

8. In order to estimate the number of people watching various TV shows, the following procedure is used. A random sample of 2200 households is obtained. The estimate is obtained from those households who agree to have their TV sets connected to an electronic device that keeps track of the programs watched for more than 8 minutes.

   (a) What was the target population?

   (b) What was the sampled population?

   (c) Comment on the accuracy of the results.

PROBLEMS

1. An experiment consists of n rolls of an unbalanced die. Let X_i be the number of spots showing on the ith roll. Does $X_1, X_2, \ldots, X_n$ constitute a random sample?

2. A random sample of size 4 is to be selected from among the integers 1 to 7, without replacement.

   (a) What is the total number of possible samples?

   (b) What is the probability of each sample?

   (c) What is the probability that the sample has at least one odd number?

   (d) What is the probability that the numbers in the sample sum to 12?

3. A random sample of size n is selected from among the integers 1 to N, without replacement. What is the probability that the sample has at least one odd number?

2.2 ESTIMATION

One of the primary purposes of a statistic is to estimate unknown properties of the population. The unknown properties that may be estimated are necessarily numerical and include items such as unknown proportions, means, probabilities, and so on. Actually, the estimate is based on a sample, a random sample if probability statements are to be made, and the estimate is an educated guess concerning some unknown property of the probability distribution of a random variable, where that random variable represents some quantity of interest in the population. For example, we might use the proportion of defective items in a sample of transistors as a statistic to estimate the unknown proportion of defective transistors in some population of transistors. A statistic that is used to estimate is called, quite naturally, an *estimator*. In this section we will discuss estimators such as the *sample mean*, the *sample variance*, and the *sample quantiles*. But first we will introduce the *empirical distribution function*, an estimator of a somewhat different kind.

Empirical Distribution Function

The true distribution function of a random variable is almost never known. Sometimes we make an educated guess as to the form of the distribution function and use our guess as an approximation of the true distribution function. One way of making a good guess is by observing several values of the random variable and constructing a graph $S(x)$ that may be used as an estimate of the entire unknown distribution function $F(x)$ of the random variable. The method of constructing the graph is best explained by an example, which follows this definition.

Definition 1 Let $X_1, X_2, \ldots, X_n$ be a random sample. The *empirical distribution function* $S(x)$ (called e.d.f. for short) is a function of x, which equals the fraction of X_is that are less than or equal to x for each x, $-\infty < x < \infty$.

EXAMPLE I

In a physical fitness study five boys were selected at random from the boys in a certain high school. They were asked to run a mile, and the time it took each of them to run the mile was recorded. The times (converted to fractions of a minute) were 6.23, 5.58, 7.06, 6.42, and 5.20, and they are presented on the horizontal axis

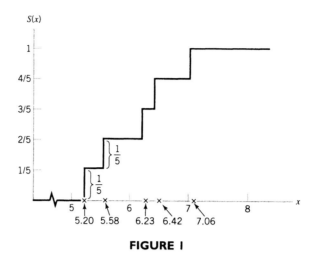

FIGURE 1

in Figure 1. The empirical distribution function $S(x)$ is the fraction of sample values less than or equal to x and, for this particular sample, is represented graphically in Figure 1. ∎

As in Example 1, the empirical distribution function is always a step function, where each step is of height $1/n$ and occurs only at the sample values. The vertical lines in Figure 1 are not part of the empirical distribution function but are included partly for appearance and partly for later convenience in determining sample quantiles. As we look at the graph of the empirical distribution function from left to right, we see that $S(x)$ equals zero until x equals the smallest value in the sample. Then $S(x)$ takes a step of $1/n$ in height. At each of the n sample values, $S(x)$ rises in height another distance of $1/n$. At the largest of the sample values, $S(x)$ reaches a height of 1.0 and remains 1.0 for all larger values of x. $S(x)$ resembles a distribution function in that it is a nondecreasing function that goes from zero to one in height. However, $S(x)$ is empirically (from a sample) determined and therefore its name.

Figure 1 represents merely one observation on $S(x)$. Another sample would have produced another and probably different graph of $S(x)$. This points out the random nature of $S(x)$. In a sense it is a random variable but, since it is a function and its observed values are entire graphs rather than single numbers, $S(x)$ is more properly called a *random function*. It is used as an estimator, since it does a reasonably good job of estimating the distribution function of the random variable, which we will call the population distribution function in order to distinguish it from the empirical (or "sample") distribution function.

In a sense, the observed value of an empirical distribution function may be considered a population distribution function. More precisely, an observed value of $S(x)$, based on the observations $x_1, x_2, \ldots, x_n$, in the sample, is identical to the distribution function of a random variable that may assume any of the numbers $x_1, x_2, \ldots, x_n$, each with probability $1/n$. The distribution function of such a random variable is a step function with jumps of height $1/n$ at each of the n numbers $x_1, x_2, \ldots, x_n$. We could find the mean, variance, and quantiles of the random variable simply by using the definitions of Chapter 1.

EXAMPLE 2

The random variable, which has a distribution function identical to the function $S(x)$ of Example 1, is the random variable X with the following probability distribution.

$$P(X = 5.20) = 0.2$$
$$P(X = 5.58) = 0.2$$
$$P(X = 6.23) = 0.2$$
$$P(X = 6.42) = 0.2$$
$$P(X = 7.06) = 0.2$$

The graph of the distribution function of X is the same as the graph in Figure 1. The median of X is 6.23 by Definition 1.4.1. The mean of X, by Definition 1.4.3, is given by

$$E(X) = \sum_x xf(x)$$

$$= (5.20)(0.2) + (5.58)(0.2) + (6.23)(0.2) + (6.42)(0.2) + (7.06)(0.2)$$

$$= 6.098 \tag{1}$$

Similarly, the variance of X may be found from Definition 1.4.4,

$$\text{Var}(X) = \sum_x (x - E(X))^2 f(x) \tag{2}$$

$$= 0.424$$

∎

Estimators

The mean, variance, and quantiles obtained from the sample, as illustrated in Example 2, will be called the *sample mean, sample variance,* and *sample quantiles* to distinguish them from the true "population" mean, variance, and quantiles. In the same way that the empirical distribution function serves as an estimator of the population distribution function, the sample mean, variance, and quantiles may be used as estimators of their population counterparts.

Definition 2 Let $X_1, X_2, \ldots, X_n$ be a random sample. The *pth sample quantile* is that number Q_p that satisfies the two conditions:

1. The fraction of the X_is that are less than Q_p is $\leq p$.
2. The fraction of the X_is that exceed Q_p is $\leq 1 - p$.

Each sample quantile may be found from the empirical distribution function in exactly the same way that the population quantile is obtained from the population distribution function. The pth sample quantile is that value of x where $S(x) = p$. If more than one number satisfies the condition that $S(x) = p$, we adopt the convention of using the average of the largest and the smallest numbers that satisfy $S(x) = p$, as we did with the population quantiles. The sample quantile Q_p depends on the random sample for its values; therefore it is a statistic. Note that for simplicity we have defined sample quantiles only for random samples.

One way to find the pth sample quantile directly from the sample and not from the graph of $S(x)$ is to multiply p by n, the sample size. Then round $p \cdot n$ upwards to the next higher integer, and find the order statistic with that rank. That observation is the pth sample quantile. If $(p \cdot n)$ is an integer, then the sample quantile is the average of the order statistic with that rank and the order statistic with the next higher rank.

EXAMPLE 3

Six married women were selected at random from among the married women in a ladies' civic club, and the number of children belonging to each was recorded. These numbers were 0, 2, 1, 2, 3, 4. The empirical distribution function is given in Figure 2. The sample median $Q_{0.5}$ is 2. The sample quartiles $Q_{0.25}$ and $Q_{0.75}$ are 1 and 3, respectively. The $1/3$ sample quantile $Q_{1/3}$ is the average of 1 and 2 by our convention, which equals 1.5. These numbers are our estimates of the unknown population quantiles. ∎

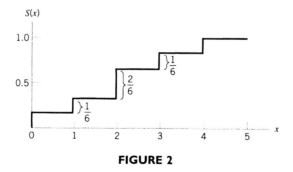

FIGURE 2

The sample mean and sample variance may be found in a simpler manner than in Example 2 by noting that $f(x) = 1/n$ in Equations 1 and 2, and may be factored out of the summation. This leaves us with simpler computation methods, which are given in the following definition.

Definition 3 Let $X_1, X_2, \ldots, X_n$ be a random sample. The *sample mean* $\overline{X}$ is defined by

$$\overline{X} = \frac{1}{n} \sum_{i=1}^{n} X_i \tag{3}$$

The *sample variance* S^2 is defined by

$$S^2 = \frac{1}{n} \sum_{i=1}^{n} (X_i - \overline{X})^2 \tag{4}$$

which is equivalent to

$$S^2 = \frac{1}{n} \sum_{i=1}^{n} X_i^2 - \overline{X}^2 \tag{5}$$

The *sample standard deviation* S is the square root of the sample variance.

EXAMPLE 4

In the random sample 0, 2, 1, 2, 3, 4 of Example 3 the sample mean is

$$\overline{X} = \tfrac{1}{6}(0 + 2 + 1 + 2 + 3 + 4)$$
$$= 2 \tag{6}$$

and the sample variance is

$$S^2 = \tfrac{1}{6}(2^2 + 0 + 1^2 + 0 + 1^2 + 2^2)$$
$$= 1\tfrac{2}{3} \tag{7}$$

Therefore our estimate of the unknown mean is 2 and our estimate of the unknown variance is $1\tfrac{2}{3}$. ∎

The estimators introduced thus far provide *point estimates* of the unknown population parameters, with the possible exception of the empirical distribution function. That is, our estimate of the unknown mean in the preceding example was provided by the statement, "Our estimate of the mean is 2." The single point "2" is the estimate.

It is usually preferred, but more difficult, to state the estimate as follows, "We are 95% confident that the unknown mean lies between 1.3 and 2.7." Such an estimate is called an *interval estimate*. An *interval estimator* consists of two statistics, one for each end of the interval, and the *confidence coefficient*, which is the probability that the interval estimator will contain the unknown population quantity. The confidence coefficient in the preceding statement is 0.95. The interval and the confidence coefficient together are usually called a confidence interval for the unknown quantity.

Point estimation is easy. To make a point estimate we need only to think of a number, any number. However, some point estimators are much better than others. Criteria for comparing point estimators, in order to determine which estimator we prefer, may be found in almost any introductory text in probability or statistics.

One of these criteria of a good estimator is called *unbiased*. In the following discussion the Greek letter θ represents a parameter in general, such as μ, σ, or ρ, while $\hat{\theta}$ represents a statistic that is used as an estimator of θ.

Definition 4 An estimator $\hat{\theta}$ is an *unbiased estimator* of a population parameter θ if $E(\hat{\theta}) = \theta$.

The following theorem shows that $\overline{X}$ is an unbiased estimator of the mean of the population from which the sample was obtained.

Theorem 1 Let $X_1, X_2, \ldots, X_n$ be independent random variables from a population with mean μ and variance σ^2. Then

$$E(\overline{X}) = \mu \tag{8}$$

and

$$\text{Var}(\overline{X}) = \sigma^2/n \tag{9}$$

Proof From Theorem 1.4.1,

$$E(X_1 + X_2 + \cdots + X_n) = E(X_1) + E(X_2) + \cdots + E(X_n) = n\mu \tag{10}$$

so

$$E(\overline{X}) = E\left(\frac{1}{n}\sum X_i\right) = \frac{1}{n}\sum E(X_i) = \frac{1}{n}n\mu = \mu \tag{11}$$

which shows that $\overline{X}$ is an unbiased estimator of the population mean. Also, from Theorem 1.4.3.

$$\text{Var}\left(\sum X_i\right) = \sum \text{Var}(X_i) = n\sigma^2 \tag{12}$$

Then, with some algebra, it follows that

$$\text{Var}(\overline{X}) = \text{Var}\left(\frac{1}{n}\sum X_i\right) = \frac{1}{n^2}\text{Var}\left(\sum X_i\right) = \frac{1}{n^2}n\sigma^2 = \sigma^2/n \tag{13}$$

which completes the proof.

Standard Error

The standard deviation of an estimator is usually called its *standard error,* so it won't be confused with the population standard deviation, which is something completely different. The standard error of $\overline{X}$ is $\sigma/\sqrt{n}$, as shown in Theorem 1.

The Unbiased Estimator s^2

It can be shown that S^2 is not an unbiased estimator for σ^2, and for this reason it has been more traditional to use the unbiased estimator s^2

$$s^2 = \frac{1}{n-1}\sum (X_i - \overline{X})^2 \tag{14}$$

as an estimator of σ^2. However, both S and s are biased estimators of the population standard deviation σ.

An Approximate Confidence Interval

From Theorem 1 and the central limit theorem (Theorem 1.5.2), if X_1, $X_2, \ldots, X_n$ are independent random variables, each with mean μ and variance σ^2, then the distribution function of

$$\frac{\sum X_i - n\mu}{\sqrt{n\sigma^2}} = \frac{\overline{X} - \mu}{\sigma/\sqrt{n}} \tag{15}$$

approaches the standard normal distribution function as n approaches infinity. For practical purposes this says that the probability

$$P\left(-z_{1-\alpha/2} < \frac{\overline{X} - \mu}{\sigma/\sqrt{n}} < z_{1-\alpha/2}\right) \tag{16}$$

is approximately $1 - \alpha$, if n is large, where $z_{1-\alpha/2}$ represents the $(1 - \alpha/2)$ quantile of a standard normal random variable.

The inequality above can be rewritten, using some algebra, as

$$P\left(\overline{X} - z_{1-\alpha/2}\frac{\sigma}{\sqrt{n}} < \mu < \overline{X} + z_{1-\alpha/2}\frac{\sigma}{\sqrt{n}}\right) \cong 1 - \alpha \tag{17}$$

which provides an approximate confidence interval for μ, valid for large n. Furthermore, since σ is seldom known, it is usually estimated using s for large n, resulting in an approximate confidence interval that is valid for random samples from all populations with finite, nonzero variances, as long as n is "large enough" for the central limit theorem to apply. For most practical purposes sample sizes larger than 30 are considered "large enough."

EXAMPLE 5

Large litters of pigs mean more profit for a hog farmer. A state experiment station is studying a new method of raising hogs that may result in large litter sizes. In 55 litters the average number of surviving pigs was 9.8, with $s = 1.4$. A 95% confidence interval for the population mean number of surviving pigs per litter has an approximate lower bound of

$$\overline{X} - z_{1-\alpha/2}\frac{s}{\sqrt{n}} = 9.8 - 1.96\,\frac{1.4}{\sqrt{55}} = 9.43 \tag{18}$$

and an approximate upper bound of

$$\overline{X} + z_{1-\alpha/2}\frac{s}{\sqrt{n}} = 9.8 + 1.96\,\frac{1.4}{\sqrt{55}} = 10.17 \tag{19}$$

They can state, with 95% confidence, that the true mean number of pigs per litter is between 9.43 and 10.17. ■

The Bootstrap

The mean and variance of $\overline{X}$ are given in Theorem 1. The mean of S^2 is also not difficult to derive. However, the variance of S^2 is not easy to find. Many statistics that are used as estimators of population parameters are difficult, or impossible, to treat theoretically as we did in Theorem 1, so other methods have been derived to estimate their means and variances. One such method is called the *bootstrap*.

The bootstrap method samples n values with replacement from the observations in the original random sample of size n. That is, some of the original observations may appear in the "bootstrap sample" once, more than once, or not at all. The bootstrap sample always has the same number of observations in it as the original random sample.

The estimator $\hat{\theta}$ of interest is calculated for each bootstrap sample. Hundreds, even thousands, of bootstrap samples are then obtained from the original random sample, using a computer of course, and for each bootstrap sample there is one value of $\hat{\theta}$. The sample mean of those hundreds, or thousands, of values of $\hat{\theta}$ is used to estimate the population mean of $\hat{\theta}$, and the sample standard deviation (using either s or S) is used to estimate the population standard deviation (the standard error) of $\hat{\theta}$. In fact, the entire empirical distribution function of these many values of $\hat{\theta}$ is used in the bootstrap method as an estimator of the true population distribution function of $\hat{\theta}$.

Clearly everything in the bootstrap procedure depends on the original sample values. A different set of sample values yields a different set of estimates.

Number of Bootstrap Replications

For simply estimating the mean and the standard error of an estimator, the number of bootstrap replications seldom needs to be more than 100 or 200, and as few as 25 replications can be very informative. However, larger numbers of replications are needed to find confidence intervals. One method of obtaining an approximate confidence interval for θ is to use the $\alpha/2$ and $1 - \alpha/2$ sample quantiles from the bootstrap sample estimator $\hat{\theta}^*$, and for this Efron and Tibshirane (1986) recommend a minimum of 250 bootstrap replications. They also present alternative, more accurate methods of obtaining confidence intervals for bootstrap samples, for which more replications, at least 1000, are recommended.

EXAMPLE 6

In Example 5 we found an approximate 95% confidence interval for the mean number of pigs per litter using the central limit theorem that applies to $\overline{X}$. Now we will use the bootstrap method to find an approximate 95% confidence interval for the population standard deviation σ of the number of pigs per litter. This parameter is useful to examine because it is better to have consistent litter sizes (small σ) than to have some litters be very small and others very large (large σ).

The original sample of 55 values is numbered from 1 to 55.

Observation Number	1	2	3	4	. . .	55		
Litter Size	9	9	8	6	. . .	11	(55 numbers)	$s = 1.4$

Now the first bootstrap sample is obtained by sampling the numbers from 1 to 55, with replacement, to obtain 55 numbers, and the estimator s^* is calculated from the bootstrap sample.

Bootstrap Sample #1:

Observation Number	4	17	4	28	. . .	16		
Litter Size	6	9	6	10	. . .	9	(55 numbers)	$s_1^* = 1.6$

This procedure is repeated 250 times. (Bootstrap procedures can replicate as many times as one desires, but 250 replications is recommended as the minimum number for finding a confidence interval.)

Bootstrap Sample #2:

Observation Number	28	23	3	16	. . .	39		
Litter Size	10	10	8	9	. . .	8	(55 numbers)	$s_2^* = 1.8$

And so on, to

Bootstrap Sample #250:

Observation Number	6	1	55	14	. . .	17		
Litter Size	10	9	11	11	. . .	9	(55 numbers)	$s_{250}^* = 1.1$

To find an approximate lower bound in the 95% confidence interval the 0.025 sample quantile of the s^*'s is found. Because 0.025(250) = 6.25, which is rounded up to 7, the 7th order statistic is found. The lower bound on this confidence interval is $s^{*(7)}$, the 7th order statistic. To find an approximate upper bound the 0.975 sample quantile is found, which is the 244th order statistic (0.975 × 250 = 243.75, which rounds up to 244), making $s^{*(244)}$ the upper bound. In this case the 250 values of s^*, ordered from smallest to largest, were

0.7,   0.8,   0.8,   0.9,   0.9,   0.9,   <u>1.0</u>,   . . . ,   <u>2.0</u>,   2.0,   2.2,   2.3,   2.3,   2.4,   2.7

so the approximate 95% confidence interval is from 1.0 to 2.0. The standard error of s is estimated by computing S on the 250 values of s^*, just as the expected value of s is estimated by computing $\overline{X}$ on the 250 values of s^*, etc.   ■

Computer Assistance

Virtually all statistics packages for the computer, and even many inexpensive hand-held calculators, compute the point estimates we have discussed. However, the bootstrap is not as easy to find. Programs to find bootstrap estimates may be found in *S-Plus, SYSTAT, Resampling Stats,* and *STATA.* See the Preface for more information about these programs. The book on resampling by Davison and Hinkley (1997) contains a library of routines for use with *S-Plus.*

Parameter Estimation in General

To summarize the process of estimating an unknown parameter θ, the two main questions are:

1. What estimator should be used? We suggest following the example of how θ is defined in terms of the population distribution function $F(x)$, and defining $\hat{\theta}$ in a similar way in terms of the empirical distribution function $S(x)$, as we did to find $\hat{\mu} = \overline{X}$, $\hat{\sigma} = S$, and the quantile estimator $\hat{x}_p = Q_p$.

2. How good is the estimator? This answer is usually found in the standard deviation of the estimator, called its *standard error.* The standard error of $\overline{X}$ is given in Theorem 1 as $\sigma/\sqrt{n}$. The standard errors of other estimators are not as easy to find, but may be estimated using the bootstrap method. The bootstrap also yields an estimate of the entire distribution function of $\hat{\theta}$, and an approximate confidence interval for θ.

□ *T h e o r y*   The theoretical justification for this approach to estimation lies in the fact that as n gets large, $S(x)$ approaches $F(x)$ in probability (the precise definition of "in probability" is not covered in this book), and thus parameters based on $F(x)$ may be estimated using $S(x)$. In most cases of interest those estimators will approach (in probability) the parameters they are estimating, making them good estimators. Repeated sampling from $F(x)$ will provide the approximate distribution function of the estimator, so repeated sampling from $S(x)$ will do nearly as well, if the sample size is large. For an introduction to the bootstrap concept see Efron and Tibshirani (1986).☐

Survival Function

The empirical distribution function $S(x)$ is a valuable link between the random sample $X_1, X_2, \ldots, X_n$ and the population distribution function $F(x)$, because it uses the relative frequency of observations less than or equal to every x to estimate $P(X \le x) = F(x)$. Equally useful in life testing, medical follow-up, and other fields is the *survival function* $P(x) = 1 - F(x)$, in which the variable of interest is the *lifetime* (time till death) of a person, an animal, an inanimate product, or simply the time until the occurrence of an event such as a cure, an arrival, a departure, and so on.

The natural estimator of $P(x)$ is the empirical survival function

$$\hat{P}(x) = 1 - S(x) \tag{20}$$

which is the relative frequency of the sample $X_1, \ldots, X_n$ that exceeds x in value.

The Kaplan-Meier Estimator

Kaplan and Meier (1958) call X the time to death, and note that the time to death may be unobservable in some cases because of the *loss* of the item from the experiment (subjects moving away, subjects entering the experiment late, the experiment ending before all subjects die, etc.). They provide a method for using some information from the lost data, namely that *death* did not occur before the loss. They use the fact that if death occurs after time x, then death also occurred after all times prior to x. Here is their reasoning.

From the definition of conditional probability (Eq. 1.2.2) we have, for $x_0 < x_1$,

$$P(X > x_1) = P(X > x_1, X > x_0) = P(X > x_1 \mid X > x_0)P(X > x_0) \tag{21}$$

Suppose that 100 items are put on test at the beginning of Year 1, and at the end of Year 1 only 30 survive. We could estimate $P(1)$ using

$$\hat{P}(1) = \hat{P}(X > 1) = 30/100 = 0.3 \tag{22}$$

where X represents the lifetime of the item.

Then at the beginning of Year 2 suppose an additional 1000 items are put on test. At the end of Year 2, 250 of the 1000 items survive, and 10 of the 30 items survive from the original 100 items. We could estimate $P(2)$ using only the original 100 items,

$$\hat{P}(2) = \hat{P}(X > 2) = 10/100 = 0.1 \tag{23}$$

However, at this point we can use the information from the 1000 items that have been on test for one year to update our estimate of $P(1)$. Since a total of 1100 items have been on test one year, with a total of $250 + 30 = 280$ survivors, an improved estimate of $P(1)$ would be

$$\hat{P}(1) = \hat{P}(X > 1) = 280/1100 = 0.255 \qquad (24)$$

Furthermore, we can use this improved estimate of $P(1)$ to get an improved estimate of $P(2)$, one which uses the fact that 280 of 1100 items survived past 1 year, with the aid of Equation 21.

$$P(2) = P(X > 2) = P(X > 2 \mid X > 1)P(X > 1) \qquad (25)$$

Our improved estimate $\hat{P}(X > 1)$ is 0.255. Unfortunately we are unable to improve our estimate of $P(X > 2 \mid X > 1)$ because we don't know what happens to the 1000 observations in the second year of the test. So we use the estimator

$$\hat{P}(X > 2 \mid X > 1) = 10/30 \qquad (26)$$

because it uses the only available information, namely, that of the 30 items for which information exists past the end of the Year 1, 10 survived the end of Year 2 also. Then our improved estimate of $P(2)$ is

$$\hat{P}(2) = \hat{P}(X > 2 \mid X > 1)\hat{P}(X > 1) = \frac{10}{30}\frac{280}{1100} = 0.085 \qquad (27)$$

Kaplan and Meier extended this line of reasoning to include the general case. Let $u_1 < u_2 < \cdots < u_k$ represent k "lifetimes," where a lifetime may be a time to death, or the time until an item is lost to the study. Also let

$$p_i = P(X > u_i \mid X > u_{i-1}) \qquad (28)$$

which is estimated using

$$\hat{p}_i = \frac{\text{Number of items known to survive past time } u_i}{\text{Number of items still being observed past time } u_{i-1}} \qquad (29)$$

Items that are lost at time u_i are considered to survive past time u_i, because at time u_i they are still known to be alive. However, items that die at time u_i do not survive past time u_i. The denominator of $\hat{p}_1$, computed at the first death or loss, is the total number of items put on test.

The Kaplan-Meier estimator of $P(x)$ is

$$\hat{P}(x) = 1 \text{ for } x < u_1$$
$$= \prod_{u_i \leq x} \hat{p}_i \text{ for } x \geq u_1 \tag{30}$$

where the product is over all lifetimes $u_i \leq x$. Note that this estimator is a decreasing step function that takes steps only at observed deaths. Also note that this method allows a more general definition of the empirical distribution function $S(x)$ also, by using the fact that $S(x) = 1 - \hat{P}(x)$, for cases where there are losses to the study.

Computer Assistance

Survival curves, and in particular Kaplan-Meier estimates, may be found in *Minitab*, *S-Plus*, and *SYSTAT*. These packages also contain methods of handling other variations of censored data that are sometimes encountered when trying to estimate a survival curve, which are not discussed in this book.

EXAMPLE 7

Ten fanbelts are tested by placing them on cars and records are kept of the mileage on each car when the fanbelt breaks. At the end of the test five fanbelts have broken, with lifetimes (in thousands of miles) of 77, 47, 81, 56, and 80. The other five fanbelts are still unbroken, and the mileages on those cars are 62, 60, 43, 71, and 37. The Kaplan-Meier estimate of the survival function is found as follows.

i	u_i	result	$\hat{p}_i$	$\hat{P}(u_i)$
1	37	lost	10/10	1
2	43	lost	9/9	1
3	47	death	7/8	0.875
4	56	death	6/7	0.75
5	60	lost	6/6	0.75
6	62	lost	5/5	0.75
7	71	lost	4/4	0.75
8	77	death	2/3	0.5
9	80	death	1/2	0.25
10	81	death	0/1	0

A graph of $\hat{P}(x)$ for all $x > 0$ is shown in Figure 3. ■

The Kaplan-Meier estimator is the same as $1 - S(x)$ if there are no losses, only deaths, starting at $\hat{P}(x) = 1$ and decreasing by steps of height $1/n$ at each death, until $\hat{P}(x) = 0$ at the final death. If there are losses as well as deaths, $\hat{P}(x)$

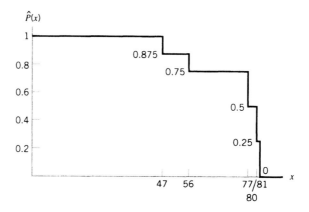

FIGURE 3   The Kaplan-Meier estimate of the Survival Function $P(x)$

starts at 1.0, but the decreasing steps may no longer be of uniform height. Also if there is a loss after the final known death, $\hat{P}(x)$ will not decrease to zero, and is not defined for x beyond the final known loss. In this case $S(x)$ is not suitable for the estimation of some parameters associated with $F(x)$ such as the mean and the variance using the usual methods described earlier in this section, but may be used to estimate some quantiles. Special methods are suggested by Kaplan and Meier (1958).

In a sense, point estimation is always a nonparametric statistical method, because no knowledge of the form of the unknown distribution function is required in order to make a point estimate. This was shown by the examples in this section, where point estimates were made without knowing anything about the unknown distribution function.

It is more difficult to tell whether the methods of forming confidence intervals are parametric or nonparametric. If no knowledge of the form of the distribution function is required in order to find a confidence interval, that method is clearly nonparametric. The approximate methods illustrated in Examples 5 and 6 are nonparametric. On the other hand, if the method requires that the unknown distribution function be a normal distribution function (see Definition 1.5.3), or some other specified form, the method is parametric. Several other nonparametric methods of forming confidence intervals will be presented later, in Sections 3.1, 3.2, 5.1, 5.5, 5.7, and 6.1.

EXERCISES

1. Ten persons are selected at random from among all persons living in a particular community. The taxable incomes for five of these persons in the previous calendar year were $8600, $15,200, $16,200, $16,400, and $29,600; there was no income for the other five people.

(a) Draw a graph of the empirical distribution function.

(b) Find the sample median income.

(c) Find the sample mean income.

(d) Find the sample variance.

(e) Find the sample standard deviation.

2.   In five consecutive games a certain basketball team had scores of 73, 68, 86, 78, and 65.

(a) Draw a graph of the empirical distribution function.

(b) Find the sample upper quartile.

(c) Find the sample interquartile range.

(d) Find the sample mean.

(e) Find the sample standard deviation.

3.   A random sample of five "12 ounce" cereal boxes actually had contents weighing 12.6, 13.0, 12.1, 11.8, and 12.1 ounces.

(a) Draw a graph of the empirical distribution function.

(b) Find the upper and lower sample quartiles, and the interquartile range.

4.   Six students took an exam and got the scores 81, 85, 89, 90, 90, and 98.

(a) Draw a graph of the empirical distribution function.

(b) Find the sample interquartile range.

5.   Using the same procedure for finding point estimators used in this section, find a point estimator for the probability $P(Y \leq c)$ for a given number c, based on a random sample $X_1, X_2, \ldots, X_n$ with the same distribution function as Y. In other words, if $X_1, X_2, \ldots, X_n$ is a random sample with the distribution function $F(x)$, estimate $F(c)$. Use your estimator to estimate the probability that the score in the next game will exceed 80 in Exercise 2.

6.   Using the same procedure for finding point estimators used in this section, find a point estimator for the range of a random variable. Will this sample range ever be larger than the population range? Will it ever be smaller? Is the expected value of the sample range smaller than the population range?

7.   The Minimum Monthly Balance is examined for a random sample of 175 checking accounts from the City Bank. The sample mean was $1156 and the sample standard deviation was $855. Find an approximate 90% confidence interval for the Average Minimum Monthly Balance of all 14,000 checking accounts at the City Bank.

8.   An academic skills test was given to a random sample of 50 seniors at Central High School. The average score was 81% and the standard deviation was 11%. Find a 95% confidence interval for the average score of all 1159 seniors at Central High School.

9.   Eighteen new members of the health club were given a fitness test. Their Percent Overweight X was recorded, and their Heartbeat Adjustment After Treadmill Exercise Y was measured. The sample correlation coefficient was $r = 0.35$. In order to estimate a 95% confidence interval for the true correlation coefficient of all new members, past, present, and future,

the bootstrap method was used. Three hundred bootstrap samples, ordered from smallest to largest, yielded the ten smallest values of r^* to be

$$-0.15, -0.06, -0.02, 0.01, 0.03, 0.03, 0.05, 0.06, 0.07, 0.09$$

and the ten largest values of r^* to be

$$0.51, 0.53, 0.53, 0.55, 0.56, 0.57, 0.59, 0.59, 0.60, 0.62.$$

Find an approximate 95% confidence interval for the population correlation coefficient.

10. Suppose that in Exercise 9, the sample mean of the 300 bootstrap values of r^* is 0.30 and the standard deviation of the 300 bootstrap values of r^* is 0.12. What is your estimate of the standard error of r? What sample size is this estimate valid for?

11. Eight light bulbs were randomly sampled from a large batch, and put on test to determine the survival function of that batch of light bulbs. The testing mechanism holds only four bulbs, so when one bulb burns out another takes its place. Bulbs burned out at 187 hours, 196 hours, 206 hours, 210 hours, and 273 hours. At the termination of the experiment, bulbs were still burning that had been on test 127 hours, 190 hours, and 194 hours. Use the Kaplan-Meier estimator to estimate the survival function for that batch of light bulbs.

12. In September 1997, 100 students were admitted to the Dental Hygienist Program, a two-year program. Seventy-two of them completed the first year, and 55 completed both years of the program. In September 1998 another 100 students were admitted to the same program, and 57 completed the first year. Use the Kaplan-Meier estimator to estimate the probability of a student, admitted in September 1999, completing the two-year program.

PROBLEMS

1. Since an estimator is a random variable, given enough information we should be able to find the probability distribution of an estimator. Suppose a finite population consists of four elements, with the respective measurements 4, 6, 7, and 10. A random sample of size 2 is drawn without replacement from the population.

   (a) How many possible random samples are there?
   (b) List the possible samples.
   (c) What is the probability of drawing each of the samples listed in part b?
   (d) What is the sample median for each of the samples listed in part b?
   (e) What is the probability of getting each of the sample medians listed in part d?
   (f) Graph the distribution function of the sample median.
   (g) Use the same procedure just outlined and obtain the probability function of the sample range.

2. A statistic is said to be an *unbiased estimator* of a population parameter if the mean of the estimator equals the parameter.

   (a) From Problem 1, find the mean of the sample median. Does this equal the population median? Is the sample median an unbiased estimator of the population median?

(b) From Problem 1, find the mean of the sample range. Does this equal the population range? Is the sample range an unbiased estimator of the population range? (Compare with Exercise 6.)

2.3 HYPOTHESIS TESTING

Statistical inference has many forms. The form that has received the most attention by the developers and users of nonparametric methods is called hypothesis testing and is treated in this section and the next.

Hypothesis testing is the process of inferring from a sample whether or not a given statement about the population appears to be true. The statement itself is called the hypothesis. Examples of hypotheses include statements such as these.

1. Women are more likely than men to have automobile accidents.
2. Nursery school helps a child achieve better marks in elementary school.
3. The defendant is guilty.
4. Toothpaste A is more effective in preventing cavities than toothpaste B.

A nonstatistical test of a particular hypothesis may be very simple to perform. We may observe a set of data related to the hypothesis, or a set of data not related to the hypothesis, or perhaps no data at all, and arrive at a decision to accept or reject the hypothesis, although that decision may be of doubtful value. However, the type of hypothesis test we will discuss is more properly called a statistical hypothesis test, and the test procedure is well defined. Here is a brief outline of the steps involved in such a test.

1. The hypotheses are stated in terms of the population. There are always two hypotheses involved. The statement that the experimenter would like to prove is called the *alternative hypothesis*, or in quality control this is the statement that says the quality of the product or service is unsatisfactory or "out of control." Typical alternative hypotheses are, "The new product is better than the old product," or, "This medication is effective in curing the illness." Sometimes the alternative hypothesis is referred to as the *research hypothesis*.

   The negation of the alternative hypothesis is then called the *null hypothesis* or the *test hypothesis*. This is the hypothesis that is tested in a hypothesis test. The null hypotheses corresponding to the above examples of alternative hypotheses are, "The new product is no better than the old product," and, "This medication is not effective in curing the illness." In quality control the null hypothesis states that the quality of the product or service is satisfactory.

   If the data in the sample strongly disagree with the null hypothesis, the null hypothesis is rejected. If the data in the sample do not conflict with the null hypothesis, or if there are insufficient data to show a conflict with the null hypothesis, the experimenter "fails to reject" the null hypothesis. Sometimes the experimenter says "the null hypothesis is accepted," which means the same thing, but this statement should not be misinterpreted to mean the data prove the null hypothesis to be true. "Accept the null hypothesis" simply means the null hypothesis has failed to be rejected.

2. A *test statistic* is selected. A good test statistic is one that tends to take on some values when the null hypothesis is true, and tends to take on other values when the null hypothesis is false. That is, a good test statistic is a sensitive indicator of whether the data agree or disagree with the null hypothesis.

3. A *decision rule* is made, in terms of possible values of the test statistic, for deciding whether to accept or reject the null hypothesis.

4. On the basis of a random sample from the population, the test statistic is evaluated, and a decision is made to accept or reject the null hypothesis.

A more precise description of the testing procedure follows Example 1.

EXAMPLE 1

A certain machine manufactures parts. The machine is considered to be operating properly if 5% or less of the manufactured parts are defective. If more than 5% of the parts are defective the machine needs remedial attention. The *null hypothesis*

$$H_0: \text{The machine is operating properly}$$

is the hypothesis to be tested. The *alternative hypothesis*

$$H_1: \text{The machine needs attention}$$

is the hypothesis we want to be able to detect, if it is true. H_0 will be tested on the basis of a random sample of 10 parts, from the population of all parts being produced by the machine. If H_0 is rejected we will need to take remedial action to restore the machine to its proper mode of operation.

The assumption is made that each part has the same probability p of being defective, independently of whether or not the other parts are defective. Therefore, in the assumed model, the original hypotheses H_0 and H_1 are equivalent to

$$H_0: p \leq 0.05$$
$$H_1: p > 0.05$$

We feel that if too many parts are defective, we should reject H_0. So let the test statistic T be the total number of defective items. Then, according to Example 1.3.5, T has the binomial distribution with parameters p, and 10 for n. From

Table A3 we see that if H_0 is true ($p \leq 0.05$), then

$$P(T \leq 2) \geq 0.9885 \tag{1}$$

equaling 0.9885 if $p = 0.05$, and therefore

$$P(T > 2) \leq 0.0115 \tag{2}$$

equaling 0.0115 if $p = 0.05$. Because the probability of rejecting H_0 when H_0 is true will be small, less than or equal to 0.0115, we decide to reject H_0 if T exceeds 2. The set of points in the sample space that correspond to values of T greater than 2 is called the *critical region*. The decision rule is this: Reject H_0 if the observed outcome is in the critical region (when T exceeds 2); otherwise, accept H_0.

Suppose a random sample consisting of 10 machined parts is observed and 4 of the parts are found to be defective. Then $T = 4$ and the null hypothesis is rejected. We conclude that the machine needs attention. ∎

In Example 1 assumptions are made concerning the conditions under which the data are collected and the type of data collected. These assumptions are tantamount to forming a model, or idealized experiment. "Under the model" means "under the assumptions." The experimenter tries to collect the data in such a way that these assumptions are met as closely as possible.

Under the model, the original hypotheses may be restated in an equivalent form, usually using statistical terminology. These hypotheses may be classified as either *simple* or *composite*.

Definition 1 The hypothesis is *simple* if the assumption that the hypothesis is true leads to only one probability function defined on the sample space. The hypothesis is *composite* if the assumption that the hypothesis is true leads to two or more probability functions defined on the sample space.

In the example, the model induces the binomial probability $\binom{10}{k} p^k (1 - p)^{10-k}$ on each sample point with k defective items and $10 - k$ nondefective items. This represents a whole class of probability functions defined on the sample space, depending on what value p has. (For each point, k is known.) Assume H_0 is true. Still, p may be any value from 0 to 0.05, so there are several possible probability functions, and H_0 is a composite hypothesis. The same is true for H_1. The hypothesis "$p = 0.05$" would be a simple hypothesis because, assuming $p = 0.05$ is true, the probability function assigns the probability $\binom{10}{k} (0.05)^k (0.95)^{10-k}$ to a point representing k defective parts, and that probability function is well defined (no unknown parameters) and the only one possible.

Definition 2 A *test statistic* is a statistic used to help make the decision in a hypothesis test.

Critical Region

A desired property of a test statistic is that it should assign real numbers to the points in the sample space in such a way that the points are arranged in some order corresponding to their ability to distinguish between a true H_0 and a false H_0. For example, the points that indicate most strongly that the experimenter should reject H_0 might be given large values by the test statistic, and the points that indicate that the experimenter should accept H_0 might be given small values by the test statistic. Then the larger the value assumed by the test statistic, the more the outcome of the experiment indicates that H_0 should be rejected. In this way, all values of the test statistic greater than a certain number might result in the decision to reject H_0. Furthermore, this enables the experimenter to determine objectively how much smaller or larger the rejection region might have been and still result in the same decision. Such a test, where the rejection region corresponds to the largest values of the test statistic, is called an *upper-tailed test*. Similarly, if the ordering is reversed so that the rejection region corresponds to the smallest values of the test statistic, the test is called a *lower-tailed test*.

These are both *one-tailed tests*. The test in the example was one-tailed. If the test statistic is selected so that the largest values of the test statistic and the smallest values of the test statistic, combined, correspond to the rejection region, the test is called a *two-tailed test,* since the rejection region corresponds to both "tails" of the test statistic's possible values.

> **Definition 3** The *critical region* is the set of all points in the sample space that result in the decision to reject the null hypothesis.

Sometimes the critical region is called the *rejection region,* and the set of all points in the sample space not in the critical region is called the *acceptance region,* for obvious reasons.

Error Types

There are two ways of making an incorrect decision in hypothesis testing. If the null hypothesis is true we might make the mistake of rejecting it, thus committing an error known as an *error of the first kind*, or a *type I error*. That is, a type I error occurs when H_0 is true and yet the outcome of our experiment is in the critical region.

> **Definition 4** A *type I error* is the error of rejecting a true null hypothesis.

The second way of committing an error in hypothesis testing is by accepting the null hypothesis when the null hypothesis is false. This error is known as an *error of the second kind,* or a *type II error*.

Definition 5 A *type II error* is the error of accepting a false null hypothesis.

Level of Significance

These two error types have associated with them certain probabilities of the errors being made. Consider first the probability of making a type I error.

> **Definition 6** The *level of significance*, or α, is the maximum probability of rejecting a true null hypothesis.

The level of significance may be found by first assuming H_0 is true and then ascertaining the probability of getting a point in the critical region. If H_0 is a simple hypothesis, the assumption that H_0 is true leads to only one probability function defined on the sample space, and α may be found by adding the probabilities of all points in the critical region. Usually, however, it is easier to find α by computing the probability that the test statistic will assume one of the values that results in rejection of H_0, under the assumption that H_0 is true.

Null Distribution

It is necessary in statistical hypothesis testing to know the probability distribution of the test statistic when the null hypothesis is true. This is called the *null distribution* of the test statistic.

> **Definition 7** The *null distribution* of the test statistic is its probability distribution when the null hypothesis is assumed to be true.

In Example 1 the null distribution of the test statistic T, the number of defective parts in the sample of 10 parts, is the binomial distribution with parameter $p \leq 0.05$. This was a result of the assumptions of independence and constant probability p that we made. The level of significance in every statistical hypothesis test is found from the null distribution of the test statistic.

If H_0 is a composite hypothesis, α is the *maximum* probability of rejecting H_0, where the maximum is obtained by considering all of the probability distributions possible when H_0 is true. In the example H_0 was composite, and the probability of rejecting a true null hypothesis was

$$P(\text{reject a true } H_0) = P(T > 2 \mid H_0 \text{ is true})$$

$$= \sum_{i=3}^{10} \binom{10}{i} p^i (1-p)^{10-i}; \qquad p \leq 0.05 \tag{3}$$

which differs for each value of p. However the probability in Equation 3 is a maximum when p is a maximum. The maximum value of p, under H_0, is 0.05, so the level of significance is given by

$$\alpha = \text{maximum } P(T > 2 \mid H_0 \text{ is true})$$
$$= P(T > 2 \mid p = 0.05)$$
$$= 0.0115 \tag{4}$$

from Table A3 or from Equation 2.

The level of significance is sometimes called the *size of the critical region,* for obvious reasons. If H_0 is true the maximum probability of rejecting H_0 is α and, therefore, the minimum probability of accepting H_0, making the correct decision, is $1 - \alpha$.

The probability of committing an error of the second kind is denoted by β. Obviously it is desirable in hypothesis testing for α and β to be close to zero. In practice the sample size helps determine how small α and β may become. Only when the sample includes all of the information contained in the population may the possibility of error be completely eliminated.

Power

If H_0 is false the decision may be to accept H_0, with a probability β, or to reject H_0, with a probability $1 - \beta$. This latter probability represents the *power of the test* to detect a false null hypothesis.

> **Definition 8**   The *power,* denoted by $1 - \beta$, is the probability of rejecting a false null hypothesis.

Unlike α, the power is not always a unique number. If H_1 is simple the assumption that H_1 is true (equivalent to "H_0 is false") leads to one probability function and hence one probability of rejecting H_0, or getting a point in the critical region. Thus, when H_1 is simple, $1 - \beta$ is unique. If H_1 is composite each probability function, under H_1, has a possibly different value for $1 - \beta$, so the power depends on the various possible probability functions.

		The Decision	
		Accept H_0	Reject H_0
The True Situation	**H_0 is true**	Correct decision probability $= 1 - \alpha$	Type I error probability $= \alpha$ (level of significance)
	H_0 is false	Type II error probability $= \beta$	Correct decision probability $= 1 - \beta$ (power)

Now that the error types have been discussed, we can return to the topic of the critical region. Although the critical region was discussed, no mention was made concerning how it is selected. If the test statistic has been chosen so as to result in a one- or two-tailed test, the selection of a critical region depends only on the experimenter's preference concerning the size of the critical region, the level of significance. Usually a desirable decrease in the level of significance α is accompanied by an undesirable increase in β. Our two objectives in hypothesis testing are to reject H_0 as seldom as possible if H_0 is true and as often as possible if H_0 is false. As a result, the critical region is usually the set of points with the largest value of $1 - \beta$, from among those sets of points of some fixed size α. By convention more than any other reason, α is usually chosen near 0.05 or 0.01, and the critical region is then selected in terms of possible values of the test statistic.

p-Value

The results of a hypothesis test are much more meaningful if the *p-value* is also stated.

Definition 9 The *p-value* is the smallest significance level at which the null hypothesis would be rejected for the given observation.

Let t_{obs} represent the observed value of a test statistic T. In an upper-tailed test the p-value is $P(T \geq t_{obs})$ computed using the null distribution of T. In a lower-tailed test the p-value is $P(T \leq t_{obs})$.

In a two-tailed test, the p-value can be stated as twice the smaller of the one-tailed p-values. Strictly speaking, this disagrees with the definition of p-value if the null distribution of T is discrete and the upper tail of the rejection region doesn't have the same probability as the lower tail of the rejection region, making it impossible to construct an exact significance level with equal probabilities in both tails. However, to avoid ambiguities, we will consider the two-tailed p-value to be twice the one-tailed probability for the tail of the null distribution in which the observed value falls.

In Example 1 the test is upper-tailed and the observed value of T is 4, so the p-value is $P(T \geq 4 \mid p = 0.05) = 0.0010$ from Table A3. The p-value is sometimes abbreviated to simply p, but in Example 1 that symbol represented the probability of a defective so there it is better to use the full term "p-value" to avoid confusion.

In many publications of research results the statistical test is abbreviated to the point where only the name of the test, the hypotheses, and the p-value are reported. The null hypothesis is rejected if the p-value is less than or equal to α, which is usually understood to be 0.05.

EXAMPLE 2

In order to see if children with nursery school experience perform differently academically than children without nursery school experience, 12 third-grade students are selected for study, 4 of whom attended nursery school. The hypothesis to be tested is

H_0: The academic performance of third-grade children does not depend on whether or not they attended nursery school

The alternative hypothesis is

H_1: There is a dependence between academic performance and attendance at nursery school

The model assumes that the 12 children are a random sample of all third-grade children, and also that the children can be ranked from 1 to 12 (best to worst) academically. The "dependence" in the hypotheses is assumed to mean either the nursery school children tend to do better as a group or they tend to do worse than the nonnursery school children. Under the model the hypotheses may be restated as

H_0: The ranks of the four children with nursery school experience are a random sample of the ranks from 1 to 12

H_1: The ranks of children with nursery school experience tend to be higher or lower as a group than a random sample of 4 ranks out of 12

We choose as a test statistic T, the sum of the ranks of the 4 children who attended nursery school. We decide to let the critical region correspond to values of T that are either very large or very small, so the test is two tailed.

Each possible outcome consists of 4 numbers from 1 to 12, corresponding to the ranks of the 4 children who attended nursery school. Therefore there are $\binom{12}{4} = 495$ points in the sample space. To decide which of these points to include in the critical region, we will assume H_0 is true and keep an eye on α as we decide on the critical region.

If H_0 is true, the ranks of the 4 children should behave as a *random sample* of 4 ranks out of the 12 possible. Therefore each selection of 4 ranks is equally likely, and so each point in the sample space has equal probability, $1/495$. Thus H_0 is a simple hypothesis. Because we decided on a two-tailed test, we examine the points that correspond to high and low values of T. The highest and lowest possible values of T are 42 and 10, corresponding to the points $(12, 11, 10, 9)$ and $(1, 2, 3, 4)$, respectively. Other high and low values of T and their corresponding experimental outcomes are given as follows.

T	Point	T	Point
10	(1, 2, 3, 4)	42	(9, 10, 11, 12)
11	(1, 2, 3, 5)	41	(8, 10, 11, 12)
12	(1, 2, 3, 6)	40	(7, 10, 11, 12)
12	(1, 2, 4, 5)	40	(8, 9, 11, 12)
13	(1, 2, 3, 7)	39	(6, 10, 11, 12)
13	(1, 2, 4, 6)	39	(7, 9, 11, 12)

(continued)

13	(1, 3, 4, 5)	39	(8, 9, 10, 12)
14	(1, 2, 3, 8)	38	(5, 10, 11, 12)
14	(1, 2, 4, 7)	38	(6, 9, 11, 12)
14	(1, 2, 5, 6)	38	(7, 8, 11, 12)
14	(1, 3, 4, 6)	38	(7, 9, 10, 12)
14	(2, 3, 4, 5)	38	(8, 9, 10, 11)

Note that there are 12 points that correspond to values of $T \leq 14$ and 12 points that correspond to values of $T \geq 38$. If the critical region consists of all points that correspond to values of $T \leq 14$ or ≥ 38, α is given by

$$\alpha = \frac{\text{number of points in critical region}}{\text{number of points in sample space}}$$

$$= \frac{24}{495}$$

$$= 0.0485 \tag{5}$$

since all points in the sample space have equal probability under H_0. Our decision rule is: If the observed value of T is ≤ 14 or ≥ 38 we reject H_0; otherwise we accept H_0.

The sample is observed, and the academic ranks of the children who attend nursery school are 2, 5, 6, and 9, providing a value of

$$T = 22 \tag{6}$$

so we accept H_0. The p-value may be approximated using the normal distribution (see Example 1.5.7). The lower-tailed p-value is the probability of getting the observed value $T = 22$ or smaller when the null hypothesis is true. The mean and variance of T are given by Theorem 1.4.5 ($n = 4$, $N = 12$) as 26 and 34.67 respectively, so the standard deviation of T is 5.888. The normal approximation gives

$$P(T \leq 22) \cong P\left(Z \leq \frac{22 - 26}{5.888}\right) = P(Z \leq -0.6794) = 0.248 \tag{7}$$

from Table A1. This is doubled to find the two-tailed p-value 0.496.

A p-value this large indicates that the observed value of T is well within the values expected under the null hypothesis, and therefore there is no reason to suspect the null hypothesis isn't true, based on this set of data. ∎

The test procedure explained in Example 2 is known as the Mann-Whitney test or the Wilcoxon test and will be discussed extensively in Chapter 5 along with its many variations. The data in Example 2 have the ordinal scale of measurement. We did not need to know the numerical value of the academic achievement for each child. In fact, such information usually has little value because each school, even each teacher, has a different interpretation of such numbers, while ranks have a universal interpretation.

Example 1 illustrated the analysis of nominal type data, "defective" or "not defective." The test of Example 1 was based on the binomial distribution. In Chapter 3 this test and other tests based on the binomial distribution will be presented formally.

Computer Assistance

Most statistics computer packages perform hypothesis tests. In some packages, the user specifies the null hypothesis and the alternative hypothesis, and the package returns the correct p-value. In other packages, the computer always returns a p-value for a two-sided test and the user must decide if that is the desired p-value, or if it must be halved to obtain a one-tailed p-value. If the p-value is less than or equal to the desired level of significance, which is selected by the user, then the null hypothesis is rejected.

Many computer packages use approximate methods for finding p-values. This is sufficient in many cases, but not all. More and more computer packages are following the example of *StatXact*, which computes exact p-values or uses monte carlo simulation to approximate the exact p-value when the exact p-values are impractical to obtain.

EXERCISES

1. A new teaching method is being tested to see if it is better than the existing teaching method.

   (a) What are the appropriate H_0 and H_1?

   (b) What does "level of significance" represent in this problem?

   (c) What does "power" represent in this problem?

2. A defendant is being tried by a judge, and it is assumed that the defendant is innocent until proven guilty.

   (a) Who is doing the hypothesis testing?

   (b) What are H_0 and H_1?

   (c) What are the sample and the population?

   (d) What do "level of significance" and "power" mean in this problem?

3. What is the appropriate H_1 for each of the following?

   (a) H_0: Fertilizer B is at least as good as fertilizer A.

   (b) H_0: My opponent is not cheating.

   (c) H_0: The occurrence of sun spots does not affect the economic cycle.

4. What is the appropriate H_0 for each of the following?

   (a) H_1: The subject has extrasensory perception.

   (b) H_1: The dowsing rod is effective in finding water.

   (c) H_1: Our average yearly temperatures are rising.

5. A coin is tossed five times, and the sequence of heads and tails observed is the outcome. The critical region is the event "at least four heads." If H_0 is true, all outcomes in the sample space are equally likely. What is α? If H_1 is true, "head" has probability 0.6 of occurring on each toss. What is the power?

6. A coin is tossed 4 times, and the critical region is "one head or less." Let $p = P(\text{Head})$ for each toss. The hypotheses are $H_0: p = 0.5$ and $H_1: p = 0.1$. Find the power of the test. What additional assumption are you making that was not stated in this problem?

7. The sample space contains 10 points, only 1 of which is in the critical region. If H_0 is true, all of the points are equally likely. If H_1 is true, the point in the critical region has probability 0.91, and the other points each have probability 0.01. What is α? What is the power?

8. Assume the sample space contains 50 points, including 2 points named A and B. If the null hypothesis is true all points in the sample space are equally likely. If the alternative hypothesis is true points A and B each carries 26 times as much probability as each of the other 48 points, which are still equally likely. The critical region consists of points A and B.

   (a) Find alpha, the level of significance.
   (b) Find the power.

PROBLEMS

1. There are 12 plastic chips in a jar, and the chips are numbered consecutively from 1 to 12. An experiment consists of drawing 2 chips with replacement. The outcome of the experiment consists of the 2 numbers on the chips, in the order in which they are drawn. Let the test statistic X be the sum of the numbers on the drawn chips and let the critical region correspond to values of X that are less than 5. Suppose that if H_0 is true the drawing of the chips is random. Also suppose that if H_1 is true chips 1, 2, and 3 are each twice as likely to be drawn as each of the other chips.

   (a) List the points in the critical region.
   (b) Find α.
   (c) What is the power?
   (d) Are H_0 and H_1 simple or composite?
   (e) Is the test one tailed or two tailed?

2. Seven chips numbered consecutively from 1 to 7 are placed independently of each other into either of two boxes A and B. The outcome of the experiment consists of numbers of the chips in box A without regard to the order in which they were placed there. Let the test statistic X be the sum of the numbers on the chips in box A and let the critical region correspond to values of X less than 6. Assume that if H_0 is true each chip has probability 0.5 of being placed in box A and if H_1 is true the corresponding probability is 0.3.

   (a) List the points in the critical region.
   (b) Find α.
   (c) Find the power.
   (d) Are H_0 and H_1 simple or composite?
   (e) Is the test one tailed or two tailed?

2.4 SOME PROPERTIES OF HYPOTHESIS TESTS

Once the hypotheses are formulated, there are usually several hypothesis tests available for testing the null hypothesis. In order to select one of these tests, we consider carefully several properties of the various tests. One of the most important questions is, "Are the assumptions of this test valid assumptions in my experiment?" If the answer is, "No," that test probably should be discarded. However, before discarding the test, one should be sure that the assumptions behind the test are understood. For example, in most parametric tests one of the stated assumptions is that the random variable being examined has a normal distribution. Further investigation sometimes reveals that if the random variable has a distribution only slightly resembling a normal distribution, the test is still approximately valid. So the implied assumption is "approximate normality," and the test should not be discarded if the assumptions are "approximately true." However, at least one black mark may be registered against the test. Another result of this criterion is that the test with the fewer assumptions in the model compares favorably with the test that has more assumptions.

The use of a test in a situation where the assumptions of the test are not valid is dangerous for two reasons. First, the data may result in rejection of the null hypothesis, not because the data indicate that the null hypothesis is false, but because the data indicate that one of the assumptions of the test is invalid. Hypothesis tests in general may be sensitive detectors not only of false hypotheses but also of false assumptions in the model. The second danger is that sometimes the data indicate strongly that the null hypothesis is false, and a false assumption in the model is also affecting the data, but these two effects neutralize each other in the test, so that the test reveals nothing and the null hypothesis is accepted.

From among the tests that are appropriate, based on the preceding criterion, the best test may be selected on the basis of other properties. These properties, which involve terms that will be defined later in this section, are as follows.

1. The test should be unbiased.
2. The test should be consistent.
3. The test should be more efficient in some sense than the other tests.

Of these the efficiency, which is related to the power, is the most important and the most widely used.

Sometimes we are content if one or two of the three criteria are met. Only rarely are all three met. The rest of this section discusses the terms unbiased, consistent, efficiency, and more about the power of test.

Power Function

If H_1 is composite, the power may vary as the probability function varies. If H_1 is stated in terms of some unknown parameter, the power usually may be given as a function of that parameter. Such a function is appropriately called a *power function* and may be represented algebraically or graphically. Unlike the power, which is the probability of rejecting H_0 when H_1 is true, the power function is usually defined for all values of the parameter under both H_0 and H_1. In that sense the power function gives more than just the power; it gives the probability of rejecting H_0 whether or not H_0 is true.

EXAMPLE I

In Example 2.3.1 the critical region consisted of all points with more than 2 defectives in the 10 items examined. Under the assumptions of the model, the probability of getting a point in the critical region, the same as the probability of rejecting H_0, is given by

$$P(\text{reject } H_0) = \sum_{i=3}^{10} \binom{10}{i} p^i (1-p)^{10-i} = 1 - \sum_{i=0}^{2} \binom{10}{i} p^i (1-p)^{10-i} \tag{1}$$

where p is the probability of a defective item. The probability of rejecting H_0 is a function of p, and a rough graph of the power function may be drawn with the aid of Table A3.

p	$P(\text{reject } H_0)$	p	$P(\text{reject } H_0)$
0	0.0000	0.50	0.9453
0.05	0.0115	0.55	0.9726
0.10	0.0702	0.60	0.9877
0.15	0.1798	0.65	0.9952
0.20	0.3222	0.70	0.9984
0.25	0.4744	0.75	0.9996
0.30	0.6172	0.80	0.9999
0.35	0.7384	0.85	1.0000
0.40	0.8327	0.90	1.0000
0.45	0.9004	1.00	1.0000

As indicated in Figure 4, the null hypothesis states that p is between zero and 0.05. The maximum value of the curve, when H_0 is true, is the level of significance, and is shown by Figure 4 and also by Equation 2.3.4 to equal 0.0115. The power is seen to range from 0.0115 for p close to 0.05 to 1.0000 for p equal to 1.0. ∎

Two tests may be compared on the basis of their power functions. This basis of comparison is discussed again later in this section when relative efficiency is defined.

Computer Assistance

The power of a test is a function of the level of significance, the simple alternative hypothesis of interest, and the sample size. The computer package *PASS* concen-

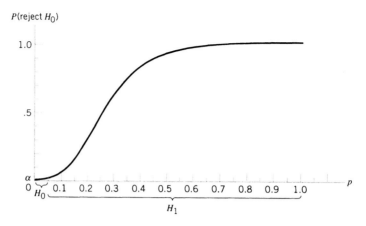

FIGURE 4   A power function.

trates on computing power of a test, given the level of significance, the range of alternatives of interest, and the sample size. It can also compute the sample size required for a given power. *Minitab* also addresses power computations for some nonparametric tests.

Unbiased Test

It is obviously desirable for a test to be more likely to reject H_0 when H_0 is false than when H_0 is true.

> **Definition 1**   An *unbiased test* is a test in which the probability of rejecting H_0 when H_0 is false is always greater than or equal to the probability of rejecting H_0 when H_0 is true.

Thus an unbiased test is one where the power is always at least as large as the level of significance. A test that is not unbiased is called a *biased* test. The test described in Example 2.3.1 and discussed further in Example 1 of this section is an unbiased test, a fact that is readily apparent from Figure 4.

Consistent Test

Another desirable property of a test is that of being *consistent*. Although we refer to a test as being "consistent" or "not consistent," the term *consistent* actually applies to a sequence of tests, because the term applies when the sample size approaches the population size. For convenience we will call the population size "infinity" even though it may be finite. Technically, for each different sample size we have a different test, because the sample space and the critical region change as the sample size changes. Thus, as the sample size increases, we consider a sequence of tests, one for each sample size.

Definition 2   A sequence of tests is *consistent against all alternatives in the class* H_1 if the power of the tests approaches 1.0 as the sample size approaches infinity, for each fixed alternative possible under H_1. The level of significance of each test in the sequence is assumed to be as close as possible to but not exceeding some constant level of significance $\alpha > 0$.

EXAMPLE 2

We wish to determine whether human births tend to produce more babies of one sex, instead of both sexes being equally likely. We are testing

H_0:   A human birth is equally likely to be male or female

against the alternative hypothesis

H_1:   Male births are either more likely, or less likely, to occur than female births

The sampled population consists of births registered in a particular country. The sample consists of the last n births registered, for some selected value of n. It is assumed that this method of sampling is equivalent to random sampling as far as the characteristics "male" and "female" are concerned. It is also assumed that the probability p (say) of a male birth remains constant from birth to birth and that the births are mutually independent as far as the events "male" and "female" go. Then the hypotheses are equivalent to the following.

$$H_0: p = 1/2$$
$$H_1: p \neq 1/2$$

Let the test statistic T be the number of male births. The critical region is chosen to correspond symmetrically to the largest values and the smallest values of T, called the upper and lower tails of T, of the largest size not exceeding 0.05.

Thus we have described an entire sequence of tests, one for each value of the sample size. Each test is two tailed and has a level of significance of 0.05 or smaller, and T has the binomial distribution. For the various tests the critical regions are given by Dixon (1953) as follows.

n	Values of T Corresponding to the Critical Region			α
5		None		0
6	$T = 0$	and	$T = 6$	0.03125
8	$T = 0$	and	$T = 8$	0.00781
10	$T \leq 1$	and	$T \geq 9$	0.02148
15	$T \leq 3$	and	$T \geq 12$	0.03516
20	$T \leq 5$	and	$T \geq 15$	0.04139
30	$T \leq 9$	and	$T \geq 21$	0.04277
60	$T \leq 21$	and	$T \geq 39$	0.02734
100	$T \leq 39$	and	$T \geq 61$	0.03520

Note that for $n \leq 20$ these same values can be obtained from Table A3. For $n > 20$ the normal approximation (Example 1.5.6) could have been used, but the exact tables are preferred.

To see if this sequence of tests is consistent, the power functions of the tests are compared. Several of these power functions are plotted on the same graph in Figure 5, from tables given by Dixon (1953). We can see that as the sample size increases, the power at each fixed value of p (except $p = 0.5$) increases toward 1.0. ∎

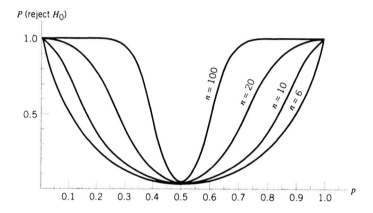

FIGURE 5 A comparison of several power functions.

This example merely demonstrates the idea behind the term consistent as it applies to a sequence of tests. This demonstration is not a proof that the sequence of tests is consistent. A rigorous proof of consistency usually requires more mathematics than we care to use in this introductory book, and so we will merely state whether or not a sequence of tests (or a "test") is consistent.

Relative Efficiency

Many other properties of statistical tests have been defined and may be found in various books (e.g., Lehmann, 1959). We will limit our discussion to one more property, that of *efficiency*. Efficiency is a relative term and is used to compare the sample size of one test with that of another test under similar conditions. Suppose two tests may be used to test a particular H_0 against a particular H_1. Also suppose that the two tests have the same α and the same β and therefore are "comparable" with respect to level of significance and power. (Note that the condition that β be the same for both tests usually excludes consideration of composite alternative hypotheses, since those usually have more than one value of β.) Then the test requiring the smaller sample size is preferred over the other test, because a smaller sample size means less cost and effort is required in the experiment. The test with the smaller sample size is said to be *more efficient* than the other test, and its *relative efficiency* is greater than one.

> **Definition 3**   Let T_1 and T_2 represent two tests that test the same H_0 against the same H_1, with the critical regions of the same size α and with the same values of β. The *relative efficiency of T_1 to T_2* (or "efficiency of T_1 relative to T_2") is the ratio n_2/n_1, where n_1 and n_2 are the sample sizes of the tests T_1 and T_2, respectively.

According to Definition 3, if n_1 is smaller than n_2, the efficiency of T_1 relative to T_2 is greater than unity, satisfying our preconceived notions.

If the alternative hypothesis is composite, the relative efficiency may be computed for each probability function defined by the alternative hypothesis, resulting in a multitude of values for relative efficiency that may then be represented in a table or, occasionally, graphically.

EXAMPLE 3

Two tests are available for testing the same H_0 against the same H_1. Both tests have $\alpha = 0.01$ and $\beta = 0.14$. The first test requires a sample size of 75. The second test requires a sample size of 50. The first test is therefore less efficient than the second test. The relative efficiency of the first test to the second test is

$$\frac{50}{75} = 0.67$$

and the efficiency of the second test relative to the first is

$$\frac{75}{50} = 1.5$$

If we know that the efficiency of the first test relative to the second test at $\alpha = 0.05$, $\beta = 0.30$, $n_1 = 40$, is 0.75, the sample size required by the second test may be obtained.

$$\text{relative efficiency} = \frac{n_2}{n_1}$$

$$0.75 = \frac{n_2}{40}$$

$$n_2 = 30$$

A sample of size 30 will provide as good an analysis using the second testing method as a sample of size 40 would using the first.   ∎

Asymptotic Relative Efficiency (A.R.E.)

The relative efficiency depends on the choice of α, the choice of β, and the particular alternative being considered if H_1 is composite. In order to provide an overall comparison of one test with another it is clear that relative efficiency leaves much to be desired. We would prefer a comparison that does not depend on our choice of α, β, or a particular alternative possible under H_1 if H_1 is composite, which it usually is. One way this sometimes may be accomplished is described briefly as follows.

Consider a sequence of tests, all with the same fixed α. If the sequence of tests is consistent, β will become smaller as the sample size n_1 gets larger. Instead of allowing β to become smaller, we could consider a different alternative each time (under the composite alternative hypothesis) for each different value of n_1 where, each time, the alternative considered is one that allows β to remain constant from test to test. Thus, as n_1 becomes larger, α and β remain fixed and the alternative being considered varies.

This may be illustrated by considering Figure 5 again. As n_1 becomes larger, the graphs in Figure 5 show that β can remain constant by considering consecutive values of the parameter p that approach closer to $p = 0.5$.

For each value of n_1, a value of n_2 is calculated so the second test has the same α and β under the alternative considered. Then there is a sequence of values of relative efficiency n_2/n_1, one for each test in the original sequence of tests. If n_2/n_1 approaches a constant as n_1 becomes large, and if that constant is the same no matter which values of α and β are being used, then that constant is called the *asymptotic relative efficiency* of the first test to the second test, or more correctly, the first sequence of tests to the second sequence of tests. Sometimes the name *Pitman's efficiency* is used for this definition of asymptotic relative efficiency to distinguish it from other definitions of asymptotic relative efficiency.

Definition 4   Let n_1 and n_2 be the sample sizes required for two tests T_1 and T_2 to have the same power under the same level of significance. If α and β remain fixed, the limit of n_2/n_1, as n_1 approaches infinity, is called the *asymptotic relative efficiency* (A.R.E.) of the first test to the second test, if that limit is independent of α and β.

In our quest to select the test with the greatest power, we usually are forced to select the test with the greatest A.R.E. because the power depends on too many factors. Thus the A.R.E. of one test relative to another is important to know.

The A.R.E. of two tests is usually difficult to calculate. A comprehensive study of A.R.E. of various pairs of tests could be the subject of a book by itself. A book by Noether (1967a) contains many of the more important results of studies of A.R.E. See also Stuart (1954) and Ruist (1955) for further discussions.

So the A.R.E. may be given instead of tables of values of relative efficiency, but of what use is A.R.E. if it considers the infinite (and thus impossible) sample size? Studies of the exact relative efficiency for very small sample sizes show that A.R.E. provides a good approximation to the relative efficiency in many situations of practical interest. Thus the A.R.E. often provides a compact summary of the relative efficiency between two tests.

Conservative Test

The term *conservative* is another term we will sometimes use when discussing a test.

Definition 5   A test is *conservative* if the actual level of significance is smaller than the stated level of significance.

At times it is difficult to compute the exact level of significance of a test, and then some methods of approximating α are used. The approximate value is then reported as being the level of significance. If the approximate level of significance is larger than the true (but unknown) level of significance, the test is conservative, and we know the risk of making a type I error is not as great as it is stated to be.

EXERCISES

1. A coin is tossed five times. At each toss the experimenter observes whether it is a head or a tail, and a blindfolded subject being tested for extrasensory perception "states" whether it is a head or a tail. The null hypothesis is that the subject's predictions have probability $p = 0.5$ of being correct, while the alternative hypothesis is that $p > 0.5$. The critical region consists of all five correct predictions.

   (a) Find α.
   (b) What is the power function?
   (c) Draw a graph of the power function.
   (d) Is the test unbiased?

2. Two types of shoe leather are being tested to see which is more durable. Eight pairs of shoes are made; the shoes seem to be identical except that one shoe is made from leather A and the other from leather B. These shoes are subjected to normal wear for a period of time and are then judged as to which leather seemed to be more durable for each pair. Let X equal the number of pairs of shoes where leather A is judged to be more durable. The null hypothesis is that $p = 0.5$, where p is the probability that the shoe made out of leather A was more durable than the other shoe, while H_1 is $p \neq 0.5$. The critical region corresponds to $X = 0, 1, 7$, and 8.

   (a) Find α.
   (b) What is the power function?
   (c) Draw a graph of the power function.
   (d) Is the test unbiased?

3. Let $T_1, T_2, \ldots$, represent a sequence of tests, and suppose the power P_n of T_n is given by $P_n = n/(n + 10)$. Is the sequence of tests consistent?

4. Let $T_1, T_2, \ldots$, represent a sequence of tests where the power of test T_n is $n/(2n + 10)$. Is the sequence consistent?

5. The hypothesis $H_0: p = 1/2$ is being tested against $H_1: p = 3/4$ using two tests T_1 and T_2 at the same level of significance. If T_1 uses a sample of size 20, then T_2 requires a sample of size 35 in order for the power of T_2 to equal the power of T_1.

   (a) What is the efficiency of T_2 relative to T_1?

   (b) What is the efficiency of T_1 relative to T_2?

6. The hypothesis $H_0: p = 1/2$ is being tested against $H_1: p \neq 1/2$ using two tests at the same level of significance. T_2 needs a sample of size 30 when T_1 has a sample of size 15 in order for their power functions to be equal at the particular alternative $p = 1/3$.

   (a) What is the efficiency of T_2 relative to T_1?

   (b) Is the efficiency necessarily the same at the alternative $p = 2/3$?

PROBLEMS

1. Suppose that the asymptotic efficiency of T_2 relative to T_1 is 0.75, and suppose that the relative efficiency for finite sample sizes is always greater than the asymptotic relative efficiency. If an experimenter prefers to use test T_2 but wishes to have at least as much power as if test T_1 were being used with a sample of size 24, what should the minimum sample size be?

2. Suppose the A.R.E. of test T_1 relative to test T_2 is $3/\pi$ and the A.R.E. of test T_3 relative to T_2 is $2/\pi$. What is the A.R.E. of test T_1 relative to T_3?

2.5 SOME COMMENTS ON NONPARAMETRIC STATISTICS

In this section we will attempt to distinguish between the terms *parametric statistics* and *nonparametric statistics,* although the distinction is not always clear even in the minds of professional statisticians. We will use the term *nonparametric* and the more descriptive term *distribution-free* interchangeably even though some statisticians distinguish between the two. Also we will provide some guidance for when to use nonparametric methods in the analysis of data, and when parametric methods should be preferred.

Good Methods to Use

First we will discuss hypothesis testing and confidence intervals. In this chapter it was pointed out that a test of hypothesis relies on a good test statistic, one that is sensitive to the difference between the null hypothesis and the alternative hypothesis, and one whose probability distribution under the null hypothesis is known. A confidence interval is the inversion of a hypothesis test in that the confidence interval is the collection of null hypotheses that are not rejected by the data, so a good (powerful) hypothesis test relates to a good (short) confidence interval, and vice versa.

For example, the sample mean $\overline{X}$ is a good test statistic for testing hypotheses about the population mean μ because it is sensitive to differences in the population mean. Similarly S and s are good statistics to use for inferences about the population standard deviation σ. However, the probability distributions of $\overline{X}$, S and s depend on the population probability distribution of the Xs, which is usually unknown.

Parametric Methods

If the population probability distribution is a normal distribution then $\overline{X}$, or some statistic based on $\overline{X}$, can be used to test hypotheses about μ or find confidence intervals for estimating μ because its null distribution is known. Similarly, hypotheses about the population standard deviation σ can be tested and confidence intervals for σ can be formed if the distribution is normal, based on either S or s, the sample standard deviation.

These are called *parametric methods* because their validity depends on knowing the population distribution function. Any hypothesis test or confidence interval that is based on the assumption that the population distribution function is known, or known except for some unknown parameters, is called a parametric method.

But how can we be sure that the population probability distribution is the normal distribution, or any other distribution for that matter? The answer is, simply, we can't. We can look at the data to see if it looks like it might have come from a normal distribution, or we can use one of the many hypotheses tests designed to detect nonnormality of the data as a preliminary step before using a test based on the normality assumption.

Most parametric methods are based on the normality assumption because the theory behind the test can be worked out with the normal population distribution. The resulting procedures are efficient and powerful procedures for normally distributed data. Other parametric procedures have been developed by assuming the population has other distributions, such as the exponential, Weibull, and so on.

Robust Methods

No population has exactly a normal distribution, or any other known distribution. If the population distribution is approximately normal, then usually (but not always) it is safe to use a method based on the normal distribution. However, if the data appear to come from a distinctly nonnormal distribution, or a distribution not covered by the parametric methods, then a nonparametric method should be considered.

A method of analysis that is approximately valid even if one of the assumptions behind the method is not true is considered to be *robust* against that assumption. Generally the term *robust* refers to methods based on the normality assumption that have approximately the same null distribution of the test statistic even when the underlying population is nonnormal.

Some parametric tests, such as the one-sample *t* test or the two-sample *t* test, are known to be robust against the assumption of normality, especially if the sample sizes are reasonably large. This means that the null distribution of the test statistic is approximately what it would be if the population were normal, and the experimenter can refer the test statistic to the tables of the *t* distribution, which are exact when the population is normal, with the confidence that the quantiles given in the table will approximate the true quantiles of the test statistic even if the population is nonnormal.

However, just because a method is robust is no assurance that the method is still powerful when the population is nonnormal. Therefore the question of which statistical method to use is answered not only by the question, Is it robust? but also by the question, Is it powerful? A statistical procedure should be robust, of course, so that the reported level of significance is close to the true level of significance. But especially it should be powerful, so that efficient use is made of the data, and the null hypothesis can be rejected when it is not true.

Nonparametric Methods

Nonparametric methods are based on some of the same assumptions on which parametric methods are based, such as the assumption that the sample is a random sample. However, nonparametric methods do not assume a particular population probability distribution, and are therefore valid for data from any population with any probability distribution, which can remain unknown.

Nonparametric methods are perfectly robust for distribution assumptions on the population, because they are equally valid for all distributions.

If the population distribution function has lighter tails than the normal distribution, such as the uniform distribution, then the parametric procedures based on the normality assumption generally have good power, equal to or greater than the power of nonparametric methods based on ranks, presented in Chapter 5. An example of data with light tails is opinion survey data, where respondents are asked to respond on a scale from 1 to 5, or 1 to 7. Even though the distribution of responses is discrete and perhaps nonsymmetric, and therefore distinctly nonnormal, because of the light tails the usual normal-based parametric methods for testing hypotheses about population means are preferred over the nonparametric methods because of their superior power.

On the other hand, if the population distribution function has heavier tails than the normal distribution, such as with the exponential distribution (presented in Chapter 6), the lognormal distribution (where the logs of the data follow a normal distribution), the chi-squared distribution (or its parent family of gamma distributions), and many other distributions that appear to be reasonable population models, then the parametric methods based on the normality assumption may have low power compared to nonparametric methods based on ranks.

Data containing *outliers*, which are observations that are much larger or much smaller than the bulk of the observations in the sample, are good examples of the type of data that comes from heavy-tailed distributions. In those cases it is important to consider using nonparametric methods, such as the rank methods introduced in Chapter 5, to analyze the data because of the superior power of those rank methods when compared with the parametric methods based on the normal assumption.

Asymptotically Distribution-Free

Many parametric tests that are robust against the assumption of nonnormality are also *asymptotically distribution-free*. This means that as the sample size gets larger the method becomes more robust, approaching the point where for an infinite sample size the method becomes exact, no matter what the population distribution may be. The central limit theorem is usually the basis for showing parametric methods based on the sample mean to be asymptotically distribution-free. This was done when presenting the approximate confidence interval for the population mean μ in Section 2.2.

A statistical procedure should not be preferred over others simply because it is nonparametric, or robust, or asymptotically distribution-free. The relative power of the parametric test, or the relative size of the confidence interval, as compared with its nonparametric alternative, usually remains good or remains bad regardless of the size of the sample, even if the procedure is asymptotically distribution-free. The above discussion, concerning which statistical procedure is preferred for various types of data, is pertinent whether the sample size is small or large.

Keep in mind that most methods we are considering are *consistent*, which means that larger sample sizes mean more absolute power. Carefully selecting the more powerful procedure may be unnecessary if the sample size is large enough to reject the null hypothesis using a less-powerful test, or if the confidence interval is small enough for the experimenter's purposes using a less efficient method. This is something else for the experimenter to consider when selecting a statistical method to analyze data.

Methods for Analyzing Nominal Data

As we stated in the introduction, when most people think of nonparametric methods they think of the methods based on ranks presented in Chapters 5 and 6, because rank methods are often the logical alternative to parametric tests such as the t test and the analysis of variance used on data with the interval or ratio scale of measurement. However, nonparametric methods also may be used on qualitative data, with the nominal scale of measurement, or with ordinal scale data.

The concept of a population probability distribution for nominal or ordinal data is difficult to imagine without treating such data as if it were at least interval, so there are no parametric methods for purely nominal or ordinal data. Methods for analyzing qualitative (nominal) data or data where only the orders or ranks are known are necessarily nonparametric. Chapters 3 and 4 present methods for analyzing qualitative data. Most of the methods in Chapters 5 and 6 are valid for ordinal data.

Definition of Nonparametric

As a definition of *nonparametric* we offer the following, which seems to work well.

Definition 1 A statistical method is nonparametric if it satisfies at least one of the following criteria.

1. The method may be used on data with a nominal scale of measurement.
2. The method may be used on data with an ordinal scale of measurement.
3. The method may be used on data with an interval or ratio scale of measurement, where the distribution function of the random variable producing the data is either unspecified or specified except for an infinite number of unknown parameters.

The test in Example 2.3.1 analyzed data with a nominal scale of measurement (defective or nondefective) and therefore the test is nonparametric by the first criterion. The test in Example 2.3.2 analyzed data with an ordinal scale of measurement and therefore, by the second criterion, it too is nonparametric. Nearly all nonparametric hypothesis tests satisfy one of these two criteria. The point estimates of Section 2.2 satisfy the third criterion, and so do the procedures that assume symmetric distributions in Section 5.7, 5.10, and 5.11. Therefore we consider them to be nonparametric.

This book is primarily concerned with hypothesis testing and the forming of confidence intervals. Unfortunately, this emphasis often gives experimenters the false impression that if they do not test some hypothesis or form some confidence interval, they are not using a statistical analysis. Other forms of statistical analysis are just as important, such as a description of the population, an interpretation of the data, prediction of unknown events, and point estimation.

These other forms of inference depend to a great extent on the experimenter's maturity and good judgment instead of on complicated probabilistic arguments; therefore we consider them too difficult to present in a book. We are attempting to assist the experimenter who already possesses maturity and good judgment by spelling out the complicated probabilistic arguments associated with hypothesis testing and confidence intervals.

Nonparametric statistical methods have been developed for several types of problems that are not covered in this book. These areas (and some references for the interested reader) include bioassay (Miller, 1973, Chmiel, 1976), survival curves (Susarla and Van Ryzin, 1976, Tarone and Ware, 1977), and longitudinal studies (Ghosh, Grizzle, and Sen, 1973). Multivariate methods are discussed by Aitchison and Aitken (1976), and Bhapkar and Patterson (1977) and are the topic of a book by Puri and Sen (1971). For discrimination analysis see Gessaman and Gessaman (1972), Broffitt, Randles, and Hogg (1976), Randles, Broffitt, Ramberg, and Hogg (1978), and Conover and Iman (1980).

Robust methods are methods that depend to some extent on the population distribution function but are not very sensitive to departures from the assumed distributional form. Robust methods are discussed briefly in Section 5.12. A more complete discussion of robust methods may be found in Govindarajulu and Leslie (1972), Hogg (1974), Pearson and Please (1975), Policello and Hettmansperger (1976), and other references cited in Section 5.12. General overviews of the field of nonparametric statistics may be obtained by reading articles by Kendall and Sundrum (1953), Blum and Fattu (1954), Savage (1969), Bell (1964), and Govindarajulu (1976), or books by Tate (1957), Fraser (1957), Walsh (1962), Noether (1967), Pierce (1970), Hollander and Wolfe (1973), Tapia (1978), Randles (1979), Buringer (1980), Henley (1981), Pratt (1981), and Manoukin (1986).

2.6 REVIEW PROBLEMS FOR CHAPTER 2

1. A box contains seven tickets. Five tickets belong to students and the other two belong to faculty. Two tickets are drawn from the box, without replacement, to determine the two winners. The null hypothesis is that the drawing is random. The alternative hypothesis is that the drawing is rigged so that the first ticket drawn belongs to a faculty member and the second ticket is then randomly selected from the remaining six tickets.

    (a) Suppose the decision rule is to reject the null hypothesis if both tickets drawn belong to faculty members. Find α. Find the power.

    (b) Suppose, instead, the decision rule is to reject the null hypothesis if the first ticket drawn belongs to a faculty member. Find α. Find the power.

    (c) Some people might prefer the test in part a because it has a smaller level of significance. Others might prefer the test in part b because it has greater power. Discuss some of the social consequences connected with using each test. Which test would you use?

2. What is the scale of measurement for the following random variables?

    (a) The number of pounds gained (or lost) while on a particular diet.
    (b) The standing of the Kansas City Royals within their league.
    (c) Your student identification number.
    (d) The scoring average for a particular basketball player.
    (e) The score a figure skater receives in an Olympic contest.

3. Two students are playing chess to see if they are equal in ability. The rules are that seven games will be played, not counting "draws" as games. If either person wins at least six of the games, they agree that they are not of equal chess-playing ability.

   (a) What is H_0?
   (b) What is H_1?
   (c) Write down any one point in the sample space.
   (d) List the points in the sample space that constitute the critical region.
   (e) Find the level of significance.
   (f) Is H_0 simple or composite?
   (g) Is H_1 simple or composite?
   (h) What is the equation of the power function?
   (i) Is the test unbiased?
   (j) What assumptions did you make here?

4. A restaurant manager adds the customers' checks at the cash register when the customers depart. Later, an auditor checks the manager's results. He finds that out of the 12 mistakes in addition, the manager made 10 mistakes in favor of the customer, and 2 mistakes in favor of the manager. Let the null hypothesis be that the manager is equally likely to make a mistake in either direction, and the alternative hypothesis be that the manager is more likely to make mistakes in one direction or the other.

   Let the critical region be the outcomes that result in 10 or more mistakes in favor of the customer, or 10 or more mistakes in favor of the manager.

   (a) Find α, the level of significance of this test.
   (b) Find the power of this test, if the manager is three times as likely to make a mistake in favor of the customer as he is in his own favor.
   (c) Draw a sketch of the power curve. Is the test unbiased? Explain.

5. Twelve percent of the cars manufactured do not pass the inspection at the end of the assembly line, and require special attention. Let X be the number of cars that do not pass inspection out of four cars manufactured. Assume independence from one car to the next.

   (a) Find the mean and the standard deviation of X.
   (b) Find the median and the interquartile range of X.

6. In a continuation of Problem 5, six groups of four cars each are manufactured, and the number of cars not passing inspection at the end of the assembly line for each group is 0, 0, 0, 1, 1, 2.

   (a) Find the sample mean and the sample standard deviation.
   (b) Find the sample median and the sample interquartile range.
   (c) Estimate the probability of a car not passing inspection from the data.

7. Discuss the measurement scales as they apply to Problem 5.

   (a) What measurement scale is used on each car?
   (b) What measurement scale is used by X?

8. In Texas, 20% of all households have no vehicles available for transportation, 30% have one vehicle, 30% have two vehicles, 10% have three vehicles, and the rest have more than three vehicles. A random sample of ten households is obtained, with the following results. Three households have no vehicles, two have one vehicle, two have two vehicles, one has three vehicles, one has four vehicles, and one has five vehicles.

   (a) Find the population median number of vehicles per household.
   (b) Find the population interquartile range.
   (c) Discuss why the population median might be preferred over the population mean in this problem.
   (d) Find the sample median number of vehicles per household.
   (e) Find the sample mean number of vehicles.
   (f) Find the sample standard deviation, using S rather than s.
   (g) Draw a graph of the population distribution function.
   (h) Draw a graph of the empirical distribution function.

9. Let X equal the number of people in an automobile traveling the freeway during rush-hour traffic. The probabilities for X are $P(X = 1) = 0.40$, $P(X = 2) = 0.30$, $P(X = 3) = 0.20$, and $P(X = 4) = 0.10$. A random sample of ten automobiles revealed the following values for X:

$$4, \quad 1, \quad 1, \quad 2, \quad 1, \quad 2, \quad 3, \quad 1, \quad 4, \quad 1$$

   Find the sample mean. Draw a graph of the empirical distribution function (Definition 2.2.1). Find the sample interquartile range. Compare these values with the population mean, the distribution function (Definition 1.3.4), and the population interquartile range, respectively.

10. If the null hypothesis is true, each customer is equally likely to select the red box or the blue box of cereal. If the alternative hypothesis is true, customers prefer the blue box 3 to 1 over the red box. Assume each customer selects one box of cereal, independently of the other customers. The null hypothesis will be rejected if 15 or more of the first 20 customers select blue boxes. Find the level of significance. Find the power. If 17 of the first 20 customers actually observed selected blue boxes, find the p-value.

11. Seven male students have interviewed for three positions as summer camp counselors. The students are ranked according to height, from 1 (tallest) to 7 (shortest). The null hypothesis is that each student is equally likely to be selected, while the alternative hypothesis is that the three taller students are twice as likely to be selected as the four shorter students. Assume the students are selected independently of each other. The test statistic is the sum of the ranks of the three students who were selected. The decision rule is to reject the null hypothesis if the test statistic is 6 or less.

   (a) Is the null hypothesis simple or composite?
   (b) Find the level of significance.
   (c) Find the power.

12. Under one theory of genetics each offspring of two particular dogs should have a 25% chance of being spotted in color. Let this be the null hypothesis. Under another theory each

puppy should have a 75% chance of being spotted. Let this be the alternative hypothesis. A litter of puppies is born. There are five spotted puppies out of the eight puppies in the litter.

(a) Using the target level of significance of 0.05, find the critical region for a conservative test.

(b) What is the exact level of significance for your test? (Use the exact formula, and then use the normal approximation, to get two slightly different results.)

(c) What is the exact power of your test? (Use the tables to get your answer.)

(d) What is the p-value in this case?

(e) Is the test unbiased? Explain.

(f) Is the alternative hypothesis simple or composite?

13. A biased coin has probability 2/3 of falling "Heads" and 1/3 of falling "Tails." The coin is tossed 10 times, with the result that "Heads" occurred 5 times and "Tails" occurred 5 times. Let X equal the number of "Heads."

(a) Sketch the graph of the population distribution function of X.

(b) Find the population median.

(c) Sketch the graph of the empirical distribution function of X.

(d) Find the sample median.

14. A student is given three multiple-choice questions with five possible responses to each. If the student has studied the subject he has an 80% chance of answering correctly on each question. If the student has not studied the subject (the null hypothesis) he has equal chances of responding each of the five possible ways on each question. The null hypothesis is rejected if the student gets all three answers correct.

(a) Find the level of significance.

(b) Find the power of the test.

(c) What assumption are you making that has not been explicitly stated?

CHAPTER 3

SOME TESTS BASED ON THE BINOMIAL DISTRIBUTION*

PRELIMINARY REMARKS

The binomial probability distribution was introduced in Chapter 1 to describe the probabilities associated with the number of heads when a coin is tossed n times. In its more general form each of n independent trials results either in "success," with probability p, or "failure," with probability $q = 1 - p$. The binomial distribution describes the probability of obtaining exactly k successes. Table A3 presents some of the binomial distribution functions.

Many experimental situations in the applied sciences may be modeled this way. Several customers enter a store and independently decide to buy or not to buy a particular product. Several animals are given a certain medicine and either they are cured or not cured. Examples can be found in almost any field. Data obtained in these situations may be analyzed using some of the simplest statistical methods known, those based on the binomial distribution. In this chapter we present a few of the available methods. The literature abounds with other procedures based on the binomial distribution. After studying the variety of tests presented in this chapter, the reader should be able to invent variations to match a given experimental situation.

* Review Problems for Chapter 3 are included in the Review Problems for Chapter 4.

3.1 THE BINOMIAL TEST AND ESTIMATION OF p

One example of the binomial test has already been presented. In Example 2.3.1 the binomial test was applied to a quality control problem. This entire chapter (Chapter 3) is little more than an elaboration of Example 2.3.1, showing the many uses and amazing versatility of that simple little binomial test. With a little ingenuity the binomial test may be adapted to test almost any hypothesis, with almost any type of data amenable to statistical analysis. In some situations the binomial test is the most powerful test; in those situations the test is claimed by both parametric and nonparametric statistics. In other situations more powerful tests are available, and the binomial test is claimed only by nonparametric statistics. However, even in situations where more powerful tests are available, the binomial test is sometimes preferred because it is usually simple to perform, simple to explain, and sometimes powerful enough to reject the null hypothesis when it should be rejected.

We will now formally present the binomial test and, at the same time, introduce the format for presenting tests. We feel that there is a need for some format in presenting tests both for the convenience of the reader and for ready review by the users of nonparametric techniques.

▶ **The Binomial Test** _____

Data   The sample consists of the outcomes of n independent trials. Each outcome is in either "class 1" or "class 2," but not both. The number of observations in class 1 is O_1 and the number of observations in class 2 is $O_2 = n - O_1$.

Assumptions

1. The n trials are mutually independent.

2. Each trial has probability p of resulting in the outcome "class 1," where p is the same for all n trials.

Test Statistic   Since we are concerned with the probability of the outcome "class 1," we will let the test statistic T be the number of times the outcome is "class 1." That is,

$$T = O_1 \tag{1}$$

Null Distribution   Let p^* be the probability specified in the null hypothesis. The null distribution of T is the binomial distribution with parameters $p = p^*$ and $n = $ the sample size. The null distribution of T is tabulated in Table A3 for $n \leq 20$ and selected values of p.

For other values of n and p the normal approximation is used. That is, approximate quantiles x_q for T are given by

$$x_q = n \cdot p + z_q \sqrt{n \cdot p \cdot (1 - p)} \qquad (2)$$

where z_q is the qth quantile of a standard normal random variable, given in Table A1.

Hypotheses   Let p^* be some specified probability, $0 \le p^* \le 1$. The hypotheses may take one of the following three forms.

A. (Two-Tailed Test)

$$H_0: p = p^*$$
$$H_1: p \neq p^*$$

The rejection region of desired size α corresponds to the two tails of the null distribution of T, where the size of the lower tail is denoted by α_1, approximately half of α, and the size of the upper tail is denoted by α_2, also approximately half of α. The true level of significance is $\alpha_1 + \alpha_2$, which is seldom a nice round number due to the discrete nature of T.

That is, from Table A3 for the particular values p^* and n, find the number t_1 such that

$$P(Y \le t_1) = \alpha_1 \qquad (3)$$

and find the number t_2 such that

$$P(Y \le t_2) = 1 - \alpha_2 \qquad (4)$$

where Y is a binomial random variable with parameters p^* and n.

If n is greater than 20 use the normal approximation. That is, use Equation 2 to approximate t_1, the $\alpha/2$ quantile, and t_2, the $(1 - \alpha/2)$ quantile, of a binomial random variable with parameters p^* and n, by letting $q = \alpha/2$ and $q = 1 - \alpha/2$ respectively.

Reject H_0 if T is less than or equal to t_1 or if T is greater than t_2. Otherwise accept the null hypothesis.

The p-value is twice the smaller of the probabilities that Y is less than or equal to the observed value of T, or greater than or equal to the observed value of T, which are found from Table A3 for $n \le 20$, using $p = p^*$, or from Table A1 for $n > 20$, using

$$P(Y \le t_{obs}) \cong P\left(Z \le \frac{t_{obs} - n \cdot p^* + 0.5}{\sqrt{n \cdot p^*(1 - p^*)}}\right) \tag{5}$$

and

$$P(Y \ge t_{obs}) \cong 1 - P\left(Z \le \frac{t_{obs} - n \cdot p^* - 0.5}{\sqrt{n \cdot p^*(1 - p^*)}}\right) \tag{6}$$

which incorporates 0.5 as a "correction for continuity" that improves the normal approximation to the binomial.

B. (Lower-Tailed Test)

$$H_0: p \ge p^*$$
$$H_1: p < p^*$$

Because small values of T indicate H_0 is false, the rejection region of size α consists of all values of T less than or equal to t, where t is obtained from Table A3, using p^* and n, so that

$$P(Y \le t) = \alpha \tag{7}$$

where Y is a binomial random variable with parameters p^* and n.

If n is greater than 20 use the normal approximation. That is, use Equation 2 to approximate t, the α quantile of a binomial random variable with parameters p^* and n, by letting $q = \alpha$.

Reject H_0 if T is less than or equal to t. Otherwise accept the null hypothesis.

The p-value is the probability that Y is less than or equal to the observed value of T, which is found from Table A3 for $n \le 20$, using $p = p^*$, or from Table A1 for $n > 20$, using

$$P(Y \le t_{obs}) \cong P\left(Z \le \frac{t_{obs} - n \cdot p^* + 0.5}{\sqrt{n \cdot p^*(1 - p^*)}}\right) \tag{8}$$

which incorporates 0.5 as a "correction for continuity" that improves the normal approximation to the binomial.

C. (Upper-Tailed Test)

$$H_0: p \le p^*$$
$$H_1: p > p^*$$

Because large values of T indicate H_0 is false, the rejection region of size α corresponds to all values of T greater than t, where t is obtained from Table A3, using p^* and n, so that

$$P(Y \leq t) = 1 - \alpha \tag{9}$$

where Y is a binomial random variable with parameters p^* and n.

If n is greater than 20 use the normal approximation. That is, use Equation 2 to approximate t, the $(1 - \alpha)$ quantile of a binomial random variable with parameters p^* and n, by letting $q = 1 - \alpha$.

Reject H_0 if T is greater than t. Otherwise accept the null hypothesis.

The p-value is the probability that Y is greater than or equal to the observed value of T, which is found from Table A3 for $n \leq 20$, using $p = p^*$, or from Table A1 for $n > 20$, using

$$P(Y \geq t_{obs}) \cong 1 - P\left(Z \leq \frac{t_{obs} - n \cdot p^* - 0.5}{\sqrt{n \cdot p^*(1 - p^*)}}\right) \tag{10}$$

which incorporates 0.5 as a "correction for continuity" that improves the normal approximation to the binomial.

Computer Assistance   Computer packages that will perform this test include *Minitab*, *S-Plus*, and *StatXact*, which finds exact p-values. *Minitab* also finds the power, and the sample size required to achieve a given level of power. ————◀

EXAMPLE I

It is estimated that at least half of the men who currently undergo an operation to remove prostate cancer suffer from a particular undesirable side effect. In an effort to reduce the likelihood of this side effect the FDA studied a new method of performing the operation. Out of 19 operations only 3 men suffered the unpleasant side effect. Is it safe to conclude the new method of operating is effective in reducing the side effect?

Let p equal the probability of the patient experiencing the side effect. Then this is a lower-tailed test of H_0: $p \geq 0.5$ versus H_1: $p < 0.5$. If the target α is 0.05, the critical region consists of $T \leq 5$, with an actual $\alpha = 0.0318$. (See Table A3, $n = 19$, $p = 0.5$.)

The observed value of T is 3, so H_0 is rejected, and it is concluded the new operation is effective in reducing the likelihood of the side effect. The p-value is

$$P(T \leq 3) = 0.0022$$

which is quite small, indicating that the sample data are in strong disagreement with the null hypothesis. ■

One should always use exact methods when exact methods are available, but to illustrate how well the normal approximation works, consider Example 1. The approximate 0.05 quantile is found from Equation 2,

$$x_{0.05} = 19(0.5) + (-1.6449) \sqrt{19(0.5)(0.5)} = 5.9$$

resulting in the same rejection region as before. The exact α is estimated from Equation 5.

$$P(Y \le 5) \cong P\left(Z \le \frac{5 - 19(0.5) + 0.5}{\sqrt{19(0.5)(0.5)}} \right) = 0.033$$

which is close to the exact α, 0.032. The exact p-value is also estimated from Equation 5

$$P(Y \le 3) \cong P\left(Z \le \frac{3 - 19(0.5) + 0.5}{\sqrt{19(0.5)(0.5)}} \right) = 0.003$$

Again, this is close to the exact p-value, 0.002.

EXAMPLE 2

Under simple Mendelian inheritance a cross between plants of two particular genotypes may be expected to produce progeny one-fourth of which are "dwarf" and three-fourths of which are "tall." In an experiment to determine if the assumption of simple Mendelian inheritance is reasonable in a certain situation, a cross results in progeny having 243 dwarf plants and 682 tall plants. If "class 1" denotes "tall," then $p^* = 3/4$ and T equals the number of tall plants. The null hypothesis of simple Mendelian inheritance is equivalent under the model to the hypothesis

$$H_0: p = 3/4$$

The alternative of interest is two-sided,

$$H_1: p \ne 3/4$$

Since $n = 925$, $(243 + 682)$, the critical region of approximate size $\alpha = 0.05$ may be obtained using the large sample approximation given by Equation 2. Thus the critical region corresponds to all values of T less than or equal to t_1, where

$$t_1 = np^* + z_{0.025} \sqrt{np^*(1 - p^*)}$$
$$= (925)(\tfrac{3}{4}) + (-1.960) \sqrt{(925)(\tfrac{3}{4})(\tfrac{1}{4})}$$
$$= 667.94 \tag{11}$$

and all values of T greater than t_2, where

$$t_2 = np^* + z_{0.975} \sqrt{np^*(1 - p^*)}$$
$$= (925)(\tfrac{3}{4}) + (1.960) \sqrt{(925)(\tfrac{3}{4})(\tfrac{1}{4})}$$
$$= 719.56 \qquad (12)$$

The value of T obtained is 682 in this experiment. Therefore the null hypothesis is accepted.

The p-value may be found from Equation 5.

$$P(Y \leq 682) \cong P\left(Z \leq \frac{682 - 693.75 + 0.5}{13.17}\right) = P(Z \leq -0.8542) = 0.196 \qquad (13)$$

where Z has the standard normal distribution as given in Table A1. This one-tailed p-value is doubled to find the two-tailed p-value 0.392.

A level of significance of at least 0.392 would be required to reject H_0. Thus the data are in good agreement with the null hypothesis. ■

The previous example illustrates the two-tailed form of the binomial test. The one-tailed binomial test was also illustrated in Example 2.3.1.

□ *Theory*   That the test statistic in the binomial test has a binomial distribution is easily seen by comparing the assumptions in the binomial test with the assumptions in Examples 1.3.5 and 1.2.8. That is, if T equals the number of trials that result in the outcome "class 1," where the trials are mutually independent and where each trial has probability p of resulting in that outcome (as stated by the assumptions), then T has the binomial distribution with parameters p and n. The size of the critical region is a maximum when p equals p^*, under the null hypothesis, and so Table A3 is entered with n and p^* to determine the exact value of α. □

As mentioned earlier, hypothesis testing is only one branch of statistical inference. We will now discuss another branch, *interval estimation*. If we are attempting to make some inferences regarding an unknown parameter associated with some population, it is reasonable to examine a random sample from that population and, on the basis of that sample, to make some statement regarding the population parameter. Such a statement might be "the population parameter lies between a and b," where a and b are two real numbers obtained from the sample. The numbers a and b are computed from the sample and are therefore realizations of two statistics. The two statistics that furnish us with the lower and upper boundary points for the interval will be denoted by L and U, respectively, for "lower" and "upper." The interval from L to U is called the *interval estimator*. The probability that the unknown population parameter lies within its interval estimate is called the *confidence coefficient*. The interval estimator together with the confidence coefficient provide us with the *confidence interval*.

A method for finding a confidence interval for p, the unknown probability of any particular event occurring, is closely related to the binomial test.

▶ **Confidence Interval for a Probability or Population Proportion** ___

Data   A sample consisting of observations on n independent trials is examined, and the number Y of times the specified event occurs is noted.

Assumptions

1. The n trials are mutually independent.
2. The probability p of the specified event occurring remains constant from one trial to the next.

Method A   For n less than or equal to 30, and confidence coefficients of 0.90, 0.95, or 0.99, use Table A4. Simply enter the table with sample size n and the observed Y. Reading across gives the exact lower and upper bounds in the columns for the desired confidence interval.

Method B   For n greater than 30, or confidence coefficients not covered in Table A4, use the normal approximation.

$$L = \frac{Y}{n} - z_{1-\alpha/2} \sqrt{Y(n - Y)/n^3} \tag{14}$$

and

$$U = \frac{Y}{n} + z_{1-\alpha/2} \sqrt{Y(n - Y)/n^3} \tag{15}$$

where $z_{1-\alpha/2}$ is the quantile of a normally distributed random variable, obtained from Table A1. The confidence coefficient is approximately $1 - \alpha$.

Computer Assistance   Computer packages that find confidence intervals for the binomial parameter p, or population proportion p, include *Minitab*, *S-Plus*, and *StatXact*. _____ ◀

For the sake of illustration, both methods of computing confidence intervals are used in the following example.

EXAMPLE 3

In a certain state 20 high schools were selected at random to see if they met the standards of excellence proposed by a national committee on education. It was found that 7 schools did qualify and accordingly were designated "excellent."

What is a 95% confidence interval for p, the proportion of all high schools in the state that would qualify for the designation "excellent"?

First, we assume that the number of high schools in the state is large enough so the high schools are classified "excellent" or "not excellent" independently of one another.

Because we assumed the selection was random, p is the same for all schools and represents the probability of a randomly selected school being designated "excellent."

Because n equals 20 and Y equals 7, Table A4 can be used. The exact 95% confidence interval is given in Table A4 as the interval from 0.154 to 0.592.

Method B, the use of the normal approximation based on the central limit theorem, gives

$$L = \frac{Y}{n} - z_{0.975} \sqrt{Y(n - Y)/n^3}$$
$$= 0.35 - (1.960) \sqrt{(7)(13)/(20)^3}$$
$$= 0.35 - 0.209$$
$$= 0.141 \tag{16}$$

and

$$U = 0.35 + 0.209$$
$$= 0.559 \tag{17}$$

The confidence interval furnished by the normal approximation is from 0.141 to 0.559, which is close to the exact confidence interval, but still different enough to show the clear advantage of using exact intervals when they are available. ∎

□ *Theory* For the exact Method A just described, the confidence interval consists of all values of p^* such that the data obtained in the sample would result in acceptance of

$$H_0: p = p^*$$

if one were using the two-tailed binomial test. More precisely, if we want to form a $(1 - \alpha)$ confidence interval, we observe the sample and determine Y. Then we ask, "For the given value of Y, which values may we use for p^* in the hypothesis

$$H_0: p = p^*$$

such that a two-tailed binomial test (at level α) would result in acceptance of H_0?" Those values of p^* would be in our confidence interval. The values of p^* that would result in rejection of H_0 would not be in the confidence interval. Since each tail of the binomial test has probability $\alpha/2$, the value of L is selected as the value of p^* that would barely result in rejection of H_0, for the given value of Y, say y, or a larger value. Thus $p_1{}^*$ is selected so that

$$P(Y \geq y \,|\, p = p_1{}^*) = \frac{\alpha}{2} = \sum_{i=y}^{n} \binom{n}{i} (p_1{}^*)^i (1 - p_1{}^*)^{n-i} \tag{18}$$

and then $L = p_1{}^*$. Next, another value of p^* is selected so the same value y is barely in the lower tail. That is, $p_2{}^*$ is selected so

$$P(Y \leq y \,|\, p = p_2{}^*) = \frac{\alpha}{2} = \sum_{i=0}^{y} \binom{n}{i} (p_2{}^*)^i (1 - p_2{}^*)^{n-i} \tag{19}$$

and we set $U = p_2{}^*$. Equations 18 and 19 are impossible to solve algebraically. They were solved using a search procedure on a computer to obtain Table A4.

More information on confidence intervals for the binomial parameter p may be found in Clopper and Pearson (1934).

The large sample approximations for L and U may be obtained by considering Example 1.5.6, which states that if Y is a binomially distributed random variable with parameters p and large n, then

$$Z = \frac{Y - np}{\sqrt{npq}} \tag{20}$$

is a random variable whose distribution may be approximated by the standard normal distribution. Then, if $z_{1-\alpha/2}$ is the $(1 - \alpha/2)$ quantile from Table A1, and because $z_{\alpha/2} = -z_{1-\alpha/2}$, we have

$$1 - \alpha = P\left(-z_{1-\alpha/2} < \frac{Y - np}{\sqrt{npq}} < z_{1-\alpha/2}\right)$$
$$= P(-z_{1-\alpha/2}\sqrt{npq} < Y - np < z_{1-\alpha/2}\sqrt{npq})$$

Multiplication by (-1) reverses the sense of the inequalities

$$1 - \alpha = P(z_{1-\alpha/2}\sqrt{npq} > np - Y > -z_{1-\alpha/2}\sqrt{npq})$$

and reversal of the reading order gives

$$1 - \alpha = P(-z_{1-\alpha/2}\sqrt{npq} < np - Y < z_{1-\alpha/2}\sqrt{npq})$$
$$= P(Y - z_{1-\alpha/2}\sqrt{npq} < np < Y + z_{1-\alpha/2}\sqrt{npq})$$

Now we divide through by n

$$1 - \alpha = P\left(\frac{Y}{n} - z_{1-\alpha/2}\sqrt{\frac{pq}{n}} < p < \frac{Y}{n} + z_{1-\alpha/2}\sqrt{\frac{pq}{n}}\right) \tag{21}$$

Using a further approximation, the estimator Y/n is used for p under the radical in Equation 21 to obtain

$$1 - \alpha \cong P\left(\frac{Y}{n} - z_{1-\alpha/2}\sqrt{\frac{Y}{n}\left(1 - \frac{Y}{n}\right)\Big/ n} < p < \frac{Y}{n} + z_{1-\alpha/2}\sqrt{\frac{Y}{n}\left(1 - \frac{Y}{n}\right)\Big/ n}\right)$$

$$\cong P(L < p < U) \tag{22}$$

where L and U are the same as in Equations 14 and 15. This latter approximation of Y/n for p results in a slight difference between the confidence interval and the hypothesis test, when the large sample approximations are used for both.

Multiplication by the sample size n in the preceding procedures gives nL and nU as the lower and upper bounds of the confidence interval for np, used to test hypotheses involving the mean of a binomial random variable, because

$$H_0: p = p^*$$

is equivalent to

$$H_0: np = np^* \qquad \qquad \square$$

Other methods of obtaining binomial confidence limits are given by Anderson and Burstein (1967 and 1968). Methods dealing with simultaneous confidence intervals for multinomial proportions are given by Quesenberry and Hurst (1964) and Goodman (1965).

3.1 EXERCISES

In each of the following exercises clearly state H_0, H_1, T, the decision rule, α, the decision, the p-value and the name of the test used, where such information is appropriate.

1. It is known that 20% of a certain species of insect exhibit a particular characteristic A. Eighteen insects of that species are obtained from an unusual environment, and none of these have characteristic A. Is it reasonable to assume that insects from that environment have the same probability of 0.20 that the species in general has? Use a two-tailed test.

2. Of 16 cars inspected during a safety campaign, 6 were found to be unsafe. Test the hypothesis that no more than 10% of the cars in the population are unsafe. (Which assumption is more likely to be false in this application?)

3. In a dice game a pair of dice were thrown 180 times. The event "seven" occurred on 38 of those times. If these dice are fair the probability of "seven" is one-sixth. If they are loaded the probability is higher.

    (a) Is the probability of "seven" what it should be if the dice were fair? Use a one-tailed test.

    (b) Find a 95% confidence interval for P(seven) using the large sample approximation.

4. In Exercise 2, what is a 90% confidence interval for the true proportion of unsafe cars in the population?

5. Twenty independent observations on a random variable X with the unknown distribution function $F(x)$ resulted in the following numbers.

142	134	98	119	131
103	154	122	93	137
86	119	161	144	158
165	81	117	128	103

    Find a 95% confidence interval for $F(100)$.

6. A civic group reported to the town council that at least 60% of the town residents were in favor of a particular bond issue. The town council then asked a random sample of 100 residents if they were in favor of the bond issue. Forty-eight said yes. Is the report of the civic group reasonable?

7. Out of 20 recent takeover attempts, 5 were successfully resisted by the companies being taken over. Assume these are independent events, and estimate the probability of a takeover attempt being successfully resisted. That is, find a 95% confidence interval.

    (a) Use Table A4.

    (b) Use Table A1.

8. An instructor is trying to adjust the level of difficulty of a continuing education class to meet the needs of his students. After teaching the course several times and giving the students a simple evaluation questionnaire each time, he found that 12 students said the course was too easy, 84 students said the course was about right, and 3 students said the course was too hard. Test the hypothesis that the students are equally likely to think the course is too easy or too hard against the two-sided alternative that the level of the course needs to be adjusted, at the 5% level of significance.

9. Use the binomial formula to find the exact p-value (critical level) when the null hypothesis is $P(\text{success}) \leq 0.3$ and the alternative hypothesis is $P(\text{success}) > 0.3$, and 3 successes are observed in 25 independent trials.

10. Twenty Texas Tech law school graduates took the bar exam and 18 of them passed. If this represents a random sample of all Texas Tech law school graduates, is this proof that the probability of a Texas Tech law school graduate passing the law exam is higher than the state average, which is 70%?

11. Twenty torpedoes are released in an underwater war exercise. Fifteen of them hit their target. Find a 90% confidence interval for the probability of a torpedo hitting the target.

    (a) Use Table A4 to solve this problem.

    (b) Use the large sample approximation to solve this problem.

    (c) Discuss the assumptions you are making in this problem.

12. Seventy chemical detection kits of one type are placed in a gas chamber together, for a fixed period of time, and a measured amount of a lethal gas is introduced into the chamber. Fifty-six kits register positive for the lethal gas, while the other 14 fail to register positive. Find a 90% confidence interval for the probability of registering positive under these conditions.

PROBLEMS

1. The *continuity correction*. It is obvious that if Y has a binomial distribution, then

$$P(Y \leq 4) = P(Y \leq 4.1) = \cdots = P(Y \leq 4.999)$$

because Y takes on only integer values, such as 4 or 5, but no values between integers. Therefore, which number should be used in the normal approximation to the binomial distribution: 4, or 4.1, or what? The *continuity correction* (because we are trying to use a continuous distribution such as the normal to approximate a discrete distribution such as the binomial) says to use the number midway between two adjacent values in the discrete distribution. That is, in the binomial distribution estimate $P(Y \leq 4)$, with

$$P(Y \leq 4) \cong P\left(Z \leq \frac{4 + 0.5 - np}{\sqrt{nqp}}\right)$$

where Z has a normal distribution, because 4.5 is halfway between 4 and 5.

Usually the continuity correction works well when using the normal distribution to approximate binomial probabilities.

(a) For $n = 20$, $p = 0.1$, find the exact value of $P(Y \leq 1)$ from Table A3. Use the normal approximation to estimate $P(Y \leq 1)$, first without the continuity correction and then with the continuity correction. Which estimate is closer?

(b) Repeat part a, but change from $p = 0.1$ to $p = 0.3$. Now which estimate is closer?

2. Let Y_1 and Y_2 be independent binomial random variables with parameters n_1 and p_1 and n_2 and p_2, respectively.

(a) Show that $Y_1/n_1 - Y_2/n_2$ has mean $p_1 - p_2$.

(b) Show that $Y_1/n_1 - Y_2/n_2$ has variance $p_1(1 - p_1)/n_1 + p_2(1 - p_2)/n_2$.

(c) Justify the use of $Y_1(n_1 - Y_1)/n_1^3 + Y_2(n_2 - Y_2)/n_2^3$ as an estimate of the variance of $Y_1/n_1 - Y_2/n_2$.

(d) If $Y_1/n_1 - Y_2/n_2$ is approximately normal, show how an approximate $1 - \alpha$ confidence interval for $(p_1 - p_2)$ is given by

$$\frac{Y_1}{n_1} - \frac{Y_2}{n_2} - z_{1-\alpha/2}s < p_1 - p_2 < \frac{Y_1}{n_1} - \frac{Y_2}{n_2} + z_{1-\alpha/2}s$$

where

$$s = \sqrt{Y_1(n_1 - Y)/n_1^3 + Y_2(n_2 - Y_2)/n_2^3}$$

and where $z_{1-\alpha/2}$ is obtained from Table A1.

3.2 THE QUANTILE TEST AND ESTIMATION OF x_p

The binomial test may be used to test hypotheses concerning the quantiles of a random variable, in which case we call it the quantile test. For example, we may examine a random sample of values of some random variable X to see if the median of X is equal to 17 (say). If the median of X is 17, then about half of the observations should fall on either side of 17, as in a binomial distribution with $p = 1/2$. If very few of the sample observations are less than 17, the median of X appears to be larger than 17. If by far most of the sample observations are less than 17, then the median of X appears to be less than 17.

The measurement scale is usually at least ordinal for the quantile test, although the binomial test only required the weaker nominal scale for its measurements. This is because quantiles have little or no meaning with nominal scale measurements. If the random variable being examined is a continuous random variable, the hypothesis being tested,

$$H_0: \text{The } p^*\text{th quantile of } X \text{ is } x^* \text{ (specified)}$$

is the same as

$$H_0: P(X \le x^*) = p^*$$

from the definition of the word *quantile*. If we represent the unknown probability $P(X \le x^*)$ by p, H_0 becomes

$$H_0: p = p^*$$

which is the same null hypothesis tested with the binomial test. The test statistic equals the number of sample values that are less than or equal to x^*, and the two-tailed binomial test may be used.

The situation is not as simple if the random variable is not assumed to be continuous. Then the null hypothesis

$$H_0: \text{The } p^*\text{th quantile of } X \text{ is } x^*$$

is the same as

$$H_0: P(X \le x^*) \ge p^* \quad \text{and} \quad P(X < x^*) \le p^*$$

Now the binomial test may be used, but the adaptation of the test to this hypothesis is a little tricky, so we will present the procedure as a separate test.

▶ **The Quantile Test** _____

Data   Let $X_1, X_2, \ldots, X_n$ be a random sample. The data consist of observations on the X_i.

Assumptions

1. The X_is are a random sample (i.e., they are independent and identically distributed random variables).
2. The measurement scale of the X_is is at least ordinal.

Test Statistic   We will use two test statistics in this test. Let T_1 equal the number of observations less than or equal to x^*, and let T_2 equal the number of observations less than x^*. Then $T_1 = T_2$ if none of the numbers in the data exactly equals x^*. Otherwise, T_1 is greater than T_2.

Null Distribution   The null distribution of the test statistics T_1 and T_2 is the binomial distribution, with parameters n = sample size, and $p = p^*$ as given in the null hypothesis. The null distribution is given in Table A3 for $n \leq 20$ and selected values of p.

For other values of n and p the normal approximation is used. That is, approximate quantiles x_q for T are given by

$$x_q = n \cdot p + z_q \sqrt{n \cdot p \cdot (1 - p)} \tag{1}$$

where z_q is the qth quantile of a standard normal random variable, given in Table A1.

Hypotheses   Let x^* and p^* represent some specified numbers, $0 < p^* < 1$. The hypotheses may take one of the following three forms.

A. (Two-Tailed Test)

$$H_0: \text{The } p^*\text{th population quantile is } x^*$$

[This is equivalent to $H_0: P(X \leq x^*) \geq p^*$, and $P(X < x^*) \leq p^*$, where X has the same distribution as the X_is in the random sample.]

$$H_1: x^* \text{ is not the } p^*\text{th population quantile}$$

The rejection region corresponds to values of T_2 that are too large [indicating that perhaps $P(X < x^*)$ is greater than p^*] and to values of T_1 that are too small [indicating that perhaps $P(X \leq x^*)$ is less than p^*]. The rejection region is found by entering Table A3 with the sample size n and the hypothesized probability p^*, as in the two-tailed binomial test. Find the number t_1 such that

$$P(Y \leq t_1) = \alpha_1 \tag{2}$$

where Y has the binomial distribution with parameters n and p^* and where α_1 is about half of the desired level of significance. Then find the number t_2 such that

$$P(Y \leq t_2) = 1 - \alpha_2 \tag{3}$$

where α_2 is chosen so that $\alpha_1 + \alpha_2$ is about equal to the desired level of significance. Reject H_0 if T_1 is less than or equal to t_1 or if T_2 is greater than t_2. Otherwise accept H_0. The level of significance equals $\alpha_1 + \alpha_2$.

For $n > 20$ or values of p^* not covered by Table A3, use Equation 1 to find $t_1 = x_{\alpha/2}$ and $t_2 = x_{1-\alpha/2}$ by letting $q = \alpha/2$ and $q = 1 - \alpha/2$, respectively.

The p-value is twice the smaller of the probabilities that a binomial random variable Y is less than or equal to the observed value of T_1, or greater than or equal to the observed value of T_2, which are found from Table A3 for $n \leq 20$, using $p = p^*$, or from Table A1 for $n > 20$, using

$$P(Y \leq T_1) \cong P\left(Z \leq \frac{T_1 - n \cdot p^* + 0.5}{\sqrt{n \cdot p^* \cdot (1 - p^*)}}\right) \tag{4}$$

and

$$P(Y \geq T_2) \cong 1 - P\left(Z \leq \frac{T_2 - n \cdot p^* - 0.5}{\sqrt{n \cdot p^* \cdot (1 - p^*)}}\right) \tag{5}$$

which incorporates 0.5 as a "correction for continuity" that improves the normal approximation to the binomial.

B. (Lower-Tailed Test)

H_0: The p^* population quantile is no greater than x^*

[Or H_0: $P(X \leq x^*) \geq p^*$.]

H_1: The p^* population quantile is greater than x^*

[Or H_1: $P(X \leq x^*) < p^*$.]

Small values of T_1 indicate H_0 is false, so enter Table A3 with the sample size n and the specified probability p^* to find t_1 such that

$$P(Y \leq t_1) = \alpha \tag{6}$$

for an acceptable level α, where Y has a binomial distribution with parameters n and p^*. Reject H_0 if T_1 is less than or equal to t_1. Accept H_0 if T_1 exceeds t_1. For $n > 20$ let $q = \alpha$ in Equation 1 to find $t_1 = x_\alpha$.

The p-value is the probability that a binomial random variable Y is less than or equal to the observed value of T_1, which is found from Table A3 for $n \leq 20$, using $p = p^*$, or from Table A1 for $n > 20$, using Equation 4.

C. (Upper-Tailed Test)

H_0: The p^*th population quantile is at least as great as x^*

[This is equivalent to H_0: $P(X < x^*) \leq p^*$.]

H_1: The p^*th population quantile is less than x^*

[This is the same as H_1: $P(X < x^*) > p^*$.]

Since large values of T_2 indicate that H_0 is false, enter Table A3 with the sample size n and the hypothesized p^* as p. Find the number t_2 such that

$$P(Y > t_2) = \alpha$$

which is the same as

$$P(Y \leq t_2) = 1 - \alpha \qquad (7)$$

for some acceptable level of significance α. Then reject H_0 if T_2 exceeds t_2. Accept H_0 if T_2 is less than or equal to t_2. For $n > 20$ let $q = 1 - \alpha$ in Equation 1 to find $t_2 = x_{1-\alpha}$.

The p-value is the probability that a binomial random variable Y is greater than or equal to the observed value of T_2, which is found from Table A3 for $n \leq 20$, using $p = p^*$, or from Table A1 for $n > 20$, using Equation 5.

Computer Assistance *Minitab* tests the null hypothesis when $p = 1/2$ under the name Median Test. ——————————————————————————◀

EXAMPLE 1

Entering college freshmen have taken a particular high school achievement examination for many years, and the upper quartile is well established at a score of 193. A particular high school sends 15 of its graduates to college, where they take the exam and get the following scores

189	233	195	160	212
176	231	185	199	213
202	193	174	166	248

It is assumed that these 15 students represent a random sample of all students from that high school who go on to college. One way of comparing college students from that high school with other college students is by testing the hypothesis that the above scores come from a population whose upper quartile is 193. That is,

H_0: The upper quartile is 193

is tested against the alternative

H_1: The upper quartile is not 193

where we are referring to the upper quartile of the test scores of all college students from that high school, past, present, or future.

The two-tailed quantile test is applied. A critical region of approximate size 0.05 is obtained by entering Table A3 with $n = 15$ and $p = 0.75$. There it is seen that, for the binomial random variable Y,

$$P(Y \leq 7) = 0.0173 \tag{8}$$

and

$$P(Y \leq 14) = 0.9866$$
$$= 1 - 0.0134 \tag{9}$$

The critical region of size

$$\alpha = 0.0173 + 0.0134$$
$$= 0.0307 \tag{10}$$

corresponds to values of T_1 less than or equal to $t_1 = 7$, and values of T_2 greater than $t_2 = 14$.

In this example $T_1 = 7$, the number of observations less than or equal to 193, and $T_2 = 6$, since one observation exactly equals 193. Therefore T_1 is too small, and H_0 is rejected. The upper quartile for students from that high school does not seem to be 193. The p-value is $2 \cdot P(Y \leq 7) = 2(0.0173) = 0.0346$. ■

The one-tailed quantile test, with the large sample approximation, is illustrated in the following example.

EXAMPLE 2

The time interval between eruptions of Old Faithful geyser is recorded 112 times to see whether the median interval is less than or equal to 60 minutes (null hypothesis) or whether the median interval is greater than 60 minutes (alternative hypothesis). If the median interval is 60, 60 is $x_{0.50}$, or the median. If the median interval is less than 60, 60 is a p quantile for some $p \geq 0.50$. Thus H_0 is $P(X \leq 60) \geq 0.50$, and H_1 is $P(X \leq 60) < 0.50$, where X is the time interval between eruptions. Assuming that the various intervals are independent and identically distributed, the lower-tailed quantile test may be used. The test statistic T_1 equals the number of intervals that are less than or equal to 60 minutes, and the critical region of size 0.05 corresponds to values of T_1 less than or equal to

$$t_1 = np^* + z_{0.05} \sqrt{np^*(1 - p^*)}$$
$$= (112)(0.50) - (1.645) \sqrt{(112)(0.50)(0.50)}$$
$$= 47.3 \tag{11}$$

Of the 112 time intervals, 8 are 60 minutes or less, so $T_1 = 8$, and H_0 is soundly rejected in favor of the alternative "the median time interval between eruptions is greater than 60 minutes." The p-value is found using Equation 4.

$$P(Y \le 8) \cong P\left[Z \le \frac{8 - (112)(0.50) + 0.5}{\sqrt{(112)(0.50)(0.50)}} \right] = P(Z \le -8.977) << 0.0001 \tag{12}$$

which is read "much less than 0.0001." ∎

□ *Theory*   First we will explain why the hypotheses within the parentheses in A, B, and C are equivalent to the hypotheses not in parentheses. Perhaps this is most easily seen by referring to the graph of an arbitrary distribution function, as in Figure 1.

The distribution function at x^* may be in one of three phases: it may be rising vertically, as at x_1; it may be in a horizontal segment, as at x_2; or it may be rising gradually, as at x_3. H_0 in the second set of hypotheses (set B) states that the p^*th population quantile (x_{p^*}) is no greater than x^*, or $x_{p^*} \le x^*$. Because every value of x^* is some sort of a quantile, we can say that x^* is the pth quantile for some p, say p_0. (We are temporarily ignoring our convention of choosing only the midpoint of the horizontal segments as the quantile and adhering directly to the definition of quantile.) Because the graph of the distribution function never descends as x gets larger, $x_{p^*} \le x^*$ implies that $p^* \le p_0$, which may be seen by imaging x^* as being in each of the three phases typified by x_1, x_2, and x_3 in Figure 1. Any value of x_{p^*} to the left of

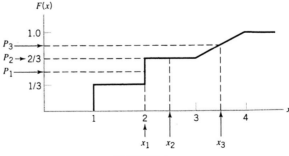

FIGURE I

x^* implies that the ordinate p^* at x_{p^*} is no greater than the ordinate p_0 of x^*. From the definition of quantile, Definition 1.4.1,

$$P(X > x^*) \leq 1 - p_0 \tag{13}$$

which is the same as

$$p_0 \leq 1 - P(X > x^*) = P(X \leq x^*) \tag{14}$$

Since $p^* \leq p_0$, this implies

$$p^* \leq P(X \leq x^*) \tag{15}$$

which is the equivalent form of H_0 in set B of hypotheses. The negation of H_0 is H_1, and the negation of Equation 15 is

$$p^* > P(X \leq x^*) \tag{16}$$

as stated in the alternative hypotheses. The same reasoning is used to show the other hypotheses to be equivalent.

Briefly, Figure 1 is used to visualize that $x_{p_0} \leq x_{p^*}$ (H_0 in C) implies that $p_0 \leq p^*$. If $x^* = x_{p_0}$ then by Definition 1.4.1,

$$P(X < x^*) \leq p_0 \leq p^* \tag{17}$$

is true, which furnishes the equivalent form of H_0.

The binomial test is applied directly to test the hypothesis in parentheses. H_0 in C is tested by defining the "class 1" of the binomial test as those observations less than x^*. H_0 in B is tested by considering "class 1" to represent those observations less than or equal to x^*. The two tests in B and C are combined to give the two-tailed test in A. The assumptions of independence and constant probability p in the binomial test are satisfied because the X_i are independent and identically distributed (respectively). □

In the previous section we showed how to find a confidence interval for a probability p. The same method is used to find a confidence interval for $F(x_0)$, the distribution function at some specified number x_0. That is, given a number x_0, we can use the method in the previous section to find a "vertical" confidence interval (referring to a graph) for the unknown probability $F(x_0)$. Suppose, however, we are given a probability, say p^*, and asked to find a "horizontal" confidence interval for the unknown quantile x_{p^*}. This second type of confidence interval, confidence interval for a quantile, is found if we wish to make a statement concerning a specified quantile such as the median, the upper quartile, or any p^* quantile where p^* is a specified constant, $0 < p^* < 1$. The statement then takes the form

$$P(X^{(r)} \leq x_{p^*} \leq X^{(s)}) = 1 - \alpha \tag{18}$$

where $1 - \alpha$ is a known *confidence coefficient* and where $X^{(r)}$ and $X^{(s)}$ are order statistics (see Definition 2.1.4) with r and s specified. The values for r and s may be determined prior to drawing the sample in the manner described next, with knowledge of only the sample size n and the desired confidence coefficient. The sample $X_1, \ldots, X_n$ needs only to be random. No restrictions are made on the distribution function of X_i. Thus this statistical method may be applied freely to any random sample from any population.

▶ Confidence Interval for a Quantile

Data   The data consist of observations on $X_1, X_2, \ldots, X_n$, which are independent and identically distributed random variables. $X^{(1)} \leq X^{(2)} \leq \cdots \leq X^{(r)} \leq \cdots \leq X^{(s)} \leq \cdots \leq X^{(n)}$ represents the ordered sample, where $1 \leq r < s \leq n$. We wish to find a confidence interval for the (unknown) p^*th quantile, where p^* is some specified number between zero and one.

Assumptions

1. The sample $X_1, X_2, \ldots, X_n$ is a random sample.
2. The measurement scale of the X_is is at least ordinal.

Method A (small samples)   For $n \leq 20$ Table A3 may be used to find r and s. Enter Table A3 with the sample size n and the probability $p = p^*$. Read down the column for $p = p^*$ until reaching an entry approximately equal to $\alpha/2$, where $1 - \alpha$ is the approximate confidence coefficient desired. Call that entry α_1, and the corresponding value for y (to the far left of α_1) is $r - 1$. Add 1 to get r. Then continue down the column for $p = p^*$ until reaching an entry approximately equal to $1 - (\alpha/2)$, which we will call $1 - \alpha_2$. The value of y corresponding to the entry $1 - \alpha_2$ is called $s - 1$, and 1 is added to obtain s. Thus we have determined α_1, α_2, r, and s. The exact confidence coefficient is $1 - \alpha_1 - \alpha_2$. The interval estimator is the interval between $X^{(r)}$ and $X^{(s)}$, whose values may be obtained from the data. Then

$$P(X^{(r)} \leq x_{p^*} \leq X^{(s)}) \geq 1 - \alpha_1 - \alpha_2 \tag{19}$$

provides the confidence interval. If we assume that the unknown distribution function is continuous, then

$$P(X^{(r)} \leq x_{p^*} \leq X^{(s)}) = 1 - \alpha_1 - \alpha_2 \tag{20}$$

as stated in Equation 18 also.

Method B (large sample approximation)   For n greater than 20 the approximation based on the central limit theorem may be used. (See Equation 1.) Compute

$$r^* = np^* + z_{\alpha/2} \sqrt{np^*(1 - p^*)} \tag{21}$$

and

$$s^* = np^* + z_{1-\alpha/2} \sqrt{np^*(1 - p^*)} \tag{22}$$

where the quantiles z_p are obtained from Table A1 and where $1 - \alpha$ is the desired confidence coefficient. In general, r^* and s^* will not be integers. Let r and s be the integers obtained by rounding r^* and s^* upward to the next higher integers. Then the approximate confidence interval is given by Equation 19, or Equation 20 if the unknown distribution function is continuous.

A one-sided confidence interval may be formed by finding only r or s as described. One-sided confidence intervals are of the form

$$P(X^{(r)} \le x_{p^*}) = 1 - \alpha_1 \tag{23}$$

and

$$P(x_{p^*} \le X^{(s)}) = 1 - \alpha_2 \tag{24}$$

if the distribution function is continuous, or

$$P(X^{(r)} \le x_{p^*}) \ge 1 - \alpha_1 \tag{25}$$

and

$$P(x_{p^*} \le X^{(s)}) \ge 1 - \alpha_2 \tag{26}$$

otherwise.

Computer Assistance   *Minitab* (under Median Test) and *StatXact* (under Sign Test) find a confidence interval for the median. _____◀

EXAMPLE 3

Sixteen transistors are selected at random from a large batch of transistors and are tested. The number of hours until failure is recorded for each one. We wish to find a confidence interval for the upper quartile, with a confidence coefficient close to 90%. Table A3 is entered with $n = 16$ and $p = 0.75$. Reading down the column for $p = 0.75$ the probability 0.0271 is selected as being close to 0.05. The value of y associated with $\alpha_1 = 0.0271$ is $y = 8$; therefore r equals 9. The probability closest to 0.95 is 0.9365 = 1 − α_2, which has a corresponding y of 14. Therefore $s = 15$. The confidence interval is

$$P(X^{(9)} \leq x_{0.75} \leq X^{(15)}) = 0.9094 \tag{27}$$

(It is reasonable to assume the time to failure is a continuous random variable, so we can use Equation 20.)

The results of the testing, arranged in increasing order, are as follows.

$X^{(1)} = 46.9$	$X^{(5)} = 56.8$	$X^{(9)} = 63.3$	$X^{(13)} = 67.1$
$X^{(2)} = 47.2$	$X^{(6)} = 59.2$	$X^{(10)} = 63.4$	$X^{(14)} = 67.7$
$X^{(3)} = 49.1$	$X^{(7)} = 59.9$	$X^{(11)} = 63.7$	$X^{(15)} = 73.3$
$X^{(4)} = 56.5$	$X^{(8)} = 63.2$	$X^{(12)} = 64.1$	$X^{(16)} = 78.5$

Because $X^{(9)} = 63.3$ and $X^{(15)} = 73.3$, we may say "the interval from 63.3 hours to 73.3 hours, inclusive, is a 90.94% confidence interval for the upper quartile."

The large sample approximation, furnished by Equations 21 and 22, yields

$$r^* = (16)(0.75) + (-1.645)\sqrt{(16)(0.75)(0.25)}$$
$$= 12 - 2.86$$
$$= 9.14 \tag{28}$$

and

$$s^* = 12 + 2.86$$
$$= 14.86 \tag{29}$$

Therefore $r = 10$ and $s = 15$, so the 90% confidence interval becomes (63.4, 73.3), slightly smaller than the more precise method used before. ∎

□ **Theory**   Consider first the simpler case where the distribution function is continuous. If x_{p^*} is the p^*th quantile, we have the exact relationship

$$P(X \geq x_{p^*}) = P(X > x_{p^*}) = 1 - p^* \tag{30}$$

where the distribution function of X is the same as that of the random sample.

The order statistic of rank 1, $X^{(1)}$, will assume a value larger than some specified constant only if the smallest value in the sample is larger than the constant. Therefore $X^{(1)}$ is greater than the constant only if all n values in the sample are greater than the constant. Choosing x_{p^*} as the constant, we may conclude

$$P(x_{p^*} < X^{(1)}) = P(\text{all sample values exceed } x_{p^*})$$
$$= P(x_{p^*} < X_1, x_{p^*} < X_2, \ldots, x_{p^*} < X_p)$$
$$= P(x_{p^*} < X_1) \cdot P(x_{p^*} < X_2) \cdot \cdots \cdot P(x_{p^*} < X_p)$$
$$= (1 - p^*)^n \tag{31}$$

because the X_is are independent, and they all have the same p^*th quantile x_{p^*}.

If x_{p^*} is less than $X^{(2)}$, then either exactly $n - 1$ observations are greater than x_{p^*}, in which case $X^{(1)} \le x_{p^*} < X^{(2)}$, or else exactly n observations are greater than the x_{p^*}, in which case $x_{p^*} < X^{(1)} < X^{(2)}$. Therefore

$$
\begin{aligned}
P(x_{p^*} < X^{(2)}) &= P(x_{p^*} < X^{(1)}) + P(X^{(1)} \le x_{p^*} < X^{(2)}) \\
&= P(\text{at least } n - 1 \text{ of the } X_i \text{ exceed } x_{p^*}) \\
&= P(1 \text{ or fewer of the } X_i \text{ are } \le x_{p^*})
\end{aligned}
\tag{32}
$$

Now the probability in Equation 32 is given by the binomial distribution function, because each X_i has probability p^* of being less than or equal to x_{p^*}, and the X_i are independent. Therefore Equation 32 leads to

$$
P(x_{p^*} < X^{(2)}) = \sum_{i=0}^{1} \binom{n}{i} (p^*)^i (1 - p^*)^{n-i}
\tag{33}
$$

With the aid of the binomial distribution function given by Equation 1.3.8, the preceding argument may be extended as follows.

$$
\begin{aligned}
P(x_{p^*} < X^{(r)}) &= P(\text{at least } n - r + 1 \text{ of the } X_i \text{ exceed } x_{p^*}) \\
&= P(r - 1 \text{ or fewer of the } X_i \text{ are } \le x_{p^*}) \\
&= \sum_{i=0}^{r-1} \binom{n}{i} (p^*)^i (1 - p^*)^{n-i}
\end{aligned}
\tag{34}
$$

The confidence coefficient is given by

$$
\begin{aligned}
1 - \alpha &\cong P(X^{(r)} \le x_{p^*} \le X^{(s)}) \\
&= P(x_{p^*} \le X^{(s)}) - P(x_{p^*} < X^{(r)})
\end{aligned}
\tag{35}
$$

Therefore r and s may be selected, with the aid of Equation 34 and Table A3, so that

$$
1 - \alpha_2 = P(x_{p^*} \le X^{(s)}) \cong 1 - \frac{\alpha}{2}
\tag{36}
$$

and

$$
\alpha_1 = P(x_{p^*} < X^{(r)}) \cong \frac{\alpha}{2}
\tag{37}
$$

Then the confidence coefficient will be $1 - \alpha_2 - \alpha_1 \cong 1 - \alpha$. Note that because the distribution function is assumed to be continuous, we have

$$
P(x_{p^*} \le X^{(s)}) = P(x_{p^*} < X^{(s)})
\tag{38}
$$

so that Table A3 may be used to find s.

If the distribution function of X, and therefore of the X_is, is not necessarily continuous, Equation 30 is not necessarily true. Instead, by Definition 1.5.1 we have

$$P(X > x_{p^*}) \leq 1 - p^* \tag{39}$$

and

$$P(X \geq x_{p^*}) \geq 1 - p^* \tag{40}$$

First we will consider how Equation 39 affects Equation 34 and, therefore, our method for finding r in Equation 37. Because Equation 39 is true, the probability of each observation exceeding x_{p^*} may be smaller than it was when X was continuous. Therefore there may be less of a tendency for each of the order statistics to exceed x_{p^*} than was formerly the case. That is, the probability $P(x_{p^*} < X^{(r)})$ may be smaller than it was in the continuous case, which was then given by Equation 34. So, in general, the following holds true instead of Equation 34.

$$P(x_{p^*} < X^{(r)}) \leq \sum_{i=0}^{r-1} \binom{n}{i} (p^*)^i (1 - p^*)^{n-i} \tag{41}$$

If Table A3 is used to find r in the manner just described, then

$$P(x_{p^*} < X^{(r)}) \leq \alpha_1 \tag{42}$$

Now we will consider how Equation 40 affects the probability $1 - \alpha_2$ resulting from our method of selecting s. Because Equation 40 is true, there may be a larger probability for each observation to be greater than or equal to x_{p^*} than in the continuous case. Therefore the number of observations exceeding or equaling x_{p^*} may tend to be larger and the probability of $X^{(s)} \geq x_{p^*}$ may be larger than in the continuous case. As a result, Equation 34 is modified in the general case to read

$$P(x_{p^*} \leq X^{(s)}) \geq \sum_{i=0}^{s-1} \binom{n}{i} (p^*)^i (1 - p^*)^{n-i} \tag{43}$$

Therefore, if Table A3 is used to find s in the manner described, we have

$$P(x_{p^*} \leq X^{(s)}) \geq 1 - \alpha_2 \tag{44}$$

Equations 42 and 44, which are true for all distributions, may be used as follows.

$$\begin{aligned} P(X^{(r)} \leq x_{p^*} \leq X^{(s)}) &= P(x_{p^*} \leq X^{(s)}) - P(x_{p^*} < X^{(r)}) \\ &\geq P(x_{p^*} \leq X^{(s)}) - \alpha_1 \\ &\geq 1 - \alpha_2 - \alpha_1 \end{aligned} \tag{45}$$

So this method may be conservative with discrete random variables, or ordinal data with ties. Thus the method of finding a confidence interval for a quantile has been justified for the case where exact tables of the binomial distribution function are available.

The large sample method of obtaining r and s is based on the use of the standard normal distribution to approximate the binomial distribution. Different arguments may be advanced for the different possible ways of converting r^* and s^* into the integers r and s, but the method given here of simply rounding upward to the next higher integer seems to provide a sufficiently close approximation.

This quantile test may be used with ordinal data and therefore is more applicable than any parametric test. If the data are at least interval in scale of measurement, and furthermore have a normal distribution, the median equals the mean, and so the quantile test of the median may be compared with the one-sample t test, where it has an A.R.E. of only $2/\pi = 0.637$. For a uniform distribution, which has light tails, its A.R.E. is only $1/3 = 0.333$. However, for a symmetric, heavy-tailed distribution known as the double exponential distribution the A.R.E. of the quantile test relative to the t test jumps to 2.0, indicating that this quantile test may have more power than the parametric test for some heavy-tailed nonnormal distributions. $\square$

The usage of one-sided confidence intervals for quantiles in life testing situations is discussed by Barlow and Gupta (1966). Tables of distribution-free confidence limits are given by Van der Parren (1970) for the median and by Van der Parren (1973) for quantiles. Confidence intervals for intervals between quantiles are discussed by Krewski (1976) and Reiss and Rüschendorf (1976).

EXERCISES

1. A random sample of tenth-grade boys resulted in the following 20 observed weights.

$$
\begin{array}{ccccc}
142 & 134 & 98 & 119 & 131 \\
103 & 154 & 122 & 93 & 137 \\
86 & 119 & 161 & 144 & 158 \\
165 & 81 & 117 & 128 & 103
\end{array}
$$

Test the hypothesis that the median weight is 103.

2. In Exercise 1 test the hypothesis that the upper quartile is at least 150.

3. In Exercise 1 test the hypothesis that the third decile is no greater than 100.

4. In Exercise 1 find an approximate 90% confidence interval for the median. What is the exact confidence coefficient? Also compare the results using the exact method with the results obtained using the large sample approximation.

5. It is desired to design a given automobile to allow enough headroom to accommodate comfortably all but the tallest 5% of the people who drive. Former studies indicate that the 95th percentile was 70.3 inches. In order to see if the former studies are still valid, a random sample of size 100 is selected. It is found that the 12 tallest persons in the sample have the following heights.

$$\begin{array}{cccc} 72.6 & 70.0 & 71.3 & 70.5 \\ 70.8 & 76.0 & 70.1 & 72.5 \\ 71.1 & 70.6 & 71.9 & 72.8 \end{array}$$

Is it reasonable to use 70.3 as the 95th percentile?

6. In Exercise 5, what is a 95% confidence interval for the 95th percentile of drivers from which the sample was selected?

7. The sergeant recalls that in the "good old days" the upper quartile for time required to complete the obstacle course was 42 minutes. He suspects that the new recruits are not as fit as the recruits in the good old days, so he times them on the obstacle course. He finds that out of 38 recruits, only ten of them complete the course within the 42 minute time period. Use the quantile test to test the hypothesis that the upper quartile is 42 minutes against the appropriate one-sided alternative.

8. Armor plating with a thickness of 10 cm is being tested to see how deeply a given projectile will penetrate armor. Fifty projectiles are fired at the armor plating, and the depth of penetration is measured. Seven of the projectiles pierced a hole through the armor plating, so their depth of penetration is recorded as 10+. All fifty values, ordered from smallest to largest, are given as follows.

5.37, 5.39, 5.42, 5.51, 5.63, 5.74, 5.82, 5.83, 5.94, 5.98, 6.07, 6.07, 6.13, 6.20, 6.21, 6.23, 6.25, 6.26, 6.26, 6.28, 6.29, 6.31, 6.35, 6.41, 6.57, 6.67, 6.81, 7.03, 7.40, 7.44, 7.82, 8.03, 8.11, 8.44, 8.51, 8.72, 8.83, 9.04, 9.33, 9.51, 9.61, 9.68, 9.82, 10+, 10+, 10+, 10+, 10+, 10+, 10+

Find a 95% confidence interval for the median penetration of the armor.

3.2 PROBLEM

1. One parametric method for finding a $1 - \alpha$ confidence interval for the median is to assume that the population is normal and use

$$\overline{X} + t_{\alpha/2}S/\sqrt{n-1} < x_{0.5} < \overline{X} + t_{1-\alpha/2}S/\sqrt{n-1}$$

where $\overline{X}$ is the sample mean, S is the sample standard deviation (Definition 2.2.3), n is the sample size, and t_p is the pth quantile from Table A21, $n - 1$ degrees of freedom. Compute the preceding confidence interval on the data in Exercise 1 and compare it with the nonparametric confidence intervals of Exercise 4, where $\alpha = 0.10$. Which confidence interval is the easier to justify? Which confidence interval is "better" (in terms of being shorter)?

3.3 TOLERANCE LIMITS

The confidence intervals of Sections 3.1 and 3.2 provide interval estimates for unknown population parameters, such as the unknown probability p or the unknown quantile x_p, and a certain probability $1 - \alpha$ (confidence coefficient) that the unknown parameter is within the interval. Tolerance limits differ from confidence intervals in that tolerance limits provide an interval within which at least a proportion q of the population lies, with probability $1 - \alpha$ or more that the stated interval does indeed "contain" the proportion q of the population. A typical application would be in a situation where we are about to draw a random sample of size n, $X_1, X_2, \ldots, X_n$, and we want to know how large n should be so that we can be 95% sure that at least 90% of the population lies between $X^{(1)}$ and $X^{(n)}$, the largest and smallest observations in our sample. We may generalize somewhat and consider the question, "How large must the sample size n be so that at least a proportion q of the population is between $X^{(r)}$ and $X^{(n+1-m)}$ with probability $1 - \alpha$ or more?" The numbers q, r, m, and $1 - \alpha$ are known (or selected) beforehand, and only n needs to be determined.

Another typical situation is when a random sample of size n is available, and we want 95% confidence (or some other value of $1 - \alpha$) that the limits we choose will contain at least q of the population. What will the population proportion q be if we choose the two extreme values in the sample, $X^{(1)}$ and $X^{(n)}$, as our limits? Or should we choose the second most extreme values as our limits, $X^{(2)}$ and $X^{(n-1)}$? What proportion of the population will lie within those limits, with 95% confidence? In this version of the problem, q is the unknown quantity and is obtained after we know, or set, values for α, n, r, and m.

The preceding tolerance limits would be two-sided tolerance limits. One-sided tolerance limits are of the form, "At least a proportion q of the population is greater than $X^{(r)}$, with probability $1 - \alpha$," or "At least a proportion q of the population is less than $X^{(n+1-m)}$, with probability $1 - \alpha$." One-sided tolerance limits are identical with one-sided confidence intervals for quantiles, as will be shown in this section.

The population referred to here is either infinite or, if the population is finite, the sample is drawn with replacement so the X_is are independent. For finite populations where the sampling is without replacement and where the sample size n is small compared to the population size N, these methods are fairly accurate. More precise methods for finite populations may be found in Wilks (1962).

▶ **Tolerance Limits** _____

Data   The data consist of a random sample $X_1, X_2, \ldots, X_n$ from a large population. Choose a confidence coefficient $1 - \alpha$ and a pair of positive integers r and m. Either we wish to determine the required sample size n after selecting a desired population proportion q (see Method A), or we wish to determine the population proportion q for a given sample size n (see Method B). We are trying to make the statement, "The probability is $1 - \alpha$ that the random interval from $X^{(r)}$ to $X^{(n+1-m)}$ inclusive contains a proportion q or more of the population." Note that we are using the convention $X^{(0)} = -\infty$ and $X^{(n+1)} = +\infty$, so that one-sided tolerance limits may be obtained by setting either r or m equal to zero.

Assumptions

1. The $X_1, X_2, \ldots , X_n$ constitute a random sample.
2. The measurement scale is at least ordinal.

Method A (to find n) If $r + m$ equals 1, that is, if either r or m equals zero as in a one-sided tolerance limit, read n directly from Table A5 for the appropriate values of α and q. If $r + m$ equals 2, read n directly from Table A6 for the appropriate values of α and q. If Tables A5 and A6 are not appropriate, use the approximation

$$n \cong \frac{1}{4} x_{1-\alpha} \frac{1+q}{1-q} + \frac{1}{2}(r + m - 1) \tag{1}$$

where $x_{1-\alpha}$ is the $(1 - \alpha)$ quantile of a chi-squared random variable with $2(r + m)$ degrees of freedom, obtained from Table A2.

Method B (to find q) For a given sample size n and selected values of α, r, and m, the approximate value of q, the proportion of the population, is given by

$$q = \frac{4n - 2(r + m - 1) - x_{1-\alpha}}{4n - 2(r + m - 1) + x_{1-\alpha}} \tag{2}$$

where $x_{1-\alpha}$ is the $(1 - \alpha)$ quantile of a chi-squared random variable with $2(r + m)$ degrees of freedom.

Tolerance Limit With a sample of size n, there is probability at least $1 - \alpha$ that at least q [or $(100)(q)\%$] of the population is between $X^{(r)}$ and $X^{(n+1-m)}$ inclusive. That is

$$P(X^{(r)} \leq \text{at least a fraction } q \text{ of the population} \leq X^{(n+1-m)}) \geq 1 - \alpha \tag{3}$$

For one-sided tolerance regions let either r or m equal zero, where $X^{(0)}$ and $X^{(n+1)}$ are considered to be $-\infty$ and $+\infty$ repectively, and proceed as described above. ◀

EXAMPLE I

Probably the most widely used two-sided tolerance limits are those where $r = 1$ and $m = 1$. Electric seat adjusters are available on a popular luxury car. The manufacturer wants to know what range of vertical adjustment is necessary to be 90% certain that at least 80% of the population of potential buyers will be able to adjust their seats to the desired height. What must n be so that $X^{(n)}$ and $X^{(1)}$ furnish our upper and lower limits?

Table A6 is entered with $q = 0.80$ and $1 - \alpha = 0.90$. The obtained value for n is 18. The approximation furnished by Equation 1 is

$$n \cong \frac{1}{4} x_{1-\alpha} \frac{1 + q}{1 - q} + \frac{1}{2}(r + m - 1)$$

$$= \frac{1}{4}(7.779) \frac{1.80}{0.20} + \frac{1}{2}$$

$$= 18.003$$

A sample of size 18 is drawn from the population of potential buyers, and the amount of adjustment from a base position is measured. The largest value in the sample is

$$X^{(18)} = 7.57 \text{ inches}$$

and the smallest value is

$$X^{(1)} = 1.21 \text{ inches}$$

Therefore there is probability 0.90 that at least 80% of the population requires a vertical seat adjustment equal to or between 1.21 and 7.57 inches. ∎

The following is an example of a one-sided tolerance limit.

EXAMPLE 2

Along with each lot of steel bars, the manufacturer guarantees that at least 90% of the bars will have a breaking point above a number specified for each lot. Because of variable manufacturing conditions the guaranteed breaking point is established separately for each lot by breaking a random sample of bars from each lot and setting the guaranteed breaking point equal to the minimum breaking point in the sample. How large should the sample be so that the manufacturer can be 95% sure the guarantee statement is correct?

Table A5 is entered with $q = 0.90$ and $1 - \alpha = 0.95$, with the result $n = 29$. In each lot a sample of size 29 is selected at random, and the smallest breaking point of these bars in the sample is stated as the guaranteed breaking point, at which at least 90% of the bars in the lot will still be intact, with probability 0.95. ∎

EXAMPLE 3

A large population of drums containing radioactive waste is being stored for safe keeping. Each drum has marked on it the amount of radioactive waste contained in the drum. Periodic audits are made where randomly selected drums are scanned externally to estimate the amount of radioactive waste contained in the drum, and the estimate is compared with the label to obtain the discrepancy X. Over a period of three months 122 drums have been examined in this way, and the results are a random sample $X_1, \ldots, X_{122}$, where each X_i is the discrepancy between the amount marked on the drum and the amount estimated by the scan.

The bounds $X^{(2)}$ and $X^{(121)}$ are selected ($r = 2$, $m = 2$), and the confidence level 95% is chosen. The population percentage falling within those bounds is found from Equation 2, using the 0.95 quantile from the chi-squared distribution with $2(r + m) = 2(2 + 2) = 8$ degrees of freedom, which is found in Table A2 as 15.51.

$$q = \frac{4n - 2(r + m - 1) - x_{1-\alpha}}{4n - 2(r + m - 1) + x_{1-\alpha}} = \frac{488 - 6 - 15.51}{488 - 6 + 15.51} = 0.938$$

We can be 95% confident that at least 93.8% of the drums have discrepancies between the second smallest and the second largest observed discrepancies in the 122 observed drums. ∎

□ *Theory*   A careful examination of the statement furnished by the one-sided tolerance limit reveals the similarity it has with one-sided confidence interval for quantiles. That is, the one-sided tolerance limit says

$$P(\text{at least } q \text{ of the population is} \leq X^{(n+1-m)}) \geq 1 - \alpha \tag{4}$$

However, "at least q of the population is $\leq X^{(n+1-m)}$" is the same as saying "the q quantile is $\leq X^{(n+1-m)}$"; the two statements are merely different ways of stating the same idea. So we have

$$\begin{aligned}
P(\text{at least } q \text{ of the population is} &\leq X^{(n+1-m)}) \\
&= P(\text{the } q \text{ quantile is} \leq X^{(n+1-m)}) \\
&= P(x_q \leq X^{(n+1-m)})
\end{aligned} \tag{5}$$

The probability in Equation 5 was given in Equation 3.2.43 as

$$P(x_q \leq X^{(n+1-m)}) \geq \sum_{i=0}^{n-m} \binom{n}{i} q^i (1 - q)^{n-i} \tag{6}$$

The right side of Equation 6 is examined to find the smallest value for n such that the right side of Equation 6 exceeds $1 - \alpha$. This may be accomplished by entering Table A3 with $y = n - m$, the parameter p equal to q, and then searching for the lowest value of n for which the entry is greater than or equal to $1 - \alpha$.

Because the value for y changes as n changes, it is more convenient to rewrite the right side of Equation 6 as

$$\sum_{i=0}^{n-m} \binom{n}{i} q^i (1-q)^{n-i} = 1 - \sum_{i=n-m+1}^{n} \binom{n}{i} q^i (1-q)^{n-i} \qquad (7)$$

which is possible because the sum of all the binomial probabilities equals unity. A change of index, $j = n - i$, on the right side of Equation 7 results in

$$\sum_{i=0}^{n-m} \binom{n}{i} q^i (1-q)^{n-i} = 1 - \sum_{j=0}^{m-1} \binom{n}{j} (1-q)^j q^{n-j} \qquad (8)$$

Equation 8 could have been obtained immediately by saying, "The probability of $n - m$ or fewer successes equals the probability of m or more failures, which equals 1 minus the probability of $m - 1$ or fewer failures." The combination of Equations 8 and 6 shows that we could find n by solving for the smallest value of n that satisfies

$$\sum_{j=0}^{m-1} \binom{n}{j} (1-q)^j q^{n-j} \leq \alpha \qquad (9)$$

which is obtained from the inequality

$$1 - \sum_{j=0}^{m-1} \binom{n}{j} (1-q)^j q^{n-j} \geq 1 - \alpha \qquad (10)$$

Then Table A3 may be entered with $y = m - 1$ and $p = 1 - q$ and the pages turned until the entry in the table is less than or equal to α. That value of n is the sample size selected.

The other one-sided tolerance limit is

$$P(X^{(r)} \leq \text{at least } q \text{ of the population}) \geq 1 - \alpha \qquad (11)$$

which is equivalent to the statement

$$P(X^{(r)} \leq x_{1-q}) \geq 1 - \alpha \qquad (12)$$

because at least $1 - q$ of the population is greater than or equal to x_{1-q}. Equation 12 becomes

$$\alpha \geq 1 - P(X^{(r)} \leq x_{1-q}) = P(x_{1-q} < X^{(r)}) \qquad (13)$$

From Equation 3.2.41 we see that the solution to Equation 13 is the smallest value of n such that

$$\alpha \geq \sum_{i=0}^{r-1} \binom{n}{i} (1 - q)^i q^{n-i} \tag{14}$$

just as in Equation 9.

In fact, it can be shown, with the aid of calculus (see Noether, 1967a), that for the two-sided tolerance limits and for both types of one-sided tolerance limits, the sample size n depends on the solution to

$$\alpha \geq \sum_{i=0}^{r+m-1} \binom{n}{i} (1 - q)^i q^{n-i} \tag{15}$$

which is somewhat surprising in that Equation 15 depends on the sum $r + m$ but does not depend on whether we wish to choose as our interval all values to the right of $X^{(r+m)}$, or all values of the left of $X^{(n+1-r-m)}$, or all values between $X^{(r)}$ and $X^{(n+1-m)}$, or any combination of two order statistics whose ranks have a difference of $n + 1 - m - r$.

The use of Table A3 to solve Equation 15 is, at best, frustrating. Therefore Tables A5 and A6 are given for the most popular values $r + m = 1$ and $r + m = 2$. The approximation in Equation 1 is furnished without proof by Scheffé and Tukey (1944). Equation 2 is obtained by solving Equation 1 for q. Graphs to aid in finding n are given by Murphy (1948) and Birnbaum and Zuckerman (1949). More extensive tables are given by Owen (1962). □

Tolerance limits may also be used with two samples (Danziger and Davis, 1964), with a single censored sample (Bohrer, 1968), or for deciding from which of two possibly multivariate populations a sample was obtained (Quesenberry and Gessaman, 1968). Usage of tolerance limits on discrete random variables is examined by Hanson and Owen (1963). An application of tolerance intervals to the regression problem is discussed by Bowden (1968). Other articles dealing with tolerance limits are given by Mack (1969) and Goodman and Madansky (1962).

3.3 EXERCISES

1. What must the sample size be to be 90% sure that at least 95% of the population lies within the sample range?

   (a) Use the exact table.

   (b) Use the approximation.

2. What must the sample size be to be 95% certain that at least 90% of the population equals or exceeds $X^{(1)}$?

   (a) Use the exact table.

   (b) Use the approximation.

3. What must the sample size be in order for there to be a probability 0.90 that at least 85% of the population is $\leq X^{(n)}$?

   (a) Use the exact table.

   (b) Use the approximation.

4. What must the sample size be in order for there to be a 95% chance that 99% of the population is $\geq X^{(2)}$?

   (a) Use the exact table.

   (b) Use the approximation.

5. What must the sample size be so there is probability 0.90 that at least 50% of the population is between $X^{(5)}$ and $X^{(n-4)}$ inclusive?

6. What must the sample size be in Exercise 5 if the probability is 0.95 instead of 0.90? How about 0.99?

7. A fitness gym has measured the percentage of fat on 86 of its members.

   (a) At least what percent of its members have fat percentages between the smallest and largest of the percentages measured on the 86 members in the sample, with 95% certainty? With 90% certainty?

   (b) At least what percent of its members have fat percentage between $X^{(2)}$ and $X^{(85)}$ with 95% certainty? With 90% certainty?

8. A mail order catalog company surveyed 146 of its customers to find out the shipping time (from order date to the date of delivery) for their recent orders using regular U.S. mail.

   (a) At least what percentage of its customers can expect delivery between $X^{(1)}$ and $X^{(142)}$ as given by its sample observations, with 95% certainty? With 90% certainty.

   (b) Notice that the endpoints of the tolerance interval are not symmetric in this case. What is an advantage of using these asymmetric endpoints in this problem?

9. An engineer writing the acceptance specs for a load of steel reinforcement rods would like to specify that at least 90% of the rods are between the sixth longest and the sixth shortest rods in a random sample she selects. In order to have 99% confidence in this statement, what should the sample size be?

10. A computer model is developed to simulate the conditions within a combat unit (e.g., a communications center) in battle conditions. One of the items of interest, determined by the computer model, is the minimum number of people required to maintain a satisfactory level of operation of the combat unit. We want to staff the combat unit with enough people so it will operate satisfactorily during 90% of the battles.

    (a) How many computer runs are necessary so we can be 99.9% sure that the number of people required is no more than $X^{(n)}$, the largest observed number in the computer runs?

    (b) How many computer runs are necessary so we can be 99.9% sure that the number of people required is between $X^{(1)}$ and $X^{(n)}$?

(c) How many computer runs are necessary so we can be 99.9% sure that the number of people required is between $X^{(2)}$ and $X^{(n-1)}$?

(d) How many computer runs are necessary so that we can be 99.9% sure that the number of people required is no more than $X^{(n-4)}$?

PROBLEM

1. Use Table A3 to solve Exercise 3. Find the exact value of α.

3.4 THE SIGN TEST

After straying from hypothesis testing somewhat, at least in the previous section, we now return to discuss the oldest of all nonparametric tests, the sign test. Actually, the sign test is just the binomial test, with $p^* = 1/2$. But the sign test deserves special consideration because of its versatility, its age (dating back to 1710), and because $p^* = 1/2 = 1 - p^*$ makes it even simpler than the binomial test. The sign test is useful for testing whether one random variable in a pair (X, Y) tends to be larger than the other random variable in the pair. Also, as we will see in Section 3.5, it may be used to test for trend in a series of ordinal measurements or as a test for correlation. In many situations where the sign test may be used, more powerful nonparametric tests are available for the same model. However, the sign test is usually simpler and easier to use, and special tables to find the critical region are sometimes not needed.

▶ **The Sign Test**

Data The data consist of observations on a bivariate random sample (X_1, Y_1), $(X_2, Y_2), \ldots , (X_{n'}, Y_{n'})$, where there are n' pairs of observations. There should be some natural basis for pairing the observations; otherwise the Xs and Ys are independent, and the more powerful Mann-Whitney test of Chapter 5 is more appropriate.

Within each pair (X_i, Y_i) a comparison is made, and the pair is classified as "+" or "plus" if $X_i < Y_i$, as "−" or "minus" if $X_i > Y_i$, or as "0" or "tie" if $X_i = Y_i$. Thus the measurement scale needs only to be ordinal.

Assumptions

1. The bivariate random variables (X_i, Y_i), $i = 1, 2, \ldots , n'$, are mutually independent.
2. The measurement scale is at least ordinal within each pair. That is, each pair (X_i, Y_i) may be determined to be a "plus," "minus," or "tie."

3. The pairs (X_i, Y_i) are internally consistent, in that if $P(+) > P(-)$ for one pair (X_i, Y_i), then $P(+) > P(-)$ for all pairs. The same is true for $P(+) < P(-)$, and $P(+) = P(-)$.

Test Statistic   Let the test statistic T equal the number of "plus" pairs; that is, T equals the number of pairs (X_i, Y_i) in which X_i is less than Y_i.

$$T = \text{total number of } +\text{'s}$$

Null Distribution   The null distribution of T is the binomial distribution with $p = 1/2$ and $n =$ the number of nontied pairs. That is, disregard all tied pairs (where $X = Y$) and let

$$n = \text{total number of } +\text{'s and } -\text{'s}$$

Hypotheses

A. (Two-Tailed Test)

$$H_0: P(+) = P(-)$$
$$H_1: P(+) \neq P(-)$$

For $n \leq 20$, use Table A3 with the proper value of n and $p = 1/2$. Select a table value of about $\alpha/2$ and call it α_1. The value of y corresponding to α_1 is called t. The critical region of size $2\alpha_1$ corresponds to values of T less than or equal to t, or greater than or equal to $n - t$. Reject H_0 if $T \leq t$ or if $T \geq n - t$, at a level of significance of $2\alpha_1$. Otherwise accept H_0.

For n larger than 20 the normal approximation at the end of Table A3 is used to obtain

$$t = \tfrac{1}{2}(n + z_{\alpha/2} \sqrt{n}) \tag{1}$$

where $z_{\alpha/2}$ is obtained from Table A1. If $\alpha = 0.05$, $z_{\alpha/2} = (-1.96)$, and Equation 1 becomes approximately

$$t = \frac{n}{2} - \sqrt{n} \tag{2}$$

which may be easily remembered.

The p-value is twice the smaller of the probabilities that Y is less than or equal to the observed value of T, or greater than or equal to the observed value of T, which are found from Table A3 for $n \leq 20$, using $p = 1/2$, or from Table A1 for $n > 20$, using

$$P(Y \leq t_{obs}) = P\left(Z \leq \frac{2 \cdot t_{obs} - n + 1}{\sqrt{n}}\right) \tag{3}$$

and

$$P(Y \geq t_{obs}) = 1 - P\left(Z \leq \frac{2 \cdot t_{obs} - n - 1}{\sqrt{n}}\right) \tag{4}$$

which incorporates 1.0 as a "correction for continuity" that improves the normal approximation to the binomial.

B. (Lower-Tailed Test)

$$H_0: P(+) \geq P(-)$$
$$H_1: P(+) < P(-)$$

Small values of T indicate that a minus is more probable than a plus, in agreement with H_1. Therefore t is found by entering Table A3 with $p = 1/2$ and n and finding the table entry that approximately equals α, say α_1. The value of y corresponding to α_1 is t. For n greater than 20, t may be found from the approximation

$$t = \tfrac{1}{2}(n + z_\alpha \sqrt{n}) \tag{5}$$

where z_α is obtained from Table A1.

The critical region of size α_1 (or α) corresponds to values of T less than or equal to t. Reject H_0 if $T \leq t$, at a level of significance of α_1 (or α in the case of $n > 20$). Otherwise accept H_0.

The p-value is the probability that Y is less than or equal to the observed value of T, which is found from Table A3 for $n \leq 20$, using $p = 0.5$, or from Table A1 for $n > 20$, using Equation 3.

C. (Upper-Tailed Test)

$$H_0: P(+) \leq P(-)$$
$$H_1: P(+) > P(-)$$

Large values of T indicate that a plus is more probable than a minus, as stated by H_1. Therefore the critical region corresponds to values of T greater than or equal to $n - t$, where t is found (as in the Lower-Tailed Test) by entering Table A3 with $p = 1/2$ and n and finding the table entry that approximately equals α. The corresponding value of y is t. For n greater than 20, t may be found from the approximation given by Equation 5. H_0 is rejected at the level of significance α if T is greater than or equal to $n - t$.

The p-value is the probability that Y is greater than or equal to the observed value of T, which is found from Table A3 for $n \leq 20$, using $p = 0.5$, or from Table A1 for $n > 20$, using Equation 4.

It should be noted that the sign test is unbiased and consistent when testing these hypotheses. The sign test is also used for testing the following counterparts of the hypotheses, in which case it is neither unbiased nor consistent unless additional assumptions concerning the distributions of (X_i, Y_i) are made.

A. (Two-Tailed Test)

The null hypothesis is interpreted as "X_i and Y_i have the same location parameter" and, therefore,

$$H_0: E(X_i) = E(Y_i) \qquad \text{for all } i$$

is tested against the alternative

$$H_1: E(X_i) \neq E(Y_i) \qquad \text{for all } i$$

to see if X_i and Y_i have different means. Similarly, the test may be a test of medians.

$$H_0: \text{The median of } X_i \text{ equals the median of } Y_i \text{ for all } i$$
$$H_1: X_i \text{ and } Y_i \text{ have different medians for all } i$$

B. (Lower-Tailed Test)

The null hypothesis in the preceding category B may be considered to indicate that X_i has a tendency to assume smaller values than does Y_i; hence this one-tailed sign test may be used to test

$$H_0: E(X_i) \leq E(Y_i) \qquad \text{for all } i$$

against the alternative

$$H_1: E(X_i) > E(Y_i) \qquad \text{for all } i$$

A similar statement may be made concerning the medians.

C. (Upper-Tailed Test)

The null hypothesis may be considered to indicate that the values of X_i tend to be larger than the values of Y_i because H_0 states that X_i may be more likely to exceed Y_i than to be less than Y_i. Therefore this one-tailed sign test is sometimes used to test

$$H_0: E(X_i) \geq E(Y_i) \qquad \text{for all } i$$

against the alternative

$$H_1: E(X_i) < E(Y_i) \qquad \text{for all } i$$

A similar set of hypotheses may be stated in terms of the median.

Computer Assistance *Minitab* and *StatXact* perform the sign test. ———— ◀

EXAMPLE I

An item A is manufactured using a certain process. Item B serves the same function as A but is manufactured using a new process. The manufacturer wishes to determine whether B is preferred to A by the consumer, so she selects a random sample consisting of 10 consumers, gives each of them one A and one B, and asks them to use the items for some period of time. The sign test (one-tailed) will be used to test

$$H_0: P(+) \leq P(-)$$

against

$$H_1: P(+) > P(-)$$

where "+" represents the event "item B is preferred over item A," and "−" represents the event "item A is preferred over item B." In words, H_0 says, "Item B does not tend to be preferred to item A," while H_1 says, "Item B tends to be preferred to item A." The test statistic T is the number of + signs, the number of consumers who prefer B over A. The critical region corresponds to values of T greater than or equal to $n - t$. However, we need to know how many ties there are before we can find n and, hence, t.

At the end of the allotted period of time the consumers report their preferences to the manufacturer. Eight consumers preferred B to A, 1 preferred A to B, and 1 reported "no preference." Therefore,

$$8 = \text{number of +'s}$$
$$1 = \text{number of −'s}$$
$$1 = \text{number of ties}$$
$$n = \text{number of +'s and −'s}$$
$$= 8 + 1 = 9$$
$$T = \text{number of +'s}$$
$$= 8$$

Table A3 is entered with $n = 9$ and $p = 1/2$ and for an entry close to 0.05. The critical region of size $\alpha_1 = 0.0195$ corresponds to values of T greater than or equal to

$$n - t = 9 - 1 = 8$$

Since $T = 8$, H_0 is rejected. The p-value is $P(Y \geq 8) = 0.0195$.

The manufacturer decides that the consumer population prefers B to A. ∎

A two-tailed sign test illustrating the use of the large sample approximation is presented in the next example.

EXAMPLE 2

In what was perhaps the first published report of a nonparametric test, Arbuthnott (1710) examined the available London birth records of 82 years and for each year compared the number of males born with the number of females born. If for each year we denote the event "more males than females were born" by "+" and the opposite event by "−," (there were no ties), we may consider the hypotheses to be

$$H_0: P(+) = P(-)$$
$$H_1: P(+) \neq P(-)$$

The test statistic T equals the number of + signs, and the critical region of size $\alpha = 0.05$ corresponds to values of T less than

$$t = 0.5(82 - (1.960)\sqrt{82})$$
$$= 32.1$$

and values of T greater than

$$n - t = 82 - 32.1$$
$$= 49.9$$

where t is calculated using Equation 1.

From the records, Arbuthnott obtained 82 plus signs, no minus signs, and no ties. So $T = 82$ and the null hypothesis is rejected. In fact, H_0 could have been rejected at an α as small as

$$P(T = 0) + P(T = 82) = (\tfrac{1}{2})^{82} + (\tfrac{1}{2})^{82} = (\tfrac{1}{2})^{81}$$

which is therefore the p-value. ∎

To see the versatility of the sign test, consider the following example, which was suggested to Batschelet (1965) by K. Schmidt-Koenig.

EXAMPLE 3

Ten homing pigeons were taken to a point 25 kilometers west of their loft and released singly to see whether they dispersed at random in all directions (the null hypothesis) or whether they tended to proceed eastward toward their loft. Field glasses were used to observe the birds until they disappeared from view, at which time the angle of the vanishing point was noted. These 10 angles are: 20, 35, 350, 120, 85, 345, 80, 320, 280, and 85 degrees. Let "+" denote directions more eastward than westward (angles from 0 to 90 degrees or from 270 to 360 degrees) and let "−" denote directions away from the loft (between 90 and 270 degrees). The hypotheses

$$H_0: P(+) \leq P(-)$$
$$H_1: P(+) > P(-)$$

match set C, so the critical region consists of large values of T, the number of "+" signs. From Table A3, for $n = 10$ and $p = 1/2$, the critical region of size $\alpha = 0.0547$ corresponds to values of T greater than or equal to $10 - 2 = 8$.

For these data $T = 9$, so the null hypothesis is rejected. The conclusion is that these pigeons tend to fly homeward instead of in random directions. The p-value is $P(T \geq 9) = 0.0107$. ∎

□ *Theory* The event "+" represents the event "$Y_i > X_i$," or "$Y_i - X_i > 0$," which says that the difference $Y_i - X_i$ is positive. Similarly, "−" and "0" represent the event $Y_i - X_i$ is negative or zero, respectively. Therefore the sign test is a test for comparing the probability of a positive difference with the probability of a negative difference. In the binomial test these were called "class 1" and "class 2" probabilities, respectively. By omitting ties we have

$$P(+) + P(-) = 1$$

and so the hypothesis

$$H_0: P(+) = P(-)$$

is the same as saying

$$H_0: P(+) = 1/2$$

which is in the same form as that of the binomial test with $p^* = 1/2$. So the same binomial test procedure is used, although a slight simplification results from the symmetry of

$$p^* = 1/2 = 1 - p^*$$

When the sign test is used with the original sets A, B, and C of hypotheses, the sign test is unbiased and consistent (Hemelrijk, 1952). Example 2.4.2 illustrated the binomial test with $p = 1/2$, which is the same as the sign test if there are no ties. Therefore the power functions graphed in Figure 2.4.4 in that example are power functions for the sign test. It is evident from those graphs that the sign test is unbiased and consistent, although such evidence is not conclusive proof. □

If, in addition to the assumptions in the sign test, we can also assume legitimately that the differences $Y_i - X_i$ are random variables with a symmetric distribution function [the distribution function of a random variable Z is symmetric about some point c if $P(Z \leq c - x) = P(Z \geq c + x)$ for all x], the Wilcoxon signed ranks test is more appropriate (see Section 5.7). Furthermore, if the differences $Y_i - X_i$ are independent and identically distributed normal random variables, the appropriate parametric test is called the paired t test. The A.R.E. compared to the paired t test under these conditions is only $2/\pi = 0.637$. Also, under these conditions the A.R.E. compared to the Wilcoxon test is $2/3$. If the differences have a uniform (light-tailed) distribution the A.R.E. of the sign test relative to the t test or the Wilcoxon test drops to $1/3 = 0.333$. For a heavy-tailed symmetric distribution known as the double-exponential distribution the A.R.E. of the sign test relative to the t test rises to 2.0, and is $4/3 = 1.333$ relative to the Wilcoxon signed ranks test.

Both small and large sample relative efficiencies have been examined by Walsh (1951), Dixon (1953), Hodges and Lehmann (1956), and Gibbons (1964), among others. Special tables for sample sizes to 1000 are given by MacKinnon (1964). Hemelrijk (1952) discusses ties.

Data that occur naturally in pairs, as in the sign test, are usually analyzed by reducing the sequence of pairs to a sequence of single values, and then the data are analyzed as if only one sample were involved. That is, bivariate samples are usually analyzed using univariate techniques. In the sign test the differences $Y_i - X_i$ were analyzed in the same manner that one would analyze a series of values to see if positive values are more likely than negative values. This principle of reducing bivariate (or even multivariate) data to a simple univariate sample is a useful one to remember.

EXERCISES

1. Six students went on a diet in an attempt to lose weight, with the following results:

Name	Abdul	Ed	Jim	Max	Phil	Ray
Weight Before	174	191	188	182	201	188
Weight After	165	186	183	178	203	181

    Is the diet an effective means of losing weight?

2. The reaction time before lunch was compared with the reaction time after lunch for a group of 28 office workers. Twenty-two workers found their reaction time before lunch was shorter, and two could detect no difference. Is the reaction time after lunch significantly longer than the reaction time before lunch?

3. Two different additives were compared to see which one is better for improving the durability of concrete. One hundred small batches of concrete were mixed under various conditions and, during the mixing, each batch was divided into two parts. One part received additive A and the other part received additive B. After the concrete hardened, the two parts in each batch were crushed against each other, and an observer determined which part appeared to be the most durable. In 77 cases the concrete with additive A was rated more durable; in 23 cases the concrete with additive B was rated more durable. Is there a significant difference between the effects of the two additives?

4. Twenty-two customers in a grocery store were asked to taste each of two types of cheese and declare their preference. Seven customers preferred one kind, 12 preferred the other kind, and 3 had no preference. Does this indicate a significant difference in preference?

5. An obstetrician claimed that more babies are born at night (6 P.M. to 6 A.M.) than during the day, while his friend the statistician said it only seemed that way. For the next year they kept track of the time of birth of all spontaneous births in that doctor's care to see who was correct. The result was:

Midnight to 3 A.M.—16 births	Noon to 3 P.M.—10 births
3 A.M. to 6 A.M.—17 births	3 P.M. to 6 P.M.—11 births
6 A.M. to 9 A.M.—12 births	6 P.M. to 9 P.M.—12 births
9 A.M. to noon—9 births	9 P.M. to midnight—15 births

Is the statistician correct?

6. In a laboratory, insects of a certain type are released in the middle of a circle drawn on a plain, flat table. A scent, intended to attract that type of insect, is located at one end of the table. Each insect is released singly and observed until it crosses the boundary of the circle. At that time it is recorded whether the insect crossed the half of the boundary "toward" the scent or the half "away" from the scent. At the conclusion of the experiment, 33 insects went "toward" the scent, 16 went "away," and 12 did not cross the boundary within a reasonable time. Does the scent attract those insects?

3.4 PROBLEM

1. If the normal approximation is used in a two-tailed test at $\alpha = 0.05$, the value of t may be computed using the approximation given by Equation 1

$$t_1 = \tfrac{1}{2}(n - 1.9600 \sqrt{n})$$

or its approximation

$$t_2 = \tfrac{1}{2}n - \sqrt{n}$$

For example, if $n = 21$, $t_1 = 6.009$ and $t_2 = 5.917$, so the first critical region includes the integer 6 while the second does not, and the two equations yield different critical regions. For which values of n between 20 and 30 do the two equations result in identical tests? Are the two results equivalent at $n = 16$?

3.5 SOME VARIATIONS OF THE SIGN TEST

Suppose now that the data are not ordinal as in the sign test but nominal, with two categories that we will call "0" and "1." That is, each X_i is either 0 or 1, and similarly for each Y_i. Then a question sometimes asked is, "Can we detect a difference between the probability of (0, 1) and the probability of (1, 0)?" Such a question arises when X_i in the pair (X_i, Y_i) represents the condition (or state) of the subject before the experiment and Y_i represents the condition of the same subject after the experiment. The same procedure as used in the sign test may be used here also, but the test is well known by a different name.

▶ **The McNemar Test for Significance of Changes** _____

Data The data consist of observations on n' independent bivariate random variables (X_i, Y_i), $i = 1, 2, \ldots, n'$. The measurement scale for the X_i and the Y_i is nominal with two categories, which we call "0" and "1"; that is, the possible values of (X_i, Y_i) are (0, 0), (0, 1), (1, 0), and (1, 1). In the McNemar test the data are usually summarized in a 2×2 *contingency table*, as follows.

		Classification of the Y_i	
		$Y_i = 0$	$Y_i = 1$
Classification	$X_i = 0$	a (the number of pairs where $X_i = 0$ and $Y_i = 0$)	b (the number of pairs where $X_i = 0$ and $Y_i = 1$)
of the X_i	$X_i = 1$	c (the number of pairs where $X_i = 1$ and $Y_i = 0$)	d (the number of pairs where $X_i = 1$ and $Y_i = 1$)

Assumptions

1. The pairs (X_i, Y_i) are mutually independent.
2. The measurement scale is nominal with two categories for all X_i and Y_i.
3. The difference $P(X_i = 0, Y_i = 1) - P(X_i = 1, Y_i = 0)$ is negative for all i, or zero for all i, or positive for all i.

Test Statistic The test statistic for the McNemar test is usually written as

$$T_1 = \frac{(b - c)^2}{b + c} \tag{1}$$

However, for $b + c \leq 20$, the following test statistic is preferred.

$$T_2 = b \tag{2}$$

Note that neither T_1 nor T_2 depends on a or d. This is because a and d represent the number of "ties," and ties are discarded in this analysis.

Null Distribution   The null distribution of T_1 is approximately the chi-squared distribution with 1 degree of freedom when $(b + c)$ is large. The exact distribution of T_2 is the binomial distribution with $p = 1/2$ and $n = b + c$.

Hypotheses

$$H_0: P(X_i = 0, Y_i = 1) = P(X_i = 1, Y_i = 0) \qquad \text{for all } i$$
$$H_1: P(X_i = 0, Y_i = 1) \neq P(X_i = 1, Y_i = 0) \qquad \text{for all } i$$

These hypotheses may take a slightly different form if we add $P(X_i = 0, Y_i = 0)$ to both sides of the equation in H_0 to get

$$H_0: P(X_i = 0, Y_i = 1) + P(X_i = 0, Y_i = 0) = P(X_i = 1, Y_i = 0) + P(X_i = 0, Y_i = 0)$$

The left side of H_0 includes all possibilities of Y_i and hence equals $P(X_i = 0)$. Similarly, the right side includes all possibilities for X_i and so equals $P(Y_i = 0)$. Therefore we have a new set of hypotheses in the form

$$H_0: P(X_i = 0) = P(Y_i = 0) \qquad \text{for all } i$$
$$H_1: P(X_i = 0) \neq P(Y_i = 0) \qquad \text{for all } i$$

Of course, these are also equivalent to

$$H_0: P(X_i = 1) = P(Y_i = 1) \qquad \text{for all } i$$
$$H_1: P(X_i = 1) \neq P(Y_i = 1) \qquad \text{for all } i$$

The latter sets of hypotheses are usually easier to interpret in terms of the experiment.

Let n equal $b + c$. If $n \leq 20$, use Table A3. If α is the desired level of significance, enter Table A3 with $n = b + c$ and $p = 1/2$ to find the table entry approximately equal to $\alpha/2$. Call this entry α_1, and the corresponding value of y is called t. Reject H_0 if $T_2 \leq t$, or if $T_2 \geq n - t$, at a level of significance of $2\alpha_1$. Otherwise accept H_0. The p-value is twice the probability of T_2 being less than or equal to the observed value, or greater than or equal to the observed value, whichever is smaller. The probabilities are found from Table A3 using $p = 1/2$ and $n = b + c$.

If n exceeds 20, use T_1 and Table A2. Reject H_0 at a level of significance α if T_1 exceeds the $(1 - \alpha)$ quantile of a chi-squared random variable with 1 degree of freedom. Otherwise accept H_0. The p-value is the probability of T_1 exceeding the observed value, as found in Table A2 for the chi-squared distribution with

1 degree of freedom. A more precise p-value can be found by comparing the negative square root of T_1 with Table A1, and doubling the lower-tailed probability.

Computer Assistance　*S-Plus, StatXact,* and *SAS* perform the McNemar Test.

◀

EXAMPLE 1

Prior to a nationally televised debate between the two presidential candidates, a random sample of 100 persons stated their choice of candidates as follows. Eighty-four persons favored the Democratic candidate, and the remaining 16 favored the Republican. After the debate the same 100 people expressed their preference again. Of the persons who formerly favored the Democrat, exactly one-fourth of them changed their minds, and also one-fourth of the people formerly favoring the Republican switched to the Democratic side. The results are summarized in the following 2×2 contingency table.

		After		Total
		Democrat	**Republican**	**before**
Before	**Democrat**	63	21	84
	Republican	4	12	16
				100

　　The McNemar test may be used to test H_0: The population voting alignment was not altered by the debate, against H_1: There has been a change in the proportion of all voters who favor the Democrat. Consider the X_i in (X_i, Y_i) to be 0 if the ith person favored the Democrat before, or 1 if the Republican was favored before. Similarly, Y_i represents the choice of the ith person after the debate. (Our choice of whether to represent the Democrat by 0 or 1 does not affect the results, as long as the X_i and the Y_i use the same representation.) The test statistic T_1 in the McNemar test becomes

$$T_1 = \frac{(b - c)^2}{b + c}$$
$$= \frac{(21 - 4)^2}{21 + 4}$$
$$= \frac{289}{25}$$
$$= 11.56 \qquad\qquad (3)$$

The critical region of size $\alpha = 0.05$ corresponds to all values of T_1 greater than 3.841, the 0.95 quantile of a chi-squared random variable with 1 degree of freedom, obtained from Table A2. Because 11.56 exceeds 3.841, the null hypothesis is rejected, and the conclusion is that the voter alignment has been altered. The p-value is less than 0.001. ∎

□ *Theory* This is a variation of the sign test, where the event $(0, 1)$ was called "+," the event $(1, 0)$ was called "−," and the events $(1, 1)$ and $(0, 0)$ were called ties. The hypothesis of the McNemar test then takes the form

$$H_0: P(+) = P(-)$$

which is the same as H_0 in the two-tailed sign test. The critical region for T_2 is found just as in the sign test for $n \leq 20$.

For n greater than 20 the sign test suggests using the normal approximation, based on the idea that

$$Z = \frac{T_2 - n(\frac{1}{2})}{\sqrt{n(\frac{1}{2})(\frac{1}{2})}} = \frac{b - n(\frac{1}{2})}{(\frac{1}{2})\sqrt{n}} \tag{4}$$

has approximately the standard normal distribution when H_0 is true (see Example 1.5.6). Because $n = b + c$, Equation 4 reduces to

$$
\begin{aligned}
Z &= \frac{b - [(b + c)/2]}{(\frac{1}{2})\sqrt{b + c}} \\
&= \frac{b - c}{\sqrt{b + c}}
\end{aligned}
\tag{5}
$$

Therefore

$$T_1 = Z^2$$

has approximately a chi-squared distribution with 1 degree of freedom (see Theorem 1.5.3). A two-tailed test involving T_2 or Z is comparable to using the upper tail of $T_1 = Z^2$ for a critical region. □

As the sign test was presented in both the two-tailed and the one-tailed forms, so could the McNemar test take both forms. The easiest way of performing a one-tailed McNemar test is just to use the one-tailed sign test. The McNemar test and its variations are discussed by Bennett and Underwood (1970), Ury (1975), Mantel and Fleiss (1975), and McKinlay (1975).

Another modification of the sign test is one introduced by Cox and Stuart (1955), and it is used to test for the presence of *trend*. A sequence of numbers is said to have trend if the later numbers in the sequence tend to be greater than the earlier numbers (upward trend) or less than the earlier numbers (downward trend). This test involves pairing the later numbers with the earlier numbers and then performing a sign test on the pairs thus formed. If there is a trend, one member of each pair will have a tendency to be higher or lower than the other member. On the other hand, if there is no trend and the sequence of numbers actually represents observations on independent and identically distributed random variables, there will be no tendency for one particular member of each pair to exceed the other one.

▶ **Cox and Stuart Test for Trend** _____

Data The data consist of observations on a sequence of random variables $X_1, X_2, \ldots, X_{n'}$, arranged in a particular order, such as the order in which the random variables are observed. It is desired to see if a trend exists in the sequence. Group the random variables into pairs $(X_1, X_{1+c}), (X_2, X_{2+c}), \ldots, (X_{n'-c}, X_{n'})$, where $c = n'/2$ if n' is even, and $c = (n' + 1)/2$ if n' is odd. (Note that the middle random variable is eliminated using this scheme if n' is odd.) Replace each pair (X_i, X_{i+c}) with a "+" if $X_i < X_{i+c}$, or a "−" if $X_i > X_{i+c}$, eliminating ties. The number of untied pairs is called n.

This test may be used to detect any specified type of nonrandom pattern, such as a sine wave or other periodic pattern. The sequence of random variables is merely rearranged so that the smallest numbers, as predicted, will be near the beginning of the sequence and the larger numbers near the end. Then the presence of an upward trend in the rearranged sequence is evidence that the predicted pattern is present in the original arrangement of the sequence.

Assumptions

1. The random variables $X_1, X_2, \ldots, X_{n'}$ are mutually independent.
2. The measurement scale of the X_is is at least ordinal.
3. Either the X_is are identically distributed or there is a trend; that is, the later random variables are more likely to be greater than instead of less than the earlier random variables (or vice versa).

Test Statistic $T = $ total number of +'s

Null Distribution The null distribution of the test statistic is the binomial distribution with $p = 1/2$ and $n = $ the number of untied pairs, where X_i is not equal to X_{i+c}.

Hypotheses The rest of the test is identical to the sign test presented in the previous section, and is not repeated here. The null hypothesis is that no trend is present. An upper-tailed test is used to detect an upward trend. A lower-tailed test is used to detect a downward trend. The two-tailed test is used if the alternative hypothesis is that any type of trend (upward or downward) exists. _____ ◀

The following is an example in which the two-tailed Cox and Stuart test for trend is applied.

EXAMPLE 2

Total annual precipitation is recorded yearly for 19 years. This record is examined to see if the amount of precipitation is tending to increase or decrease. The precipitation in inches was 45.25, 45.83, 41.77, 36.26, 45.37, 52.25, 35.37, 57.16, 35.37, 58.32, 41.05, 33.72, 45.73, 37.90, 41.72, 36.07, 49.83, 36.24, and 39.90. Because $n' = 19$ is odd, the middle number 58.32 is omitted. The remaining numbers are paired.

(45.25, 41.05)	(45.37, 41.72)
(45.83, 33.72)	(52.25, 36.07)
(41.77, 45.73)	(35.37, 49.83)
(36.26, 37.90)	(57.16, 36.24)
	(35.37, 39.90)

There are no ties, so $n = 9$. The test statistic T equals the number of pairs in which the second number exceeds the first number. The critical region of size 0.0390 corresponds to values of T less than or equal to 1 and values of T greater than or equal to $9 - 1 = 8$.

For the data obtained, $T = 4$, well within the region of acceptance. The p-value is 1.0. Therefore the null hypothesis "no trend exists" is accepted. ■

In Example 2 the assumptions of the model on which the test is valid are reasonable assumptions. Thus the test is reasonably valid. However, the assumptions listed are not all necessary. We need only to assume enough to satisfy the model for the sign test. That is, we need only assume:

1. The bivariate random variables (X_i, X_{i+c}) are mutually independent.
2. The probabilities $P(X_i < X_{i+c})$ and $P(X_i > X_{i+c})$ have the same relative size for all pairs.
3. Each pair (X_i, X_{i+c}) may be judged to be a $+$, a $-$, or a tie.

These assumptions are not as readily understood as the set of assumptions given in the test, but they may prove more useful in some applications such as the following.

EXAMPLE 3

On a certain stream the average rate of water discharge is recorded each month (in cubic feet per second) for a period of 24 months. The hypothesis to be tested is

H_0: The rate of discharge is not decreasing

against the alternative

H_1: The rate of discharge is decreasing

The rate of discharge is known to follow a yearly cycle, so that nothing is learned by pairing stream discharges for two different months. However, by pairing the same months in two successive years the existence of a trend can be investigated. The following data were collected.

Month	First Year	Second Year	Month	First Year	Second Year
Jan	14.6	14.2	Jul	92.8	88.1
Feb	12.2	10.5	Aug	74.4	80.0
Mar	104	123	Sep	75.4	75.6
Apr	220	190	Oct	51.7	48.8
May	110	138	Nov	29.3	27.1
Jun	86.0	98.1	Dec	16.0	15.7

The test statistic T equals the number of pairs where the second year had a higher discharge than the first year, which is 5 in this example. Because the test is to detect a downward trend, the critical region of size 0.0730 corresponds to all values of T less than or equal to 3 (from Table A3, $n = 12$, $p = 1/2$). Therefore H_0 is accepted. The p-value is given by

$$P(T \leq 5 \,|\, H_0 \text{ is true}) = 0.3872$$

which is too large to be an acceptable α.   ■

The examples presented in this section represent only a few of the many ways the sign test may be applied to test different types of hypotheses. Two more applications conclude this section. In the first the sign test is used as a simple method of detecting correlation, that is, detecting whether high values of one random variable tend to be paired with high values of a second random variable and low values with low values (positive correlation), or whether high values of one random variable tend to be paired with low values of the second random variable and low values with high values (negative correlation). The test involves arranging the pairs (the pairs remain intact) so that one member of the pair (usually the variable with the fewer ties, which may be either the first member or second) is arranged in increasing order. If there is correlation the other member of the pair will exhibit a trend, upward if the correlation is positive, and downward if the correlation is negative. The Cox and Stuart test for trend may be used on the sequence formed by the other member of the pair.

EXAMPLE 4

Cochran (1937) compares the reactions of several patients with each of two drugs, to see if there is a positive correlation between the two reactions for each patient.

Patient	Drug 1	Drug 2	Patient	Drug 1	Drug 2
1	+0.7	+1.9	6	+3.4	+4.4
2	−1.6	+0.8	7	+3.7	+5.5
3	−0.2	+1.1	8	+0.8	+1.6
4	−1.2	+0.1	9	0.0	+4.6
5	−0.1	−0.1	10	+2.0	+3.4

Ordering the pairs according to the reaction from drug 1 gives

Patient	Drug 1	Drug 2	Patient	Drug 1	Drug 2
2	−1.6	+0.8	1	+0.7	+1.9
4	−1.2	+0.1	8	+0.8	+1.6
3	−0.2	+1.1	10	+2.0	+3.4
5	−0.1	−0.1	6	+3.4	+4.4
9	0.0	+4.6	7	+3.7	+5.5

The one-tailed Cox and Stuart test for trend is applied to the newly arranged sequence of observations on drug 2. The five resulting pairs are (+0.8, +1.9), (+0.1, +1.6), (+1.1, +3.4), (−0.1, +4.4), and (+4.6, +5.5). Because we are testing.

$$H_0: \text{There is no positive correlation}$$

against the alternative

$$H_1: \text{There is positive correlation}$$

we are, in essence, testing for the presence of an upward trend (H_1). The test statistic T equals 5, because in all five pairs the second observation on drug 2 exceeds the first observation on drug 2. The critical region of size 0.0312 (obtained from Table A3 for $n = 5$, $p = 1/2$, and hence $t = 0$) corresponds to the single value $T = 5$. Therefore the null hypothesis is rejected, and we may conclude that there is a positive correlation between reactions to the two drugs. The p-value in this example is also 0.0312. ∎

The final example illustrates how the sign test, or rather the Cox and Stuart test for trend, may be used to test for the presence of a predicted pattern.

EXAMPLE 5

The number of eggs laid by a group of insects in a laboratory is counted on an hourly basis during a 24-hour experiment, to test

H_0: The 24 egg counts constitute observations on 24 identically distributed random variables against the alternative

H_1: The number of eggs laid tends to be a minimum at 2:15 P.M., increasing to a maximum at 2:15 A.M., and decreasing again until 2:15 P.M.

The hourly counts are as follows.

Time	Number of Eggs	Time	Number of Eggs	Time	Number of Eggs
9 A.M.	151	5 P.M.	83	1 A.M.	286
10 A.M.	119	6 P.M.	166	2 A.M.	235
11 A.M.	146	7 P.M.	143	3 A.M.	223
Noon	111	8 P.M.	116	4 A.M.	176
1 P.M.	63	9 P.M.	163	5 A.M.	176
2 P.M.	84	10 P.M.	208	6 A.M.	174
3 P.M.	60	11 P.M.	283	7 A.M.	139
4 P.M.	109	Midnight	296	8 A.M.	137

If the alternative hypothesis is true, the egg counts nearest 2:15 P.M. should tend to be the smallest and those nearest 2:15 A.M. should tend to be the largest. Therefore the number of eggs is rearranged according to the times, from the time nearest 2:15 P.M. to the times nearest 2:15 A.M.

Time	Number of Eggs	Time	Number of Eggs
2 P.M.	84	8 A.M.	137
3 P.M.	60	9 P.M.	163
1 P.M.	63	7 A.M.	139
4 P.M.	109	10 P.M.	208
Noon	111	6 A.M.	174
5 P.M.	83	11 P.M.	283
11 A.M.	146	5 A.M.	176
6 P.M.	166	Midnight	296
10 A.M.	119	4 A.M.	176
7 P.M.	143	1 A.M.	286
9 A.M.	151	3 A.M.	223
8 P.M.	116	2 A.M.	235

If H_1 is true these numbers should exhibit an upward trend. The Cox and Stuart one-tailed test for trend is used. The first half of the sequence (first column) is paired with the last half of the sequence (second column), with the result that the two egg counts on each line form a pair. In all 12 pairs the number in the second column exceeds the number in the first column, so $T = 12$. For $n = 12$, $p = 1/2$, Table A3 shows that the critical region of size $\alpha = 0.0193$ corresponds to values of T greater than or equal to $12 - 2 = 10$. Therefore H_0 is rejected, and we conclude that the predicted pattern does seem to be present. The p-value is given as

$$P(T \geq 12) = 0.0002$$

Therefore H_0 would have been rejected at any reasonable level of significance.   ■

□ *Theory*   The Cox and Stuart test for trend is an obvious modification of the sign test and, therefore, the distribution of the test statistic when H_0 is true is obviously binomial. Also, the test is unbiased and consistent when the first sets A, B, and C of hypotheses are being used, but not necessarily so when the later sets are used. Stuart (1956) shows that the test, when applied to random variables known to be normally distributed, has an A.R.E. of 0.78 with respect to the best parametric test, a test based on the regression coefficient. Under the same conditions it has an A.R.E. of 0.79 compared to Spearman's or Kendall's rank correlation tests used as tests of randomness, which will be presented in Chapter 5.

If the test is altered so that the middle one-third of the observations are eliminated and only the first one-third of the observations are paired with the last one-third of the observations, the A.R.E. increases to 0.83 when compared to the best parametric test, under ideal conditions for the parametric test. Apparently the loss of data is small as compared with the gain in larger differences. This suggests another variation, that of pairing from the ends of the sequence. That is, by forming the pairs (X_1, X_n), (X_2, X_{n-1}), and so forth, using all the data, perhaps the larger differences may be preserved, along with no loss in data. The test may still be performed as just described, because the distribution of the test statistic under the null hypothesis remains unchanged.

The test for correlation, described in Example 4, has not been investigated to see what its properties are. One of the difficulties in applying the test for correlation is that if many observations equal each other, there is more than one way of arranging the observations so that the test for trend can be applied. Therefore it is recommended that the original data pairs be arranged using the pair member that has the smallest number of ties. Of the arrangements still possible due to ties, the conservative approach is to choose the arrangement that will be least likely to result in rejection of H_0. □

A bivariate sign test for location is discussed by Chatterjee (1966). Other modifications of the sign test may be used to test for trends in dispersion (Ury, 1966) or to compare several treatments with a control (Rhyne and Steel, 1965). Rao (1968) uses the Cox and Stuart test for testing trend in dispersion. The power of the test for trend is discussed further by Mansfield (1962). Olshen (1967) presents tests for testing quadratic trend versus linear trend. Other variations of the sign test appear in Woodbury, Manton, and Woodbury (1977) and Altham (1971). A paper by Schaafsma (1973) examines the consequences of order dependence on the sign test; that is, a customer preferring one brand over another may be influenced by which brand he or she was exposed to first.

3.5 EXERCISES

1. One hundred thirty-five citizens were selected at random and were asked to state their opinion regarding U.S. foreign policy. Forty-three were opposed to the U.S. foreign policy. After several weeks, during which they received an informative newsletter, they were again asked their opinion; 37 were opposed, and 30 of the 37 were persons who originally were not opposed to the U.S. foreign policy. Is the change in numbers of people opposed to the U.S. foreign policy significant?

2. In Exercise 1, suppose all 37 of the persons opposed to the foreign policy after the experiment were also among those opposed to the U.S. foreign policy before the experiment. Is the change in the number of people opposed to the U.S. foreign policy significant?

3. In a certain city the mortality rate per 100,000 citizens due to automobile accidents for each of the last 15 years was 17.3, 17.9, 18.4, 18.1, 18.3, 19.6, 18.6, 19.2, 17.7, 20.0, 19.0, 18.8, 19.3, 20.2, and 19.9. Is there any basis for the statement that the mortality rate is increasing?

4. For each of the last 34 years a small Midwestern college recorded the average heights of male freshmen. The averages were 68.3, 68.6, 68.4, 68.1, 68.4, 68.2, 68.7, 68.9, 69.0, 68.8, 69.0, 68.6, 69.2, 69.2, 68.9, 68.6, 68.6, 68.8, 69.2, 68.8, 68.7, 69.5, 68.7, 68.8, 69.4, 69.3, 69.3, 69.5, 69.5, 69.0, 69.2, 69.2, 69.1, and 69.9. Do these averages indicate an increasing trend in height?

5. A manufacturer computes the average cost in dollars of producing a certain item for each of 44 months with the resulting averages 13.65, 13.41, 13.53, 13.23, 13.58, 13.43, 13.73, 13.40, 13.70, 13.58, 13.80, 13.40, 13.63, 13.69, 13.92, 13.68, 13.72, 13.42, 13.66, 13.98, 13.81, 13.60, 13.32, 13.45, 13.27, 13.26, 13.28, 13.29, 13.10, 13.09, 13.36, 13.40, 13.35, 13.53, 13.66, 13.10, 13.28, 13.33, 13.02, 13.09, 13.12, 13.16, 12.96, and 12.95. Is there a statistically significant trend in these averages?

6. In an experiment to determine the influence of suggestion, 20 straight lines of varying lengths were shown one at a time to subjects A and B, and the subjects estimated aloud the length of each line. Subject A stated her preference first and, unknown to subject B, was under instructions to overestimate the first 10 lines and underestimate the last 10 lines. After hearing subject A's estimate, subject B stated his estimate. The errors of the estimates, measured by subtracting the true lengths of the lines from the estimated lengths of the lines, were as follows.

	Line									
	1	2	3	4	5	6	7	8	9	10
Error by A	+0.3	+1.1	+0.9	+0.6	+1.0	+1.3	+0.8	+1.6	+1.2	+0.8
Error by B	−0.1	+0.6	+1.0	+0.7	+0.2	+0.9	−0.1	+0.2	0.0	+0.5

					Line				
	11	**12**	**13**	**14**	**15**	**16**	**17**	**18**	**19**	**20**
Error by A	−1.3	−1.1	−1.3	−0.7	−1.4	−1.1	−0.8	−0.5	−1.2	−1.0
Error by B	−0.6	−1.2	−1.0	−0.7	−1.0	−0.1	−0.5	0.0	−0.4	−0.3

Is there a significant positive correlation between subject A's errors and subject B's errors?

7. A major league baseball player had compiled the following record over 12 years.

	1988	**1989**	**1990**	**1991**	**1992**	**1993**
Number of Home Runs	7	14	17	15	9	19
Batting Averages	0.212	0.232	0.234	0.210	0.201	0.256

	1994	**1995**	**1996**	**1997**	**1998**	**1999**
Number of Home Runs	16	17	22	17	13	10
Batting Averages	0.261	0.247	0.255	0.241	0.238	0.235

Is there significant correlation between the number of home runs he hit and his batting average for that year?

8. Test the following data to see if there is a significant correlation between the yearly income of a family and the number of children in that family.

Income	Number of Children	Income	Number of Children	Income	Number of Children
$17,440	3	$23,320	3	$28,940	3
17,664	2	23,569	4	29,300	1
17,721	4	23,950	2	29,371	3
17,883	3	24,023	3	29,512	1
18,000	4	24,330	5	29,662	1
18,332	2	24,545	2	29,804	2
18,653	0	24,922	5	30,167	2
18,781	3	25,571	4	30,634	3
19,087	6	25,624	4	31,235	1
19,686	5	25,873	2	31,797	3
19,832	2	26,010	1	31,880	4
20,100	1	26,145	3	32,363	1
20,222	6	26,513	2	32,946	3
20,435	3	26,660	4	33,586	2
20,961	5	26,984	5	34,000	2
21,382	2	27,463	0	34,443	3
21,957	0	27,702	1	35,693	1
22,190	8	27,914	4	39,247	1
22,212	1	28,244	2	40,540	1
22,635	4	28,698	4	55,686	2

9. Two landing fields 50 miles apart are observed for a one-year period to determine whether there is a significant difference in availability due to weather conditions. On 286 days both

fields were open all day. On 62 days both fields were closed at least part of the day due to inclement weather. On 14 days Field A was closed while Field B was open, and on 3 days the reverse was true. Is there a significant difference in availability due to weather conditions?

PROBLEMS

1. A barber shop is considering raising the price of haircuts one dollar and then giving the customers a coupon worth one free refreshing drink at a nearby pub. A survey was conducted, and 200 people selected at random from the population of real and potential customers were given an explanation of this proposal. Ten percent of the customers in the sample said they would go elsewhere for their haircuts. Five percent of the noncustomers in the sample said they would become customers at that barber shop. Test the null hypothesis that the proposed change will not increase the total number of customers who receive haircuts in that shop, if only 20 people in the sample are presently customers. How does the answer change if there are 60 customers in the sample instead of 20?

2. Data for the McNemar test may be written as bivariate observations X_i, Y_i, where each observation is 0 or 1 "before" and 0 or 1 "after." A parametric test called the "paired t test" is often applied to data of this type. The paired t test uses the differences $D_i = X_i - Y_i$, $i = 1, 2, \ldots , n'$. The sample mean $\overline{D}$ and sample standard deviation S are used in the statistic

    $$t = \overline{D}\sqrt{n' - 1}/S$$

    which is compared with the quantiles in Table A21, $n' - 1$ degrees of freedom. This test is only approximate if the D_is are not normally distributed.

    Show that the following relationship holds between t and T_1

    $$t^2 = \frac{(n' - 1)T_1}{n' - T_1} \qquad \text{or} \qquad T_1 = \frac{n't^2}{n' - 1 + t^2}$$

    where T_1 is given by Equation 1. That is, as T_1 gets larger, t^2 also gets larger, so the two tests that reject H_0 for large T_1 or large t^2 are equivalent if their critical regions correspond to each other's.

CHAPTER 4

CONTINGENCY TABLES

PRELIMINARY REMARKS

A *contingency table* is an array of natural numbers in matrix form where those natural numbers represent counts, or frequencies. For example, an entomologist observing insects may say he observed 37 insects, or he may say he observed:

Moths	Grasshoppers	Others	Total
12	22	3	37

using a 1 × 3 (one by three) contingency table. This is a one-way contingency table because it has only one row.

The entomologist may wish to be more specific and use a 2 × 3 contingency table, as follows.

	Moths	Grasshoppers	Others	Total
Alive	3	21	3	27
Dead	9	1	0	10
Total	12	22	3	37

The totals, consisting of two *row totals*, three *column totals*, and one *grand total*, are optional and are usually included only for the reader's convenience. This is a two-way contingency table, and may be extended to include several rows (r) and several columns (c) as an $r \times c$ contingency table. Contingency tables with three or more dimensions also may occur, but are discussed only briefly in this chapter.

4.1   THE 2 × 2 CONTINGENCY TABLE

In general an $r \times c$ contingency table is an array of natural numbers arranged into r rows and c columns and thus has rc *cells* or places for the numbers. This section is concerned only with the case where $r = 2$ and $c = 2$, the 2×2 contingency table. Because there are four cells, the 2×2 contingency table is also called the *fourfold* contingency table.

One application of the 2×2 contingency table arises when N objects (or persons), possibly selected at random from some population, are classified into one of two categories before a treatment is applied or an event takes place. After the treatment is applied the same N objects are again examined and classified into the two categories. The question to be answered is, "Does the treatment significantly alter the proportion of objects in each of the two categories?" This use of the contingency table was introduced in Section 3.5, and the appropriate statistical procedure was seen to be a variation of the sign test known as the McNemar test. The McNemar test is often able to detect subtle differences, primarily because the same sample is used in the two situations (such as "before" and "after"). Another way of testing the same hypothesis tested with the McNemar test is by drawing a random sample from the population before the treatment and then comparing it with another random sample drawn from the population after that treatment. The additional variability introduced by using two different random samples is undesirable because it tends to obscure the changes in the population caused by the treatment. However, there are times when it is not practical, or even possible, to use the same sample twice. Then the procedures to be described in this section may be used.

In the first procedure, two random samples are drawn, one from each of two populations, to test the null hypothesis that the probability of event A (some specified event) is the same for both populations. The null hypothesis may also be stated as "the proportion of the population with characteristic A is the same for both populations."

▶   **The Chi-squared Test for Differences in Probabilities, 2 × 2** _____

Data   A random sample of n_1 observations is drawn from one population (or before a treatment is applied) and each observation is classified into either class 1 or class 2, the total numbers in the two classes being O_{11} and O_{12}, respectively, where $O_{11} + O_{12} = n_1$. A second random sample of n_2 observations is drawn from a second population (or the first population after some treatment is applied) and the number of observations in class 1 or class 2 is O_{21} or O_{22} respectively, where $O_{21} + O_{22} = n_2$. The data are arranged in the following 2 × 2 contingency table.

	Class 1	Class 2	Total
Population 1	O_{11}	O_{12}	n_1
Population 2	O_{21}	O_{22}	n_2
Total	C_1	C_2	$N = n_1 + n_2$

The total number of observations is denoted by N.

Assumptions

1. Each sample is a random sample.
2. The two samples are mutually independent.
3. Each observation may be categorized either into class 1 or class 2.

Test Statistic If any column total is zero, the test statistic is defined as $T_1 = 0$. Otherwise,

$$T_1 = \frac{\sqrt{N}(O_{11}O_{22} - O_{12}O_{21})}{\sqrt{n_1 n_2 C_1 C_2}} \tag{1}$$

Null Distribution The exact distribution of T_1 is difficult to tabulate because of all the different combinations of values possible for O_{11}, O_{12}, O_{21}, and O_{22}. Therefore the large sample approximation is used, which is the standard normal distribution whose quantiles are given in Table A1.

Hypotheses Let the probability that a randomly selected element will be in class 1 be denoted by p_1 in population 1 and p_2 in population 2. Note that it is not necessary for p_1 and p_2 to be known. The hypotheses merely specify a relationship between them.

A. (Two-Tailed Test)

$$H_0: p_1 = p_2$$
$$H_1: p_1 \neq p_2$$

Reject H_0 at the approximate level α if T_1 is less than the $\alpha/2$ quantile of a standard normal random variable Z, or if T_1 is greater than the $1 - \alpha/2$ quantile of Z, where the quantiles of Z are given in Table A1.

The p-value is twice the smaller of the probabilities that Z is less than the observed value of T_1 or greater than the observed value of T_1, from Table A1.

Note that for the above hypotheses, T_1^2 is often used instead of T_1 as the test statistic. Then the rejection region is the upper tail of the chi-squared distribution with 1 degree of freedom, given in Table A2.

B. (Lower-Tailed Test)

$$H_0: p_1 \geq p_2$$
$$H_1: p_1 < p_2$$

Reject H_0 at the approximate level α if T_1 is less than the α quantile of a standard normal random variable Z, where the quantiles of Z are given in Table A1.

The p-value is the probability that Z is less than the observed value of T_1, found in Table A1.

C. (Upper-Tailed Test)

$$H_0: p_1 \leq p_2$$
$$H_1: p_1 > p_2$$

Reject H_0 at the approximate level α if T_1 is greater than the $1 - \alpha$ quantile of a standard normal random variable Z, where the quantiles of Z are given in Table A1.

The p-value is the probability that Z is greater than the observed value of T_1, obtained from Table A1.

Computer Assistance   *Minitab, S-Plus, SAS,* and *StatXact* perform this test, and also find a confidence interval for the difference between two probabilities, as introduced in Problem 3.1.2. ───────────────────────────── ◀

EXAMPLE 1

Two carloads of manufactured items are sampled randomly to determine if the proportion of defective items is different for the two carloads. From the first carload 13 of the 86 items were defective. From the second carload 17 of the 74 items were considered defective.

	Defective	Nondefective	Totals
Carload 1	13	73	86
Carload 2	17	57	74
Totals	30	130	160

The assumptions are met, and so the two-tailed test is used to test

H_0: The proportion of defectives is equal in the two carloads

using the test statistic

$$T_1 = \frac{\sqrt{N}(O_{11}O_{22} - O_{12}O_{21})}{\sqrt{n_1 n_2 C_1 C_2}}$$

$$= \frac{\sqrt{160}((13)(57) - (73)(17))}{\sqrt{(86)(74)(30)(130)}}$$

$$= -1.2695$$

The 0.975 quantile of a standard normal random variable is found from Table A1 to be 1.9600. Therefore the rejection region of approximate size 0.05 consists of all values of T_1 greater than 1.9600, or less than -1.9600. The observed value is -1.2695, so the null hypothesis is accepted at the $\alpha = 0.05$ level of significance.

The p-value is twice the probability of Z being less than the observed value -1.2695, which is found from Table A1 as 0.102, so the p-value is approximately 0.204. Therefore the decision to accept H_0 seems to be a fairly safe one. ∎

The following example illustrates the use of the one-tailed test.

EXAMPLE 2

At the U.S. Naval Academy a new lighting system was installed throughout the midshipmen's living quarters. It was claimed that the new lighting system resulted in poor eyesight due to a continual strain on the eyes of the midshipmen. Consider a (fictitious) study to test the null hypothesis,

H_0: The probability of a graduating midshipman having 20-20 (good) vision is the same or greater under the new lighting system than it was under the old lighting system

against the one-sided alternative

H_1: The probability of good vision is less now than it was

Let p_1 be the probability that a randomly selected graduating midshipman had good vision under the old lighting system and let p_2 be the corresponding probability with the new lights. Then the preceding hypotheses may be restated as

$$H_0: p_1 \leq p_2$$
$$H_1: p_1 > p_2$$

which matches the set C of hypotheses. The random samples are taken to be the entire graduation class just prior to the installation of the new lights for population 1, and the first graduation class to spend 4 years using the new lighting system for population 2. It is hoped that these samples will behave the same as would random samples from the entire population of graduating seniors, real and potential.

Suppose the results were as follows.

	Good vision	Poor vision	
Old lights	$O_{11} = 714$	$O_{12} = 111$	$n_1 = 825$
New lights	$O_{21} = 662$	$O_{22} = 154$	$n_2 = 816$
Totals	1376	265	$N = 1641$

Decision rule C defines the critical region for $\alpha = 0.05$ to be all values of T_1 greater than 1.6449 from Table A1. Computation of T_1 gives

$$T_1 = \frac{\sqrt{N}(O_{11}O_{22} - O_{12}O_{21})}{\sqrt{n_1 n_2 C_1 C_2}}$$

$$= \frac{\sqrt{1641}[(714)(154) - (111)(662)]}{\sqrt{(825)(816)(1376)(265)}}$$

$$= 2.982$$

so the null hypothesis is clearly rejected. From Table A1 we see that the null hypothesis could have been rejected at a level of significance as small as about 0.002, so the p-value is 0.002.

We may therefore conclude that the populations represented by the two graduation classes do differ with respect to the proportions having poor eyesight, and in the direction predicted. That is, population 2 (with the new lights) has poorer eyesight than population 1 (with the old lights). Whether the poorer eyesight is a *result* of the new lights has not been shown. However, an association of poor eyesight with the new lights has been shown in this hypothetical example. ■

□ *Theory*   The 2×2 contingency table just presented is actually a special case of the $r \times c$ contingency table presented in the next section, and so the theory involved is a special case of the theory behind the $r \times c$ case. However, the exact distribution of the test statistic is difficult to find unless r and c are very small, so the exact distribution of T_1 is presented now.

The exact probability distribution of T_1 when H_0: $p_1 = p_2 = p$ (say) is true may be calculated as illustrated in the following. For the sample from population 1, the probability of exactly x_1 items in class 1 and $n_1 - x_1$ items in class 2 is given by the binomial probability distribution.

$$P\left(\begin{array}{c}\\ \textbf{Population I}\end{array}\begin{array}{|c|c|}\hline x_1 & n_1 - x_1 \\\hline\end{array}\right) = \binom{n_1}{x_1} p^{x_1}(1-p)^{n_1-x_1} \tag{2}$$

Similarly the probability of the sample from population 2 having exactly x_2 items in class 1 and $n_2 - x_2$ items in class 2 is given by

$$P\left(\begin{array}{c}\\ \textbf{Population 2}\end{array}\begin{array}{|c|c|}\hline x_2 & n_2 - x_2 \\\hline\end{array}\right) = \binom{n_2}{x_2} p^{x_2}(1-p)^{n_2-x_2} \tag{3}$$

Because the two samples are independent the probability of the joint event may be obtained by multiplying the right sides of Equations 2 and 3. Thus

$$P\left(\begin{array}{c}\textbf{Population I}\\ \textbf{Population 2}\end{array}\begin{array}{|c|c|}\hline x_1 & n_1 - x_1 \\\hline x_2 & n_2 - x_2 \\\hline\end{array}\right) = \binom{n_1}{x_1}\binom{n_2}{x_2} p^{x_1+x_2}(1-p)^{N-x_1-x_2} \tag{4}$$

In the simple case where $n_1 = 2$ and $n_2 = 2$ there are nine different points in the sample space, corresponding to the nine possible tables that appear on this and the following page.

Probabilities if H_0 Is True

Tables		($p = 1/2$)	($p = 1$)	T_1
$\begin{array}{\|c\|c\|}\hline 2 & 0 \\\hline 2 & 0 \\\hline\end{array}$	p^4	1/16	1	Undefined
$\begin{array}{\|c\|c\|}\hline 2 & 0 \\\hline 1 & 1 \\\hline\end{array}$	$2p^3(1-p)$	1/8	0	1.1547

Table	Probability			T_1
$\begin{array}{\|c\|c\|}\hline 2 & 0 \\\hline 0 & 2 \\\hline\end{array}$	$p^2(1-p)^2$	$1/16$	0	2.0000
$\begin{array}{\|c\|c\|}\hline 1 & 1 \\\hline 2 & 0 \\\hline\end{array}$	$2p^3(1-p)$	$1/8$	0	-1.1547
$\begin{array}{\|c\|c\|}\hline 1 & 1 \\\hline 1 & 1 \\\hline\end{array}$	$4p^2(1-p)^2$	$1/4$	0	0
$\begin{array}{\|c\|c\|}\hline 1 & 1 \\\hline 0 & 2 \\\hline\end{array}$	$2p(1-p)^3$	$1/8$	0	1.1547
$\begin{array}{\|c\|c\|}\hline 0 & 2 \\\hline 2 & 0 \\\hline\end{array}$	$p^2(1-p)^2$	$1/16$	0	-2.0000
$\begin{array}{\|c\|c\|}\hline 0 & 2 \\\hline 1 & 1 \\\hline\end{array}$	$2p(1-p)^3$	$1/8$	0	-1.1547
$\begin{array}{\|c\|c\|}\hline 0 & 2 \\\hline 0 & 2 \\\hline\end{array}$	$(1-p)^4$	$1/16$	0	Undefined

The undefined values for T_1 arise from the indeterminate form $0/0$. However, since the two outcomes that result in undefined values for T_1 are strongly indicative that H_0 is true, just as the fifth outcome is strongly indicative that H_0 is true, we may arbitrarily define T_1 to be 0 for the first and last outcomes in agreement with the fifth outcome. Then T_1 has the following probability distribution.

$$p = 1/2 \qquad\qquad p = 1$$
$$P(T_1 = -2) = 1/16 \qquad P(T_1 = 0) = 1$$
$$P(T_1 = -1.1547) = 1/4$$
$$P(T_1 = 0) = 3/8$$
$$P(T_1 = 1.1547) = 1/4$$
$$P(T_1 = 2) = 1/16$$

Similarly for any sample sizes n_1 and n_2 the exact probability distributions may be found after the appropriate defining of the undefined values of T_1. However, the probability function is not unique even when H_0 is assumed to be true, as is seen in the previous example, but, instead, it depends on p. Hence the null hypothesis in the preceding test is a composite hypothesis. It is not easy to show, but the size of the critical region in the prior small sample case is a maximum when $p = 1/2$. Therefore α may be found by setting p equal to $1/2$. If the critical region corresponds to the largest value of T_1 (i.e., $T_1 = 2$), then $\alpha = 0.0625$.

To justify the normal distribution as the large sample approximation, consider that the mean of $O_{11}/n_1 - O_{21}/n_2$ is $p_1 - p_2$ and the variance is $p_1 q_1/n_1 + p_2 q_2/n_2$. Under H_0: $p_1 = p_2$ the mean is zero and the variance can be estimated using C_1/N as an estimate for the common value of p, and C_2/N as an estimate for the common value of $q = 1 - p$. By the central limit theorem both O_{11} and O_{21} are approximately normal, so $O_{11}/n_1 - O_{21}/n_2$ is also approximately normal. After subtracting the mean under H_0 (zero) and dividing by the estimated standard deviation we get

$$\frac{O_{11}/n_1 - O_{21}/n_2}{\sqrt{\dfrac{C_1 C_2}{NN}\left(\dfrac{1}{n_1} + \dfrac{1}{n_2}\right)}} \tag{5}$$

which is now approximately a standard normal random variable when H_0 is true. However, the expression in Equation 5 is simply the test statistic T_1, after some algebra, thus showing that the null distribution of T_1 is approximately standard normal. $\square$

Another use for the 2 × 2 contingency table appears when each observation in a single sample of size N is classified according to two properties, where each property may take one of two forms. Then there are $(2)(2) = 4$ different combinations of the two properties, and the 2 × 2 contingency table is a convenient means of tabulating the number of observations in each category. However, this use of the 2 × 2 contingency table is a special case of the $r \times c$ contingency table and does not have any special variation (such as the one-sided test of this section) that would warrant a separate presentation. Therefore it is presented in the next section.

The primary difference between this type of contingency table and the first one is that in this contingency table the row totals are random variables whose values are unknown until after the data are examined. In the first table the row totals represented sample sizes for the two samples, which are known prior to the examination of the data and are therefore not random. In both tables the column totals are random variables.

The third type of contingency table is one with nonrandom row and column totals. That is, both row totals and both column totals are known prior to an examination of the data. This situation does not occur as often as the first two types of contingency tables, but the following statistical procedure is often employed, no matter which of the three types of contingency tables actually occurs, because the exact p-value can be determined fairly easily. The procedure was developed almost simultaneously in the mid 1930's by R. A. Fisher (1935), J. O. Irwin (1935), and F. Yates (1934). It is widely known as Fisher's exact test.

▶ **Fisher's Exact Test** _____

Data   The N observations in the data are summarized in a 2×2 contingency table as in the previous test, except both of the row totals, r and $N - r$, and both of the column totals, c and $N - c$, are determined beforehand, and are therefore fixed, not random.

	Col 1	Col 2	
Row 1	x	$r - x$	r
Row 2	$c - x$	$N - r - c + x$	$N - r$
	c	$N - c$	N

Assumptions

1. Each observation is classified into exactly one cell.
2. The row and column totals are fixed, not random. (However, see the comment at the end for random totals in rows, columns, or both.)

Test Statistic   The test statistic T_2 is the number of observations in the cell in row 1, column 1.

Null Distribution   The exact distribution of T_2 when H_0 is true is given by the hypergeometric distribution (see Equation 1.3.17).

$$P(T_2 = x) = \frac{\binom{r}{x}\binom{N-r}{c-x}}{\binom{N}{c}} \qquad x = 0, 1, \ldots, \min(r, c)$$

$$= 0 \qquad \text{for all other values of } x \tag{6}$$

For a large sample approximation use

$$T_3 = \frac{x - \dfrac{rc}{N}}{\sqrt{\dfrac{rc(N-r)(N-c)}{N^2(N-1)}}} \tag{7}$$

which has the standard normal distribution given in Table A1 as an approximation. If row totals or column totals, or both, are random it is more accurate to use T_1 given by Equation 1 in the large sample approximation.

Hypotheses   Let p_1 be the probability of an observation in row 1 being classified into column 1. Let p_2 be the probability of an observation in row 2 being classified into column 1. Let t_{obs} be the observed value of T_2.

A. (Two-Tailed Test)

$$H_0: p_1 = p_2$$
$$H_1: p_1 \neq p_2$$

First find the p-value using Equation 6. The p-value is twice the smaller of $P(T_2 \leq t_{obs})$ or $P(T_2 \geq t_{obs})$. Reject H_0 at the level of significance α if the p-value is $\leq \alpha$.

B. (Lower-Tailed Test)

$$H_0: p_1 \geq p_2$$
$$H_1: p_1 < p_2$$

Find the p-value $P(T_2 \leq t_{obs})$ using Equation 6. Reject H_0 at the level of significance α if $P(T_2 \leq t_{obs}) \leq \alpha$.

C. (Upper-Tailed Test)

$$H_0: p_1 \leq p_2$$
$$H_1: p_1 > p_2$$

Find the p-value $P(T_2 \geq t_{obs})$ using Equation 6. Reject H_0 at the level of significance α if $P(T_2 \geq t_{obs}) \leq \alpha$.

Computer Assistance   Fisher's exact test is found in *S-Plus, SAS,* and *StatXact.* ◀

Comment

This test is valid for contingency tables with random row totals, random column totals, or both. That is, this exact test finds the p-value for one subset of the sample space, the one with the given row and column totals. Each different set of row and column totals represents another mutually exclusive subset, thus partitioning the entire sample space into mutually exclusive subsets. If the critical region in each subset has a conditional probability $\leq \alpha$ under H_0, then the union of all critical regions has an unconditional probability $\leq \alpha$ under H_0, and the test is valid. However, the power of this exact test is usually less than the power of a more appropriate, approximate, test in those cases where row totals, or column totals, or both, are random.

Comment (continuity correction)

The large sample approximation for T_3 is improved by using a continuity correction. That is, for lower-tailed probabilities, add 0.5 to the numerator of T_3 before looking up the p-value in Table A1. For upper-tailed probabilities subtract 0.5 from the numerator. The resulting probability will be more accurate in most cases.

EXAMPLE 3

Fourteen newly hired business majors, 10 males and 4 females, all equally qualified, are being assigned by the bank president to their new jobs. Ten of the new jobs are as tellers, and four are as account representatives. The null hypothesis is that males and females have equal chances at getting the more desirable account representative jobs. The one-sided alternative of interest is that females are more likely than males to get the account representative jobs.

Only one female is assigned a teller position. Can the null hypothesis be rejected? The information given is sufficient to fill in the following 2×2 contingency table, because the row totals and column totals are already known (nonrandom).

	Account Representative	Teller	
Males	1	9	10
Females	3	1	4
	4	10	14

$$H_0: p_1 \geq p_2$$
$$H_1: p_1 < p_2$$

The exact lower-tailed p-value is given by Equation 6 as

$$P(T_2 \leq 1) = P(T_2 = 0) + P(T_2 = 1)$$

$$= \frac{\binom{10}{0}\binom{4}{4}}{\binom{14}{4}} + \frac{\binom{10}{1}\binom{4}{3}}{\binom{14}{4}}$$

$$= \frac{1}{1001} + \frac{40}{1001} = 0.041$$

The null hypothesis is rejected at $\alpha = 0.05$. ∎

Comment

For comparison purposes, the exact p-value in Example 3 is compared with the exact p-value if the column totals were random. That is, suppose the column totals were random in a problem that resulted in the same contingency table given above, and the experimenter wanted to find the exact p-value using Equation 4. The maximum lower-tailed probability for T_1 is obtained by substituting $p = 0.3$ into Equation 4 (see Problem 3). The exact p-value is 0.012, which is much lower than the value 0.041 given above using Fisher's exact test. The normal approximation for $T_1 = -2.4321$, obtained for the contingency table in the example, is found from Table A1 to be 0.008, which is close to the true p-value. This illustrates that Fisher's exact test is exact only if the row and column totals are nonrandom. In other cases it is still valid but has a tendency to become quite conservative.

□ *T h e o r y*   To show that T_2 has the hypergeometric distribution, start with a contingency table with fixed row totals, whose probability is given in Equation 4 as

$$\binom{r}{x}\binom{N-r}{c-x} p^c (1-p)^{N-c} \tag{8}$$

after switching to the new notation. The probability of getting the column totals c and $N - c$ is given by the binomial probability

$$\binom{N}{c} p^c (1-p)^{N-c} \tag{9}$$

The conditional probability of getting the table results, given the column totals, is found by dividing Equation 8 by Equation 9 to get Equation 6, as we did earlier in Example 1.3.8. The large sample normal approximation is obtained by subtracting the mean and dividing by the standard deviation of the hypergeometric distribution to obtain T_3, and invoking the central limit theorem on T_2. □

Sometimes it is necessary to combine results of several 2 × 2 contingency tables into one overall analysis. This situation occurs when the overall experiment consists of several smaller experiments conducted in various environments, the common probability p under H_0 may be different from one environment to another, and each sub-experiment results in its own 2 × 2 contingency table. The tables cannot be combined into a single 2 × 2 contingency table because of the different environments in which the tables were obtained.

One method for combining several 2 × 2 contingency tables was presented by Mantel and Haenszel (1959).

▶ **The Mantel-Haenszel Test** _____

Data   The data are summarized in several 2×2 contingency tables, each with nonrandom row and column totals. Let the number of tables be $k \geq 2$, and let the ith table be represented with the following notation.

	Col 1	Col 2	
Row 1	x_i	$r_i - x_i$	r_i
Row 2	$c_i - x_i$	$N_i - r_i - c_i + x_i$	$N_i - r_i$
	c_i	$N_i - c_i$	N_i

Assumptions   The assumptions for each contingency table are the same as for the Fisher exact test. In addition, the several contingency tables are obtained from independent experiments.

Test Statistic

$$T_4 = \frac{\sum x_i - \sum \frac{r_i c_i}{N_i}}{\sqrt{\sum \frac{r_i c_i (N_i - r_i)(N_i - c_i)}{N_i^2 (N_i - 1)}}}$$

Null Distribution   The distribution of T_4 is approximately the standard normal distribution given in Table A1, when H_0 is true. The exact probabilities are improved by using a continuity correction. That is, for lower-tailed probabilities add 0.5 to the numerator of T_4 before looking up the p-value in Table A1. For upper-tailed probabilities subtract 0.5 from the numerator before looking up the p-value. The resulting probabilities will be more accurate in most cases.

Hypotheses   Let p_{1i} be the probability of an observation in row 1 being classified into column 1, in the ith contingency table, and let p_{2i} be the corresponding probability for row 2.

A. (Two-Tailed Test)

$H_0: p_{1i} = p_{2i}$ for all $i = 1, \ldots, k$

$H_1:$ Either $p_{1i} > p_{2i}$ for some i, or $p_{1i} < p_{2i}$ for some i, but not both

Reject H_0 at the level α if T_4 exceeds $z_{1-\alpha/2}$, or if T_4 is less than $z_{\alpha/2}$, where z_p represents the pth quantile of the standard normal distribution, given in Table A1.

The p-value is twice the probability of a standard normal random variable being less than the observed T_4, or greater than the observed T_4, whichever is smaller.

B. (Lower-Tailed Test)

$$H_0: p_{1i} \geq p_{2i} \quad \text{for all } i = 1, \ldots, k$$
$$H_1: p_{1i} \leq p_{2i} \quad \text{for all } i, \text{ and } p_{1i} < p_{2i} \text{ for some } i$$

Reject H_0 at the level α if T_4 is less than z_α, from Table A1. The p-value is the probability of a standard normal random variable being less than the observed T_4.

C. (Upper-Tailed Test)

$$H_0: p_{1i} \leq p_{2i} \quad \text{for all } i = 1, \ldots, k$$
$$H_1: p_{1i} \geq p_{2i} \quad \text{for all } i, \text{ and } p_{1i} > p_{2i} \text{ for some } i$$

Reject H_0 at the level α if T_4 is greater than $z_{1-\alpha}$, from Table A1. The p-value is the probability of a standard normal random variable being greater than the observed T_4.

Computer Assistance The Mantel-Haenszel test is found in *S-Plus* and *SAS*. ——————————————————————————————————◄

Comment

As with the Fisher exact test this test is valid even though the row totals or column totals are random. However, in that case it is more accurate to use the test statistic

$$T_5 = \frac{\sum x_i - \sum \frac{r_i c_i}{N_i}}{\sqrt{\sum \frac{r_i c_i (N_i - r_i)(N_i - c_i)}{N_i^3}}}$$

instead of T_4. It is compared with the normal distribution as described above for T_4, but the continuity correction should not be used to find p-values when T_5 is used.

EXAMPLE 4

From Li, Simon, and Gart (1979), three groups of cancer patients are given an experimental treatment to see if the success rate is improved. The numbers of successes and failures are summarized as follows.

	Group 1		Group 2		Group 3	
	Success	Failure	Success	Failure	Success	Failure
Treatment	10	1	9	0	8	0
Control	12	1	11	1	7	3

Because p_{1i} represents the probability of success in a treated patient in group i, the upper-tailed test is used.

$$T_4 = \frac{(10 + 9 + 8) - \left(\dfrac{11 \cdot 22}{24} + \dfrac{9 \cdot 20}{21} + \dfrac{8 \cdot 15}{18}\right)}{\sqrt{\dfrac{11 \cdot 22 \cdot 13 \cdot 2}{24^2 \cdot 23} + \dfrac{9 \cdot 20 \cdot 12 \cdot 1}{21^2 \cdot 20} + \dfrac{8 \cdot 15 \cdot 10 \cdot 3}{18^2 \cdot 17}}}$$

$$= \frac{1.6786}{1.1719} = 1.4323$$

Because T_4 does not exceed 1.6449, the 0.95 quantile from Table A1, the null hypothesis is accepted. The p-value is found by subtracting 0.5 from the numerator of T_4 to get a continuity corrected value of T_4 to be 1.0057. Table A1 shows the upper-tailed p-value to be 0.157.

Because the column totals are obviously random in this example, a more accurate test involves the use of $T_5 = 1.4690$, which has an upper-tailed p-value of 0.071. This demonstrates the greater power associated with using the more appropriate T_5 in the case of random row totals or column totals, or both. However, the null hypothesis is still accepted at $\alpha = 0.05$. ∎

□ *Theory*   The numerator of T_4 is obtained by summing the cell counts in row 1, column 1 of each table, i.e., the Fisher exact test statistic T_2. Then, as in T_3, the mean is subtracted, and the result is divided by the standard deviation, which is the square root of the sum of variances. By the central limit theorem, T_4 has approximately the standard normal distribution.

The statistic T_5 is obtained by rearranging T_1 to look like T_3, and noting that the only difference between T_1 and T_3 is the term N instead of $N - 1$ in the denominator. Then the same reasoning used in the previous paragraph is used to justify the normal approximation for T_5.

The correction for continuity is, typically, half the distance between adjacent values of the discrete random variable. With T_4 and T_3 the row and column totals never change, so the denominator remains constant while the numerator takes successive values one unit apart. Therefore a continuity correction of one-half unit in the numerator is appropriate.

However, when row totals or column totals are random, the denominator of T_1, and T_5, assumes many different values. Likewise the numerator assumes numerous possible values, not evenly spaced. A correction for continuity becomes almost impossible to calculate, and is much smaller than 0.5, so it is reasonable to omit attempting any correction in that case. Certainly a correction of 0.5, as recommended by some, is entirely too large and results in a worse estimate of the true probability in most cases. See Pearson (1947), Plackett (1964), Grizzle (1967) and Conover (1974) for support of this statement. More information on the Mantel-Haenszel test may be found in Li, Simon, and Gart (1997) and Breslow and Liang (1982). □

Confidence intervals may be formed for any unknown probabilities associated with the 2 × 2 contingency table or any contingency table, for that matter, by applying the procedure described in Section 3.1. Similarly, the test in Section 3.1 may be used on contingency tables, whenever the hypotheses are pertinent and the assumptions of the test are met.

A shortcut rule for the one-sided test is given by Ott and Free (1969). Further discussion of the continuity correction may be found in Mantel and Greenhouse (1968), Pirie and Hamdan (1972), and Maxwell (1976). The power of the test is discussed by Harkness and Katz (1964). The exact test is considered by Gail and Gart (1973), Garside and Mack (1976), and McDonald, Davis, and Milliken (1977). For methods of combining the test statistics in several 2 × 2 contingency tables see Radhakrishna (1965), Nelson (1966), Meeker (1978), and Zelen (1971). Possible errors in the marginal totals because of misclassification is the subject of many papers, including ones by Chiacchierini and Arnold (1977) and Plackett (1977). Other related papers are by Fienberg and Gilbert (1970), Upton and Lee (1976), and Ray (1976). An excellent book by Fleiss (1973) is concerned primarily with a discussion of 2 × 2 contingency tables.

EXERCISES

1. A random sample of 135 people was drawn from each of two populations to gauge reaction to pending legislation. In the first sample there were 43 responses of "opposed"; in the second sample there were 37 "opposed." Is there a difference in the proportion of people opposed in the two populations? Does a comparison of this problem with Exercises 1 and 2 in Section 3.5 suggest an advantage in using the same persons in both samples whenever possible, such as in a "before" and "after" situation?

2. Sixty students were divided into two classes of 30 each and taught how to write a program for a computer. One class used the conventional method of learning, and the other class used a new, experimental method. At the end of the courses, each student was given a test that consisted of writing a computer program. The program was either correct or incorrect, and the results were tabulated as follows.

	Correct Program	Incorrect Program
Conventional Class	23	7
Experimental Class	27	3

Is there reason to believe the experimental method is superior? Or could the preceding differences be due to chance fluctuations?

3. One hundred men and 100 women were asked to try a new toothpaste and to state whether they liked or did not like the new taste. Thirty-two men and 26 women said they did not like the new taste. Does this indicate a difference in preferences between men and women in general?

4. Contingency tables may be used to present data representing scales of measurement higher than the nominal scale. For example, a random sample of size 20 was selected from the graduate students who are U.S. citizens, and their grade point averages were recorded.

3.42	3.54	3.21	3.63	3.22
3.80	3.70	3.20	3.75	3.31
3.86	4.00	2.86	2.92	3.59
2.91	3.77	2.70	3.06	3.30

Also, a random sample of 20 students was selected from the non-U.S. citizen group of graduate students at the same university. Their grade point averages were as follows.

3.50	4.00	3.43	3.85	3.84
3.21	3.58	3.94	3.48	3.76
3.87	2.93	4.00	3.37	3.72
4.00	3.06	3.92	3.72	3.91

Test the null hypothesis that the proportion of graduate students with averages of 3.50 or higher is the same for both the U.S. citizens and the non-U.S. citizens.

5. Fisher's exact test can be used as a quick test for correlation between two variables X and Y, each of which has at least an ordinal scale of measurement. Divide the scatterplot of the N values of (X, Y) with a vertical line at the median of X, and a horizontal line at the median of Y, and count the number of observations in each of the four quadrants. Note that the row and column totals are $N/2$, and are not random.

   Suppose that 16 observations of X = age of marriage of a husband, and Y = age of marriage of his father, resulted in 7 pairs where both ages were above the median. Are the two variables positively correlated?

6. Use the Fisher exact test as a test for positive correlation, as explained in Exercise 5, for the data in Example 3.5.4. There the reaction of a patient to Drug 1 (X) and the reaction of the same patient to Drug 2 (Y) was given for 10 patients as $(0.7, 1.9)$, $(-1.6, 0.8)$, $(-0.2, 1.1)$, $(-1.2, 0.1)$, $(-0.1, -0.1)$, $(3.4, 4.4)$, $(3.7, 5.5)$, $(0.8, 1.6)$, $(0.0, 4.6)$, $(2.0, 3.4)$. Compare the p-value using the Fisher exact test to 0.0312, the p-value obtained in Section 3.5 using the Cox and Stuart test for trend as a test for correlation.

7. Exposure to nitrous oxide, an anesthetic, is suspected as a cause for miscarriages among pregnant nurses and dental assistants who sustained prolonged periods of exposure in

their occupation. Data are collected from three different groups of pregnant females and it is recorded how many have miscarriages and how many full-term deliveries.

	Dental Assistants		Operating Room Nurses		Out-Patient Nurses	
	Mis-carriage	Full Term	Mis-carriage	Full Term	Mis-carriage	Full Term
Exposed	8	32	3	18	0	7
Not Exposed	26	210	3	21	10	75

(a) Use T_4, with a correction for continuity when finding the p-value, to investigate this theory.

(b) Use T_5 to test the hypothesis of no miscarriage effect due to exposure to nitrous oxide. Compare the p-value with part a.

(c) Which analysis, using T_4 or using T_5, seems more appropriate in this case?

8. (a) A university received faculty applications from 21 males last year and hired 10, or 48%. At the same time it received faculty applications from 63 female applicants and hired 14, or 22%. Does this university appear to have a higher probability of hiring male applicants than female applicants? (Use Fisher's exact test.)

(b) A breakdown of the above data by colleges showed the following.

Applicants	Education		Bus. Admin.		Engineering	
	Hired	Not Hired	Hired	Not Hired	Hired	Not Hired
Male	2	8	5	0	3	3
Female	12	48	1	0	1	1

Now does this university appear to have a higher probability of hiring male applicants than female applicants? (Use the Mantel-Haenszel test.)

(c) What is the reason for the two different conclusions in part a and part b? Discuss.

PROBLEMS

1. In the test for differences in probabilities, find the exact probability distribution of the test statistic when $n_1 = 2$, $n_2 = 3$. Also, let the largest value of T_1 correspond to the critical region and find α.

2. The data in this section may be considered as two independent samples, $X_1, X_2, \ldots, X_{n_1}$ from population 1, and $Y_1, Y_2, \ldots, Y_{n_2}$ from population 2. Each X_i or Y_i equals 0 if the observation is in class 1 and equals 1 if it is in class 2. Thus each sample is a set of zeros

and ones. The parametric approach to the problem of two independent samples uses the "two-sample t test" with the test statistic

$$t = \frac{\overline{X} - \overline{Y}}{\sqrt{n_1 S_x^2 + n_2 S_y^2}} \sqrt{\frac{n_1 n_2 (n_1 + n_2 - 2)}{n_1 + n_2}}$$

where $\overline{X}, \overline{Y}, S_x^2$, and S_y^2 are the respective sample means and variances. Show that t and T_1 (Equation 1) are related as follows:

$$t = T_1 \sqrt{\frac{n_1 + n_2 - 2}{n_1 + n_2 - T_1^2}}$$

or its equivalent

$$T_1 = t \sqrt{\frac{n_1 + n_2}{n_1 + n_2 - 2 + t^2}}$$

As a result of this relationship, a test that rejects H_0 for large T_1 is equivalent to a test that rejects H_0 for large t, if their critical regions coincide.

3. Consider two random samples, of sizes $n_1 = 10$ and $n_2 = 4$, in a test of $p_1 = p_2$, against the one-sided alternative $p_1 \leq p_2$ in a lower-tailed test. The observed 2×2 contingency table is

	Class 1	**Class 2**	
Pop. 1	1	9	10
Pop. 2	3	1	4
	4	10	14

(a) Calculate the test statistic T_1.

(b) Find the four contingency tables with $n_1 = 10$ and $n_2 = 4$ that produce smaller (more negative) values of T_1. Can you find more than four?

(c) Use Equation 4 to find the exact p-value, that is, the probability of the observed table and the four more extreme tables from part b, as a function of the common probability $p = p_1 = p_2$ under H_0.

(d) Substitute $p = 0.3$ into the probability equation you obtained from part c to find the p-value associated with the observed table in a lower-tailed test. (*Hint.* You should get $p = 0.012$, as we did in this section).

(e) Substitute a value of $p < 0.30$ and a value of $p > 0.30$ into the probability equation you obtained from part c to show that the maximum p-value occurs at $p = 0.30$.

4. Show that T_1 and T_3 are related as follows.

$$T_3 = \frac{\sqrt{N-1}}{\sqrt{N}} T_1$$

4.2 THE $r \times c$ CONTINGENCY TABLE

As an immediate generalization of the 2×2 contingency table of the previous section, we have the contingency table with r rows and c columns, called the *$r \times c$ contingency table*. This contingency table may be used, as in the previous section, to present a tabulation of the data contained in several samples, where the data represent at least a nominal scale of measurement, and to test the hypothesis that the probabilities do not differ from sample to sample. Another use for the $r \times c$ contingency table is with the single sample, where each element in the sample may be classified into one of r different categories according to one criterion and, at the same time, into one of c different categories according to a second criterion. Both of these applications are treated the same in the statistical analysis, but basic differences between the two applications justify separate discussions of the two situations. A third application, similar to the other two, will also be discussed.

First we will consider the extension of the two-sample application presented in the previous section. Now, instead of only two samples, we have r samples, where each sample is tabulated in one of the r rows. Instead of each sample furnishing two categories (formerly called class 1 and class 2), we now consider c categories, corresponding to the c columns. Thus the entry in the (i, j) cell (ith row and jth column) is the number of observations from the ith sample that belong to the jth category. Because of the many rows and columns, the one-sided hypotheses of the previous section are no longer appropriate. Therefore we consider only the two-sided alternative hypothesis, and the test statistic is the square of T_1 from the previous section, generalized to the $r \times c$ case.

▶ The Chi-squared Test for Differences in Probabilities, $r \times c$ _____

Data There are r populations in all, and one random sample is drawn from each population. Let n_i represent the number of observations in the ith sample (from the ith population) for $1 \le i \le r$. Each observation in each sample may be classified into one of c different categories. Let O_{ij} be the number of observations from the ith sample that fall into category j, so

$$n_i = O_{i1} + O_{i2} + \cdots + O_{ic} \qquad \text{for all } i \tag{1}$$

The data are arranged in the following $r \times c$ contingency table.

	Class 1	Class 2	$\cdots$	Class c	Totals
Population 1	O_{11}	O_{12}	$\cdots$	O_{1c}	n_1
Population 2	O_{21}	O_{22}	$\cdots$	O_{2c}	n_2
$\cdots$	$\cdots$	$\cdots$	$\cdots$	$\cdots$	$\cdots$
Population r	O_{r1}	O_{r2}	$\cdots$	O_{rc}	n_r
Totals	C_1	C_2	$\cdots$	C_c	N

The total number of observations from all samples is denoted by N.

$$N = n_1 + n_2 + \cdots + n_r \tag{2}$$

The number of observations in the jth column is denoted by C_j. That is, C_j is the total number of observations in the jth category, or class, from all samples combined.

$$C_j = O_{1j} + O_{2j} + \cdots + O_{rj}, \quad \text{for } j = 1, 2, \ldots, c \tag{3}$$

Assumptions

1. Each sample is a random sample.
2. The outcomes of the various samples are all mutually independent (particularly among samples, because independence within samples is part of the first assumption).
3. Each observation may be categorized into exactly one of the c categories or classes.

Test Statistic   The test statistic T is given by

$$T = \sum_{i=1}^{r} \sum_{j=1}^{c} \frac{(O_{ij} - E_{ij})^2}{E_{ij}}, \quad \text{where} \quad E_{ij} = \frac{n_i C_j}{N} \tag{4}$$

While the term O_{ij} represents the observed number in cell (i, j), the term E_{ij} represents the *expected* number of observations in cell (i, j), if H_0 is really true. That is, if H_0 is true the number of observations in cell (i, j) should be close to the ith sample size n_i multiplied by the proportion C_j/N of all observations in category j. Note that in the 2×2 case this T equals T_1^2 from the previous section, because only the two-sided alternative hypothesis is considered.

An equivalent expression for T, more suited for hand calculator use, is given by

$$T = \sum_{i=1}^{r} \sum_{j=1}^{c} \frac{O_{ij}^2}{E_{ij}} - N \tag{5}$$

Null Distribution The null distribution of T is given approximately by the chi-squared distribution with $(r - 1)(c - 1)$ degrees of freedom, whose quantiles are given in Table A2. The exact distribution of T is very difficult to find and therefore is almost never used.

This chi-squared approximation is satisfactory if the E_{ij}s in the test statistic are not too small. The chi-squared approximation appears to be satisfactory in most cases if all E_{ij}s are greater than 0.5 and at least half are greater than 1.0. However, if any E_{ij}s are less than 0.5, or if most E_{ij}s are less than 1.0, the chi-squared approximation may not be accurate, and thought should be given to combining several similar rows or columns to eliminate the very small row or column totals, or omitting a row or column with very few observations in it.

Hypotheses Let the probability of a randomly selected value from the ith population being classified in the jth class be denoted by p_{ij}, for $i = 1, 2, \ldots, r$, and $j = 1, 2, \ldots, c$.

> H_0: All of the probabilities in the same column are equal to each other
> (i.e., $p_{1j} = p_{2j} = \cdots = p_{rj}$, for all j)
>
> H_1: At least two of the probabilities in the same column are not equal to
> each other (i.e., $p_{ij} \neq p_{kj}$ for some j, and for some pair i and k)

Note that it is not necessary to stipulate the various probabilities. The null hypothesis merely states the probability of being in class j is the same for all populations, no matter what that probability might be (and no matter which category we are considering).

Because of the difficulties involved in tabulating the exact distribution of T, the approximation based on the large sample distribution (where the E_{ij}s are large) is used to find the critical region. The critical region of approximate size α corresponds to values of T larger than $x_{1-\alpha}$, the $1 - \alpha$ quantile of a chi-squared random variable with $(r - 1)(c - 1)$ degrees of freedom obtained from Table A2. Reject H_0 if T exceeds $x_{1-\alpha}$. Otherwise accept H_0.

The p-value is the probability of a chi-squared random variable with $(r - 1)(c - 1)$ degrees of freedom exceeding the observed value of T. This can also be found in Table A2.

Computer Assistance *Minitab, S-Plus, SAS,* and *StatXact* all perform this test. Some of these programs also analyze more complex contingency tables than we cover in this book. ———————————————————————◀

Comment (small E_{ij}s)

Because the asymptotic distribution is used, the approximate value for α, as found here, is a good approximation to the true value of α if the E_{ij}s are fairly large. However, if some of the E_{ij}s are small, the approximation may be very poor.

Cochran (1952) states that if any E_{ij} is less than 1 or if more than 20% of the E_{ij}s are less than 5, the approximation may be poor. This seems to be overly conservative according to unpublished studies by various researchers, including students of Oscar Kempthorne and students of B. L. van der Waerden, and an article by Roscoe and Byars (1971). If r and c are not too small, I feel that some of the E_{ij}s may be as small as 0.5, as long as most are greater than 1.0, without endangering the validity of the test. If some of the E_{ij}s are too small, several categories should be combined to eliminate the E_{ij}s that are too small. Just which categories should be combined is a matter of judgment. Generally, categories are combined only if they are similar in some respects, so that the hypotheses retain their meaning.

EXAMPLE I

A sample of students randomly selected from private high schools and a sample of students randomly selected from public high schools were given standardized achievement tests with the following results.

	Test Scores				
	0–275	**276–350**	**351–425**	**426–500**	**Totals**
Private School	6	14	17	9	46
Public School	30	32	17	3	82
Totals	36	46	34	12	128

To test the null hypothesis that the distribution of test scores is the same for private and public high school students, the test for differences in probabilities is used. A critical region of approximate size $\alpha = 0.05$ corresponds to values of T greater than 7.815, obtained from the chi-squared distribution in Table A2 with $(r - 1)(c - 1) = (2 - 1)(4 - 1) = 3$ degrees of freedom.

The values of E_{ij} are computed using Equation 4 and are given as follows.

	Column			
E_{ij}:	**1**	**2**	**3**	**4**
Row 1	12.9	16.5	12.2	4.3
Row 2	23.1	29.5	21.8	7.7

Note that the E_{ij} satisfy Cochran's criteria. Also note that the row and column sums for the E_{ij} are always the same as those for the O_{ij}. This may be used as a check on the calculations.

For the cell in row 1, column 1, we have

$$\frac{(O_{ij} - E_{ij})^2}{E_{ij}} = \frac{(O_{11} - E_{11})^2}{E_{11}} = \frac{(6 - 12.9)^2}{12.9} = \frac{47.61}{12.9} = 3.69$$

A similar calculation is made for each cell and the result is

$$T = 3.69 + 0.38 + 1.89 + 5.14$$
$$+ 2.06 + 0.21 + 1.06 + 2.87$$
$$= 17.3$$

Since 17.3 is greater than 7.815, the null hypothesis is rejected. In fact, the null hypothesis could have been rejected using a level of significance as small as 0.001, so the p-value is approximately 0.001.

The conclusion is that test scores are distributed differently among public and private high school students. ■

In Example 1 the data (the test scores before grouping) possessed at least an ordinal scale of measurement, a stronger scale than the nominal scale of measurement considered to be more appropriate for the test used. If the alternative hypothesis of interest was that students from the private schools tended to score higher (or lower) than students from the public schools, then a more powerful test based on ranks could have been used, such as the Mann-Whitney test presented in the next chapter. However, the alternative hypothesis in this example included differences of all types, such as higher scores, lower scores, a smaller variance in scores, a larger variance in scores, and so forth, so this chi-squared test is more appropriate.

□ *T h e o r y*   The exact distribution of T in the $r \times c$ case may be found in exactly the same way as it was found in the previous section for the 2×2 case. That is, the row totals (sample sizes) are held constant, and then all possible contingency tables having those same row totals are listed, and their probabilities are calculated using the multinomial distribution for each row. The column totals may vary freely from one table to the next, but the row totals may not change. This is the essential difference between this application of the contingency table and the next to be described. In the next application the row totals are not fixed and, therefore, a greater number of different contingency tables are possible. The only requirement is that the total number of observations N remains the same for all tables. Also, a third variation will be presented. In that application the row totals and the column totals are all fixed and do not vary from table to table. The number of possible tables is greatly reduced, and the exact distribution is then much easier to find.

In all three applications of contingency tables in this section, the asymptotic distribution of T is the same, namely chi-squared with $(r - 1)(c - 1)$ degrees of freedom. Therefore this distribution is used to provide an approximate value for

α, so that the exact tables are not needed. The asymptotic distribution is derived in Cramér (1946). □

The second application of the $r \times c$ contingency table involves a single random sample of size N, where each observation may be classified according to two criteria. There are r categories (rows) resulting from the first criterion, and c categories (columns) resulting from the second criterion. Each observation is classified according to both criteria and thus ends up being assigned to a particular cell in the $r \times c$ contingency table. The cell entries represent the number of observations belonging to that cell. A nominal scale of measurement is all that is required, although higher scales may be used. The hypothesis tested is one of independence; loosely stated, the null hypothesis says that the rows and columns represent two independent classification schemes. A more precise description is now given.

▶ The Chi-squared Test for Independence

Data A random sample of size N is obtained. The observations in the random sample may be classified according to two criteria. Using the first criterion each observation is associated with one of the r rows, and using the second criterion each observation is associated with one of the c columns. Let O_{ij} be the number of observations associated with row i and column j simultaneously. The cell counts O_{ij} may be arranged in an $r \times c$ contingency table. The total number of observations in row i is designated by R_i, (instead of n_i as the previous test, to emphasize that the row totals are now random rather than fixed), and in column j by C_j. The sum of the numbers in all of the cells is N.

	Column					
	1	**2**	**3**	$\cdots$	**c**	**Totals**
Row 1	O_{11}	O_{12}	O_{13}	$\cdots$	O_{1c}	R_1
2	O_{21}	O_{22}	O_{23}	$\cdots$	O_{2c}	R_2
$\cdots$	$\cdots$	$\cdots$	$\cdots$	$\cdots$	$\cdots$	$\cdots$
r	O_{r1}	O_{r2}	O_{r3}	$\cdots$	O_{rc}	R_r
Totals	C_1	C_2	C_3	$\cdots$	C_c	N

Assumptions

1. The sample of N observations is a random sample. (Each observation has the same probability as every other observation of being classified in row i and column j, independently of the other observations.)

2. Each observation may be classified into exactly one of r different categories according to one criterion and into exactly one of c different categories according to a second criterion.

Test Statistic   Let E_{ij} equal $R_i C_j / N$. Then the test statistic is given by

$$T = \sum_{i=1}^{r} \sum_{j=1}^{c} \frac{(O_{ij} - E_{ij})^2}{E_{ij}} \tag{6}$$

or, for more convenient use with hand calculators,

$$T = \sum_{i=1}^{r} \sum_{j=1}^{c} \frac{O_{ij}^2}{E_{ij}} - N \tag{7}$$

where the summation is taken over all cells in the contingency table. Note that these are identical to the two forms of the test statistic given in the previous test as Equations 4 and 5.

Null Distribution   As in the previous test, the null distribution of the test statistic is given approximately by the chi-squared distribution with $(r - 1)(c - 1)$ degrees of freedom, found in Table A2. The exact distribution of T is very difficult to find and therefore is almost never used.

As in the previous test, the chi-squared approximation is usually satisfactory if all of the E_{ij}s are greater than 0.5, and at least half are greater than 1.0. Small E_{ij}s can be eliminated by combining rows or columns that have small totals with other rows or columns that represent similar characteristics, or by simply eliminating rows or columns that have very few observations in them.

Hypotheses

H_0: The event "an observation is in row i" is independent of the event "that same observation is in column j," for all i and j

By the definition of independence of events, H_0 may be stated as follows.

H_0: P(row i, column j) = P(row i) $\cdot$ P(column j),   for all i, j

The negation of H_0 is conveniently stated as

H_1: P(row i, column j) $\neq$ P(row i) $\cdot$ P(column j),   for some i, j

Reject H_0 if T exceeds the $1 - \alpha$ quantile of a chi-squared random variable with $(r - 1)(c - 1)$ degrees of freedom, obtained from Table A2. The approximate level of significance is then α. The p-value is also obtained from Table A2, as the probability of a chi-squared random variable with $(r - 1)(c - 1)$ degrees of freedom exceeding the observed value of T.

Computer Assistance   *Minitab, S-Plus, SAS,* and *StatXact* all perform this test. Some of these programs also analyze more complex tables than we cover in this book. ◀

EXAMPLE 2

A random sample of students at a certain university were classified according to the college in which they were enrolled and also according to whether they graduated from a high school in the state or out of the state. The results were put into a 2 × 4 contingency table.

	Engineering	Arts and Sciences	Home Economics	Other	Totals
In State	16	14	13	13	56
Out of State	14	6	10	8	38
Totals	30	20	23	21	94

In order to test the null hypothesis that the college in which each student is enrolled is independent of whether high school training was in state or out of state, the chi-squared test for independence is selected. The rejection region of approximate size 0.05 corresponds to values of T greater than 7.815, the 0.95 quantile of a chi-squared random variable with $(r - 1)(c - 1) = 3$ degrees of freedom, obtained from Table A2. Therefore α is approximately 0.05.

Using Equation 6, T is computed for these data, resulting in

$$T = 1.52$$

Therefore H_0 is accepted. From Table A2 we may say that the p-value exceeds 0.25.

■

□ **Theory**   The exact distribution of T is illustrated here for the relatively simple case where $N = 4$. Let p_{ij} be the probability of an observation being classified in row i and column j (cell i, j). (Note that this p_{ij} is not the same as the p_{ij} of the previous test. Here the sum of the p_{ij} in all cells is one. In the previous test the p_{ij} in each row added to unity.) Then the probability of the particular outcome

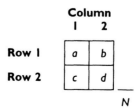

is found, using the multinomial distribution, to be

$$\frac{N!}{a!b!c!d!}(p_{11})^a(p_{12})^b(p_{21})^c(p_{22})^d \tag{8}$$

because the number of ways N objects can result in the preceding cell counts is given by the multinomial coefficient $N!/a!b!c!d!$, and each result has probability

$$(p_{11})^a(p_{12})^b(p_{21})^c(p_{22})^d \tag{9}$$

The null hypothesis says that each p_{ij} equals the product of its row probability times its column probability. The maximum size of the upper tail of T, when H_0 is true, is found by setting all of the p_{ij}s equal to each other, $1/4$ in this case (we will not prove this). Therefore α is found by computing

$$P\left(\begin{array}{|c|c|} \hline a & b \\ \hline c & d \\ \hline \end{array}\right) = \frac{N!}{a!b!c!d!}\left(\frac{1}{4}\right)^N \tag{10}$$

for each possible arrangement. For $N = 4$ there are 35 different contingency tables. These are listed in Figure 1 along with their probabilities and the corresponding values of T. As before we define zero divided by zero to be zero.

$$P(T = 0) = 84/256 = 0.33$$
$$P(T = 4/9) = 48/256 = 0.19$$
$$P(T = 4/3) = 96/256 = 0.37$$
$$P(T = 4) = 28/256 = 0.11$$

If the critical region corresponds to the largest value of T, $T = 4$, then $\alpha = 0.11$. This compares with $\alpha = 0.125$ for the two-tailed version of the exact distribution discussed in the previous section. A comparison of the preceding distribution, where only N is fixed, with the distribution derived in the previous section where the row totals are also fixed, shows that in this case the distribution of T is more complicated to obtain because of the many more possible tables. Also, additional values of T are now possible, and the probability distribution is altered somewhat. $\square$

Even though the exact distributions of T under the two applications differ somewhat, the asymptotic distributions are both chi-squared with $(r - 1)(c - 1)$ degrees of freedom.

In the third application of the contingency table, not only are the row totals fixed, as in the first application, but the column totals are fixed as well. Thus the exact distribution of T is easier to find than in both applications previously introduced. However, easier or not, the exact distribution is still too complicated

T = 0

Outcome	Probability
$\begin{array}{cc}4&0\\0&0\end{array}$	$(1/4)^4$
$\begin{array}{cc}0&4\\0&0\end{array}$	$(1/4)^4$
$\begin{array}{cc}0&0\\0&4\end{array}$	$(1/4)^4$
$\begin{array}{cc}0&0\\4&0\end{array}$	$(1/4)^4$
$\begin{array}{cc}3&1\\0&0\end{array}$	$4(1/4)^4$
$\begin{array}{cc}0&3\\0&1\end{array}$	$4(1/4)^4$
$\begin{array}{cc}0&0\\1&3\end{array}$	$4(1/4)^4$
$\begin{array}{cc}1&0\\3&0\end{array}$	$4(1/4)^4$
$\begin{array}{cc}1&3\\0&0\end{array}$	$4(1/4)^4$
$\begin{array}{cc}0&1\\0&3\end{array}$	$4(1/4)^4$
$\begin{array}{cc}0&0\\3&1\end{array}$	$4(1/4)^4$
$\begin{array}{cc}3&0\\1&0\end{array}$	$4(1/4)^4$
$\begin{array}{cc}2&2\\0&0\end{array}$	$6(1/4)^4$
$\begin{array}{cc}0&2\\0&2\end{array}$	$6(1/4)^4$
$\begin{array}{cc}0&0\\2&2\end{array}$	$6(1/4)^4$
$\begin{array}{cc}2&0\\2&0\end{array}$	$6(1/4)^4$
$\begin{array}{cc}1&1\\1&1\end{array}$	$24(1/4)^4$

Total = 84/256

T = 4/9

Outcome	Probability
$\begin{array}{cc}2&1\\1&0\end{array}$	$12(1/4)^4$
$\begin{array}{cc}1&2\\0&1\end{array}$	$12(1/4)^4$
$\begin{array}{cc}0&1\\1&2\end{array}$	$12(1/4)^4$
$\begin{array}{cc}1&0\\2&1\end{array}$	$12(1/4)^4$

Total = 48/256

T = 4/3

Outcome	Probability
$\begin{array}{cc}2&1\\0&1\end{array}$	$12(1/4)^4$
$\begin{array}{cc}0&2\\1&1\end{array}$	$12(1/4)^4$
$\begin{array}{cc}1&0\\1&2\end{array}$	$12(1/4)^4$
$\begin{array}{cc}1&1\\2&0\end{array}$	$12(1/4)^4$
$\begin{array}{cc}2&0\\1&1\end{array}$	$12(1/4)^4$
$\begin{array}{cc}1&2\\1&0\end{array}$	$12(1/4)^4$
$\begin{array}{cc}1&1\\0&2\end{array}$	$12(1/4)^4$
$\begin{array}{cc}0&1\\2&1\end{array}$	$12(1/4)^4$

Total = 96/256

T = 4

Outcome	Probability
$\begin{array}{cc}3&0\\0&1\end{array}$	$4(1/4)^4$
$\begin{array}{cc}0&3\\1&0\end{array}$	$4(1/4)^4$
$\begin{array}{cc}1&0\\0&3\end{array}$	$4(1/4)^4$
$\begin{array}{cc}0&1\\3&0\end{array}$	$4(1/4)^4$
$\begin{array}{cc}2&0\\0&2\end{array}$	$6(1/4)^4$
$\begin{array}{cc}0&2\\2&0\end{array}$	$6(1/4)^4$

Total = 28/256

FIGURE I   The exact distribution of T, when all p_{ij}s equal $1/4$

for practical purposes, unless extensive tables or a computer are available. The chi-squared approximation is recommended for finding the critical region and α.

▶ The Chi-squared Test with Fixed Marginal Totals _____

Data The data are summarized in an $r \times c$ contingency table, as in the two previous applications, except that the row and column totals are determined beforehand, and are therefore fixed, not random.

	Column				
	1	**2**	$\cdots$	**c**	**Totals**
Row 1	O_{11}	O_{12}	$\cdots$	O_{1c}	n_1
2	O_{21}	O_{22}	$\cdots$	O_{2c}	n_2
$\cdots$	$\cdots$	$\cdots$	$\cdots$	$\cdots$	$\cdots$
r	O_{r1}	O_{r2}	$\cdots$	O_{rc}	n_r
Totals	c_1	c_2	$\cdots$	c_c	N

The row and column totals are denoted by n_i and c_j, respectively, to emphasize the fact that they are given and not random. The total number of observations is N.

Assumptions

1. Each observation is classified into exactly one cell.
2. The observations are observations on a random sample. Each observation has the same probability of being classified into cell (i, j) as any other observation.
3. The row and column totals are given, not random.

Test Statistic Let $E_{ij} = n_i c_j / N$ be the expected number of observations in cell (i, j). Then the test statistic, as before, is given by

$$T = \sum_{i=1}^{r} \sum_{j=1}^{c} (O_{ij} - E_{ij})^2 / E_{ij} \tag{11}$$

where the summation is over all rc cells. Notice that this is the same test statistic that was used in the previous two tests, and therefore the more computationally friendly form given by Equation 5 may be used instead.

Null Distribution   As in the two previous tests, the null distribution of the test statistic is given approximately by the chi-squared distribution with $(r - 1)$ $(c - 1)$ degrees of freedom, found in Table A2. The exact distribution of T is easier to find in this case than in the two previous tests, but it is still very difficult to find and therefore is almost never used.

As in the previous test, the chi-squared approximation is usually satisfactory if all of the E_{ij}s are greater than 0.5, and at least half are greater than 1.0. Small E_{ij}s can be eliminated by combining rows or columns that have small totals with other rows or columns that represent similar characteristics, or by simply eliminating rows or columns that have very few observations in them.

Hypotheses   The hypotheses may be either of the two sets of hypotheses introduced in the two previous applications in this section, under the condition that the row and column totals are fixed. Or the hypotheses may be tailored to fit the particular experimental situation. Usually the hypotheses are variations of the independence hypotheses of the previous test. See Examples 3 and 4 for particular modifications that are dictated by the experiment.

Reject H_0 if T exceeds the $1 - \alpha$ quantile of a chi-squared random variable with $(r - 1) (c - 1)$ degrees of freedom, obtained from Table A2. The approximate level of significance is then α. The p-value is also obtained from Table A2, as the probability of a chi-squared random variable with $(r - 1) (c - 1)$ degrees of freedom exceeding the observed value of T.

Computer Assistance   *Minitab, S-Plus, SAS,* and *StatXact* all perform this test. Exact p-values may be found in some cases by *StatXact.* ————————————◄

EXAMPLE 3

The chi-squared test with fixed marginal totals may be used to test the hypothesis that two random variables X and Y are independent. Starting with a scatter diagram of 24 points, which represent independent observations on the bivariate random variable (X, Y), a contingency table may be constructed. The x-coordinate of each point is the observed value of X and the y-coordinate is the observed value of Y in each observation on (X, Y). Assume the observed pairs (X, Y) are mutually independent. We wish to test

$$H_0\text{: } X \text{ and } Y \text{ are independent of each other}$$

against the alternative hypothesis of dependence.

To form the contingency table so that all E_{ij}s are equal, we note that 3 and 4 both are factors of the sample size 24. Therefore we divide the points into 3 rows of 8 points each, and 4 columns of 6 points each, using dotted lines as in Figure 2. (It is recommended that if the E_{ij}s are small, they should be very nearly equal to each other. One way of accomplishing this is by having equal row totals and equal column totals.) This resulting contingency table of counts is given as follows.

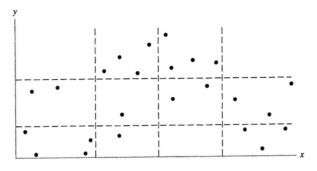

FIGURE 2

	Column				
	1	2	3	4	Totals
Row 1	0	4	4	0	8
2	2	1	2	3	8
3	4	1	0	3	8
Totals	6	6	6	6	24

The critical region of approximate size 0.05 corresponds to values of T greater than 12.59, the 0.95 quantile of a chi-squared random variable with $(r - 1)$ $(c - 1) = (2)(3) = 6$ degrees of freedom, obtained from Table A2.

The test statistic is evaluated using Equation 11 and $E_{ij} = (6)(8)/24 = 2$.

$$T = \sum_{i=1}^{r} \sum_{j=1}^{c} \frac{(O_{ij} - E_{ij})^2}{E_{ij}} = \sum_{i=1}^{3} \sum_{j=1}^{4} \frac{(O_{ij} - 2)^2}{2}$$

$$= 14 \tag{12}$$

Because T exceeds 12.59, H_0 is rejected, and we conclude that X and Y are not independent. H_0 could have been rejected at a level of significance as small as 0.03, so the p-value is 0.03.   ■

EXAMPLE 4

A psychologist asks a subject to learn 25 words. The subject is given 25 blue cards, each with one word on it. Five of the words are nouns, 5 are adjectives, 5 are adverbs, 5 are verbs, and 5 are prepositions. She must pair these blue cards with 25 white cards, each with one word on it and also containing the different parts of speech, 5 words each. The subject is allowed 5 minutes to pair the cards (1 white card with each blue card) and 5 minutes to study the pairs thus formed.

Then she is asked to close her eyes, and the words on the white cards are read to her one by one. When each word is read to her, she tries to furnish the word on the blue card associated with the word read.

The psychologist is not interested in the number of correct words but, instead, in examining the pairing structure to see if it represents an ordering of some sort. The hypotheses are as follows.

H_0: There is no organization of pairs according to parts of speech

H_1: The subject tends to pair particular parts of speech on the blue cards with particular parts of speech (not necessarily the same) on the white cards

The pairings are summarized in a 5×5 contingency table.

	Noun	Adjective	Adverb	Verb	Preposition	Totals
Noun		3			2	5
Adjective	4	1				5
Adverb				5		5
Verb			5			5
Preposition	1	1			3	5
Totals	5	5	5	5	5	25

The chi-squared test with fixed marginal totals is selected. The experimenter feels that large values of T indicate H_1 is true. The marginal totals represent the number of words in each category, which was fixed in advance of the actual experiment. The critical region of approximate size 0.05 corresponds to values of T greater than 26.30, the 0.95 quantile of a chi-squared random variable with $(r - 1)(c - 1) = (4)(4) = 16$ degrees of freedom, obtained from Table A2. The observed value of T is obtained using Equation 11.

$$E_{ij} = \frac{(5)(5)}{25} = 1 \qquad \text{for all } i \text{ and } j$$

$$T = \sum_{i=1}^{5} \sum_{j=1}^{5} \frac{(O_{ij} - 1)^2}{1}$$

$$= 66 \tag{13}$$

Because $T = 66$, H_0 is soundly rejected in favor of H_1. The p-value is less than 0.001.

∎

□ *Theory* The exact distribution of T was found in the previous section for the 2×2 case with nonrandom row and column totals. The distribution is the hypergeometric distribution, with probabilities given by Equation 4.1.6. We will now illustrate the exact distribution of T for a simple 2×2 contingency table.

If the row totals and column totals all equal 2, there are three possible contingency tables.

Table	Probability	T
$\begin{array}{\|c\|c\|} \hline 2 & 0 \\ \hline 0 & 2 \\ \hline \end{array}$	$\dfrac{\binom{2}{2}\binom{2}{0}}{\binom{4}{2}} = 1/6$	4
$\begin{array}{\|c\|c\|} \hline 1 & 1 \\ \hline 1 & 1 \\ \hline \end{array}$	$\dfrac{\binom{2}{1}\binom{2}{1}}{\binom{4}{2}} = 2/3$	0
$\begin{array}{\|c\|c\|} \hline 0 & 2 \\ \hline 2 & 0 \\ \hline \end{array}$	$\dfrac{\binom{2}{0}\binom{2}{2}}{\binom{4}{2}} = 1/6$	4

Because the probability distribution of T is unique, H_0 is simple in this application.

Fixed row totals and fixed column totals greatly reduce the number of contingency tables possible, and so the exact distribution of T is more feasible in this case than in the previous two applications. When $r = 2$ and $c = 2$, the test is known as "Fisher's exact test," and extensive exact tables of probabilities are available (Finney, 1948). Programming Fisher's exact test is discussed by Robertson (1960).

For r and c in general, the exact probability of the table

	Column				
	1	2	$\cdots$	c	**Totals**
Row 1	O_{11}	O_{12}	$\cdots$	O_{1c}	n_1
2	O_{21}	O_{22}	$\cdots$	O_{2c}	n_2
$\cdots$	$\cdots$	$\cdots$	$\cdots$	$\cdots$	$\cdots$
r	O_{r1}	O_{r2}	$\cdots$	O_{rc}	n_r
Totals	c_1	c_2	$\cdots$	c_c	N

with fixed marginal totals is given by

$$\text{probability} = \frac{\begin{bmatrix} n_1 \\ O_{1i} \end{bmatrix}\begin{bmatrix} n_2 \\ O_{2i} \end{bmatrix} \cdots \begin{bmatrix} n_r \\ O_{ri} \end{bmatrix}}{\begin{bmatrix} N \\ c_i \end{bmatrix}} \tag{14}$$

where the multinomial coefficients are as defined by Rule 3 in Section 1.1. □

The contingency tables of this section and Section 4.1 could be called *two-way contingency tables* because the observations are classified two ways, by rows and by columns. An immediate extension may be made to include the situation where observations are classified according to three or more criteria, and thus the data are presented in the form of a three- (or more) way contingency table.

For convenience in extending the chi-squared contingency table test, the two-way test statistic is rewritten as

$$T = \sum_{i,j} \frac{\left[O_{ij} - N\dfrac{R_i}{N}\dfrac{C_j}{N} \right]^2}{N\dfrac{R_i}{N}\dfrac{C_j}{N}} \tag{15}$$

which has $(r - 1)(c - 1)$ degrees of freedom. In a three-way contingency table with r rows, c columns, and t blocks, denote the block totals by B_k, $k = 1, 2, \ldots,$ t, to correspond to the row totals R_i and the column totals C_j. Let N still denote the total number of observations. Then

$$R_i = \sum_{j,k} O_{ijk} \tag{16}$$

$$C_j = \sum_{i,k} O_{ijk} \tag{17}$$

$$B_k = \sum_{i,j} O_{ijk} \tag{18}$$

where O_{ijk} represents the total number of observations classified in row i, column j, and block k. Then E_{ijk}, the expected number of observations in row i, column j, and block k, assuming the null hypothesis of row-column-block independence is true, may be estimated from

$$E_{ijk} = N\frac{R_i}{N}\frac{C_j}{N}\frac{B_k}{N} \tag{19}$$

and the test statistic may be computed using

$$T = \sum_{i,j,k} \frac{(O_{ijk} - E_{ijk})^2}{E_{ijk}} \tag{20}$$

where the summation is over all $r \cdot c \cdot t$ cells. The test statistic is then tested for significance using the chi-squared distribution with $rct - r - c - t + 2$ degrees of freedom. The extension of the test to contingency tables of any dimension should be apparent.

So-called "loglinear models" have been used successfully to analyze multidimensional contingency tables and are discussed in the final section of this chapter. More detailed discussion of the analysis of multidimensional contingency tables is found in Goodman (1970), Ireland, Ku, and Kullback (1969), Ku, Varner, and Kullback (1971), Kullback (1971), Goodman (1971), Koch, Johnson, and Tolley (1972), Darroch (1974), and Halperin et al. (1977). A short book on analyzing contingency tables has been written by Maxwell (1961). The important subject of estimation in contingency tables is addressed by Fienberg (1970a), McNeil and Tukey (1975), and Quade and Salama (1975). If some of the data are only partially classified, see Chen and Fienberg (1974) or Hocking and Oxspring (1974). Contingency tables where one or both categories have a natural ordering are discussed in Section 5.2, as well as by Williams and Grizzle (1972), Simon (1974), and Clayton (1974).

The exact distribution of the 2×3 contingency table test statistic is given for equal column totals and fixed row totals by Bennett and Nakamura (1963, 1964) and is discussed by Healy (1969). Ireland and Kullback (1968) give a different test for contingency tables with given row and column totals. The power of chi-squared tests for contingency tables is examined by Chapman and Meng (1966).

An excellent and readable survey article on contingency tables is one by Mosteller (1968). Haynam and Leone (1965) give an approximation to the exact distribution of T. Misclassification of data is the subject of an article by Mote and Anderson (1965). Tables with small or zero cell frequencies are discussed by Ku (1963) and Sugiura and Otake (1968). See Goodman (1964, 1968) and Bhapkar and Koch (1968) for information concerning tests for interaction. A class of bivariate contingency-type distributions is discussed by Plackett (1965), Mardia (1967b), and Steck (1968). Other methods for examining contingency tables are given by Ishii (1960), Gregory (1961), Claringbold (1961), Kullback, Kupperman, and Ku (1962), Diamond (1963), Mielke and Siddiqui (1965), Hoeffding (1965), Gart (1966), and Chacko (1966). The many papers on contingency tables illustrate the usefulness and versatility of this type of analysis; see, for example, Elston (1970), Crowley and Breslow (1975), Light and Margolin (1971), Margolin and Light (1974), and Shuster and Downing (1976). Mantel and Haenszel (1959) present a useful procedure for testing homogeneity in row-column distributions on N independent $2 \times r$ tables. Other applications of contingency tables are given in the following sections of this chapter.

EXERCISES

1. Test whether the following observations indicate a dependence between the two variables observed: (3.6, 13), (4.7, 19), (1.4, 9), (5.5, 15), (4.8, 27), (4.3, 14), (3.0, 6), (4.2, 11), (6.0, 24), (6.8, 26), (4.1, 18), (3.2, 9), (4.0, 8), (1.9, 6), (0.4, 7), (4.9, 14), (5.6, 18), and (5.6, 20). Which test of this section is being used?

2. One horse was selected at random from each of 80 races and categorized according to post position (the position assigned to the horse for the start of the race) and the position in which the horse crossed the finish line (first, second, etc.).

		Finish			
		1	**2**	**3**	**Other**
Post	**1–4**	8	6	8	16
Position	**5–9**	3	6	5	28

Is the horse's position at the end of the race dependent on post position? Which test of this section is being used?

3. In another study, all of the horses in all of the races for three days were classified by post position and by the order in which they finished.

		Finish			
		1	**2**	**3**	**Other**
Post	**1–4**	15	14	15	52
Position	**5–9**	9	10	9	72

Is the horse's position at the end of the race dependent on post position? Which test of this section is being used?

4. Three professors are teaching large classes in introductory statistics. At the end of the semester, they compare grades to see if there are significant differences in their grading policies.

Professor	Grade						
	A	**B**	**C**	**D**	**F**	**WP**	**WF**
Smith	12	45	49	6	13	18	2
Jones	10	32	43	18	4	12	6
White	15	19	32	20	6	9	7

Are these differences significant? Which test are you using? Are the grades assigned by Professors Jones and White significantly different? How would the results be interpreted?

5. A random sample of stocks from the American Stock Exchange is compared with a random sample of stocks from the New York Stock Exchange to see if there is any difference in the rating percentages of the two exchanges. Of the 23 stocks from the ASE there were 11 As, 11 Bs, and 1 C. Of the 35 stocks from the NYSE there were 24 As, 11 Bs, and no Cs. What does your analysis look like?

6. A team of observers moves through a wooded area and reports all sightings, false and true, of camouflaged equipment. Two types of camouflage are used, plain and patterned. The team report includes the type of camouflage used, and the location of the equipment. The team is being monitored by a person who knows which sightings are true and which are false. The results of true and false sightings are as follows.

	Types of Camouflage	
	Patterned	Plain
Number of False Detections	14	4
Number of True Detections	27	32

Is there a significant difference in the probability of a reported sighting being incorrect? (Note that this study does not address undetected equipment, nor sightings that misidentify the type of camouflage.) Which type of contingency table is this?

7. A random sample of 30 graduating university seniors was categorized by college and religious preference as follows. Is there a relationship between college and religious preference?

Arts & Sciences	Business	Engineering	Other
Catholic	Protestant	Catholic	Protestant
Catholic	Catholic	Protestant	Catholic
Jewish	Jewish	Protestant	Catholic
Catholic	Jewish	Protestant	Other
Protestant	Protestant	Other	Catholic
Protestant	Protestant		Other
Other	Other		Catholic
Protestant			Jewish
Other			Other

8. A television marketing firm kept track of the telephone responses to its various products in order to determine if the time the ad appeared on television was related to the product being sold. The number of responses was as follows. What does your analysis look like? What assumptions are necessary here?

	Product			
	Fishing Rod	Kitchen Tool	Music CD	Exercise Machine
Daytime	6	73	55	7
Nighttime	14	65	82	8
Weekend	21	58	48	8

PROBLEMS

1. Show that the two forms of T given by Equations 4 and 5 are equivalent.

2. Show that if $r = 2$ and $c = 2$, Equation 4 is equivalent to Equation 4.1.1 squared

$$T = \frac{N(O_{11}O_{22} - O_{12}O_{21})}{n_1 n_2 C_1 C_2}$$

3. A different method of analyzing contingency tables uses the statistic

$$T' = 2 \sum_{i=1}^{r} \sum_{j=1}^{c} O_{ij} \ln (O_{ij} / E_{ij})$$

instead of T, where ln refers to natural logarithm, found on most calculators. Otherwise the two test procedures are exactly the same. Use T' in Exercise 3 to see if the result of the analysis is similar to the result using T. (The two tests are not equivalent in general, even though they may produce similar results in particular cases.)

4.3   THE MEDIAN TEST

The median test is designed to examine whether several samples came from populations having the same median. Actually, the median test is not new to this chapter; it is merely a special application of the chi-squared test with fixed marginal totals introduced in the previous section. It is a very useful application, however, and we consider it worth special treatment.

To test whether several (c) populations have the same median, a random sample is drawn from each population. (The scale of measurement is at least ordinal, or else the term "median" is without meaning.) A $2 \times c$ contingency table is constructed, and the two entities in the ith column are the numbers of observations in the ith sample that are above and below the grand median (the median of all observations combined). The usual chi-squared test is then applied to the contingency table.

▶ **The Median Test** _____

Data   From each of c populations a random sample of size n_i is obtained, $i = 1$, $2, \ldots, c$. The combined sample median is determined; that is, the number that is exceeded by about half of the observations in the entire array of N ($= n_1 + n_2 + \cdots + n_c$) sample values is determined. This is called the *grand median*. Let O_{1i} be the number of observations in the ith sample that exceed the grand median and let O_{2i} be the number in the ith sample that are less than or equal to the grand median. Arrange the frequency counts into a $2 \times c$ contingency table as follows.

Sample	1	2	$\cdots$	c	Totals
>Median	O_{11}	O_{12}	$\cdots$	O_{1c}	a
≤Median	O_{21}	O_{22}	$\cdots$	O_{2c}	b
Totals	n_1	n_2	$\cdots$	n_c	N

Let a equal the total number of observations above the grand median in all samples and let b equal the total number of values less than or equal to the grand median. Then $a + b = N$, the total number of observations.

Assumptions

1. Each sample is a random sample.
2. The samples are independent of each other.
3. The measurement scale is at least ordinal.
4. If all populations have the same median, all populations have the same probability p of an observation exceeding the grand median.

Test Statistic   The test statistic is obtained by a rearrangement of the test statistic given in the previous section, noting that $O_{2i} = n_i - O_{1i}$ in the special case of two rows.

$$T = \frac{N^2}{ab} \sum_{i=1}^{c} \frac{\left(O_{1i} - \frac{n_i a}{N}\right)^2}{n_i} \tag{1}$$

If a calculator is being used, the following form is more convenient.

$$T = \frac{N^2}{ab} \sum_{i=1}^{c} \frac{O_{1i}^2}{n_i} - \frac{Na}{b} \tag{2}$$

If a is approximately equal to b, as it should be unless there are many values equal to the grand median, the following simplification of the test statistic may be used.

$$T = \sum_{i=1}^{c} \frac{(O_{1i} - O_{2i})^2}{n_i} \tag{3}$$

The simplified form for T given in Equation 3 is exact if $a = b$. It is only approximate otherwise.

Null Distribution   The exact distribution of T is difficult to tabulate, so the large sample approximation is used to approximate the distribution of T. (See the discussion of the theory, later in this section, for the exact distribution of T.) The approximate null distribution is the chi-squared distribution with $c - 1$ degrees of freedom.

Hypotheses

H_0: All c populations have the same median

H_1: At least two of the populations have different medians

The critical region of approximate size α corresponds to values of T greater than $x_{1-\alpha}$, the $(1 - \alpha)$ quantile of a chi-squared random variable with $c - 1$ degrees of freedom, obtained from Table A2. If T exceeds $x_{1-\alpha}$, reject H_0. Otherwise accept H_0.

The approximate p-value is the probability of a chi-squared random variable exceeding the observed value of T, which is obtained from Table A2.

If some of the sample sizes n_i are too small, the preceding approximation may not be very accurate (the true α may vary considerably from the approximate α determined above). The same rule given in the previous section may be used as a rule of thumb. In this case the rule will be met if all samples of size 1 are dropped from the analysis.

Multiple Comparisons   If the null hypothesis is rejected, pairwise multiple comparisons may be made between populations by using the median test repeatedly on 2×2 contingency tables. In each comparison the median of the two samples is found, and the number above or below that median is used in the 2×2 contingency table. If the test statistic T, computed using Equation 1, 2, or 3 on the 2×2 table, exceeds the $1 - \alpha$ quantile of a chi-squared random variable with 1 degree of freedom from Table A2, then the medians of those two populations are considered to be unequal.

Computer Assistance   The median test can be found in *Minitab* and *Stat-Xact*. _____ ◀

EXAMPLE I

Four different methods of growing corn were randomly assigned to a large number of different plots of land and the yield per acre was computed for each plot.

Method

1	2	3	4
83	91	101	78
91	90	100	82
94	81	91	81
89	83	93	77
89	84	96	79
96	83	95	81
91	88	94	80
92	91		81
90	89		
	84		

In order to determine whether there is a difference in yields as a result of the method used, the median test was employed because it was felt that a difference in population medians could be interpreted as a difference in the value of the method used. The hypotheses may be stated as follows.

H_0: All methods have the same median yield per acre

H_1: At least two of the methods differ with respect to the median yield per acre

A quick count reveals there are 34 observations in all, so the average of the seventeenth and eighteenth smallest observations is the grand median, and is seen to be 89. Then, for each method (sample), the number of values that exceed 89 and the number that are less than or equal to 89 are recorded in the following form.

	Method				
	1	**2**	**3**	**4**	**Totals**
>89	6	3	7	0	16
≤89	3	7	0	8	18
Totals	9	10	7	8	34

The critical region corresponds to values of T greater than 7.815, the 0.95 quantile of a chi-squared random variable with $c - 1 = 3$ degrees of freedom, obtained from Table A2. T is computed using Equation 1.

$$T = \frac{(34)^2}{(16)(18)} \left\{ \frac{\left[6 - \frac{(9)(16)}{34}\right]^2}{9} + \cdots + \frac{\left[0 - \frac{(8)(16)}{34}\right]^2}{8} \right\}$$

$$= 4.01(0.34 + 0.29 + 1.97 + 1.78)$$

$$= 17.6 \tag{4}$$

Use of the more convenient Equation 3 gives

$$T = \tfrac{8}{9} + \tfrac{16}{10} + \tfrac{49}{7} + \tfrac{64}{8}$$
$$= 17.6 \tag{5}$$

which is identical to the rounded-off value previously obtained.

Because the T of 17.6 exceeds the critical value 7.815, H_0 is rejected. Inspection of Table A2 shows the p-value to be slightly less than 0.001. Because the null hypothesis of equal medians is rejected for the four methods of growing corn, it is appropriate to make multiple pairwise comparisons. A comparison between Method 1 and Method 2 reveals a sample median of the 19 observations to be 89. Method 1 has 6 of its 9 observations above 89, and Method 2 has 3 of its 10 observations above 89. The 2×2 contingency table test statistic is 2.55, which is less than the 0.95 quantile of a chi-squared random variable with 1 degree of freedom, 3.841, therefore the medians of Methods 1 and 2 cannot be considered different. However, all other pairwise comparisons are significant at $\alpha = 0.05$.

Methods	Median	T
1 and 2	89	2.55
1 and 3	92.5	6.35
1 and 4	83	13.43
2 and 3	91	13.25
2 and 4	82.5	14.40
3 and 4	82	15.00

Multiple comparisons of the populations by repeatedly using the same test on subgroups of the original data always distorts the true level of significance of all tests but the first. Such repeated testing procedures are for one's personal satisfaction or for use as an objective "yardstick" for separating the various populations, but they cannot receive the same interpretation as may legitimately be given to the first, overall test. For a further discussion of repeated testing procedures see Gabriel (1966) or Knoke (1976).

In Example 1 the experiment has been arranged in a so-called "completely randomized design," which assumes that the different methods are assigned to the different plots in some random manner (or a manner equivalent to a random manner). The usual parametric method of analyzing the data is called a "one-way analysis of variance." The A.R.E. of the median test relative to the one-way analysis of variance depends on the form of the population distribution function. For normal distributions the A.R.E. is only $2/\pi = 64\%$. However, for the double exponential distribution (which has heavier tails) it is 200%.

The median test may be extended to become a "quantile test" for testing the null hypothesis that several populations have the same quantile, for any particular quantile chosen, merely by altering the data section of the test so that the observations are classified as being above or not above the grand quantile for the entire array of values. The remainder of this test remains the same, except the approximation Equation 3 will not be applicable. The exact distribution of T, as given in the theory to follow, remains the same. An extension of the median test to allow for two-stage sampling is given by Wolfe (1977a).

□ *Theory*   The row totals a and b are fixed, as in the third test of the previous section, because of the objective set of rules that is used to determine which observations are to be counted in the upper or lower cells of the contingency table. For example, if the test is an "upper quartile" test, then a is about $N/4$ and b is about $3N/4$, with allowances for tied, or equal, sample values. Thus the exact distribution of T is a conditional distribution that depends on the row and column totals. The probability of obtaining the table

$$
\begin{array}{|c|c|c|c|}
\hline
O_{11} & O_{12} & \cdots & O_{1c} \\
\hline
O_{21} & O_{22} & \cdots & O_{2c} \\
\hline
\end{array}
\begin{array}{c} a \\[2mm] b \end{array}
\qquad (6)
$$

$$
\begin{array}{cccc}
n_1 & n_2 & \cdots & n_c
\end{array}
$$

with the column totals fixed, is the product of the binomial probabilities

$$
P\left(\begin{array}{|c|}\hline O_{1i} \\ \hline O_{2i} \\ \hline \end{array}\right) = \binom{n_i}{O_{1i}} p^{O_{1i}}(1-p)^{O_{2i}}; \qquad i = 1, 2, \ldots, c \qquad (7)
$$

where p is the probability of an observation exceeding the grand median. Now H_0 merely states that all populations have the same median, and this does not necessarily imply that all populations have the same probability p of exceeding the grand (sample) median. On the other hand, H_0 does not preclude this latter situation, and the two statements (H_0 and the previous situation) are very similar in intent. To find the distribution of T when H_0 is true, we need to require that the probabilities of exceeding the grand median be the same for all populations. This is why the fourth assumption was placed in the model.

Because the samples are independent of each other, we can obtain the joint probability by multiplication, using Equation 7;

$$
P\left(\begin{array}{|c|c|c|c|}\hline O_{11} & O_{12} & \cdots & O_{1c} \\ \hline O_{21} & O_{22} & \cdots & O_{2c} \\ \hline \end{array}\right) = \binom{n_1}{O_{11}}\binom{n_2}{O_{12}} \cdots \binom{n_c}{O_{1c}} p^a(1-p)^b \qquad (8)
$$

where

$$a = O_{11} + O_{12} + \cdots + O_{1c}$$

and

$$b = O_{21} + O_{22} + \cdots + O_{2c}$$

The probability of the event in Equation 8, given the row totals a and b, is found by dividing the probability in Equation 8 by the probability of getting the row totals a and b, as in the latter part of the previous section. The result is

$$P\left(\begin{array}{|c|c|c|c|}\hline O_{11} & O_{12} & \cdots & O_{1c} \\\hline O_{21} & O_{22} & \cdots & O_{2c} \\\hline n_1 & n_2 & \cdots & n_c \end{array}\, \begin{array}{c} a \\ b \\ N \end{array} \right) = \frac{\binom{n_1}{O_{11}}\binom{n_2}{O_{12}}\cdots\binom{n_c}{O_{1c}}}{\binom{N}{a}} \qquad (9)$$

which can be written in terms of multinomial coefficients as

$$\text{Probability} = \frac{\begin{bmatrix} a \\ O_{1i} \end{bmatrix}\begin{bmatrix} b \\ O_{2i} \end{bmatrix}}{\begin{bmatrix} N \\ n_i \end{bmatrix}} \qquad (10)$$

in agreement with Equation 14 of Section 4.2, which was given without derivation.

Thus the exact distribution of T can be, but almost never is, found using Equation 9 or 10. Instead, the chi-squared distribution with $(c - 1)$ degrees of freedom is used (because the number of rows is two), as in the previous section. □

▶ **An Extension of the Median Test** _____

The preceding median test may be extended so that more complex experiments may be analyzed. Because of the cumbersome notation involved in the extension of the median test, we will introduce the test by presenting an example of its application. _____ ◀

EXAMPLE 2

Four different fertilizers are used on each of six different fields, and the entire experiment is replicated using three different types of seed. The yield per acre is calculated at the conclusion of the experiment under each of the $(4)(6)(3) = 72$ different conditions with the following results.

		Seed 1				Seed 2				Seed 3			
						Fertilizer							
		1	2	3	4	1	2	3	4	1	2	3	4
Field	1	80.5	90.1	87.0	88.0	79.1	87.0	82.6	81.5	85.4	92.3	92.0	89.3
	2	87.0	83.4	89.1	90.3	77.6	82.0	81.4	87.9	89.2	90.1	90.2	93.6
	3	86.1	82.4	91.0	86.1	84.1	80.6	89.0	80.4	90.0	88.1	87.2	90.8
	4	82.1	84.9	84.4	83.1	83.3	79.5	86.3	83.1	83.4	85.3	94.3	87.6
	5	79.3	87.1	92.2	90.8	76.6	86.2	84.0	87.4	87.1	86.3	88.4	93.7
	6	84.2	89.3	85.3	84.7	81.0	84.1	88.1	85.0	82.3	92.9	95.1	82.9

To test the null hypothesis

H_0: There is no difference in median yields due to the different fertilizers

let $x_{i_1i_2i_3}$ denote the observed yield using fertilizer i_1 in field i_2 with seed i_3. For example, x_{213} is the yield using fertilizer 2 in field 1 with seed 3, which is 92.3. Then x_{213} is compared with the median of x_{113}, x_{213}, x_{313}, and x_{413}, the four yields obtained under identical circumstances except for fertilizers (which H_0 claims to have no effect). Thus x_{213} is compared with the median of 85.4, 92.3, 92.0, and 89.3, which is

$$(\tfrac{1}{2})(89.3 + 92.0) = 90.65$$

If x_{213} exceeds 90.65, it is replaced in the table by a one; otherwise it is replaced by a zero.

Similarly, each $x_{i_1i_2i_3}$ is compared with the median of $x_{1i_2i_3}$, $x_{2i_2i_3}$, . . . , $x_{ci_2i_3}$, the observations obtained under similar conditions, except for the c different fertilizers. In our example each yield is compared with the median of the yields in the same row (field) and same block (seed) and replaced by one or zero according to whether it exceeds or does not exceed its respective median. The results are as follows.

		Seed 1				Seed 2				Seed 3			
						Fertilizer							
		1	2	3	4	1	2	3	4	1	2	3	4
Field	1	0	1	0	1	0	1	1	0	0	1	1	0
	2	0	0	1	1	0	1	0	1	0	0	1	1
	3	0	0	1	0	1	0	1	0	1	0	0	1
	4	0	1	1	0	1	0	1	0	0	0	1	1
	5	0	0	1	1	0	1	0	1	0	0	1	1
	6	0	1	1	0	0	0	1	1	0	1	1	0

Let O_j be the number of fields in which fertilizer j was used and where the yield exceeded its respective median. Then O_j is the total number of "ones" under fertilizer j in the preceding tables. The O_j are given in the following table for $j = 1, 2, . . . , c$.

	Fertilizer				
O_j = Number of	**1**	**2**	**3**	**4**	**Total**
"Ones"	3	8	14	10	$a = 35$
Number of "Zeros"	15	10	4	8	$b = 37$
	$n_1 = 18$	$n_2 = 18$	$n_3 = 18$	$n_4 = 18$	$N = 72$

The usual median test is then applied to this table. Using Equation 3, we obtain

$$T = \frac{(144 + 4 + 100 + 4)}{18}$$

$$= 14.0 \tag{11}$$

A comparison of this value of T with the 0.95 quantile of a chi-squared random variable with $c - 1 = 3$ degrees of freedom, obtained from Table A2 as $x_{0.95} = 7.815$, results in rejection of H_0. The p-value in this experiment is about 0.004. ∎

EXERCISES

1. Test the hypothesis that the following samples were obtained from populations having the same medians.

   Sample 1: 35, 42, 42, 30, 15, 31, 29, 29, 17, 21
   Sample 2: 34, 38, 26, 17, 42, 28, 35, 33, 16, 40
   Sample 3: 17, 29, 30, 36, 41, 30, 31, 23, 38, 30
   Sample 4: 39, 34, 22, 27, 42, 33, 24, 36, 29, 25

2. A number of oil leases were auctioned to the highest bidder. Each lease received one or more sealed bids. Test the hypothesis that the leases that eventually became producers of oil have the same medium number of bids as the leases that never produced oil. A random sample of each type of lease is given below.

   Number of Bids on Each Lease

   Producers   6, 3, 1, 14, 8 9, 12, 1, 3, 2, 1, 7
   Nonproducers   6, 2, 1, 1, 3, 1, 2, 4, 8, 1, 2

3. Do the experimental results of Example 2 indicate a difference among seeds?

4. Do the experimental results of Example 2 indicate a difference among fields?

5. A random sample of 30 stocks was selected from each of the three major U.S. exchanges,

and their performance over the previous year was noted. The median performance for all 90 stocks was noted and the following table constructed.

Exchange	Above Median
New York	18
American	17
NASDAQ	10

Was there a significant difference in the performance of stocks on the three exchanges during the previous year?

6. One hundred army recruits were randomly assigned to four drill sergeants in boot camp. At the end of boot camp 84 recruits remained, and their performance in the obstacle course was timed. For Sergeant Adams, 11 of his 20 recruits performed above the median. For Sergeant Baker, 8 of her 22 recruits performed above the median. Sergeant Callahan had 8 of his 20, and Sergeant Davis had 15 perform above the median. Is there a significant difference in the performance results for the four drill sergeants?

PROBLEMS

1. Show that if a equals b, then Equation 1 becomes Equation 3.

2. Show that Equation 1 is the same as Equation 4.2.11 when r equals 2.

3. The usual parametric test for the design in this section (the one-way layout) assumes that each observation has a normal distribution instead of being merely a zero or one, depending on whether it is below or above the median. If the observations in each sample are called 0s when they are below the grand median and 1s if they are equal to or above the grand median, the statistic for the previous parametric test, computed on the 0s and 1s, simplifies to

$$F = \frac{\left(\sum_{j=1}^{c} \frac{O_{1j}^2}{n_j} - \frac{a^2}{N} \right)(N - c)}{\left(a - \sum_{j=1}^{c} \frac{O_{1j}^2}{n_j} \right)(c - 1)}$$

Show that F may be written as the following function of T,

$$F = \frac{T(N - c)}{(N - T)(c - 1)}$$

and that therefore rejecting H_0 for large T is equivalent to rejecting H_0 for large F.

4.4 MEASURES OF DEPENDENCE

The contingency table is a convenient form for examining data to see if there is some sort of dependence inherent in the data. The particular type of dependence revealed by a contingency table is a row-column dependence. If the different rows represent samples from different populations and the columns represent different categories of classification of the data from the samples, a row-column dependence is synonymous with a functional dependence of the probabilities of being in the various categories on the population from which the sample was obtained. Similarly, if the observations from one random sample are classified into rows and columns according to each of two different criteria, a row-column dependence has an obvious interpretation as a dependence between the two criteria of classification.

Suppose that instead of testing hypotheses, as we have been doing so far in this chapter, we merely wish to express the degree of dependence shown in a particular contingency table. Ideally, we would like to be able to express the degree of dependence in some simple form, and in a form that easily conveys to other people the exact degree of dependence exhibited by the table.

As the first approach, we could use the test statistic of the previous sections

$$T = \sum_{i=1}^{r} \sum_{j=1}^{c} \frac{(O_{ij} - E_{ij})^2}{E_{ij}} \tag{1}$$

as a measure of dependence, with the philosophy, "If it is good enough to test for dependence, it is good enough to measure dependence." The use of T seems satisfactory as far as convenience and simplicity are concerned. However, in order to convey the degree of dependence to other people, the number of degrees of freedom should also be stated along with the T value because, without knowing the number of degrees of freedom, it is not possible to tell the degree of dependence conveyed by the value of T. Even if the number of degrees of freedom is known, the nonexpert must consult a chi-squared table in order to interpret T.

EXAMPLE I

In Example 4.2.1 the contingency table was as follows.

	Scores				
	0–275	**276–350**	**351–425**	**426–500**	**Totals**
Private School	6	14	17	9	46
Public School	30	32	17	3	82
Totals	36	46	34	12	128

For this contingency table T was found to equal 17.3. Now, 17.3 is approximately the 0.999 quantile of a chi-squared random variable with 3 degrees of freedom, so

$$p\text{-value} = 1 - p$$
$$= 0.001$$

as computed in the example. Such a small value of the p-value indicates that the data strongly disagree with the null hypothesis of independence between the row classification (type of school) and the column classification (test scores), but it does little toward measuring the level of dependence. ▪

▶ Cramér's Contingency Coefficient _____

An approach to the problem of providing an easily interpreted measure of dependence consists of modifying the value of T in Equation 1 in such a way that the result does not depend as much on the number of degrees of freedom as T does. One such modification considers dividing T by the maximum value T may attain. We know by now that large values of T arise from contingency tables that have a pronounced unevenness among the cell counts. By examining extremely uneven contingency tables we may find, by trial and error, that T is greatest (for a given number of rows r and columns c and total sample size N) when there are zeros in every cell except for one cell in each row and in each column. (If r does not equal c some rows or columns may be all zeros.) That is, T is a maximum in a contingency table resembling the following.

	| Column					
	1	2	3	4	5	Totals
Row 1	3	0	0	0	0	3
2	0	3	0	0	0	3
3	0	0	3	0	0	3
Totals	3	3	3	0	0	9

For this table, $T = 18$, after $0/0$ is defined as 0 or, equivalently, after omitting columns (and rows) with all zero cells.

In general, the maximum value of T is $N(q - 1)$, where q is either r or c, whichever is smaller, and N is the total number of observations. Division of T by its maximum gives

$$R_1 = \frac{T}{N(q - 1)} \tag{2}$$

where q is the smaller of r and c. R_1 is close to 1.0 if the table indicates a strong row-column dependence and close to 0 if the numbers across each row are in the same proportions to each other as the column totals are to each other. This measure was suggested by Cramér (1946, p. 443). Many modern computer packages, such as *SAS* and *StatXact*, use the square root of R_1 and call it "Cramér's coefficient."

$$\text{Cramér's coefficient} = \sqrt{\frac{T}{N(q-1)}} \tag{3}$$

This is currently the most widely used measure of dependence for $r \times c$ contingency tables. ──────────────────────────────◄

EXAMPLE 2

In the previous example the 2×4 contingency table furnished a value of $T = 17.3$. Because $N = 128$ and $q = 2$, we have R_1 given by

$$R_1 = \frac{T}{N(q-1)} = \frac{17.3}{128}$$
$$= 0.135$$

and Cramér's coefficient is $\sqrt{0.135} = 0.368$.   ■

Cramér's coefficient, like all good measures of dependence, is "scale invariant." That is, if the scale of the experiment becomes much larger, such as 10 times as many school children in Example 1, the measure of dependence doesn't change as long as all the observations change the same relative to each other. If 10 times as many school children were examined, and the result was that each cell was 10 times as large, the resulting table would be as follows.

	Scores				
	0–275	**276–350**	**351–425**	**426–500**	**Totals**
Private School	60	140	170	90	460
Public School	300	320	170	30	820
Totals	360	460	340	120	1280

The test statistic T also increases 10-fold to 173, but Cramér's coefficient remains unchanged at 0.368. This is because the degree of dependence is the same for the smaller scaled experiment as it is for the larger scaled experiment.

▶ **Pearson's Contingency Coefficient** ———————————————

Two other coefficients are sometimes used. The first is called *Pearson's coefficient of mean square contingency* by Yule and Kendall (1950, p. 53) and is given as

$$R_2 = \sqrt{\frac{T}{N + T}} \tag{4}$$

We stated that the maximum value of T is $N(q - 1)$, and so the maximum value of R_2 is

$$R_2(\text{max}) = \sqrt{\frac{N(q - 1)}{N + N(q - 1)}} = \sqrt{\frac{q - 1}{q}} \tag{5}$$

which is close to one in many cases. The smallest possible value of T is zero, and so

$$0 \leq R_2 \leq \sqrt{\frac{q - 1}{q}} < 1.0 \tag{6}$$

R_2 is also called the *contingency coefficient* by McNemar (1962, p. 198) and Siegel (1956, p. 196). ———————————————————————————————— ◀

EXAMPLE 3

In the contingency table of the two previous examples we have $T = 17.3$ and $N = 128$, so

$$R_2 = \sqrt{\frac{T}{N + T}} = \sqrt{\frac{17.3}{128 + 17.3}}$$
$$= 0.345$$ ■

▶ **Pearson's Mean-Square Contingency Coefficient** ——————————

We present a third measure of dependence, R_3, also attributed to Pearson (by Cramér, 1946, p. 282) and also called the *mean-square contingency* (by Yule and Kendall, 1950, p. 53). R_3 is defined as

$$R_3 = \frac{T}{N} \tag{7}$$

From the preceding discussions we may conclude that

$$0 \leq R_3 \leq q - 1$$

and that knowledge of r and c is necessary in order to interpret accurately the degree of independence from the value of R_3. ──────────────────◀

EXAMPLE 4

For the same contingency table used in the previous example we have

$$R_3 = \frac{17.3}{128} = 0.135$$

■

Finally, we just mention *Tschuprow's coefficient*, given by Yule and Kendall (1950), as

$$R_4 = \sqrt{\frac{T}{N\sqrt{(r-1)(c-1)}}} \tag{8}$$

The choice of a measure of dependence is largely a personal decision, motivated primarily by local traditions instead of by statistical considerations. See Stuart (1953) for further discussion.

For the 2×2 contingency table,

	Column		
	1	**2**	
Row 1	a	b	r_1
Row 2	c	d	r_2
	c_1	c_2	N

the preceding measures simplify somewhat. We know from Problem 4.2.2 that T reduces to

$$T = \frac{N(ad - bc)^2}{r_1 r_2 c_1 c_2} \tag{9}$$

Therefore R_3 and R_1 (because $q = 2$) reduce to

$$R_1 = R_3 = \frac{T}{N} = \frac{(ad - bc)^2}{r_1 r_2 c_1 c_2} \tag{10}$$

and Cramér's coefficient becomes

$$\sqrt{R_1} = \sqrt{\frac{T}{N(q-1)}} = \sqrt{\frac{(ad-bc)^2}{r_1 r_2 c_1 c_2}} \tag{11}$$

R_2 may be written as

$$R_2 = \sqrt{\frac{T}{N+T}} = \sqrt{\frac{(ad-bc)^2}{r_1 r_2 c_1 c_2 + (ad-bc)^2}} \tag{12}$$

In a four-fold contingency table, unlike the general $r \times c$ contingency table, it is sometimes meaningful to distinguish between a positive association and a negative association, such as when the two criteria of classification have corresponding categories.

EXAMPLE 5

Forty children are classified according to whether their mothers have dark hair or light hair and as to whether their fathers have dark or light hair. The results may show a positive association (positive correlation)

		Father Dark	Father Light	
Mother	Dark	28	0	28
	Light	5	7	12
		33	7	40

or a negative association (negative correlation)

		Father Dark	Father Light	
Mother	Dark	21	7	28
	Light	12	0	12
		33	7	40

according to whether $(ad - bc)$ is positive or negative. A lack of association (zero correlation) is indicated by the following.

		Father		
		Dark	Light	
Mother	Dark	23	5	28
	Light	10	2	12
		33	7	40

∎

▶ The Phi Coefficient

If the type of association is of interest, care must be taken to set up the table so that a and d represent the number of similar classifications (dark-dark and light-light), while b and c represent the number of unlike classifications (dark-light and light-dark). One measure of association that preserves direction is the *phi coefficient*, given by

$$R_5 = \frac{ad - bc}{\sqrt{r_1 r_2 c_1 c_2}} \tag{13}$$

which may vary from $+1$, when all items are classified in the "alike" cells (both b and c equal zero), to -1, when all items are classified as "unlike" (both a and d equal zero). The phi coefficient is merely Cramér's coefficient (see Equation 11), with the sign of $(ad - bc)$ being preserved. One reason for the popularity of the phi coefficient is because it is a special case of the *Pearson product moment correlation coefficient* (presented in the next chapter), computed by representing the classes by numbers. Notice also the close relationship between the phi coefficient and the test statistic T_1 used in the 2×2 contingency table analysis. Equation 4.1.1 shows that the phi coefficient equals T_1 divided by the square root of N. ___ ◀

EXAMPLE 6

For the first table in Example 5 we have

$$
\begin{array}{ll}
a = 28 & r_1 = 28 \\
b = 0 & r_2 = 12 \\
c = 5 & c_1 = 33 \\
d = 7 & c_2 = 7
\end{array}
$$

so that R_5 is computed as

$$R_5 = \frac{ad - bc}{\sqrt{r_1 r_2 c_1 c_2}} = \frac{(28)(7) - 0}{\sqrt{(28)(12)(33)(7)}}$$
$$= 0.703 \qquad (14)$$

For the second table in Example 5,

$$R_5 = \frac{(21)(0) - (7)(12)}{\sqrt{(28)(12)(33)(7)}}$$
$$= -0.302 \qquad (15)$$

which reflects the negative association of hairtypes. ∎

Other measures of association for the four-fold contingency table include one proposed by Yule and Kendall (1950, p. 30)

$$R_6 = \frac{ad - bc}{ad + bc} \qquad (16)$$

and one proposed by Ives and Gibbons (1967)

$$R_7 = \frac{(a + d) - (b + c)}{a + b + c + d} \qquad (17)$$

There is no end to the possible measures that may be defined. One's choice of a coefficient is solely a result of personal preferences.

Sometimes the question arises, "How can I test the null hypothesis of independence, using R_1 (or R_2, R_3, etc.) as a test statistic?" The answer is that you can find the exact small sample distribution of any of these measures in the same laborious manner that the exact small sample distribution of T was found in Section 4.2. Therefore, theoretically a test may be devised. But it is much easier, and just as effective, to use the tests presented in Sections 4.1 and 4.2 for the same hypotheses.

In particular, the coefficients

$$R_1 = \frac{T}{N(q - 1)}$$

$$R_2 = \sqrt{\frac{T}{N + 1}}$$

$$R_3 = \frac{T}{N}$$

and

$$R_4 = \sqrt{\frac{T}{N\sqrt{(r - 1)(c - 1)}}}$$

as well as Cramér's coefficient will all be "too large" whenever T is "too large," because they increase or decrease when T increases or decreases. The tests of Section 4.2 use T as a test statistic, and when T is significant we may conclude that R is significant.

A one-tailed test, appropriate only for the 2×2 contingency table, may be based on the phi coefficient R_5, because of the relationship

$$R_5 = \frac{T_1}{\sqrt{N}} \tag{18}$$

where T_1 is given by Equation 4.1.1. T_1 is approximately normally distributed and therefore $\sqrt{N} \cdot R_5$ is approximately normally distributed. Then the null hypothesis

$$H_0: \text{There is no positive correlation}$$

may be rejected if $\sqrt{N} \cdot R_5$ is too large (exceeds $z_{1-\alpha}$ from Table A1, for a level of α), and the null hypothesis

$$H_0: \text{There is no negative correlation}$$

may be rejected if $\sqrt{N} \cdot R_5$ is too small (smaller than z_α from Table A1, for a level of α). This is exactly the same as the test described in Section 4.1 based on T_1.

A one-tailed test based on the phi coefficient is illustrated in the following example.

EXAMPLE 7

In order to see if seat belts help prevent fatalities, records of the last 100 automobile accidents to occur along a particular highway were examined. These 100 accidents involved 242 persons. Each person was classified as using or not using seat belts when the accident occurred and as injured fatally or a survivor.

		Injured Fatally?		
		Yes	No	Totals
Wearing Seat Belts?	**Yes**	7	89	96
	No	24	122	146
	Totals	31	211	242

The statement we wish to prove is, "Seat belts help prevent fatalities." However, a test for correlation does not automatically imply a cause and effect relationship.

While a cause and effect relationship between two variables usually results in correlation, a significant correlation may be the result of both variables being influenced by a third variable, which might be the reckless nature of the driver in this case. Therefore the null hypothesis is

H_0: There is no negative correlation between wearing seat belts and being killed in an automobile accident

and the alternative hypothesis is

H_1: There is a negative correlation between wearing seat belts and being killed in an automobile accident

This example is slightly different than the situation we have been describing in that the two criteria of classification do not have the same corresponding classes. ("Yes" and "no" mean one thing in the rows and another in the columns.) Therefore we need to stop and assess the situation. If H_1 were true we would expect b and c to be larger than a and d. Therefore the inequality

$$ad - bc < 0$$

tends to support H_1 and also causes R_5 to be negative. So we reject H_0 if $\sqrt{N} \cdot R_5$ is less than -1.645, the 0.05 quantile of the standard normal distribution given in Table A1. In this example the test statistic,

$$\sqrt{N} \cdot R_5 = \frac{\sqrt{N}(ad - bc)}{\sqrt{r_1 r_2 c_1 c_2}}$$
$$= \frac{\sqrt{242}[(7)(122) - (89)(24)]}{\sqrt{(96)(146)(31)(211)}}$$
$$= -2.0829$$

is less than -1.645, so H_0 is rejected. We may conclude that the use of seat belts is associated with fewer fatalities. (Whether the relationship is a causal relationship remains an open question.) The p-value is found from Table A1 to be about 0.019. ∎

Several other measures of association between variables that are classified according to two criteria are introduced in classical papers by Goodman and Kruskal (1954, 1959, 1963). A partial coefficient is introduced by Davis (1967).

EXERCISES

1. One hundred married couples were interviewed, and the husband and wife were asked separately for their first choice for the next U.S. president, with the following results.

Wife's Choice

		A	B	Other
Husband's Choice	A	12	22	6
	B	25	21	4
	Other	3	7	0

Compute the following.

(a) T (b) Cramér's coefficient

(c) R_1 (d) R_2

(e) R_3 (f) R_4

2. Fifty factory workers reported to the nurse complaining of soreness due to arthritis. Twenty-five of them were given aspirin, and the rest were given a placebo without their knowledge. One hour later they were asked if the pill they took helped them to feel better. Seventeen in the aspirin group and 12 in the placebo group said it did.

   (a) Use R_5 to see if there is a positive correlation between taking aspirin and feeling better.

   (b) Compute R_6.

   (c) Compute R_7.

3. A traffic study was conducted for a short time on a well-traveled city street. Of the 64 cars observed, 16 were exceeding the limit and 48 were not. Also, 24 of the drivers had passengers and the rest did not. Twelve of the speeders were driving alone. Assume that the observed traffic behaves the same as a random sample of all traffic would.

   (a) Use R_5 to see if there is a positive correlation between speeding and driving alone.

   (b) Compute R_6.

   (c) Compute R_7.

4. A certain type of insect that is found in lakes in the southwestern United States is studied to see if the chromosomal structure is significantly different among states. The number of insects of various chromosomal types is recorded as follows.

Type	Texas	New Mexico	Arizona	California
A	54	72	83	96
B	20	6	18	6
C	17	3	12	0
D	0	12	14	1
E	0	10	0	0

Compute the following.

(a) T (b) Cramér's coefficient

(c) R_1 (d) R_2

(e) R_3 (f) R_4

PROBLEMS

1. Show that T equals $N(q - 1)$ for the following contingency table (here $r < c$).

	1	2	$\cdots$	r	$r + 1$	$\cdots$	c
1	O_{11}	0	$\cdots$	0	0	$\cdots$	0
2	0	O_{22}	$\cdots$	0	0	$\cdots$	0
$\cdots$	$\cdots$	$\cdots$	$\cdots$	$\cdots$	$\cdots$	$\cdots$	$\cdots$
r	0	0	$\cdots$	O_{rr}	0	$\cdots$	0

2. Think of another $r \times c$ contingency table, other than the one in Problem 1, that you would suspect of having a large value of T. Compute T for your contingency table. Is it greater than $N(q - 1)$?

3. Prove the identities:

   (a) $R_2^2 = \dfrac{R_3}{1 + R_3}$.

   (b) $R_3 = \dfrac{R_2^2}{1 - R_2^2}$.

   (c) When $r = 2$ and $c = 2$, $R_3 = R_5^2 = R_1$.

4. Show that the phi coefficient is merely Pearson's product moment correlation coefficient

$$r = \frac{\sum\limits_{i=1}^{N} (X_i - \overline{X})(Y_i - \overline{Y})}{\left[\sum\limits_{i=1}^{N} (X_i - \overline{X})^2 \sum\limits_{j=1}^{N} (Y_j - \overline{Y})^2 \right]^{1/2}}$$

computed on the pairs (X_i, Y_i) where (X_i, Y_i) equals $(0, 0)$, $(0, 1)$, $(1, 0)$, or $(1, 1)$, depending on which cell each observation belongs to. Then show that the same result holds true if the numbers 0 and 1 just used are replaced by any two numbers p and q.

4.5 THE CHI-SQUARED GOODNESS-OF-FIT TEST

Often the hypotheses being tested are statements concerning the unknown probability distribution of the random variable being observed. Examples include "The median is 4.0" and "The probability of being in class 1 is the same for both populations." More comprehensive hypotheses than the ones we have been examining would include "The unknown distribution function is the normal distribution function with mean 3.0 and variance 1.0" or "The distribution function of this random variable is the binomial, with parameters $n = 10$ and $p = 0.2$." These latter hypotheses are more comprehensive because they include statements concerning all of the quantiles simultaneously, rather than just the median, and all of the probabilities simultaneously, instead of an isolated statement about some of the probabilities. Hypotheses such as these may be tested with a "goodness-of-fit" test, that is, with a test designed to compare the sample obtained with the hypothesized distribution to see if the hypothesized distribution function "fits" the data in the sample.

The oldest and best-known goodness-of-fit test is the chi-squared test for goodness of fit, first presented by Pearson (1900).

▶ The Chi-squared Test for Goodness of Fit

Data The data consist of N independent observations of a random variable X. These N observations are grouped into c classes, and the numbers of observations in each class are presented in the form of a $1 \times c$ contingency table.

	Class				
	I	**2**	$\cdots$	**c**	**Total**
Observed Frequencies	O_1	O_2	$\cdots$	O_c	N

Let O_j denote the number of observations in class j, for $j = 1, 2, \ldots, c$.

Assumptions

1. The sample is a random sample.
2. The measurement scale is at least nominal.

Test Statistic Let p_j^* be the probability of a random observation on X being in class j, under the assumption that the null hypothesis is true. Then define E_j as

$$E_j = p_j^* N, \qquad j = 1, 2, \ldots, c \tag{1}$$

where E_j represents the expected number of observations in class j when H_0 is true. The test statistic T is given by

$$T = \sum_{j=1}^{c} \frac{(O_j - E_j)^2}{E_j} \qquad (2)$$

An equivalent expression, more convenient for use with a hand calculator, is

$$T = \sum_{j=1}^{c} \frac{O_j^2}{E_j} - N \qquad (3)$$

Null Distribution   The exact distribution of T is too difficult to work with, so the chi-squared distribution with $c - 1$ degrees of freedom is used as an approximation.

Hypotheses

$$H_0: P(X \text{ is in class } j) = p_j^* \text{ for } j = 1, \dots, c$$

$$H_1: P(X \text{ is in class } j) \neq p_j^* \text{ for at least one class}$$

Reject H_0 if T is greater than the $1 - \alpha$ quantile from the chi-squared distribution with $c - 1$ degrees of freedom as given in Table A2. The p-value is approximately the probability of a chi-squared random variable with $c - 1$ degrees of freedom exceeding the observed value of T, also obtained from Table A2.

Computer Assistance   *Minitab, S-Plus,* and *StatXact* contain versions of the chi-squared goodness-of-fit test. ⎯⎯⎯⎯⎯⎯⎯⎯⎯⎯⎯⎯⎯⎯⎯⎯⎯⎯ ◄

Comment

If some of the E_js are small, the asymptotic chi-squared distribution (described in the following material) may not be appropriate, but just how small the E_js may become is not clear. While Cochran (1952) suggests that none of the E_js should be less than 1 and no more than 20% should be smaller than 5, more recent results indicate that this rule can be relaxed. Yarnold (1970) says, "If the number of classes s is 3 or more, and if r denotes the number of expectations less than 5, then the minimum expectation may be as small as $5r/s$." Slakter (1973) feels that the number of classes can exceed the number of observations, which means the average expected value can be less than 1. A more recent study by Koehler and Larntz (1980) finds the chi-squared approximation to be adequate as long as $N \geq 10$, $c \geq 3$, $N^2/c \geq 10$, and all $E_j \geq 0.25$. The user may wish to combine some cells with this discussion in mind if many of the E_js are small.

EXAMPLE I

A certain computer program is supposed to furnish random digits. If the program is accomplishing its purpose, the computer prints out digits (2, 3, 7, 4, etc.) that seem to be observations on independent and identically distributed random variables, where each digit 0, 1, 2, . . . , 8, 9 is equally likely (probability 0.1) to be obtained. One way of testing

H_0: The numbers appear to be random digits

against the alternative

H_1: Some digits are more likely than others

is to count how many times each digit appears. Three hundred digits are generated with the following results.

1578748416	4705188926	6936349612
4653843213	0282868892	3928057043
5101259393	9837006785	3011679938
7122863085	6528271107	2956427027
2671728075	9759178719	9373309535
8363265100	2546793732	2212122529
9453087720	3976759377	9593511031
5605373242	1819898287	3872181027
3494768396	9296177240	8620774591
4659773922	9246724287	8326143939

Each digit is equally likely under the null hypothesis, so the expected number of occurrences is 30 for each of the ten digits. But the digit 2 occurs 41 times while the digit 4 occurs only 19 times. Is this what one could expect from random fluctuation? The complete list of observed counts is as follows:

Digit:	0	1	2	3	4	5	6	7	8	9	Total
Observed Frequency	22	28	41	35	19	25	25	40	30	35	300
Expected Frequency	30	30	30	30	30	30	30	30	30	30	300

The test statistic

$$T = \sum_{i=1}^{10} \frac{O_i^2}{E_i} - N = 317 - 300 = 17 \tag{4}$$

is compared with the 0.95 quantile of a chi-squared random variable with 9 degrees of freedom, which is given in Table A2 as 16.92. Therefore the null hypothesis of randomness is rejected at the 0.05 level of significance in favor of the alternative that the digits are not all equally likely to be generated by this computer program. The p-value is slightly less than 0.05. ∎

Comment

If the probability distribution of X is completely specified except for a number k of parameters, it is first necessary to estimate the parameters and then to proceed with the test as just outlined. The only change is in the distribution of T, which now may be approximated using a chi-squared distribution with $c - 1 - k$ degrees of freedom. That is, 1 degree of freedom is subtracted for each parameter estimated. However, subtraction of degrees of freedom is a privilege accorded only when the parameters are estimated in the proper manner. For example, in a goodness-of-fit test with four classes, H_0 is usually rejected (at $\alpha = 0.05$) if T exceeds 7.815 (see Table A2). However, if one parameter is estimated from the data before the test is applied, the hypothesized distribution has already been modified so that it will fit the data better. [This is true if the estimate is a "good" estimate. A poor estimate may be used deliberately to result in a poor fit, but then the goodness-of-fit test is no longer valid. Chase (1972) discusses the chi-squared test when the parameters are estimated independently of the data.]

The goodness-of-fit test will then be more likely to result in acceptance of H_0; the test becomes conservative and therefore less powerful. We would like to enlarge the critical region so that α again becomes 0.05 and the test regains some (or all) of the power that was lost. If we subtract 1 degree of freedom, using 2 degrees of freedom instead of 3, the critical region is enlarged and H_0 is rejected if T exceeds 5.991 instead of 7.815 as before. The question is, "Are we justified in subtracting 1 degree of freedom, as we did?"

Cramér (1946, p. 424; or see Birnbaum, 1962, p. 258) shows that 1 degree of freedom may be subtracted for each parameter estimated by the *minimum chi-squared* method. The minimum chi-squared method simply involves using the value of the parameter that results in the smallest value of the test statistic, for the given observations. From a practical standpoint this means trying all possible values of the parameter, or all possible combinations if several parameters are unknown, computing a set of E_js and then T for each value of the parameter, and selecting the value of the parameter that results in the smallest T. However, such a procedure is impractical. Therefore Cramér also presents a more usable *modified minimum chi-squared method*. Even that procedure is tedious, and so Cramér and Birnbaum, in their examples, actually use a modification of the modified minimum chi-squared method, which asymptotically still permits subtracting 1 degree of freedom for each parameter estimated. The method eventually used consists of estimating the k unknown parameters by computing the first k sample moments of the grouped data. (Each observation is assumed to be at the midpoint of its class interval, where all intervals containing observations are of finite length.) Then the first k population moments are set equal to the first k sample moments of the grouped data, and the resulting k equations are solved simultaneously for the k unknown parameters. The following example and the subsequent comment section should help to clarify the above procedure.

EXAMPLE 2

Efron and Morris (1975) presented data on the first 18 major league baseball players to have 45 times at bat in 1970. The players' names and the number of hits they got in their 45 times at bat are given as follows.

Clemente	18	Kessinger	13	Scott	10
F. Robinson	17	L. Alvarado	12	Petrocelli	10
F. Howard	16	Santo	11	E. Rodriguez	10
Johnstone	15	Swoboda	11	Campaneris	9
Berry	14	Unser	10	Munson	8
Spencer	14	Williams	10	Alvis	7

We will test the null hypothesis that these data follow a binomial distribution with $n = 45$. But first we need to estimate $p = P(\text{hit})$ for each time at bat.

A good estimate for p is the overall relative frequency of getting a hit based on these data.

$$\hat{p} = \frac{\text{total number of hits}}{\text{total number of at-bats}} = \frac{215}{810} = 0.2654 \tag{5}$$

Using $n = 45$ and $p = 0.2654$ the binomial probabilities are calculated,

$$P(X = i) = \binom{45}{i}(0.2654)^i(0.7346)^{45-i} \qquad i = 0, 1, \dots, 45 \tag{6}$$

The expected cell counts are

$$E_i = 18 \cdot P(X = i) \qquad i = 0, 1, \dots, 45 \tag{7}$$

Cells with expected values less than 0.5 are combined to avoid problems of having a poor approximation by the chi-squared distribution. The resulting cells, after combining, are given as follows.

| | | | | | No. of hits | | | | | | | | |
	≤7	8	9	10	11	12	13	14	15	16	17	≥18	Total
Observed	1	1	1	5	2	1	1	2	1	1	1	1	18
Expected	1.10	1.06	1.57	2.04	2.35	2.40	2.20	1.82	1.36	0.92	0.57	0.61	18

The test statistic is

$$T = \sum_{i=1}^{12} \frac{O_i^2}{E_i} - N = 24.73 - 18 = 6.73 \tag{8}$$

which is compared with the 0.95 quantile from the chi-squared distribution with $12 - 1 - 1 = 10$ degrees of freedom, which is given by Table A2 as 18.31, so the null hypothesis is accepted at $\alpha = 0.05$. In fact, a comparison of the observed value 6.73 with quantiles from Table A2, 10 degrees of freedom, shows the p-value to be much larger than 0.25, so the fit to the binominal distribution is quite good. Note that 1 degree of freedom was subtracted because the parameter p was estimated from the data.   ∎

Comment

The parameter p in Example 2 was estimated using the total number of hits divided by the total number of times at bat for all 18 players. This is a good estimator, but it may not be the one that minimizes the value of T, in accordance with the asymptotic theory that allows 1 degree of freedom to be subtracted because one parameter is estimated. Two comments need to be considered here.

First, the p-value is already much larger than 0.25 in Example 2, and the null hypothesis is easily accepted. There is no need to find the minimum value of T, which will further increase the p-value. The conclusion will remain the same, that the data follow the binomial distribution fairly well. Therefore the extra work required to find the minimum chi-squared statistic is not necessary unless the p-value is small and the decision is in doubt.

Second, the theory that justifies subtraction of 1 degree of freedom for each parameter estimated using the minimum chi-squared method is an asymptotic theory, as the sample size goes to infinity and the expected values in each cell also go to infinity. This by itself is no guarantee that the minimum chi-squared method results in a more accurate approximation for small sample sizes, the kind we encounter in real samples from the real world. Therefore we can be comfortable using the usual estimators for unknown parameters, such as moment estimators, or maximum likelihood estimators, knowing that for the sample being examined the chi-squared approximation may be as good as if the minimum chi-squared method were being used. For more discussion of this topic see Yule and Kendall (1950), Chernoff and Lehmann (1954), and Berkson (1980).

The chi-squared goodness-of-fit test is not limited to discrete random variables, as the previous two examples might suggest. It can also be used to test whether the data come from a specified continuous distribution, where some of the unknown parameters may be estimated from the data as in Example 2. The first step is to "discretize" the continuous random variable by forming intervals, which then become the classes described in the test. The number of observations in each interval O_j is compared with the expected number in each interval

$$E_j = N \cdot P(X \text{ is in interval } j) \tag{9}$$

when the null hypothesis is true.

The following example illustrates the chi-squared goodness-of-fit test to a continuous distribution where two parameters are estimated from the data. Note that the formation of intervals is somewhat arbitrary, and is therefore a weakness in applying the chi-squared goodness-of-fit test to any continuous distribution.

EXAMPLE 3

Fifty two-digit numbers were drawn at random from a telephone book, and the chi-squared test for goodness of fit is used to see if they could have been observations on a normally distributed random variable. The numbers, after being arranged in order from the smallest to the largest, are as follows.

23	23	24	27	29	31	32	33	33	35
36	37	40	42	43	43	44	45	48	48
54	54	56	57	57	58	58	58	58	59
61	61	62	63	64	65	66	68	68	70
73	73	74	75	77	81	87	89	93	97

The null hypothesis is

H_0: These numbers are observations on a normally distributed random variable

The normal distribution has two parameters (Definition 1.5.3), both of which are unspecified by H_0, and must be estimated before the goodness-of-fit test may be applied. For illustration, the procedure is divided into steps.

Step 1   *Divide the observations into intervals of finite length.* We arbitrarily choose the intervals, 20 to 40, 40 to 60, 60 to 80, and 80 to 100, not including the upper limit of each interval.

	Interval				
	20 to 40	40 to 60	60 to 80	80 to 100	Total
Number of Observations	12	18	15	5	50

Step 2   *Estimate μ and σ with the sample mean $\overline{X}$ and sample standard deviation S of the grouped data.* The 12 observations in the interval 20 to 40 are treated as if they all equal the middle point 30. The 18 observations from 40 to 60 are all considered to be 50, and so on. These are the numbers used for computing $\overline{X}$ and S, using the equations of Definition 2.2.3.

$$\overline{X} = \frac{1}{N} \sum_{i=1}^{N} X_i$$

$$= \frac{1}{50} [12(30) + 18(50) + 15(70) + 5(90)]$$

$$= 55.2 \tag{10}$$

$$S = \sqrt{S^2} = \sqrt{\frac{1}{N} \sum_{i=1}^{N} X_i^2 - \overline{X}^2}$$

$$= \left\{ \frac{1}{50} [12(30)^2 + 18(50)^2 + 15(70)^2 + 5(90)^2] - (55.2)^2 \right\}^{1/2}$$

$$= 18.7 \tag{11}$$

Therefore our estimates of μ and σ are 55.2 and 18.7, respectively.

Step 3   *Using the estimated parameters from Step 2, compute the E_js for the groups in Step 1 and for the "tails."*

Class Boundaries b_j	$(b_j - \overline{X})/S = x_p$	$F(x_p)$	Interval	$p_j{}^*$
$b_1 = 20$	-1.88	0.03	<20	0.03
$b_2 = 40$	-0.813	0.21	20 to 40	0.18
$b_3 = 60$	$+0.256$	0.60	40 to 60	0.39
$b_4 = 80$	$+1.33$	0.91	60 to 80	0.31
$b_5 = 100$	$+2.40$	0.99	80 to 100	0.08
			≥ 100	0.01

To find the hypothesized probabilities of being in the various classes, when the hypothesized distribution is the normal distribution with mean 55.2 and standard deviation of 18.7, the class boundaries (column 1 in the table) are considered to be the quantiles of the hypothesized distribution. These quantiles are converted to quantiles of a standard normal random variable (column 2) by Equation 1.5.3 in order to find out which quantile the boundaries represent (column 3). Subtraction of the items in column 3 then yields the probabilities $p_j{}^*$ of being in the various intervals under the hypothesized distribution. The E_js equal 50 $p_j{}^*$, from Equation 1, and are given below.

			Class			
	<20	20–40	40–60	60–80	80–100	≥100
Expected Number E_j	1.5	9.0	19.5	15.5	4	0.5
Observed Number O_j	0	12	18	15	5	0

Because of the small E_js, the first and last cells are combined with the cells adjacent to them.

	Class			
	<40	**40–60**	**60–80**	**≥80**
Expected Number E_i	10.5	19.5	15.5	4.5
Observed Number O_i	12	18	15	5

Step 4 *Compute* T. The test statistic is now computed using Equation 2.

$$T = \frac{(12 - 10.5)^2}{10.5} + \frac{(18 - 19.5)^2}{19.5} + \frac{(15 - 15.5)^2}{15.5} + \frac{(5 - 4.5)^2}{4.5}$$
$$= 0.401 \tag{12}$$

The critical region of size 0.05 corresponds to values of T greater than 3.841, the 0.95 quantile of a chi-squared random variable with $c - 1 - k = 4 - 1 - 2 = 1$ degree of freedom. Therefore H_0 is accepted, with a *p*-value well above 0.25.

Usually a modification called Sheppard's correction is used when the variance is being estimated from grouped data and when the interior intervals are of equal width, say h. Sheppard's correction consists of subtracting $h^2/12$ from S^2 in order to obtain a better estimate of variance. In this example $h = 20$ (the width of each interval), so $(20)^2/12 = 33.33$ could have been subtracted in Step 2 before extracting the square root. The result is $S = 17.8$, a smaller estimate for σ. This smaller estimate of σ results in a larger value of T in this example and, since our objective is to obtain estimates that give the smallest possible value for T, the correction was not used. In most situations we can expect a smaller T when the correction is used.

Another peculiarity of this example is the fact that a smaller value of T (0.279) may be obtained by using $\overline{X} = 55.04$ and $s = 19.0$ as estimates of μ and σ. These estimates are the sample moments obtained from the original observations, before grouping. No matter how they are obtained, the estimates to use are the estimates that result in the smallest value of T. The procedure described in this example can be relied on to provide a value of T not far from its minimum value in most cases. Therefore it is the recommended procedure.   ■

In the preceding example the test statistic just happened to be smaller when μ and σ were estimated using the sample mean and standard deviation based on the original, rather than the grouped, observations. This procedure may be used and is even recommended by Yule and Kendall (1950), but it is usually inferior in results to the other method, which uses the grouped data (Chernoff and Lehmann, 1954).

□ *Theory*   If the null hypothesis completely specifies the hypothesized distribution function, once the classes are defined there is a known probability p_j associated with each class if H_0 is true. The probability of any particular arrangement $O_1, O_2, \ldots, O_c$ of the N sample values is then given by the multinomial probability distribution,

$$P(O_1, O_2, \cdots, O_c | N) = \frac{N!}{O_1! O_2! \cdots O_c!} p_1^{O_1} p_2^{O_2} \cdots p_c^{O_c} \tag{13}$$

which is an immediate extension of the binomial distribution to c classes instead of only two classes. With the probability function given by Equation 13 the probability distribution of T may be determined, although the calculation becomes laborious for large N and c. There seems to be no theory developed to find the exact distribution of T when several parameters are first estimated from the sample. Therefore the large sample approximation is both practical and necessary in order to apply this goodness-of-fit test. The theory behind the large sample chi-squared approximation may be found in Cramér (1946). □

Usage of the chi-squared goodness-of-fit test with small expected frequencies is discussed by Slakter (1966, 1968), Dahiya (1971), Dahiya and Gurland (1972, 1973), Pahl (1969), and Koehler and Larntz (1980). Exact tables when all $E_j s = 1$ appear in Zahn and Roberts (1971). If the sample is grouped according to time of observation instead of numerical value, the usage of the test may require some modification (Putter, 1964). Chernoff (1967) discusses the adjustment of the degrees of freedom when parameters are estimated. The test is discussed further by Efron and Morris (1975), Molinari (1977), and Hewett and Tsutakawa (1972) and compared with other goodness-of-fit tests by Holst (1972), Cohen and Sackrowitz (1975), and Horn (1977).

EXERCISES

1. Test the following data to see if they could have come from a population whose values are uniformly distributed between 0.0000 and 0.9999. (Continued on next page.)

0.4755	0.5233	0.5440	0.5456	0.9056
0.2186	0.7500	0.2484	0.5101	0.8283
0.5112	0.5484	0.5758	0.3607	0.4352
0.3826	0.6454	0.9145	0.3943	0.5381
0.5758	0.8620	0.6687	0.3979	0.5646
0.4274	0.5482	0.3007	0.4438	0.4102
0.4295	0.5926	0.6521	0.6328	0.5689

0.7297	0.3768	0.8403	0.2925	0.2113
0.8757	0.4403	0.4993	0.3900	0.5166
0.8230	0.8522	0.8312	0.7979	0.4632
0.8432	0.4004	0.4295	0.9763	0.5590
0.4396	0.2595	0.3003	0.3003	0.5836
0.5337	0.8008	0.4887	0.2172	0.9329
0.5498	0.3686	0.4067	0.5274	0.4579
0.9096	0.4995	0.2172	0.6793	

2. A die was cast 600 times with the following results.

Occurrence	1	2	3	4	5	6
Frequency	87	96	108	89	122	98

Is the die balanced?

3. Use the number of hits for each player in Example 2 to test the null hypothesis that all 18 players have the same probability of getting a hit. Note that one assumption of the binomial distribution is that the probability is the same for all trials, and this is one way of testing that assumption in Example 2.

4. Without the aid of books or tables, attempt to write 300 random digits. Then apply the test of randomness described in Example 2 to see if you are a good random digit generator.

5. The number of babies born in Methodist Hospital last year was as follows. In the Winter there were 36 babies born, in the Spring 45, in the Summer 42, and in the Fall 55. Test the hypothesis that the number of births is uniformly distributed over the four seasons of the year.

6. Twenty-six observations were obtained, and the question arose as to whether they followed a normal distribution with mean 12 and standard deviation 3. None of the observations were below the lower quartile of this distribution, and 12 were above the upper quartile. Six were below the median, and 8 were between the median and the upper quartile. Do these observations appear to have come from the distribution described?

4.6 COCHRAN'S TEST FOR RELATED OBSERVATIONS

Sometimes the use of a treatment, or condition, results in one of two possible outcomes. For example, the response to a salesperson's technique may be classified as "sale" or "no sale," or a certain treatment may result in "success" or "failure." Of course, if several treatments, c in number, are each applied in several different independent trials, the results may be given in the form of a $2 \times c$ contingency table, where one row represents the number of successes and the other row represents the number of failures, and the null hypothesis of no treatment differences may be tested using a chi-squared contingency table test, as described in Section 4.2. However, it is often possible to detect more subtle differences between treatments, that is, increase the power of the test, by applying all c treatments independently to the same blocks, such as by trying all c sales techniques on each of several persons in an experimental situation and then recording for each person the results of each technique. Thus each block, or person, acts as its own control, and the treatments are more effectively compared with each other. Such an experimental technique is called "blocking," and the experimental design is called a "randomized complete block design." If the treatment result may be classified into one of two categories, the following test, proposed by Cochran (1950), may be an appropriate method of analysis.

▶ **The Cochran Test** _____

Data Each of c treatments is applied independently to each of r blocks, or subjects, and the result of each treatment application is recorded as either 1 or 0, to represent "success" or "failure," or any other dichotomization of the possible treatment results. The results are then given in the form of a table with r rows representing the blocks and c columns representing the c treatments, with entries that are either zeros or ones. Let R_i represent the row totals, $i = 1, 2, \ldots, r$, and let C_j represent the column totals, $j = 1, 2, \ldots, c$. Then the data appear as follows, where the X_{ij} are either 0 or 1, and N represents the total number of ones in the table.

	Treatments				
Blocks	1	2		c	Row Totals
1	X_{11}	X_{12}	$\ldots$	X_{1c}	R_1
2	X_{21}	X_{22}	$\ldots$	X_{2c}	R_2
$\ldots$	$\ldots$	$\ldots$	$\ldots$	$\ldots$	$\ldots$
r	X_{r1}	X_{r2}	$\ldots$	X_{rc}	R_r
Column Totals	C_1	C_2		C_c	N = Grand Total

Assumptions

1. The blocks were randomly selected from the population of all possible blocks.
2. The outcomes of the treatments may be dichotomized in a manner common to all treatments within each block, so the outcomes are listed as either "0" or "1."

Test Statistic The test statistic T may be written as

$$T = c(c-1) \frac{\sum_{j=1}^{c} \left(C_j - \frac{N}{c} \right)^2}{\sum_{i=1}^{r} R_i \left(c - R_i \right)} \tag{1}$$

The following form is more suitable for use with a hand calculator.

$$T = \frac{c(c-1)\sum_{j=1}^{c} C_j^2 - (c-1)N^2}{cN - \sum_{i=1}^{r} R_i^2} \tag{2}$$

Null Distribution   The exact distribution of T is difficult to tabulate, so the large sample approximation is used instead. The number of blocks r is assumed to be large. Then the approximate null distribution is the chi-squared distribution with $c - 1$ degrees of freedom.

Hypotheses

H_0: The treatments are equally effective

H_1: There is a difference in effectiveness among treatments

In more mathematical terms, let $p_j = P$ (scoring a "1" in column j). Then equal effectiveness among treatments implies

$H_0: p_1 = p_2 = \cdots = p_c$ within each block

where a difference in effectiveness among treatment implies

$H_1: p_i \neq p_j$ for some treatments i and j

Reject H_0 at the level α if T is greater than the $1 - \alpha$ quantile of the chi-squared distribution with $c - 1$ degrees of freedom, obtained from Table A2. The p-value is approximately the probability of a chi-squared random variable with $c - 1$ degrees of freedom exceeding the observed value of T, also obtained from Table A2.

Multiple Comparisons   If H_0 is rejected, pairwise comparisons may be made between treatments using the McNemar test, which is the two-tailed sign test, as described in Section 3.5. ⎯⎯⎯⎯⎯⎯⎯⎯⎯⎯⎯⎯⎯⎯⎯⎯⎯⎯⎯⎯◄

Computer Assistance   The Cochran test is found in *StatXact*.

EXAMPLE I

Each of three basketball enthusiasts had devised his own system for predicting the outcomes of collegiate basketball games. Twelve games were selected at random, and each sportsman presented a prediction of the outcome of each game. After the games were played, the results were tabulated, using 1 for successful prediction and 0 for unsuccessful prediction.

		Sportsman		
Game	1	2	3	Totals
1	1	1	1	3
2	1	1	1	3
3	0	1	0	1
4	1	1	0	2
5	0	0	0	0
6	1	1	1	3
7	1	1	1	3
8	1	1	0	2
9	0	0	1	1
10	0	1	0	1
11	1	1	1	3
12	1	1	1	3
Totals	8	10	7	25

The assumptions of the Cochran test were met, because the games (blocks) were selected at random from among all college basketball games being played. Therefore the Cochran test was used to test the null hypothesis

H_0: Each sportsman is equally effective in his ability to predict the outcomes of the basketball games

The test statistic is computed using Equation 1.

$$T = c(c-1) \frac{\sum_{j=1}^{c} \left(C_j - \frac{N}{c} \right)^2}{\sum_{i=1}^{r} R_i (c - R_i)}$$

$$= \frac{(3)(2)[(-\tfrac{1}{3})^2 + (\tfrac{5}{3})^2 + (-\tfrac{4}{3})^2]}{2 + 2 + 2 + 2 + 2}$$

$$= 2.8 \tag{3}$$

The critical region of approximate size 0.05 corresponds to values of T greater than 5.99, the 0.95 quantile of a chi-squared random variable with 2 degrees of freedom obtained from Table A2. Therefore, in this example H_0 is accepted and we conclude that no significant differences among prediction methods were detected. H_0 could have been rejected using an α of about 0.25, so the p-value equals 0.25. ∎

□ *Theory*   Each X_{ij}, as defined, is a random variable that follows the point binomial distribution (the binomial distribution with $n = 1$) with parameter p, which is the same within each row under the null hypothesis, but can be different from block to block. The column total, C_j, defined by

$$C_j = \sum_{i=1}^{r} X_{ij} \tag{4}$$

is therefore a random variable also. Because the random variable C_j is the sum of r independent random variables, the central limit theorem applies, and for large r the distribution of C_j is approximately normal. This implies that the distribution function of

$$\frac{C_j - E(C_j)}{\sqrt{\text{Var}(C_j)}}$$

may be approximated by the standard normal distribution function, and, according to Theorem 1.5.3, the sum

$$\sum_{j=1}^{c} \left[\frac{C_j - E(C_j)}{\sqrt{\text{Var}(C_j)}} \right]^2 = \sum_{j=1}^{c} \frac{[C_j - E(C_j)]^2}{\text{Var}(C_j)} \tag{5}$$

may be approximated by the chi-squared distribution with c degrees of freedom. However, the parameters $E(C_j)$ and $\text{Var}(C_j)$ are unknown, and the following method of estimating those parameters is shown in Blomqvist (1951) to result in the loss of 1 degree of freedom.

The mean of C_j may be estimated by the sample mean

$$\frac{1}{c} \sum_{j=1}^{c} C_j = \frac{N}{c} = \text{estimate of } E(C_j) \tag{6}$$

The same estimate is used for the mean of every C_j. The variance of C_j equals the sum of the variances of the X_{ij} in the jth column,

$$\text{Var}(C_j) = \sum_{i=1}^{r} \text{Var}(X_{ij}) \tag{7}$$

because of the block to block independence of the X_{ij} (see also Theorem 1.4.3). The variance of X_{ij} is given by Equation 1.4.8 as

$$\text{Var}(X_{ij}) = p(1 - p) \tag{8}$$

Under H_0 the probability of a "success" is the same for all columns within a row, and therefore it is natural to estimate p in each row by the average number of successes in row i, R_i/c. That is,

$$\text{estimate of } p \text{ in row } i = R_i/c \tag{9}$$

and, from Equation 8,

$$\text{estimate of Var}(X_{ij}) = \frac{R_i}{c}\left(1 - \frac{R_i}{c}\right) \tag{10}$$

However, such an estimate tends to be too small and is improved by multiplication by $c/(c-1)$. Then Var (X_{ij}) is estimated by

$$\text{estimate of Var } (X_{ij}) = \frac{R_i (c - R_i)}{c(c-1)} \tag{11}$$

and Var (C_j) is estimated, from Equation 7, as

$$\text{estimate of Var } (C_j) = \frac{1}{c(c-1)} \sum_{i=1}^{r} R_i(c - R_i) \tag{12}$$

which does not depend on j and so is used for all C_js. Substitution of the estimates for $E(C_j)$ (Equation 6) and Var (C_j) (Equation 12) into Equation 5 gives

$$T = \sum_{i=1}^{c} \frac{\left(C_j - \dfrac{N}{c}\right)^2}{\displaystyle\sum_{i=1}^{r} \dfrac{R_i(c - R_i)}{c(c-1)}}$$

$$= c(c-1) \frac{\displaystyle\sum_{j=1}^{c} \left(C_j - \dfrac{N}{c}\right)^2}{\displaystyle\sum_{i=1}^{r} R_i(c - R_i)} \tag{13}$$

which provides some insight into the use of the chi-squared distribution with $(c-1)$ degrees of freedom for the test statistic T. $\square$

Cochran's test is considered by Berger and Gold (1973) and Bhapkar and Somes (1977). Patel (1975) discusses the exact distribution of the test statistic. For another model in which Cochran's test is valid, see Fleiss (1965). The asymptotic approximation is considered by Tate and Brown (1970).

Comment

If only two treatments are being considered, such as "before" and "after" observations on the same block, with r blocks, the experimental situation is the same as that analyzed by the McNemar test for significance of changes. That is, in each situation the null hypothesis is that the proportion of the population in class 1 is the same using treatment 1 (before) as it is using treatment 2 (after). Thus it appears that if $c = 2$, the experimenter has a choice of using the Cochran test or the McNemar test. In fact, there is no choice because if $c = 2$, the Cochran test is identical with the McNemar test (Section 3.5), as shown in the following.

For $c = 2$ the Cochran test statistic reduces to

$$T = 2 \frac{\left(C_1 - \dfrac{C_1 + C_2}{2}\right)^2 + \left(C_2 - \dfrac{C_1 + C_2}{2}\right)^2}{\sum\limits_{i=1}^{r} R_i(2 - R_i)}$$

$$= 2 \frac{\left(\dfrac{C_1}{2} - \dfrac{C_2}{2}\right)^2 + \left(\dfrac{C_2}{2} - \dfrac{C_1}{2}\right)^2}{\sum\limits_{i=1}^{r} R_i(2 - R_i)}$$

$$= \frac{(C_1 - C_2)^2}{\sum\limits_{i=1}^{r} R_i(2 - R_i)} \tag{14}$$

If a block has ones in both columns, then $R_i = 2$ and $R_i (2 - R_i) = 0$. Similarly, if both columns have zeros, then $R_i = 0$ and $R_i (2 - R_i) = 0$. If there is a change from zero to one or one to zero in a given row, then $R_i = 1$ and $R_i (2 - R_i) = 1$. Thus the denominator of Equation 14 is merely the total number of rows that go from 0 to 1 or 1 to 0, which is $b + c$ in the notation of the McNemar test. Also, C_1 is the total number of ones in column one, or "before," which is $c + d$ in the notation of the McNemar test. Similarly, $C_2 = b + d$. Therefore we have

$$C_1 - C_2 = c + d - b - d$$

and Equation 14 becomes

$$T = \frac{(c - b)^2}{b + c} = \frac{(b - c)^2}{b + c}$$

which is identical with the form of McNemar's test statistic given in Equation 3.5.1. Both the McNemar test statistic and the Cochran test statistic with $c = 2$ are approximated by a chi-squared random variable with 1 degree of freedom.

EXERCISES

1. The relative effectiveness of two different sales techniques was tested on 12 volunteer housewives. Each housewife was exposed to each sales technique and asked to buy a certain product, the same product in all cases. At the end of each exposure, each housewife rated the technique with a 1 if she felt she would have agreed to buy the product and a 0 if she probably would not have bought the product.

	Housewife											
	1	**2**	**3**	**4**	**5**	**6**	**7**	**8**	**9**	**10**	**11**	**12**
Technique 1	1	1	1	1	1	0	0	0	1	1	0	1
Technique 2	0	1	1	0	0	0	0	0	1	0	0	1

(a) Use Cochran's test.

(b) Rearrange the data and use McNemar's test in the large sample form suggested by Equation 3.5.1.

(c) Ignore the blocking effect in this experiment and treat the data as if 24 different housewives were used. Analyze the data using the test for differences in probabilities given in Section 4.1. Compare with Cochran's test and discuss.

2. On a ship, 12 groups with three sailors in each group were chosen in a random manner, where the sailors in each group did similar work and were in the same division aboard the ship. In a random manner the sailors in each group were given treatment 1, 2, or 3, no two sailors from the same group receiving the same treatment. Treatment 1 was a "flu shot," treatment 2 was a "flu pill," and treatment 3 was a promise of 2 weeks extra leave if they did not catch the flu. As each sailor reported to sick bay with the flu, a report to the experimenter was made. At the end of the winter, these were the results.

Group	Sailors with the Flu (by Treatment Number)
1	2
2	1, 2
3	1, 2, 3
4	2, 3
5	2
6	None
7	1, 2
8	1, 2
9	1
10	2
11	1, 2, 3
12	2

Do these results indicate a significant difference between the various treatments?

3. In an attempt to compare the relative power of three statistical tests, 100 sets of artificial data were generated using a computer. On each set of data the three statistical tests were used, with $\alpha = 0.05$, and the decision to accept or reject H_0 was recorded. The results were as follows. (Continued on next page.)

Test 1	Test 2	Test 3	Number of Sets of Data
Accept	Accept	Accept	26
Accept	Accept	Reject	6
Accept	Reject	Accept	12
Reject	Accept	Accept	4

Reject	Reject	Accept	18
Reject	Accept	Reject	5
Accept	Reject	Reject	7
Reject	Reject	Reject	22

Is there a difference in the power of the three tests when applied to populations from which the simulated data were obtained?

PROBLEMS

1. Suppose that instead of just one observation in every treatment-block combination, we now have m independent observations in each cell. Let C_j, R_i, and N represent the treatment sum, row sum, and overall sum as before. Then justify comparing statistic

$$T' = mc(c-1)\frac{\sum_{j=1}^{c}\left(C_j - \frac{N}{c}\right)^2}{\sum_{i=1}^{r} R_i (mc - R_i)}$$

with the chi-squared distribution, $c - 1$ degrees of freedom, in the same way the distribution of T is justified in this section.

2. The usual parametric test for the design in this section assumes that the observations are made on normal random variables instead of point binomial random variables and uses the "F statistic." If the F statistic is computed on the zeros and ones, it simplifies to

$$F = (r-1)\frac{c\sum_{j=1}^{c} C_j^2 - N^2}{rcN - r\sum_{i=1}^{r} R_i^2 - c\sum_{j=1}^{c} C_j^2 + N^2}$$

Show that this F is the following function of T

$$F = \frac{(r-1)T}{r(c-1) - T}$$

and that rejecting H_0 for large T is equivalent to rejecting H_0 for large F.

4.7 SOME COMMENTS ON ALTERNATIVE METHODS OF ANALYSIS

The Likelihood Ratio Statistic

The methods described in this chapter are not the only methods available for analyzing contingency tables. They may be summarized by saying that they use the test statistic

$$T_1 = \sum_{\substack{all \\ cells}} \frac{(O_i - E_i)^2}{E_i} \tag{1}$$

where O_i is the observed count in cell i and E_i is the expected count in cell i. The statistic T_1 was introduced by Pearson (1900, 1922) and is often called "Pearson's chi-squared statistic" as opposed to other chi-squared statistics, one of which we will now introduce.

A different method of analysis, called the likelihood ratio test and mentioned in Problem 4.2.3, employs the statistic

$$T_2 = 2 \sum_{\substack{all \\ cells}} O_i \ln \left(\frac{O_i}{E_i} \right) \tag{2}$$

instead of T_1, where ln refers to natural logarithms, easily obtained on most calculators. The statistic T_2 is compared with the chi-squared distribution, just as for T_1 with the same number of degrees of freedom as used for T_1. Although the two statistics T_1 and T_2 have the same asymptotic distribution, their values for a particular contingency table will differ, possibly by a large amount. The choice of whether to use T_1 or T_2 depends largely on the user's preference.

The statistic T_2 is called the "likelihood ratio chi-squared statistic" because its derivation stems from likelihood ratio theory in statistics. It is attributed to Wilks (1935, 1938), and is often preferred because of its origins in likelihood ratio theory. However, a serious disadvantage in using T_2 is that the chi-squared approximation is usually poor if $N/rc < 5$, while the chi-squared approximation for T_1 holds up for much smaller values of N. Agresti (1990) states that the chi-squared approximation can be decent for T_1 when N/rc is as small as 1, if r or c is large, and "if the contingency table does not contain both very small and moderately large expected frequencies." Agresti's findings do not disagree with our recommendations that all expected values be at least 0.5, and most larger than 1.0.

Loglinear Models

Another popular method of analysis is the "loglinear model." This method works well in analyzing contingency tables with three or more dimensions (three-way or more) if the proper computer programs are available. Analysis by hand computation is not recommended. The same statistics T_1 and T_2 just given are used with loglinear models; the difference is in the method used for obtaining the E_is. Usually iterative methods are used, which require a computer. Computer programs such as *SAS*, *StatMost*, and *SYSTAT* will conduct a complete loglinear analysis of multidimensional contingency tables.

The name loglinear model arises for the following reason. In a two-way contingency table the null hypothesis of independence may be expressed as

$$H_0: p_{ij} = p_{i+} \cdot p_{+j}, \quad \text{all } i \text{ and } j$$

where p_{ij} is the probability of an observation being classified in cell (i, j) and where p_{i+} and p_{+j} are the row and column marginal probabilities. Taking the logarithm of p_{ij} gives

$$H_0: \log p_{ij} = \log p_{i+} + \log p_{+j}$$

which is a linear equation. The test then amounts to a test of whether or not the model for the logarithms of the cell probabilities is a linear function of the logarithms of the marginal probabilities. A complete presentation of loglinear models and their analyses is found in Bishop, Fienberg, and Holland (1975). An elementary treatment of the subject by Ku and Kullback (1974) or Lee (1978) is recommended for beginners. See also a book by Fienberg (1977).

The interested reader may pursue the topic of loglinear models by reading articles by Bishop (1969, 1971), Fienberg (1970b, 1972), Fienberg and Larntz (1976), Chen and Fienberg (1976), Grizzle, Starmer, and Koch (1969), Koch and Reinfurt (1971), Grizzle and Williams (1972), Koch, Imrey, and Reinfurt (1972), Wagner (1970), Odoroff (1970), Goodman (1971), Gart (1972), Haberman (1973), and Read (1977). See also the book by Agresti (1990).

4.8 REVIEW PROBLEMS FOR CHAPTERS 3 AND 4

1. In a Danish research project to see if alcoholism is hereditary, five psychiatrists studied men who had been separated from their biological parents since early infancy. Fifty-five of the men had one parent who had been diagnosed as an alcoholic, and 10 of these 55 were found to be alcoholic. These were compared with 78 men whose biological parents had no history of alcoholism, and 4 of these 78 were alcoholic. The study found that "significantly more" of the first group had a history of drinking problems. What would your statistical analysis look like? (Associated Press, February 21, 1973.)

2. Prior to an election, a random sample of 200 voters were asked which candidate they preferred. Candidate A was preferred by 85 voters, candidate B by 111 voters, and 4 voters were undecided. How would you predict the election results? (Discuss the items that need discussion.)

3. Several nonfreshman students at a community college were asked some questions, including how they felt about legalizing marijuana, with the following (fictitious) results.

Student Number	Sex	Commuting Distance from Home to College	Political Party Preference	Marijuana Question[a]	Freshman GPA
1	M	32	N	1	2.66
2	M	10	D	1	3.18
3	M	28.5	R	1	2.15
4	M	3.5	R	2	1.61
5	M	4	D	4	1.54
6	F	7	N	5	2.12
7	M	3.5	R	3	1.35
8	M	10	N	4	2.26
9	F	6	R	4	2.70
10	M	32	D	4	2.84
11	M	22.5	D	4	2.60
12	M	7	D	1	1.13
13	M	6.5	N	4	0.81
14	M	5	D	1	3.11
15	M	35	R	1	2.47
16	M	5.5	D	5	3.15
17	M	26.5	D	4	2.33
18	F	24	N	5	2.46
19	M	32	D	5	3.59
20	F	5	R	1	2.00
21	M	5.5	R	1	2.90
22	M	11.5	N	5	3.26
23	M	9.5	R	4	2.71
24	M	25.5	O	3	2.22
25	M	15	R	1	3.00
26	F	9	N	4	2.06
27	M	15	D	1	1.75
28	M	24	R	3	2.42

[a] 1 = strongly disagree, . . . , 5 = strongly agree.

(a) Test the hypothesis that political party preference is independent of the attitude toward legalizing marijuana.

(b) Does the GPA appear to be related to the commuting distance?

(c) Estimate the median commuting distance.

(d) Estimate the percent of female students.

(e) Does political party preference appear to be independent of whether students are male or female?

(f) Test the hypothesis that male and female students have the same freshman GPA.

4. Ten students were given a special course, and for each student a precourse text score X and a postcourse test score Y were recorded as follows:

Student Number

	1	2	3	4	5	6	7	8	9	10
X	92	80	74	85	71	68	81	82	80	81
Y	94	86	72	91	70	77	89	91	86	94

(a) Do the postcourse scores seem to be significantly better than the precourse scores?

(b) Let p = probability of a student's postcourse score being better than the precourse score. Find a confidence interval for p.

(c) Find a confidence interval for the median precourse score.

(d) Is the upper quartile of Y significantly greater than 75?

5. Consider the following experiment in pathology to determine the effectiveness of trying to detect whether ethionine was in an animal's diet by measuring, in an autopsy, the amount of iron absorbed by the liver. Thirty-four animals are randomly placed into one of two groups; 17 received ethionine in their diets and 17 did not. The animals are paired (1 from each group) and fed the same amounts within each pair. After a period of time, each liver was extracted and treated with radioactive iron in a solution that is either warm (37°C) or cool (25°C). The data consist of the amount of iron absorbed by the various livers.

	Warm			**Cool**	
Pair	Ethionine	None	Pair	Ethionine	None
1	2.59	1.40	9	6.77	4.71
2	1.54	1.51	10	4.97	1.60
3	3.68	2.49	11	1.46	0.67
4	1.96	1.74	12	0.96	0.71
5	2.94	1.59	13	5.59	5.21
6	1.61	1.36	14	9.56	5.12
7	1.23	3.00	15	1.08	0.95
8	6.96	4.81	16	1.58	1.56
			17	8.09	1.68

(a) Do livers from ethionine-fed animals seem to absorb more iron in warm solutions than livers from the other group? How about in cool solutions?

(b) Does the cool solution (rather than the warm solution) significantly enhance the absorption of iron in the livers of the ethionine-fed animals? How about the animals in the other group?

(c) Find a two-sided tolerance limit for the amount of iron absorbed by the liver of an ethionine-fed animal when treated with a cool solution of radioactive iron.

(d) Does there seem to be a correlation between the amounts of iron absorbed in the livers of the two animals in the same pair?

6. An experiment consists of sampling the air in 20 Eskimo homes in Bethel, Alaska, a native village on the Kuskokwima River in southwestern Alaska. Ten of the homes were new homes built as part of a housing development project; the other 10 were standard houses in Bethel. The objective is to compare the old houses with the new houses to see if there is a difference in the number of bacterial colonies per cubic foot of air. The measurements in the houses, as they were estimated from streptocel plates, were as follows.

Old House Number	Bacterial Colonies per Cubic Foot	New House Number	Bacterial Colonies per Cubic Foot
1	37.0	1N	1.0
2	2.6	2N	5.3
3	48.6	3N	3.4
4	47.8	4N	2.3
5	99.3	5N	5.1
6	1.4	6N	38.7
7	2.3	7N	5.0
8	3.1	8N	50.6
9	3.0	9N	1.6
10	0.3	10N	22.7

(a) Analyze the data.

(b) Find a confidence interval for the median bacterial count in the old houses.

7. Dr. G. Noether has suggested the following test for trend. Group the sequence of observations into nonoverlapping groups of three adjacent observations. Let T equal the number of monotonic groups (either increasing or decreasing). For example:

$$\underbrace{42, 44, 63,}_{\text{Increasing}} \ \ 61, 44, 52, \ \ \underbrace{73, 72, 46,}_{\text{Decreasing}} \ \ 48, 42, 53$$

For these numbers $T = 2$.

(a) What is the distribution of T under the null hypothesis of independent and identically distributed random variables?

(b) How does one find the critical region?

8. A random sample consisting of 12 computers owned by the College of Business Administration showed that 8 were IBM computers. A random sample consisting of 36 computers owned by the School of Medicine showed that 30 were IBM computers.

(a) Is this difference significant?

(b) Find a confidence interval for the overall proportion of computers owned by the School of Medicine that are IBM computers.

9. The intramural department would like to add javelin throwing to its curriculum and must decide how many yard markers are needed in the practice field to mark the length of throw. They decide to select randomly several students to throw the javelin and mark the field only between the shortest throw and the longest from that group of students. How

many students should they select so they can be 90% sure that at least 95% of the students will be throwing their javelins within the marked boundaries?

10. A broker noted the number of municipal bonds he sold each month for a 24-month period

	January	February	March	April	May	June
1997	12	16	14	18	18	14
1998	19	22	20	17	18	20

	July	August	September	October	November	December
1997	10	21	12	18	17	17
1998	20	16	16	21	24	25

Does this record indicate an increasing trend?

11. To see if a chimpanzee can learn to recognize letters, five different letters are placed randomly on five buttons. When the light goes on and the chimpanzee presses the letter E, he gets a banana, which he likes. Each day five trials are run, and the letters are changed randomly after each trial. The experiment continues for six days, with the following results.

	Number of Presses until E				
Trial Number	1	2	3	4	5
Monday	6	4	4	2	3
Tuesday	7	8	6	1	3
Wednesday	4	2	1	2	3
Thursday	1	4	3	2	2
Friday	5	2	3	1	2
Saturday	4	2	1	2	1

(a) Does the chimpanzee seem to improve during the course of the five trials within each day?

(b) Does the chimpanzee seem to be improving through the week?

(c) If the chimpanzee is pressing the buttons randomly, the number of presses should follow the geometric distribution given by $P(X = k) = (0.2)(0.8)^{k-1}$, for $k = 1, 2, 3, \ldots$. Test the hypothesis that the chimpanzee is pressing the buttons randomly.

12. Four different systems for finding flaws in a manufactured item are being tested simultaneously on an assembly line. Each item is inspected by all four systems for flaws. There are no type I errors; that is, no system says there is a flaw if there really is no flaw. There are only type II errors, where flaws may go undetected. One set of data looks like this.

	System 1	System 2	System 3	System 4
Item 1	Flaw		Flaw	Flaw
Item 2			Flaw	
Item 3				
Item 4	Flaw	Flaw	Flaw	
Item 5	Flaw			Flaw
Item 6	Flaw			
Item 7				
Item 8	Flaw		Flaw	

Is there a difference in the abilities of the four systems to detect flaws?

13. A bank obtains a random sample of customer deposits at automatic teller machines, and wishes to find a 90% confidence interval for the upper quartile. The deposits are as follows.

748	320	45	1,237	5,883	170
65	30	83	186	598	8,500
1,500	4,857	300	100	395	2,450
349	50	25	637	600	260
67	200	45	400	57	580

Find the desired confidence limits.

14. For the random sample in Problem 13, test the hypothesis that the probability of a deposit being less than $100 is equal to the probability of a deposit exceeding $1000, against the one-sided alternative that it is greater.

15. A random sample of 50 teenage boys with fathers shows that 17 boys smoke, and 27 have fathers who smoke. Twelve of the boys who smoke have fathers who smoke. Is there a significant (positive) correlation between the smoking habits of teenage boys and their fathers? What type of contingency table is this?

16. Pizza Hut on 50th Street has recently hired 12 boys and 8 girls. They needed 6 drivers and 14 counter-persons. They assigned 6 boys to the 6 driver positions. Test the null hypothesis that job assignments are made independently of sex, against the one-sided alternative that boys tend to be assigned to the driver positions. Find the exact p-value.

17. In the gambling game of roulette, the probability of red is $18/38$, the probability of black is also $18/38$, and the probability of green is $2/38$. Five hundred plays resulted in 35 greens, 241 reds, and the rest blacks. Are the observations consistent with the theoretical probabilities?

18. Suppose that the Environmental Protection Agency requires that at least 95% of the automobiles your company manufacturers should be getting more than 20 miles per gallon. You test 60 automobiles, and all of them are getting more than 20 miles per gallon. Are you 95% confident that at least 95% of the cars your company maufactures are getting more than 20 miles per gallon?

(a) Use a hypothesis test at $\alpha = 0.05$.

(b) Find the one-sided 95% confidence interval that answers the same question. (You can find a two-sided 90% confidence interval and disregard one end of it, if you don't know how to find a one-sided confidence interval.)

19. In 1997 Joe's restaurant customers used MasterCard 14 times, Visa 10 times, Discover 4 times, and American Express once. In 1998 they used MasterCard 22 times, Visa 23 times, Discover 10 times, and no one used American Express. Is there a significant shift in the pattern of credit card usage from 1997 to 1998? What is the name of the hypothesis test you are using?

20. One hundred forty freshman students were randomly selected from the incoming freshman class at a large university. Eighty of these freshmen arrived with private automobiles, and the rest did not bring a car. Forty of these freshmen failed to make their grades their freshman year and the rest made their grades. The phi coefficient for these data was 0.32. Does there appear to be a significant relationship between bringing their car and making their grades for these freshmen students? Use a two-sided test.

21. Twenty pieces of mail were used in a test to see if putting the zip code on the address made the mail likely to get to its destination earlier than if no zip code were included. Two similar pieces of mail were mailed at the same time to the same address, but the two pieces of mail differed in that one had the zip code on the address and the other had no zip code. Ten different destinations all over the United States were used. These are the results.

	Zip Code	No Zip Code
Atlanta	3 days	4 days
Baltimore	3 days	4 days
Chicago	4 days	4 days
Detroit	4 days	5 days
Elgin	3 days	5 days
Philadelphia	5 days	4 days
Gary	3 days	3 days
Hartford	5 days	7 days
Indianapolis	4 days	5 days
Los Angeles	3 days	4 days

(a) Use a hypothesis test to try to show that zip-coded mail tends to arrive earlier than non-zip-coded mail.

(b) Find a 95% confidence interval for the probability that a piece of mail will arrive sooner by including the zip code.

(c) Test the hypothesis that the upper quartile for the time it takes a letter to reach its destination with the zip code in the address is 3 days, against the one-sided alternative that it is longer than 3 days.

22. Eight domestic doctoral BA students and eight foreign doctoral BA students were selected at random from their respective populations to see if the foreign doctoral BA students have higher analytical scores on the GMAT than their domestic counterparts. The scores (percentiles) are as follows:

Foreign:	79,	86,	93,	96,	97,	97,	99,	99
Domestic:	76,	80,	87,	88,	89,	92,	94,	96

Use a procedure from Chapter 3 or 4 to analyze these data. Give an approximate *p*-value using the usual techniques. Also find the exact *p*-value, and compare it with the approximate value.

23. Forty students took a standardized exam and obtained the following scores. Assume this group of students behaves as a random sample would.

100	92	90	83	75	69	64	47
100	92	89	81	73	68	60	44
97	91	88	80	71	65	58	40
92	91	85	78	70	65	56	38
92	90	83	75	70	64	49	36

(a) In the past the upper quartile has been 85, but the instructor suspects that this quartile is now higher than 85. Test the appropriate hypothesis.

(b) Find a 95% confidence interval for the upper quartile.

(c) Test the hypothesis that the probability of scoring 90 or above equals the probability of scoring 40 or below, against the one-sided alternative that it is greater than the probability of scoring 40 or below.

24. An experimenter in psychology is trying to prove that husbands are more likely than wives to prefer the company of their mothers-in-law over their own mothers. The data from a random sample of couples is given as follows. Use the appropriate hypothesis test to analyze the data.

Couple	Husband's Preference	Wife's Preference
Adams	mother-in-law	mother-in-law
Baker	mother-in-law	mother
Chase	mother-in-law	mother
Dodge	mother	mother
Evans	mother-in-law	mother
Forrester	mother-in-law	mother-in-law
Graves	mother	mother
Holland	mother	mother
Islip	mother-in-law	mother-in-law
Jacobs	mother-in-law	mother
Kraft	mother-in-law	mother
Lewis	mother-in-law	mother
Morris	mother	mother
Noonan	mother-in-law	mother
O'Neil	mother-in-law	mother
Procter	mother	mother

Quincy	mother	mother
Reed	mother-in-law	mother-in-law
Smith	mother	mother-in-law
Tracy	mother	mother
Unseld	mother-in-law	mother
Victor	mother-in-law	mother
Williams	mother	mother
Young	mother-in-law	mother
Zyskind	mother-in-law	mother

25. The manager of a Quik-Stop Convenience Store believes that customers from out of state (those with out-of-state license plates on their cars) are more likely to use credit cards for their purchases. He keeps a tally on all of his customers one afternoon, with the following results.

	Card	Cash
Out of State	14	10
In State	6	18

Is the manager correct in his thinking? Which type of contingency table is this?

CHAPTER 5

SOME METHODS BASED ON RANKS

PRELIMINARY REMARKS

Most of the statistical procedures introduced in the previous chapters can be used on the data that have a nominal scale of measurement. In Chapter 3 several statistical methods were presented for analyzing data that were naturally dichotomous, that is, the zero-one or success-failure types of data. In Chapter 4 the discussion centered around the analysis of data that may be classified according to two or more different criteria and into two or more separate classes by each criterion. All of those procedures may also be used where more than nominal information concerning the data is available but, for various reasons such as speed and ease of calculation, abundance of data, or the particular interpretation desired of the data, some of the information contained in the data is disregarded and the data are reduced to nominal-type data for analysis. Such a loss of information usually results in a corresponding loss of power. In this chapter several statistical methods are presented that utilize more of the information contained in the data, if the data have at least an ordinal scale of measurement.

Data may be nonnumeric ("good, better, best") or numeric (7.36, 4.91, etc.). If data are nonnumeric but are ranked as in ordinal-type data, the methods of this chapter are often the most powerful ones available. If data are numeric and, furthermore, are observations on random variables that have the normal distribution so that all of the assumptions of the usual parametric tests are met, the loss of efficiency caused by using the methods of this chapter is surprisingly small. In those situations the relative efficiency of tests using only the ranks of the observation is frequently about 0.95, depending on the situation.

The rank tests of this chapter are valid for all types of populations, whether continuous, discrete, or mixtures of the two. Earlier results in nonparametric statistics required the assumption of continuous random variables in order for the tests based on ranks to be valid. Research results by Conover (1973a) and others have shown that the continuity assumption is not necessary. It can be replaced by the trivial assumption that $P(X = x) < 1$ for each x. Since it is unlikely that any experimenter will be sampling from a population consisting entirely of a single number, we will not list this assumption for the tests in this chapter.

Thus data with many ties (two observations are said to be tied if they equal each other) may be analyzed using rank tests if the data are ordinal. A word of caution should be offered here; the so-called "exact tables" for small samples are exact only when there are no ties in the data. Otherwise they are approximate. Exact tables for a given set of ties can be obtained in the same way as in the case of no ties, but it is not practical to have a set of tables for each possible configuration of ties. If there are extensive ties in the data, the large sample approximation and not the small sample tables in this book should be used.

A convenient method for arranging observations in increasing order is the stem-and-leaf method presented by Tukey (1977). Perhaps the simplest way of explaining the stem-and-leaf method is by way of an example. Suppose a class of 28 students obtained the following scores on an exam.

74	63	88	69	81	91	75
82	91	87	77	86	86	87
96	84	93	73	74	93	78
70	84	90	97	79	89	93

The tens digit in each score is the stem in this case. There are four different stems: 6, 7, 8, and 9. The units digit is considered the leaf. First the stems are listed

9
8
7
6

Then each leaf is written to the right of the appropriate stem. That is, the first score, 74, is written as a 4 next to the 7 stem. Each score is written in this manner, with the following result.

9	6	1	3	0	7	1	3	3		
8	2	4	4	8	7	1	6	6	9	7
7	4	0	7	3	4	9	5	8		
6	3	9								

A picture of the distribution of exam scores immediately emerges. But, more important for our purposes, the scores may be arranged from smallest to largest quite easily now. In this way the ranks may be assigned to the observations. This simple stem-and-leaf method can make the methods in this chapter easier to use.

5.1 TWO INDEPENDENT SAMPLES

The test presented in this section is known as the Mann-Whitney test and also as the Wilcoxon test. Equivalent forms of the same test appeared in the literature under various names at about the same time, probably partly because of the intuitive appeal of the test procedure.

The usual two-sample situation is one in which the experimenter has obtained two samples from possibly different populations and wishes to use a statistical test to see if the null hypothesis that the two populations are identical can be rejected. That is, the experimenter wishes to detect differences between the two populations on the basis of random samples from those populations. An equivalent situation is where one random sample is obtained, but it is randomly subdivided into two samples. One sample receives one treatment and the other sample receives a different treatment, such as when one group of patients in a medical experiment receives a new medication and another group receives the current standard medication or no medication at all. This type of two-sample experiment is analyzed the same as the first type.

If the samples consist of ordinal-type data, the most interesting difference is a difference in the locations of the two populations. Does one population tend to yield larger values than the other population? Are the two medians equal? Are the two means equal?

An intuitive approach to the two-sample problem is to combine both samples into a single ordered sample and then assign ranks to the sample values from the smallest value to the largest, without regard to which population each value came from. Then the test statistic might be the *sum* of the ranks assigned to those values from one of the populations. If the sum is too small (or too large), there is some indication that the values from that population tend to be smaller (or larger, as the case may be) than the values from the other population. Hence the null hypothesis of no differences between populations may be rejected if the ranks associated with one sample tend to be larger than those of the other sample.

Ranks may be considered preferable to the actual data for several reasons. First, if the numbers assigned to the observations have no meaning by themselves but attain meaning only in an ordinal comparison with the other observations, the numbers contain no more information than the ranks contain. Such is the nature of ordinal data. Second, even if the numbers have meaning but the distribution function is not a normal distribution function, the probability theory is usually beyond our reach when the test statistic is based on the actual data. The probability theory of statistics based on ranks is relatively simple and does not depend on the distribution in many cases. A third reason for preferring ranks is that the A.R.E. of the Mann-Whitney test is never too bad when compared with the two-sample t test, the usual parametric counterpart. And yet the contrary is not true; the A.R.E. of the t test compared to the Mann-Whitney test may be as small as zero, or "infinitely bad." So the Mann-Whitney test is a safer test to use.

▶ The Mann-Whitney Test

Data   The data consist of two random samples. Let $X_1, X_2, \ldots, X_n$ denote the random sample of size n from population 1 and let $Y_1, Y_2, \ldots, Y_m$ denote the random sample of size m from population 2. Assign the ranks 1 to $n + m$ to the observations from smallest to largest. Let $R(X_i)$ and $R(Y_j)$ denote the rank assigned to X_i and Y_j for all i and j. For convenience, let $N = n + m$.

If several sample values are exactly equal to each other (tied), assign to each the average of the ranks that would have been assigned to them had there been no ties (see Example 1).

Assumptions

1. Both samples are random samples from their respective populations.
2. In addition to independence within each sample, there is mutual independence between the two samples.
3. The measurement scale is at least ordinal.

Test Statistic   If there are no ties, or just a few ties, the sum of the ranks assigned to the sample from population 1 can be used as a test statistic.

$$T = \sum_{i=1}^{n} R(X_i) \tag{1}$$

If there are many ties, subtract the mean from T and divide by the standard deviation to get

$$T_1 = \frac{T - n\frac{N+1}{2}}{\sqrt{\frac{nm}{N(N-1)}\sum_{i=1}^{N} R_i^2 - \frac{nm(N+1)^2}{4(N-1)}}} \tag{2}$$

where ΣR_i^2 refers to the sum of the squares of all N of the ranks or average ranks actually used in both samples.

Null Distribution Selected lower quantiles of the null distribution of T are given in Table A7, for $n \leq 20$ and $m \leq 20$. Upper quantiles w_p for T are obtained using the relation

$$w_p = n(n + m + 1) - w_{1-p} \tag{3}$$

where the lower quantile w_{1-p} is obtained from Table A7.

As an alternative to using upper quantiles, the statistic T', defined as

$$T' = n(N + 1) - T \tag{4}$$

may be used with the lower quantiles whenever an upper-tailed test is desired. The use of T' is more convenient than finding upper quantiles, and upper-tailed p-values, from Equation 3.

The quantiles in Table A7 are exact only if there are no ties in the data and therefore no average ranks are used.

The approximate quantiles in the case of no ties, and n or m greater than 20, are found from the normal approximation,

$$w_p \cong \frac{n(N+1)}{2} + z_p \sqrt{\frac{nm(N+1)}{12}} \tag{5}$$

where the quantile z_p is obtained from Table A1.

If there are many ties in the data, then T_1 is used instead of T, and T_1 is approximately a standard normal random variable whose quantiles are given in Table A1.

Hypotheses

A. (Two-Tailed Test) Let $F(x)$ and $G(x)$ be the distribution functions corresponding to X and Y, respectively. Then the hypotheses may be stated as follows.

$$H_0: F(x) = G(x) \quad \text{for all } x$$
$$H_1: F(x) \neq G(x) \quad \text{for some } x$$

The test is sensitive for $H_1: E(X) \neq E(Y)$, and can be used as a test for means. In many real situations any difference between distributions implies that $P(X > Y)$ is no longer equal to $1/2$. Therefore $H_1: P(X > Y) \neq P(X < Y)$ is often used instead of the above.

Reject H_0 at the level of significance α if T is less than its $\alpha/2$ quantile or greater than its $1 - \alpha/2$ quantile obtained from Table A7 and Equation 3 for $n \leq 20$ and $m \leq 20$, or obtained from Table A1 and Equation 5 in the large sample approximation. If T_1 is used instead of T the quantiles are obtained directly from Table A1.

The approximate two-tailed p-value may be found using Table A1. In the case of T, substitute the smaller of T or T' into the following

$$p\text{-value} = 2 \cdot P\left(Z \leq \frac{T \,(\text{or } T') + \frac{1}{2} - n\frac{N+1}{2}}{\sqrt{\frac{nm(N+1)}{12}}} \right) \tag{6}$$

where Z is a standard normal random variable. In the case of T_1 the p-value is twice the smaller of $P(Z \leq T_1)$ or $P(Z \geq T_1)$.

B. (Lower-Tailed Test) The null hypothesis is

$$H_0: F(x) = G(x)$$

while the alternative hypothesis in a lower-tailed test takes one of the forms

$$H_1: F(x) > G(x)$$
$$H_1: E(X) < E(Y)$$

or

$$H_1: P(X > Y) < P(X < Y)$$

all of which convey the thought, "X tends to be less than Y" in various ways.

Reject H_0 if T is less than its α quantile given in Table A7 for $n \leq 20$ and $m \leq 20$, or by Equation 5 for larger sample sizes. If T_1 is used, reject H_0 at the level α if $T_1 < z_\alpha$, where z_α is obtained from Table A1.

The p-value is approximately the probability

$$p\text{-value} \cong P\left(Z \leq \frac{T + \frac{1}{2} - n\frac{N+1}{2}}{\sqrt{\frac{nm(N+1)}{12}}} \right) \tag{7}$$

obtained from Table A1. In the case of T_1 the p-value is approximately $P(Z \leq T_1)$ obtained directly from Table A1.

C. (Upper-Tailed Test) The null hypothesis is

$$H_0: F(x) = G(x)$$

while the alternative in an upper-tailed test takes one of the forms

$$H_1: F(x) < G(x)$$
$$H_1: E(X) > E(Y)$$

or

$$H_1: P(X > Y) > P(X < Y)$$

All three forms for H_1 convey the thought "X tends to be greater than Y" in various forms.

Reject H_0 at the level α if T is greater than its $1 - \alpha$ quantile obtained from Table A7 and Equation 3. It may be easier to find $T' = n(N + 1) - T$ and reject H_0 if T' is less than the α quantile from Table A7. Equation 5 is used to find quantiles if $n > 20$ or $m > 20$.

If T_1 is used, reject H_0 if $T_1 > z_{1-\alpha}$ where $z_{1-\alpha}$ is obtained from Table A1. The p-value is approximately the probability, found from Table A1,

$$p\text{-value} \cong P\left(Z \geq \frac{T - \frac{1}{2} - n\frac{N+1}{2}}{\sqrt{\frac{nm(N+1)}{12}}} \right) \tag{8}$$

which is the same as

$$p\text{-value} \cong P\left(Z \leq \frac{T' + \frac{1}{2} - n\frac{N+1}{2}}{\sqrt{\frac{nm(N+1)}{12}}} \right) \tag{9}$$

If T_1 is used the p-value is simply

$$p\text{-value} = P(Z \geq T_1) = 1 - P(Z \leq T_1) \tag{10}$$

which is obtained directly from Table A1.

Computer Assistance   Computer programs containing the Mann-Whitney Test include *Minitab, S-Plus, SAS,* and *StatXact.* _____ ◄

Comment

The Mann-Whitney test is unbiased and consistent when testing the preceding hypotheses involving $P(X > Y)$. However, the same is not always true for the hypotheses involving $E(X)$ and $E(Y)$. To insure that the test remains consistent and unbiased for hypotheses involving $E(X)$ it is sufficient to add another assumption to the previous model.

Assumption 4. If there is a difference between population distribution functions, that difference is a difference in the location of the distributions. That is, if $F(x)$ is not identical with $G(x)$, then $F(x)$ is identical with $G(x + c)$, where c is some constant.

EXAMPLE 1

The senior class in a particular high school had 48 boys. Twelve boys lived on farms and the other 36 lived in town. A test was devised to see if farm boys in general were more physically fit than town boys. Each boy in the class was given a physical fitness test in which a low score indicates poor physical condition. The scores of the farm boys (X_i) and the town boys (Y_j) are as follows.

X_i: Farm Boys		Y_j: Town Boys					
14.8	10.6	12.7	16.9	7.6	2.4	6.2	9.9
7.3	12.5	14.2	7.9	11.3	6.4	6.1	10.6
5.6	12.9	12.6	16.0	8.3	9.1	15.3	14.8
6.3	16.1	2.1	10.6	6.7	6.7	10.6	5.0
9.0	11.4	17.7	5.6	3.6	18.6	1.8	2.6
4.2	2.7	11.8	5.6	1.0	3.2	5.9	4.0

Neither group of boys is a random sample from any population. However, it is reasonable to assume that these scores resemble hypothetical random samples from the populations of farm and town boys in that age group, at least for similar localities. The other assumptions of the model seem to be reasonable, such as independence between groups. Therefore the Mann-Whitney test is selected to test

H_0: Farm boys do not tend to be more fit, physically, than town boys
H_1: Farm boys tend to be more fit than town boys

These hypotheses suggest an upper-tailed test as stated in set C of hypotheses.

The scores are ranked as follows.

X	Y	Rank	X	Y	Rank	X	Y	Rank
	1.0	1		6.2	17		11.3	33
	1.8	2	6.3		18	11.4		34
	2.1	3		6.4	19		11.8	35
	2.4	4		6.7	20.5 ⌉	12.5		36
	2.6	5		6.7	20.5 ⌋		12.6	37
2.7		6	7.3		22		12.7	38
	3.2	7		7.6	23	12.9		39
	3.6	8		7.9	24		14.2	40
	4.0	9		8.3	25		14.8	41.5 ⌉
4.2		10	9.0		26	14.8		41.5 ⌋
	5.0	11		9.1	27		15.3	43
	5.6	13 ⌉		9.9	28		16.0	44
	5.6	13		10.6	30.5 ⌉	16.1		45
5.6		13 ⌋		10.6	30.5		16.9	46
	5.9	15	10.6		30.5		17.7	47
	6.1	16		10.6	30.5 ⌋		18.6	48

There are four groups of tied scores, as indicated by the square brackets. Within each group the ranks that should have been assigned are averaged, and the average rank is assigned instead, as illustrated.

The test is one tailed. The critical region corresponds to large values of T_1. Note that this is not a large number of ties, so it is probably acceptable to use T instead of T_1. Both methods will be compared later in the example. From Table A1 we see that a critical region of size $\alpha = 0.05$ corresponds to values of T_1 greater than 1.6449.

Here we have $n = 12$, $m = 36$, so $N = 12 + 36 = 48$. The sum of the ranks assigned to the Xs is

$$T = \sum_{i=1}^{n} R(X_i)$$
$$= 6 + 10 + 13 + 18 + 22 + 26 + 30.5 + 34 + 36 + 39 + 41.5 + 45 = 321$$

The sum of the squares of all 48 ranks is

$$\sum_{i=1}^{N} R_i^2 = 38{,}016$$

which is slightly less than the sum 38,024, of the squares of all the ranks from 1 to 48 if there had been no ties (using Lemma 1.4.2.). Now we can compute T_1.

$$T_1 = \frac{T - n\dfrac{N+1}{2}}{\sqrt{\dfrac{nm}{N(N-1)}\displaystyle\sum_{i=1}^{N} R_i^2 - \dfrac{nm(N+1)^2}{4(N-1)}}}$$

$$= \frac{321 - 12\dfrac{49}{2}}{\sqrt{\dfrac{(12)(36)}{(48)(47)}(38{,}016) - \dfrac{(12)(36)(49)^2}{4(47)}}}$$

$$= 0.6431$$

which is not in the critical region, so H_0 is accepted and we conclude that these data do not show that farm boys are more physically fit than town boys. A comparison of $T_1 = 0.6431$ with Table A1 shows that 0.6431 is close to the 0.74 quantile, so the null hypothesis could have been rejected at an α level of about $1 - 0.74 = 0.26$, and therefore the p-value is 0.26.

If we had ignored the few ties and used the large sample approximation in Equation 5 we would have obtained an approximate 0.95 quantile for T as

$$w_{0.95} = n\frac{N+1}{2} + (1.6449)\sqrt{nm(N+1)/12}$$

$$= 294 + (1.6449)(42)$$

$$= 363.1$$

and H_0 would have been accepted as before.   ■

The next example illustrates a situation in which no random variables are defined explicitly. The pieces of flint are ranked according to hardness by direct comparison with each other. A random variable that assigns a measure of hardness to each piece of flint is conceivable but unnecessary in this case.

EXAMPLE 2

A simple experiment was designed to see if flint in area A tended to have the same degree of hardness as flint in area B. Four sample pieces of flint were collected in area A and five sample pieces of flint were collected in area B. To determine which of two pieces of flint was harder, the two pieces were rubbed against each other. The piece sustaining less damage was judged the harder of the two. In this manner all nine pieces of flint were ordered according to hardness. The rank 1 was assigned to the softest piece, rank 2 to the next softest, and so on.

Origin of Piece	Rank
A	1
A	2
A	3
B	4
A	5
B	6
B	7
B	8
B	9

The hypothesis to be tested is

H_0: The flints from areas A and B are of equal hardness

against the alternative

H_1: The flints are not of equal hardness

The Mann-Whitney two-tailed test is used where $n = 4$, $m = 5$, and

$$T = \text{sum of the ranks of pieces from area A}$$
$$= 1 + 2 + 3 + 5$$
$$= 11$$

The two-tailed critical region of approximate size 0.05 corresponds to values of T less than 12 and values of T greater than $(4)(10) - 12 = 28$. Because T in this example is less than 12, the null hypothesis is rejected, and it is concluded that flints from the two areas differ in degree of hardness. Because of the direction of the difference, the further conclusion that the flint in area B is harder than the flint in area A may also be drawn. The p-value may be found approximately from Equation 6 and Table A1.

$$p\text{-value} \cong 2 \cdot P\left(Z \leq \frac{11 + \frac{1}{2} - 4\frac{10}{2}}{\sqrt{\frac{4 \cdot 5 \cdot 10}{12}}} \right)$$
$$= 2 \cdot P(Z \leq -2.0821)$$
$$= 2(0.019) = 0.038 \qquad \blacksquare$$

□ *Theory*   The null distribution of T is found by assuming that X_i and Y_j are identically distributed. If the X_i and Y_j are independent and identically distributed, every arrangement of the Xs and Ys in the ordered combined sample is equally likely. This is the basic principle behind many rank tests. A formal proof of this statement requires calculus and is therefore beyond the scope of this book. However, the truth of the statement may seem to be intuitively obvious after one attempts to furnish a reason for some arrangements being more probable than others. There is no valid reason for this and, therefore, we can accept the fact that all ordered arrangements are equally likely as an intuitively obvious but unproved (here) statement.

If the X_i and Y_j are independent and identically distributed, the ranks assigned to the X_i in the combined sample should resemble a random selection of n of the integers from 1 to $n + m$. That is, there is no reason why any particular rank should have a better chance than any other rank of being assigned to a value of X_i. Because each number from 1 to $n + m$ is equally likely to be assigned to X_i as its rank and because n different numbers are selected as ranks for the Xs, the probability distribution of T, the sum of ranks, may be obtained by considering the probability distribution of the sum of n integers selected at random, without replacement, from among the integers from 1 to $n + m$.

The number of ways of selecting n integers from a total number of $n + m$ integers is $\binom{n + m}{n}$, and each way has equal probability according to the basic premise just stated. Hence the probability that $T = k$ may be found by counting the number of different sets of n integers from 1 to $n + m$ that add up to the value k and then dividing that number by $\binom{n + m}{n}$.

For example, if the sample sizes are $n = 4$ and $m = 5$ as in Example 2, the number of ways of selecting four out of nine ranks is

$$\binom{n + m}{n} = \frac{9!}{4!5!} = 126$$

The smallest value that T may assume is 10, which occurs if the four ranks 1, 2, 3, 4 are selected. The next value that T may assume is 11, which occurs only one way: 1, 2, 3, 5. The value $T = 12$ may be assumed two ways, with the ranks 1, 2, 3, 6 or with 1, 2, 4, 5. Therefore,

$$P(T = 10) = 1/126 \qquad P(T \leq 10) = 0.0079$$
$$P(T = 11) = 1/126 \qquad P(T \leq 11) = 0.0159$$
$$P(T = 12) = 2/126 \qquad P(T \leq 12) = 0.0317$$
$$\text{etc.} \qquad\qquad\qquad \text{etc.}$$

Notice that the exact p-value in Example 2 is $2 \cdot P(T \leq 11)$, which is given above as $4/126 = 0.0317$, not far from the approximation 0.038 used in the example.

Because T is the sum of the ranks of the n Xs, for large n and m the central limit theorem may be applied to obtain an approximate distribution for T. This was done in Example 1.5.7. The results of Example 1.5.7 state that T is approximately normal, with mean and variance given by Theorem 1.4.5 as

$$E(T) = \frac{n(n + m + 1)}{2} \tag{11}$$

and

$$\text{Var}\,(T) = \frac{n(n + m + 1)m}{12} \tag{12}$$

Therefore the quantiles of T may be approximated with the aid of Theorem 1.5.1:

$$w_p = E(T) + z_p \sqrt{\text{Var}\,(T)} \tag{13}$$

where z_p is the pth quantile of the standard normal distribution. The justification for using the normal approximation for T_1 is similar to the preceding justification for using the normal approximation for T, except that the term Var (T) must be based on the actual ranks and average ranks used in the two samples. The details are deferred until Section 5.3. $\square$

The Mann-Whitney test may be used for testing

$$H_0: E(X) = E(Y) + d, \quad \text{or} \quad E(X) - E(Y) = d \tag{14}$$

where d is some specified number. We simply add the number d to each Y_i and then use the Mann-Whitney test on the original Xs and the newly adjusted Ys.

By collecting all the values of d that would result in acceptance of the preceding H_0, we have a confidence interval for $E(X) - E(Y)$, the difference between the two means. This confidence interval is sometimes more meaningful to an experimenter than merely testing whether the two means are equal. We will now describe a method for obtaining the confidence interval without having to use the Mann-Whitney test over and over again.

▶ Confidence Interval for the Difference Between Two Means _____

Data The data consist of two random samples $X_1, \ldots, X_n$ and $Y_1, \ldots, Y_m$ of size n and m, respectively. Let X and Y denote random variables with the same distribution as the X_i and the Y_j, respectively.

Assumptions

1. Both samples are random samples from their respective populations.
2. In addition to independence within each sample, there is mutual independence between the two samples.
3. The two population distribution functions are identical except for a possible difference in location parameters. That is, there is a constant d (say) such that X has the same distribution function as $Y + d$.

Note that no assumption of continuity needs to be made here. Noether (1967b) shows that if the confidence coefficient of a confidence interval is $1 - \alpha$ when sampling from a continuous population then, for general populations, the true confidence coefficient of the same interval including its end points is at least $1 - \alpha$ and without its end points is at most $1 - \alpha$. We will include the end points.

Method   Determine the $\alpha/2$ quantile $w_{\alpha/2}$ for n and m from Table A7 or Equation 5 if n and m are large, where $(1 - \alpha)$ is the desired confidence coefficient. Note that Table A7 and Equation 5 are used even if there are many ties. Then calculate k, given by

$$k = w_{\alpha/2} - n(n + 1)/2 \qquad (15)$$

From all of the possible pairs (X_i, Y_j), find the k largest differences $X_i - Y_j$ and find the k smallest differences. To find the largest and smallest differences, it is convenient to order each sample first, from smallest to largest, and then to form a matrix of differences $X_i - Y_j$ using the Xs as rows and the Ys as columns. The kth largest difference is the upper limit U and the kth smallest difference is the lower limit L. That is, counting toward the middle of the ordered array of all mn possible differences, the kth differences from each end of the array are the points L and U. Then the confidence interval is given by

$$P[L \le E(X) - E(Y) \le U] \ge 1 - \alpha \qquad (16)$$

Computer Assistance   Exact nonparametric confidence intervals for the difference between two means (or medians) are obtained in *Minitab* and *StatXact*. These are also known as the Hodges-Lehmann estimates of shift. _____◄

EXAMPLE 3

A cake batter is to be mixed until it reaches a specified level of consistency. Five batches of the batter are mixed using mixer A, and another five batches are mixed using mixer B. The required times for mixing are given as follows (in minutes).

Mixer A	Mixer B
7.3	7.4
6.9	6.8
7.2	6.9
7.8	6.7
7.2	7.1

A 95% confidence interval is sought for the mean difference in mixing times, more specifically for $E(X) - E(Y)$, where X refers to mixer A and Y refers to mixture B.

For $n = 5$, $m = 5$, $\alpha = 0.05$, Table A7 yields $w_{0.025} = 18$, so $k = 18 - (5)(6)/2 = 3$. The two samples are ordered from smallest to largest and the Xs are used as rows and the Ys are used as columns in forming a matrix of differences $X_i - Y_j$ as follows.

X_i \ Y_j	6.7	6.8	6.9	7.1	7.4
6.9	0.2	0.1	0.0	−0.2	−0.5
7.2	0.5	0.4	0.3	0.1	−0.2
7.2	0.5	0.4	0.3	0.1	−0.2
7.3	0.6	0.5	0.4	0.2	−0.1
7.8	1.1	1.0	0.9	0.7	0.4

Then the largest and smallest differences are found.

Smallest Differences	Largest Differences
$6.9 - 7.4 = -0.5$	$7.8 - 6.7 = 1.1$
$6.9 - 7.1 = -0.2$	$7.8 - 6.8 = 1.0$
$7.2 - 7.4 = -0.2 = L$	$7.8 - 6.9 = 0.9 = U$

The resulting 95% confidence interval (L, U) for $E(X) - E(Y)$ is $(-0.2, 0.9)$. ∎

□ *Theory*   Note that there are mn pairs (X_i, Y_j). Let k denote the number of pairs where $X_i > Y_j$, that is, where $X_i - Y_j > 0$. Then T in Equation 1, the sum of ranks of Xs, is $k + n(n + 1)/2$ (see Problem 1). That is, if no Ys are smaller than any of the Xs, $T = 1 + 2 + \cdots + n = n(n + 1)/2$ (from Lemma 1.4.1). The effect of having k pairs (X_i, Y_j) where Y is less than X is to increase T by k units.

The "borderline" value of T, where H_0 is barely accepted, is given in Table A7 as $w_{\alpha/2}$. By subtracting $n(n + 1)/2$ from $w_{\alpha/2}$, we find the borderline value of k. Now we want to find the value of d that we can add to the Ys to achieve barely this borderline value of k, that is, so that exactly k of the pairs $(X_i, Y_j + d)$ satisfy $X_i > Y_j + d$ or $X_i - Y_j > d$.

If we add the maximum of all of the differences $X_i - Y_j$ to each of the Ys, obviously none of the Xs will be greater than the adjusted Ys because the Ys are too large. By adding the kth largest difference $X_i - Y_j$ to each of the Ys, we achieve the borderline case: fewer than k pairs satisfy $X_i > Y_j + d$, and at least k pairs satisfy $X_i \geq Y_j + d$. In this way we obtain the largest value d that results in acceptance of H_0: $E(X) = E(Y) + d$. By reversing the procedure and working from the lower end, we obtain the smallest value of d that results in acceptance of the same hypothesis. This collection of values of d gives us the confidence interval we desire. □

Comparison with Other Procedures

The natural procedure to compare with the Mann-Whitney test is the two-sample t test, as mentioned earlier. This version of the t test involves the sample means $\overline{X}$ and $\overline{Y}$ of the two samples in the following formula.

$$t = \frac{(\overline{X} - \overline{Y})\sqrt{mn(N - 2)/N}}{\sqrt{\sum_{i=1}^{n} (X_i - \overline{X})^2 + \sum_{j=1}^{m} (Y_j - \overline{Y})^2}} \tag{17}$$

The value of t is compared with quantiles obtained from Table A21, with $N - 2$ degrees of freedom. In order for these quantiles to be accurate the additional assumption must be made that both populations have a normal distribution. With this assumption the t test is the most powerful test. Some types of nonnormal distributions result in very little power when using the t test as compared with the Mann-Whitney test. This is especially true when one or both samples contain unusually large or small observations, called "outliers."

If a computer program for the t statistic is available, it can be used to facilitate calculations in the Mann-Whitney test, especially if there are many ties. Merely compute the t statistic on the ranks $R(X_i)$ and $R(Y_j)$, instead of the data X_i and Y_j, and compare the result with quantiles from Table A21, $N - 2$ degrees of freedom. Although this approximation is not quite the same as the usual normal approximation, it is slightly more accurate in most cases. For an even better approximation, find the average of T_1 given by Equation 2 and the t statistic computed on the ranks, and compare this with the average of the two quantiles obtained from Tables A1 and A21. For more details on this method see Iman (1976).

The A.R.E. of the Mann-Whitney test as compared with the t test is computed under the assumption that the distributions of X and Y are identical except for their means. If the populations are normal the A.R.E. is 0.955, if the populations are uniform the A.R.E. is 1.0, and if the populations have a symmetric distribution known as the double exponential distribution the A.R.E. is 1.5. If the two populations differ only in their location parameters the A.R.E. is never lower than 0.864 but may be as high as infinity (Hodges and Lehmann, 1956).

The median test also may be used for data of this type. The A.R.E. of the Mann-Whitney test relative to the median test is 1.5 for normal populations, 3.0 for uniform distributions, but only 0.75 in the double exponential case. Remember that this is *asymptotic* relative efficiency. For small samples the Mann-Whitney test may have much more power than the median test in the case of double exponential distribution (see Conover, Wehmanen, and Ramsey, 1978). On the other hand, the median test does not require that the populations be identical when H_0 is true. It only requires that they have the same median. Hence the median test may be applied in situations where the Mann-Whitney test is not valid.

The Mann-Whitney test was first introduced for the case $n = m$ by Wilcoxon (1945). Wilcoxon's test was extended to the case of unequal sample sizes by White (1952) and van der Reyden (1952). A test equivalent to Wilcoxon's was also developed independently and introduced by Festinger (1946). Mann and Whitney (1947) seem to be the first to consider unequal sample sizes and to furnish tables suitable for use with small samples. It is largely the work of Mann and Whitney that led to widespread use of the test. Because the test is attributed to various authors, it is the user's prerogative as to which name to call it by.

The modification of the Mann-Whitney test to examine differences in dispersion or variance or scale, introduced by Siegel and Tukey in 1960, is similar in principle to an earlier test devised by Freund and Ansari (1957). The relationship between the two tests is described on page 126 of Hájek and Sidák (1967).

More extensive tables for the Mann-Whitney test are given by Verdooren (1963) for n and $m \leq 25$ and by Milton (1964) for $n \leq 20$ and $m \leq 40$. Other tables and a bibliography are found in Jacobson (1963). The distribution of the Mann-Whitney test statistic is discussed by Klotz (1966) and Buckle, Kraft, and van Eeden (1969). Other articles are by Zaremba (1965) and Serfling (1968).

The efficiency of the Mann-Whitney test and other closely related tests is the subject of papers by Chanda (1963), Noether (1963), Hayman and Govindarajulu (1966), McNeil (1967), R.A. Shorack (1967), Stone (1967), and Conover and Kemp (1976). Justification for the treatment of ties is given by Conover (1973a). Modifications for sequential testing are given by Alling (1963), Woinsky and Kurz (1969), Bradley, Martin, and Wilcoxon (1965), Bradley, Merchant, and Wilcoxon (1966), Sen and Ghosh (1974), and Spurrier and Hewitt (1976). The problem of testing circular distributions, as discussed by Batschelet (1965) is approached by Beran (1969) and Schach (1969b).

If the two samples are censored (i.e., if some of the largest and/or smallest sample values are not observable) the data may be analyzed with modifications of the Mann-Whitney test, such as those discussed by Gastwirth (1965a), Gehan (1965a, 1965b), Gehan and Thomas (1969), Saw (1966), Basu (1968), Hettmansperger (1968), and Shorack (1968). A rank test for the bivariate two-sample problem is given by Mardia (1967a, 1968). Other two-sample nonparametric tests are presented and discussed by Hudimoto (1959), Haga (1960), Tamura (1963), Potthoff (1963), Wheeler and Watson (1964), Gastwirth (1965b), Bhattacharyya and Johnson (1968), Mielke (1972), and Pettitt (1976). The efficiency of some of these tests is examined by Mikulski (1963), Basu (1967a), Hollander (1967a), and Gibbons and Gastwirth (1970). Other related papers include Hollander, Pledger, and Lin (1974), Bickel and Lehmann (1975), Hettmansperger and Malin (1975), Doksum and Sievers (1976), and Fligner, Hogg, and Killeen (1976).

The method for finding confidence intervals is discussed by Noether (1967a), and a related graphical procedure is described by Moses in Walker and Lev (1953). An algorithm that may be useful when sample sizes are large is given by McKean and Ryan (1977). Other estimates of location differences are discussed by Hodges and Lehmann (1963), Høyland (1965), Rao, Schuster, and Littell (1975), and Switzer (1976). Related papers are by Moses (1965), Govindarajulu (1968), Bauer (1972), Ury (1972), and Kraft and van Eeden (1972).

EXERCISES

1. Test the following data to see if the mean high temperature in Des Moines is higher than the mean high temperature in Spokane, for randomly sampled days in the summer.

Des Moines	Spokane
83	78
91	82
94	81
89	77
89	79
96	81
91	80
92	81
90	

2. In a controlled environment laboratory, 10 men and 10 women were tested to determine the room temperature they found to be the most comfortable. The results were as follows.

Men	Women
74	75
72	77
77	78
76	79
76	77
73	73
75	78
73	79
74	78
75	80

Assuming that these temperatures resemble a random sample from their respective populations, is the average comfortable temperature the same for men and women?

3. Seven students were taught algebra using the present method, and six students learned

algebra according to a new method. Find a 90% confidence interval for the difference in achievement scores expected from the two methods.

Method	Students' Achievement Scores						
Present	68	72	79	69	84	80	78
New	64	60	68	73	72	70	

4. Diet A was given to four overweight girls and diet B was given to five other overweight girls, with the following observed weight losses. Find a 90% confidence interval for mean difference in effectiveness of the two diets.

Diet	Weight Losses (pounds)
A	7, 2, −1, 4
B	6, 5, 2, 8, 3

5. Eight volunteers for an experiment are divided randomly into two groups, to see if a telescopic sight improves the ability to hit a target under twilight conditions. Group A is given rifles with telescopic sights, while group B has the same kind of rifle but with open sights. After a learning period they are given a shooting test at twilight. These are their scores (100 is perfect).

Group A:	96	93	88	85
Group B:	89	83	80	77

What do you conclude?

6. Ten tents using plain camouflage and 10 tents using patterned camouflage are set up in a wooded area, and a team of observers is sent out to find them. The team reports the distance at which they first sight each tent (true sightings only) until all 20 tents are found. The purpose of the study is to determine whether the patterned camouflage is more difficult to detect than the plain camouflage. The distances at which each tent is detected are reported as follows.

Type of camouflage	Distance (meters)
Plain	25, 28, 16, 34, 38, 21, 29, 43, 32, 36
Patterned	26, 12, 16, 21, 20, 14, 10, 18, 22, 20

(a) Perform a hypothesis test.

(b) Find a nonparametric 95% confidence interval for the difference in mean detection distances.

PROBLEMS

1. Let S equal the number of pairs (X_i, Y_j) in which X_i exceeds Y_j (counting ties as one-half). Note that there are mn pairs in all. Show that S and T satisfy the relationship

$$S = T - \frac{n(n+1)}{2}$$

What statistic seems reasonable to use as an estimate of $P(X > Y)$?

2. In the case where $n = 3$, $m = 2$, and H_0 is true, find the exact distribution of T and compare it with Table A7.

3. Compute the two-sample t statistic (Equation 17) on the data in Exercise 2 and compare the results with those obtained using the Mann-Whitney test.

5.2 SEVERAL INDEPENDENT SAMPLES

The Mann-Whitney test for two independent samples, presented in Section 5.1, was extended to the problem of analyzing k independent samples, for $k > 2$, by Kruskal and Wallis (1952). The experimental situation is one where k random samples have been obtained, one from each of k possibly different populations, and we want to test the null hypothesis that all of the populations are identical against the alternative that some of the populations tend to furnish greater observed values than other populations. The term "greater" applies to observations on random variables, but actually any observations that may be arranged in increasing order according to some property such as quality, value, and the like may be analyzed using the Kruskal-Wallis test in a manner analogous to the analysis of nonnumeric data using the Mann-Whitney test, as in Example 5.1.2.

▶ **The Kruskal-Wallis Test** _____

Data　The data consist of k random samples of possibly different sizes. Denote the ith random sample of size n_i by $X_{i1}, X_{i2}, \ldots, X_{in_i}$. Then the data may be arranged into columns.

Sample 1	Sample 2	$\cdots$	Sample k
$X_{1,1}$	$X_{2,1}$		$X_{k,1}$
$X_{1,2}$	$X_{2,2}$		$X_{k,2}$
$\cdots$	$\cdots$		$\cdots$
X_{1,n_1}	X_{2,n_2}		X_{k,n_k}

Let N denote the total number of observations

$$N = \sum_{i=1}^{k} n_i \tag{1}$$

Assign rank 1 to the smallest of the totality of N observations, rank 2 to the second smallest, and so on to the largest of all N observations, which receives rank N. Let $R(X_{ij})$ represent the rank assigned to X_{ij}. Let R_i be the sum of the ranks assigned to the ith sample.

$$R_i = \sum_{j=1}^{n_i} R(X_{ij}) \qquad i = 1, 2, \ldots, k \tag{2}$$

Compute R_i for each sample.

If the ranks may be assigned in several different ways because several observations are equal to each other, assign the average rank to each of the tied observations, as in the previous test of this chapter.

Assumptions

1. All samples are random samples from their respective populations.
2. In addition to independence within each sample, there is mutual independence among the various samples.
3. The measurement scale is at least ordinal.
4. Either the k population distribution functions are identical, or else some of the populations tend to yield larger values than other populations do.

Test Statistic The test statistic T is defined as

$$T = \frac{1}{S^2}\left(\sum_{i=1}^{k} \frac{R_i^2}{n_i} - \frac{N(N+1)^2}{4}\right) \tag{3}$$

where N and R_i are defined in Equations 1 and 2, respectively, and where

$$S^2 = \frac{1}{N-1}\left(\sum_{\substack{\text{all} \\ \text{ranks}}} R(X_{ij})^2 - N\frac{(N+1)^2}{4}\right) \tag{4}$$

If there are no ties S^2 simplifies to $N(N + 1)/12$, and the test statistic reduces to

$$T = \frac{12}{N(N+1)} \sum_{i=1}^{k} \frac{R_i^2}{n_i} - 3(N+1) \tag{5}$$

If the number of ties is moderate there will be very little difference between Equations 3 and 5, so the simpler Equation 5 may be preferred.

Null Distribution The exact distribution of T is given by Table A8 for $k = 3$ and all $n_i \le 5$, but in general the exact distribution is too cumbersome to work with. Therefore the chi-squared distribution with $k - 1$ degrees of freedom is used as an approximation to the null distribution of T.

Hypotheses

H_0: All of the k population distribution functions are identical

H_1: At least one of the populations tends to yield larger observations than at least one of the other populations

Because the Kruskal-Wallis test is designed to be sensitive against differences among means in the k populations, the alternative hypothesis is sometimes stated as follows.

H_1: The k populations do not all have identical means

Reject H_0 at the level α if T is greater than its $1 - \alpha$ quantile from the null distribution. If $k = 3$, all of the sample sizes are 5 or less, and there are no ties, the exact quantile may be obtained from Table A8. More extensive exact tables are given in Iman, Quade, and Alexander (1975). When there are ties, or when exact tables are not available, the approximate quantiles may be obtained from Table A2, the chi-squared distribution with $k - 1$ degrees of freedom. Reject H_0 at the level α if T exceeds the $1 - \alpha$ quantile thus obtained. The p-value is approximately the probability of a chi-squared random variable with $k - 1$ degrees of freedom exceeding the observed value of T.

Multiple Comparisons   If, and only if, the null hypothesis is rejected, we may use the following procedure to determine which *pairs* of populations tend to differ. We can say that populations i and j seem to be different if the following inequality is satisfied:

$$\left| \frac{R_i}{n_i} - \frac{R_j}{n_j} \right| > t_{1-(\alpha/2)} \left(S^2 \frac{N-1-T}{N-k} \right)^{1/2} \left(\frac{1}{n_i} + \frac{1}{n_j} \right)^{1/2} \tag{6}$$

where R_i and R_j are the rank sums of the two samples, $t_{1-\alpha/2}$ is the $(1 - \alpha/2)$ quantile of the t distribution obtained from Table A21 with $N - k$ degrees of freedom, S^2 comes from Equation 4, and T comes from Equation 3 or 5. This procedure is repeated for all pairs of populations. The same α level is usually used here as in the Kruskal-Wallis test.

Computer Assistance   Computer programs that include the Kruskal-Wallis Test include *Minitab*, *S-Plus*, *SAS*, and *StatXact*. These and other programs will also convert the data to ranks and perform a one-way analysis of variance on the ranks, as discussed in Problem 5. This method automatically corrects for ties and usually includes a wider variety of multiple comparisons procedures to choose from. _____◀

EXAMPLE I

Data from a completely randomized design were given in Example 4.3.1, where four different methods of growing corn resulted in various yields per acre on various plots of ground where the four methods were tried. Ordinarily, only one statistical analysis is used, but here we will use the Kruskal-Wallis test so that a rough comparison may be made with the median test, which previously furnished a p-value of slightly less than 0.001.

The hypotheses may be stated as follows.

H_0: The four methods are equivalent

H_1: Some methods of growing corn tend to furnish higher yields than others

The observations are ranked from the smallest, 77, of rank 1 to the largest, 101, of rank $N = 34$. Tied values receive the average ranks. The ranks of the observations, with the sums R_i, are given next.

Method

	1		2		3		4	
Observation	Rank	Observation	Rank	Observation	Rank	Observation	Rank
83	11	91	23	101	34	78	2
91	23	90	19.5	100	33	82	9
94	28.5	81	6.5	91	23	81	6.5
89	17	83	11	93	27	77	1
89	17	84	13.5	96	31.5	79	3
96	31.5	83	11	95	30	81	6.5
91	23	88	15	94	28.5	80	4
92	26	91	23			81	6.5
90	19.5	89	17				
		84	13.5				
R_i:	196.5		153.0		207.0		38.5
n_i:	9		10		7		8
$N = 34$							

The critical region of approximate size $\alpha = 0.05$ corresponds to values of T greater than the 0.95 quantile of a chi-squared random variable with $k - 1 = 3$ degrees of freedom, which is given in Table A2 as 7.815. (Note that the median test also used the chi-squared distribution with $k - 1$ degrees of freedom, so the critical regions of the two tests will seem to be the same although the test statistics are different.)

The value of T obtained using Equation 5 is

$$T = 25.46$$

which clearly leads to rejection of H_0. A rough idea of the power of the Kruskal-Wallis test as compared with the median test may be obtained by comparing the value of the test statistics in both tests. Both test statistics have identical asymptotic distributions, the chi-squared distribution with 3 degrees of freedom. However, the value 25.46 attained in the Kruskal-Wallis test is somewhat larger than the value 17.6 computed in the median test, indicating more sensitivity to the sample differences.

Because H_0 is rejected, the multiple comparisons procedure may be used. We can ignore the few ties and use the simpler form

$$S^2 = N(N + 1)/12 = 99.167 \tag{7}$$

so that

$$\frac{S^2(N - 1 - T)}{N - k} = \frac{(99.167)(33 - 25.464)}{34 - 4} = 24.911 \tag{8}$$

and the remaining calculations are as follows:

Populations	$\left\lvert \dfrac{R_i}{n_i} - \dfrac{R_j}{n_j} \right\rvert$	$2.041(24.911)^{1/2} \left(\dfrac{1}{n_i} + \dfrac{1}{n_j} \right)^{1/2}$
1 and 2	6.533	4.681
1 and 3	7.738	5.134
1 and 4	17.021	4.950
2 and 3	14.271	5.020
2 and 4	10.488	4.832
3 and 4	24.759	5.272

In every case the second column exceeds the third column, so we may state that the multiple comparisons procedure shows every pair of populations to be different. ∎

There should be no hesitation to apply the rank tests of this chapter to situations that have many ties. In fact, the Kruskal-Wallis test is an excellent test to use in a contingency table, where the rows represent ordered categories and the columns represent the different populations, as in the following.

	Population 1	2	3	$\cdots$	k	Row Totals	$\overline{R}_i$ = Average Rank
Category 1	O_{11}	O_{12}	O_{13}	$\cdots$	O_{1k}	t_1	$(t_1 + 1)/2$
2	O_{21}	O_{22}	O_{23}	$\cdots$	O_{2k}	t_2	$t_1 + (t_2 + 1)/2$
3	O_{31}	O_{32}	O_{33}	$\cdots$	O_{3k}	t_3	$t_1 + t_2 + (t_3 + 1)/2$
$\cdots$	$\cdots$	$\cdots$	$\cdots$	$\cdots$	$\cdots$		
c	O_{c1}	O_{c2}	O_{c3}	$\cdots$	O_{ck}	t_c	$\sum_{i=1}^{c-1} t_i + (t_c + 1)/2$
Column Totals	n_1	n_2	n_3	$\cdots$	n_k	N = Grand Total	

O_{ij} is the number of observations in population j that fall into the ith category. The average rank for row i is $\overline{R}_i$, which is computed from the row totals, as shown. The difference between this structure and ordinary contingency tables is that the categories (rows) are ordered. That is, all of the observations in row 1 are considered equal to each other but less than the observations in row 2, and so on. To compute the test statistic, the following form is recommended. Let the sum of the ranks in population (column) j be denoted by R_j,

$$R_j = \sum_{i=1}^{c} O_{ij}\overline{R}_i \tag{9}$$

and compute S^2 from the following equation.

$$S^2 = \frac{1}{N-1}\left[\sum_{i=1}^{c} t_i\overline{R}_i^2 - N(N+1)^2/4\right] \tag{10}$$

Then the test statistic T is computed by substituting Equations 9 and 10 into Equation 3, as before. Note that Equations 10 and 4 yield the same value of S^2, but Equation 10 is easier to use in this situation. If the null hypothesis is rejected the multiple comparisons procedure may be used, as described, to pinpoint differences where they exist.

EXAMPLE 2

Three instructors compared the grades they assigned over the past semester to see if some of them tended to give lower grades than others. The null hypothesis is:

H_0: The three instructors grade evenly with each other

and the alternative of interest is

H_1: Some instructors tend to grade lower than others

The grades being examined are as follows.

Grades	Instructor 1	Instructor 2	Instructor 3	Row Totals	Average Ranks
A	4	10	6	20	10.5
B	14	6	7	27	34
C	17	9	8	34	64.5
D	6	7	6	19	91
F	2	6	1	9	105
Total number of students	43	38	28	109	

The column rank sums are found from Equation 9.

$$R_1 = 2370.5 \qquad R_2 = 2156.5 \qquad R_3 = 1468$$

Checking on our calculations so far, the sum of the R_j should equal $N(N + 1)/2 = 5995$ for $N = 109$, and it does. From Equation 10 we compute $S^2 = 941.71$ and, finally, Equation 3 yields $T = 0.3209$.

The critical region of size 0.05, from Table A2 for 2 degrees of freedom, corresponds to all values of T greater than 5.991. The null hypothesis is clearly accepted. None of the instructors can be said to grade higher or lower than the others on the basis of the evidence presented.   ∎

□ *Theory*   The exact distribution of T is found under the assumption that all observations were obtained from the same or identical populations. The method is that of randomization, which was also used in finding the distribution of the Mann-Whitney test statistic. That is, under the preceding assumption, each arrangement of the ranks 1 to N into groups of sizes $n_1, n_2, \ldots, n_k$, respectively, is equally likely, and occurs with probability $n_1!n_2! \cdots n_k!/N!$, which is the reciprocal of the number of ways the N ranks may be divided into groups of sizes $n_1, n_2, \ldots, n_k$. The value of T is computed for each arrangement. The probabilities associated with equal values of T are then added to give the probability distribution of T.

For example, if $n_1 = 2$, $n_2 = 1$, and $n_3 = 1$ in the three-sample case, there are 12 equally likely arrangements of the four ranks; thus each arrangement has probability $1/12$. The 12 arrangements, with the associated values of T, are given as follows.

		Sample		
Arrangement	1	2	3	T
1	1, 2	3	4	2.7
2	1, 2	4	3	2.7
3	1, 3	2	4	1.8
4	1, 3	4	2	1.8
5	1, 4	2	3	0.3
6	1, 4	3	2	0.3
7	2, 3	1	4	2.7
8	2, 3	4	1	2.7
9	2, 4	1	3	1.8
10	2, 4	3	1	1.8
11	3, 4	1	2	2.7
12	3, 4	2	1	2.7

Therefore the probability function $f(x)$ and the distribution function $F(x)$ are given as follows for $n_1 = 2$, $n_2 = 1$, and $n_3 = 1$.

x	$f(x) = P(T = x)$	$F(x) = P(T \leq x)$
0.3	$2/12 = 1/6$	$1/6$
1.8	$4/12 = 1/3$	$1/2$
2.7	$6/12 = 1/2$	1.0

The large sample approximation for the distribution of T is based on the fact that R_i in Equation 2 is the sum of n_i random variables, and for large n_i the central limit theorem may be used. Thus

$$\frac{R_i - E(R_i)}{\sqrt{\text{Var}(R_i)}}$$

is approximately distributed as a standardized normal random variable when H_0 is true. From Theorem 1.4.5 the mean and variance of R_i may be expressed by

$$E(R_i) = \frac{n_i(N+1)}{2} \tag{11}$$

and

$$\text{Var}(R_i) = \frac{n_i(N+1)(N-n_i)}{12} \tag{12}$$

Therefore,

$$\left[\frac{R_i - E(R_i)}{\sqrt{\text{Var}(R_i)}}\right]^2 = \frac{\{R_i - [n_i(N+1)/2]\}^2}{n_i(N+1)(N-n_i)/12} \tag{13}$$

is approximately distributed as a chi-squared random variable with 1 degree of freedom. If the R_i were independent of each other the distribution of the sum

$$T' = \sum_{i=1}^{k} \frac{\{R_i - [n_i(N+1)/2]\}^2}{n_i(N+1)(N-n_i)/12} \tag{14}$$

could be approximated using the chi-squared distribution with k degrees of freedom. However, the sum of the R_is is $N(N+1)/2$, so there is a dependence among the R_is. Kruskal (1952) showed that if the ith term in T' is multiplied by $(N - n_i)/N$ for $i = 1, 2, \ldots, k$, then the result

$$T = \sum_{i=1}^{k} \frac{\{R_i - [n_i(N+1)/2]\}^2}{n_i(N+1)N/12} \tag{15}$$

is asymptotically distributed as a chi-squared random variable with $k - 1$ degrees of freedom. Equation 15 is merely a rearrangement of the terms in Equation 5, which originally defined the test statistic T. Therefore we have rationalized the use of the chi-squared approximation for the distribution of the Kruskal-Wallis test statistic. $\square$

Kruskal and Wallis (1952) found that for small α (less than about 0.10) and for selected small values of n_1, n_2, and n_3, the true level of significance is smaller than the stated level of significance associated with the chi-squared distribution, which indicates that the chi-squared approximation furnishes a conservative test in many if not most situations. Gabriel and Lachenbruch (1969) show that the chi-squared approximation is good even though the sample sizes may be small. The chi-squared approximation is compared with other approximations by Iman and Davenport (1976).

For two samples the Kruskal-Wallis test is equivalent to the Mann-Whitney test. Recall that in the Mann-Whitney test (Section 5.1) one sample was called X_1, . . . , X_n, while the other was Y_1, . . . , Y_m. The statistic T was defined by Equation 5.1.1 as

$$T = \sum_{i=1}^{n} R(X_i) \tag{16}$$

the sum of the ranks of the Xs in the combined sample corresponding to R_1 in the Kruskal-Wallis test. The Mann-Whitney two-tailed test consisted of rejecting H_0 if the statistic T was too large or too small. Because T is approximately normal for large sample sizes, one could reject H_0 if the quantity

$$\frac{T - E(T)}{\sqrt{\text{Var}(T)}} \tag{17}$$

is above or below the appropriate standardized normal quantiles or if its square,

$$\frac{[T - E(T)]^2}{\text{Var}(T)} \tag{18}$$

is above the $1 - \alpha$ quantile in a chi-squared distribution with 1 degree of freedom, according to Theorem 1.5.3. So the chi-squared distribution with 1 degree of freedom could have been used in the Mann-Whitney two-tailed test, with the quantity in Equation 18 as a test statistic. The Kruskal-Wallis test, with two samples, also uses the chi-squared distribution with 1 degree of freedom to test the same hypotheses as in the Mann-Whitney two-tailed test and, in fact, the Kruskal-Wallis test statistic is identical to the form of the Mann-Whitney test statistic given in Equation 18. Showing this is left as an exercise for the reader.

Justification of the usage of the rank tests presented thus far in the case of noncontinuous distributions is given by Conover (1973a). The exact distribution of the test statistic when ties are present is discussed by Klotz and Teng (1977). The multiple comparisons procedure is simply the usual parametric procedure, called Fisher's least significant difference, computed on the ranks rather than the data, as described by Conover and Iman (1979).

The usual parametric procedure is called the "one-way analysis of variance," or sometimes simply the one-way F test. The statistic used is given by

$$F = \frac{\left(\sum_{i=1}^{k} T_i^2 / n_i - C\right)/(k-1)}{\left(\sum_{i=1}^{k}\sum_{j=1}^{n_i} X_{ij}^2 - \sum_{i=1}^{k} T_i^2 / n_i\right)/(N-k)} \tag{19}$$

where T_i is the sum of the observations in the ith sample, and $C = T^2/N$ where T is the total of all of the observations. If the assumptions of the Kruskal-Wallis test are valid and, in addition, if the populations have in common a normal distribution, then the quantiles of the F statistic are given in Table A22. Look in the column for $k_1 = k - 1$ and the row marked $k_2 = N - k$ for N and k given by the experiment. Violation of the normality assumption usually has little effect on the distribution of the F statistic when the null hypothesis is true. However, the power of the F test may be considerably less than the Kruskal-Wallis test for certain types of nonnormality when H_0 is false. Data containing outliers are better suited for the Kruskal-Wallis test, for example.

The A.R.E. of the Kruskal-Wallis test relative to the F test is never less than 0.864 but may be as high as infinity if the distribution functions have identical shapes but differ only in their means. If the populations are normal the A.R.E. is $3/\pi = 0.955$. For uniform distributions the A.R.E. relative to the F test is 1.0; for double exponential distributions it is 1.5. Compared with the median test the A.R.E. of the Kruskal-Wallis test is 1.5, 3.0, and 0.75, respectively, for the three distributions just mentioned.

Rank sum tests similar to the Kruskal-Wallis test have been adapted by Steel (1960), Sherman (1965), and McDonald and Thompson (1967) for making multiple comparisons. Some tables for making multiple comparisons are provided by Tobach, Smith, Rose, and Richter (1967). Procedures for selecting the best populations are described by Rizvi and Sobel (1967), Sobel (1967), Rizvi, Sobel, and Woodworth (1968), and Puri and Puri (1969). Rank tests are presented for censored data by Basu (1967b) and Breslow (1970); for testing against ordered alternatives by G.R. Shorack (1967), Odeh (1971, 1972), and Tryon and Hettmansperger (1973); and for analysis of covariance by Puri and Sen (1969a). For other work concerned with rank tests and several independent samples see Sen (1962, 1966), Matthes and Truax (1965), Quade (1966), Crouse (1966), Sen and Govindarajulu (1966), Odeh (1967), Deshpande (1970), and Bhapkar and Deshpande (1968). Analysis of covariance is discussed by Quade (1967). Brunden (1972) considers using ranks to analyze 2×3 contingency tables.

EXERCISES

1. Random samples from each of three different types of light bulbs were tested to see how long the light bulbs lasted, with the following results.

Brand		
A	**B**	**C**
73	84	82
64	80	79
67	81	71
62	77	75
70		

Do these results indicate a significant difference between brands? If so, which brands appear to differ?

2. Four job training programs were tried on 20 new employees, where 5 employees were randomly assigned to each training program. The 20 employees were then placed under the same supervisor and, at the end of a certain specified period, the supervisor ranked the employees according to job ability, with the lowest ranks being assigned to those employees with the lowest job ability.

Program	Ranks
1	4, 6, 7, 2, 10
2	1, 8, 12, 3, 11
3	20, 19, 16, 14, 5
4	18, 15, 17, 13, 9

Do these data indicate a difference in the effectiveness of the various training programs? If so, which ones seem to be different?

3. The amount of damage to the soil on a farm caused by water and wind is examined for many different farms. At the same time the type of farming practiced on each farm is noted, with the following results.

	Type of Farming			
Amount of Damage	Minimum Tillage	Contour	Terrace	Other
		Number of Farms		
No damage	17	19	4	21
Slight damage	3	10	4	42
Moderate damage	0	2	2	34
Severe damage	0	0	2	6

Does the type of farming affect the degree of damage? If so, which types of farming are significantly different?

4. Three different types of radios, manufactured by the same company, all carry 1-year guarantees. A record is kept of how many radios needed to be replaced, were repairable, or were not returned under warranty.

	Type		
	A	B	C
Replaced	12	3	6
Repaired	10	8	7
Not Returned	82	96	58

Does there seem to be a significant difference among the reliabilities of the different radio types? If so, which ones seem to be different?

5. The amount of iron present in the livers of white rats is measured after the animals had been fed one of five diets for a prescribed length of time. There were 10 animals randomly assigned to each of the five diets.

Diet A	Diet B	Diet C	Diet D	Diet E
2.23	5.59	4.50	1.35	1.40
1.14	0.96	3.92	1.06	1.51
2.63	6.96	10.33	0.74	2.49
1.00	1.23	8.23	0.96	1.74
1.35	1.61	2.07	1.16	1.59
2.01	2.94	4.90	2.08	1.36
1.64	1.96	6.84	0.69	3.00
1.13	3.68	6.42	0.68	4.81
1.01	1.54	3.72	0.84	5.21
1.70	2.59	6.00	1.34	5.12

Do the different diets appear to affect the amount of iron present in the livers?

6. Twelve volunteers were assigned to each of three weight-reducing plans. The assignment of the volunteers to the plans was at random, and it was assumed that the 36 volunteers in all would resemble a random sample of people who might try a weight-reducing program. Test the null hypothesis that there is no difference in the probability distributions of the amount of weight lost under the three programs against the alternative that there is a difference. The results are given as the number of pounds lost by each person.

Plan A		Plan B		Plan C	
2	17	17	5	29	5
12	4	15	6	3	25
5	25	3	19	25	32
4	6	19	4	28	24
26	21	5	9	11	36
8	6	14	7	7	20

PROBLEMS

1. Show that Equations 3 and 5 are equivalent when there are no ties.

2. Find the exact distribution of the Kruskal-Wallis test statistic when H_0 is true, $n_1 = 3$,

$n_2 = 2$, $n_3 = 1$, and there are no ties. Compare your results with the quantiles given in Table A8.

3. In the two-sample case, what are some of the reasons why we might prefer to use the Mann-Whitney test instead of the Kruskal-Wallis test?

4. Show that Equations 10 and 4 are equivalent.

5. Suppose the F statistic in Equation 19 is computed on the ranks $R(X_{ij})$ instead of the observations X_{ij}. Then show the relationship

$$F = \frac{T/(k-1)}{(N-1-T)/(N-k)}$$

holds between F and T, given by Equation 3. Therefore the test that rejects H_0 for large T is equivalent to the test that rejects H_0 for large F, if F is computed on the ranks.

5.3　A TEST FOR EQUAL VARIANCES

The usual standard of comparison for several populations is based on the means or other measures of location of the populations. However, in some situations the variances of the populations may be the quantity of interest. For example, it has been claimed that the effect of seeding clouds with silver iodide is to increase the variance of the resulting rainfall. Such a claim may be tested by the method presented in this section.

The test for variances is analogous to the test just presented for means. That is, to test H_0: $E(X) = E(Y)$, the two independent samples were combined, ranked, and the sum of the ranks of the Xs was used as a test statistic. Recall that the variance is defined as the expected value of $(X - \mu)^2$ where μ is the mean of X. Thus to test H_0: $E[(X_i - \mu_x)^2] = E[(Y_j - \mu_y)^2]$ it seems reasonable to record the values of $(X_i - \mu_x)^2$ and $(Y_j - \mu_y)^2$ from two independent samples, assign ranks to them, and use the sum of the ranks of the $(X_i - \mu_x)^2$s as the test statistic. Talwar and Gentle (1977) studied this method. Although this technique could be used, more power is obtained when the ranks are squared first and then summed. This section contains a more accurate description of such a test.

▶ **The Squared Ranks Test for Variances** _____

Data　The data consist of two random samples. Let $X_1, X_2, \ldots, X_n$ denote the random sample of size n from population 1 and let $Y_1, Y_2, \ldots, Y_m$, denote the random sample of size m from population 2. Convert each X_i and Y_j to its absolute deviation from the mean using

$$U_i = |X_i - \mu_1|, i = 1, \ldots, n \tag{1}$$

and

$$V_j = |Y_j - \mu_2|, j = 1, \ldots, m \qquad (2)$$

where μ_1 and μ_2 are the means for populations 1 and 2. If μ_1 and μ_2 are unknown, use $\overline{X}$ for μ_1 and $\overline{Y}$ for μ_2, and the following test is still approximately valid.

Assign the ranks 1 to $n + m$ to the combined sample of Us and Vs in the usual way. If several values of U and/or V are exactly equal to each other (tied), assign to each the average of the ranks that would have been assigned to them had there been no ties. Let $R(U_i)$ and $R(V_j)$ denote the ranks and average ranks thus assigned. Note that ranking the U_is and V_js achieves the same results and is easier than ranking the values of $(X_i - \mu_1)^2$ and $(Y_j - \mu_2)^2$.

Assumptions

1. Both samples are random samples from their respective populations.
2. In addition to independence within each sample there is mutual independence between the two samples.
3. The measurement scale is at least interval.

Test Statistic   If there are no values of U tied with values of V, the sum of the squares of the ranks assigned to population 1 can be used as the test statistic.

$$T = \sum_{i=1}^{n} [R(U_i)]^2 \qquad (3)$$

If there are ties, subtract the mean from T and divide by the standard deviation to get

$$T_1 = \frac{T - n\overline{R^2}}{\left[\frac{nm}{N(N-1)} \sum_{i=1}^{N} R_i^4 - \frac{nm}{N-1} (\overline{R^2})^2 \right]^{\frac{1}{2}}} \qquad (4)$$

where $N = n + m$, $\overline{R^2}$ represents the average of the squared ranks of both samples combined:

$$\overline{R^2} = \frac{1}{N} \left\{ \sum_{i=1}^{n} [R(U_i)]^2 + \sum_{j=1}^{m} [R(V_j)]^2 \right\} \qquad (5)$$

and ΣR_i^4 represents the sum of the ranks raised to the fourth power:

$$\sum_{i=1}^{N} R_i^4 = \sum_{j=1}^{n} [R(U_j)]^4 + \sum_{j=1}^{m} [R(V_j)]^4 \qquad (6)$$

Null Distribution   Quantiles of the exact null distribution of T are given in Table A9 for the case of no ties and $n \leq 10$, $m \leq 10$. For sample sizes larger than 10 the following large-sample approximation, based on the standard normal quantiles z_p given in Table A1, can be used to obtain approximate quantiles w_p for T.

$$w_p = \frac{n(N + 1)(2N + 1)}{6} + z_p \sqrt{\frac{mn(N + 1)(2N + 1)(8N + 11)}{180}} \tag{7}$$

Recall that $N = n + m$.

The approximate null distribution of T_1 is the standard normal distribution, Table A1.

Hypotheses

A. (Two-Tailed Test)

H_0: X and Y are identically distributed, except for possibly different means
H_1: Var $(X) \neq$ Var (Y)

Reject H_0 at the level α if T (or T_1 in the case of ties) is greater than its $1 - \alpha/2$ quantile or less than its $\alpha/2$ quantile, found from Table A9 or Equation 7 in the case of T, or from Table A1 if T_1 is used.

The two-tailed p-value is twice the smaller of $P(Z \leq T_1)$ or $P(Z \geq T_1)$, obtained directly from Table A1 if T_1 is used. If T is used the approximate p-value can be found using Table A9, to find the smallest two-tailed test that results in rejection of H_0, or from the normal approximation

$$p\text{-value} = 2 \cdot (\text{smaller of the one-tailed } p\text{-values}) \tag{8}$$

where the lower-tailed p-value is approximately

$$\text{lower-tailed } p\text{-value} = P\left(Z \leq \frac{T - n(N + 1)(2N + 1)/6}{\sqrt{mn(N + 1)(2N + 1)(8N + 11)/180}}\right) \tag{9}$$

and the upper-tailed p-value is approximately

$$\text{upper-tailed } p\text{-value} = P\left(Z \geq \frac{T - n(N + 1)(2N + 1)/6}{\sqrt{mn(N + 1)(2N + 1)(8N + 11)/180}}\right) \tag{10}$$

B. (Lower-Tailed Test)

H_0: X and Y are identically distributed, except for possibly different means
H_1: Var $(X) <$ Var (Y)

Reject H_0 at the level α if T (or T_1 in the case of ties) is less than its α quantile, found from Table A9 or Equation 7 in the case of T, or from Table A1 if T_1 is used. The p-value is the probability of being less than or equal to T (or T_1) in the null distribution, which is given approximately by Equation 9 for T, or by $P(Z \le T_1)$, using Table A1.

C. (Upper-Tailed Test)

> H_0: X and Y are identically distributed, except for possibly different means
>
> H_1: Var (X) > Var (Y)

Reject H_0 at the level α if T (or T_1 in the case of ties) is greater than its $1 - \alpha$ quantile, given by Table A9 or Equation 7 in the case of T, or from Table A1 if T_1 is used. The p-value is the probability of being greater than or equal to the observed value of T (or T_1) in the null distribution. It is given approximately by Equation 10 for T, or directly from Table A1 for T_1.

Computer Assistance This test is performed by *StatXact*, where it is called Conover's test. ──────────────────────────────────── ◄

A Test For More Than Two Samples

If there are three or more samples, this test is modified easily to test the equality of several variances. From each observation subtract its population mean (or its sample mean when μ_i is unknown) and convert the sign of the resulting difference to +, as just described for two samples. Rank the combined absolute differences from smallest to largest, assigning average ranks in case of ties, again as described. Compute the sum of the squares of the ranks of each sample, letting S_1, $S_2, \ldots, S_k$ denote the sums for each of the k samples. Thus S_1 corresponds to T in the preceding two-sample case.

> H_0: All k populations are identical, except for possibly different means
>
> H_1: Some of the population variances are not equal to each other

The test statistic is

$$T_2 = \frac{1}{D^2}\left[\sum_{j=1}^{k}\frac{S_j^2}{n_j} - N(\bar{S})^2\right]$$

(11)

where n_j = number of observations in sample j

> $N = n_1 + n_2 + \cdots + n_k$
>
> S_j = the sum of the squared ranks in sample j

$$\overline{S} = \frac{1}{N}\sum_{j=1}^{k} S_j = \text{the average of all the squared ranks}$$

$$D^2 = \frac{1}{N-1}\left[\sum_{i=1}^{N} R_i^4 - N(\overline{S})^2\right]$$

and ΣR_i^4 represents the sum resulting after raising each rank to the fourth power. If there are no ties D^2 and $\overline{S}$ simplify to

$$D^2 = N(N+1)(2N+1)(8N+11)/180 \qquad (12)$$

and

$$\overline{S} = (N+1)(2N+1)/6 \qquad (13)$$

The null distribution is approximately the chi-squared distribution with $k-1$ degrees of freedom, whose upper quantiles are given in Table A2.

The null hypothesis is rejected if T_2 exceeds the $1-\alpha$ quantile of the chi-squared distribution with $k-1$ degrees of freedom, obtained from Table A2.

The p-value is approximately the probability of a chi-squared random variable with $k-1$ degrees of freedom being greater than the observed value of T_2. If H_0 is rejected, multiple comparisons may be made as described in the previous section. In this case the variances of populations i and j are said to differ if the following inequality is satisfied.

$$\left|\frac{S_i}{n_i} - \frac{S_j}{n_j}\right| > t_{1-\alpha/2}\left(D^2\frac{N-1-T_2}{N-k}\right)^{\frac{1}{2}}\left(\frac{1}{n_i}+\frac{1}{n_j}\right)^{\frac{1}{2}} \qquad (14)$$

where $t_{1-\alpha/2}$ is the $1-\alpha/2$ quantile of the t distribution obtained from Table A21 with $N-k$ degrees of freedom.

EXAMPLE I

A food packaging company would like to be reasonably sure that the boxes of cereal it produces do in fact contain at least the number of ounces of cereal stamped on the outside of the box. In order to do this it must set the average amount per box a little above the advertised amount, because the unavoidable variation caused by the packaging machine will sometimes put a little less or a little more cereal in the box. A machine with smaller variation would save the company money because the average amount per box could be adjusted to be closer to the advertised amount.

A new machine is being tested to see if it is less variable than the present machine, in which case it will be purchased to replace the old machine. Several boxes are filled with cereal using the present machine and the amount in each box is measured. The same is done for the new machine to test:

H_0: Both machines have equal variability

versus

H_1: The new machine has a smaller variance

The measurements and calculations are as follows.

Original Measurements		Absolute Deviation		Rank		Squared Rank	
Present (X)	New (Y)	Present (U)	New (V)	Present	New	Present	New
10.8	10.8	.06	.01	4	2 (tie)	16	4
11.1	10.5	.36	.29	10	8	100	64
10.4	11.0	.34	.21	9	7	81	49
10.1	10.9	.64	.11	12	6	144	36
11.3	10.8	.56	.01	11	2 (tie)	121	4
	10.7		.09		5		25
	10.8		.01		2 (tie)		4
$\overline{X} = 10.74$	$\overline{Y} = 10.79$					$T = $	462

$T = $ sum of squared ranks (present) $= 462$

$$\overline{R^2} = \frac{1}{12}(16 + 100 + \cdots + 25 + 4) = 54$$

$$\sum_{i=1}^{N} R_i^4 = (16)^2 + (100)^2 + \cdots + (25)^2 + (4)^2 = 60{,}660$$

$$T_1 = \frac{462 - 5(54)}{\left[\frac{(5)(7)}{(12)(11)}(60{,}660) - \frac{(5)(7)}{11}(54)^2\right]^{\frac{1}{2}}} = 2.3273$$

The preceding hypotheses match set C, because H_1 specifies the new machine (Y) has a smaller variance. The critical region corresponds to values of T_1 greater than 1.6449, the 0.95 quantile from Table A1, for an approximate α of 0.05. In this case T_1 exceeds 1.6449, so H_0 is rejected. A comparison of the observed $T_1 = 2.3273$ with the quantiles from Table A1 reveals a p-value of about 0.01.

Considerable simplification of the computations results whenever none of the values of U are tied with values of V, as in this example. Then ranks rather than average ranks can be used and the exact tables consulted. That is, in this example the only tie is among three values of V, so instead of using $2^2 = 4$ in the column on the far right, the values $1^2 = 1$, $2^2 = 4$, and $3^2 = 9$ can be used where the three tied values occur and the remainder of the test conducted as if there were no ties. The value of T happens to remain unchanged this time, is greater than the 0.95 quantile 410 from Table A9 for $n = 5$, $m = 7$, and shows the p-value to be about 0.01 as with the approximate test. ∎

□ *Theory*   Whenever two random variables X and Y are identically distributed except for having different means μ_1 and μ_2, $X - \mu_1$ and $Y - \mu_2$ not only have zero means, but they are identically distributed also. This means $U = |X - \mu_1|$ has the same distribution as $V = |Y - \mu_2|$, and $U^2 = (X - \mu_1)^2$ has the same distribution as $V^2 = (Y - \mu_2)^2$. So random samples of Xs and Ys furnish Us and Vs that are independent and identically distributed. Thus every assignment of ranks of the Us is equally likely, as in the Mann-Whitney test, and the distribution of any function of the ranks can be found as in Section 5.1.

Note that the ranks of the Us are the same as the ranks of the corresponding values of U^2, since $U_1 < U_2$ if and only if $U_1^2 < U_2^2$. Since we are interested in comparing $E(U^2)$ with $E(V^2)$, we should be looking at the ranks of the values of U^2 and V^2; however, it is equivalent and easier to consider the ranks of the Us and Vs.

Another important difference that distinguishes this rank test from the previous rank test is that we are using the squared ranks and not the ranks themselves. We say we are using scores instead of ranks, the scores being denoted by a function of R, $a(R)$, which is used in the test statistic instead of the rank R. Let T equal the sum of the scores associated with one sample, as in this test where the scores are $a(R) = R^2$. The distribution of T is found just as in Section 5.1. If the sample sizes are $n = 3$ and $m = 4$, there are 35 ways of selecting three out of the seven ranks. The three ranks 1, 2, 3 have corresponding scores $a(1)$, $a(2)$, and $a(3)$, which in turn yield some number for the test statistic T (depending on which scores are being used), which has probability $1/35$. The 35 ways of selecting three ranks out of seven give 35 values of T, possibly not all different, and the probability function of T is then obtained quite simply as in Section 5.1.

To use the large sample normal approximation for T it is necessary to find the mean and variance of T when H_0 is true. We have

$$T = \sum_{i=1}^{n} a(R_i) \tag{15}$$

where $R_1, \ldots, R_n$ represent the ranks of $U_1, \ldots, U_n$ in the combined sample of Us and Vs. We will find $E(T)$ and Var (T) for general scores $a(R)$ and substitute $a(R) = R^2$ at the end.

The mean of T is written as

$$E(T) = E\left[\sum_{i=1}^{n} a(R_i)\right] = \sum_{i=1}^{n} E[a(R_i)] \tag{16}$$

with the aid of Theorem 1.4.1. Because $P(R_i = j) = 1/N$ for each $j = 1, 2, \ldots,$ N, we have

$$E[a(R_i)] = \sum_{j=1}^{N} a(j) \cdot \frac{1}{N} = \frac{1}{N} \sum_{j=1}^{N} a(j) = \bar{a} \tag{17}$$

say. This is the same for all $i = 1$ to n, so Equation 16 becomes

$$E(T) = n\bar{a} \tag{18}$$

where $\bar{a}$ is the average of all the scores.

For the variance of T we use Theorem 1.4.3 to get

$$\text{Var}(T) = \sum_{i=1}^{n} \text{Var}[a(R_i)] + \sum_{\substack{i=1 \\ i \neq j}}^{n} \sum_{j=1}^{n} \text{Cov}[a(R_i), a(R_j)] \tag{19}$$

where, by the definition of variance,

$$\text{Var}[a(R_i)] = E\{[a(R_i) - \bar{a}]^2\} = \sum_{k=1}^{N} [a(k) - \bar{a}]^2 \cdot \frac{1}{N} = A \tag{20}$$

and where, by the definition of covariance,

$$\begin{aligned}
\text{Cov}[a(R_i), a(R_j)] &= E\{[a(R_i) - \bar{a}][a(R_j) - \bar{a}]\} \\
&= \sum_{\substack{k=1 \\ k \neq l}}^{N} \sum_{l=1}^{N} [a(k) - \bar{a}][a(l) - \bar{a}] \frac{1}{N(N-1)}
\end{aligned} \tag{21}$$

because $P(R_i = k, R_j = l) = 1/[N(N-1)]$ for all $k \neq l$. The expression in Equation 21 is simplified by adding and subtracting the term where $k = l$;

$$\begin{aligned}
\text{Cov}[a(R_i), a(R_j)] &= \sum_{k=1}^{N} [a(k) - \bar{a}] \sum_{l=1}^{N} [a(l) - \bar{a}] \frac{1}{N(N-1)} \\
&\quad - \sum_{k=1}^{N} [a(k) - \bar{a}]^2 \frac{1}{N(N-1)}
\end{aligned} \tag{22}$$

But the first summation equals zero because of the way $\bar{a}$ was defined in Equation 17, so Equation 22 simplifies to

$$\text{Cov}[a(R_i), a(R_j)] = -\frac{1}{N-1} A \tag{23}$$

where A is defined in Equation 20. Now the variance and covariance terms of Equations 20 and 23 are substituted into Equation 19 to get

$$\begin{aligned}
\text{Var}(T) &= \sum_{i=1}^{n} A - \sum_{\substack{i=1 \\ i \neq j}}^{n} \sum_{j=1}^{n} \frac{1}{N-1} A \\
&= nA - n(n-1)\frac{1}{N-1} A \\
&= \frac{n(N-n)}{N-1} A \\
&= \frac{nm}{(N-1)N} \sum_{i=1}^{N} [a(i) - \bar{a}]^2
\end{aligned} \tag{24}$$

because $N - n = m$. These Equations 18 and 24 were used in Sections 5.1 and 5.2 when ties were present and will be useful later in this chapter. For now we are interested in the case where $a(R) = R^2$ for the squared ranks test. In that case

$\bar{a}$ is written as $\overline{R^2}$ in Equation 4, and the denominator of Equation 4 is what the square root of Equation 24 becomes when the identity

$$\sum_{i=1}^{N} [a(i) - \bar{a}]^2 = \sum_{i=1}^{N} [a(i)]^2 - N(\bar{a})^2 \qquad (25)$$

is employed for ease in computation.

The extension of the two-sample case to the k-sample case is completely analogous to the extension of the two-sample Mann-Whitney test to the k-sample Kruskal-Wallis test. That is, the sums of scores are found for each of the k samples. Call these $S_1, S_2, \ldots, S_k$. The mean and variance of S_i are found from Equations 18 and 24 to be

$$E(S_i) = n_i \bar{a} \qquad (26)$$

and

$$\text{Var}(S_i) = \frac{n_i(N - n_i)}{(N - 1)N} \sum_{i=1}^{N} [a(i) - \bar{a}]^2 \qquad (27)$$

The terms $[S_i - E(S_i)]^2/\text{Var}(S_i)$ are multiplied by $(N - n_i)/N$ for $i = 1$ to k as in the Kruskal-Wallis test and are added together to get

$$T_2 = \sum_{i=1}^{k} \frac{(S_i - n_i \bar{a})^2}{n_i D^2} \qquad (28)$$

where

$$D^2 = \frac{1}{N-1} \sum_{i=1}^{N} [a(i) - \bar{a}]^2 = \frac{1}{N-1} \left\{ \sum_{i=1}^{N} [a(i)]^2 - N(\bar{a})^2 \right\} \qquad (29)$$

Equation 28 simplifies to

$$T_2 = \frac{1}{D^2} \left[\sum_{j=1}^{k} \frac{S_j^2}{n_j} - N(\bar{a})^2 \right] \qquad (30)$$

which matches Equation 11 when the scores are the squared ranks. The suggested multiple comparisons procedure is an approximate procedure that becomes exact as the sample sizes get large. $\square$

If the populations of X and Y have normal distributions the appropriate statistic to use is the ratio of the two "sample variances,"

$$F = \frac{\dfrac{1}{n-1} \sum_{i=1}^{n} (X_i - \overline{X})^2}{\dfrac{1}{m-1} \sum_{j=1}^{m} (Y_j - \overline{Y})^2} \qquad (31)$$

which has the F distribution. The upper quantiles of F are given in Table A22 in column $k_1 = n - 1$ and row $k_2 = m - 1$. The lower quantiles are not given but may be found by taking the reciprocal of the upper quantile found in column $k_1 = m - 1$ and row $k_2 = n - 1$. Appropriate one-tailed or two-tailed tests are then obtained.

The F test is very sensitive to the assumption of normality, as pointed out by Siegel and Tukey (1960). The true distribution may be symmetric and resemble somewhat the normal distribution, such as the double exponential distribution, and yet the true level of significance may be two or three times as large as it is supposed to be. For this reason the F test is not a very safe test to use unless one is sure that the populations are normal.

If the squared ranks test is used instead of the F test when the populations are normal the A.R.E. is only $15/(2\pi^2) = 0.76$. However, for the double exponential distribution the A.R.E. is 1.08 and for the uniform distribution the A.R.E. is 1.00. These same efficiencies apply to the k-sample case as well. Thus the sensitivity of the F test to the assumption of normality, coupled with its lack of power in some reasonably nonnormal situations, encourages the consideration of a nonparametric test for variances.

The result of using $\overline{X}$ and $\overline{Y}$ in place of the true means of X and Y in the squared ranks test is to make the test approximate rather than exact and the exact distribution of the test statistic dependent on the true population distribution. This causes a problem with highly skewed populations, as shown in an extensive simulation study of 56 tests for variances, conducted by Conover, Johnson, and Johnson (1981). The level of significance can become very large due to the skewness of the populations, so the sample medians of X and Y are recommended for the adjustment instead of the sample means in that case. Either procedure gives a test that is asymptotically distribution free, however, which means that the approximation becomes exact as the sample sizes get large.

Another popular nonparametric test for the two-sample scale problem is based on the statistic

$$T = \sum_{i=1}^{n} \left[R(X_i) - \frac{n + m + 1}{2} \right]^2 \tag{32}$$

where $R(X_i)$ is the rank of X_i, as in the Mann-Whitney test. This was proposed by Mood (1954). Exact tables are given by Laubscher, Steffens, and DeLange (1968) under the null hypothesis of identical distribution functions. The null distribution tables of a related statistic

$$T = \sum_{i=1}^{n} [R(X_i) - \overline{R}_x]^2 \tag{33}$$

where

$$\overline{R}_x = \frac{1}{n} \sum_{i=1}^{n} R(X_i)$$

are given by Hollander (1963). Ansari and Bradley (1960) discuss the A.R.E. of the Mood test and others.

Further discussion of the squared ranks test and more extensive tables are given by Conover and Iman (1978a). A slight variation of the test is examined by Talwar and Gentle (1977). Other tests for scale are considered by Sen (1963), Puri (1965), Mielke (1967), Duran and Mielke (1968), Shorack (1969), Hwang and Klotz (1975) and Fligner and Killeen (1976) for two samples, and Tsai, Duran, and Lewis (1975) for several samples. A comprehensive comparison of 56 tests for variances appears in Conover, Johnson, and Johnson (1981). Tests designed to detect both location and scale differences are presented by Lepage (1971, 1973, 1977), Mielke (1972), and Duran, Tsai, and Lewis (1976). The correlation between rank tests for scale and rank tests for location is studied by Gibbons (1967) and Hollander (1968). Estimation of scale parameters is considered further by Moses (1963), van Eeden (1964), Basu and Woodworth (1967), Bauer (1972), Laubscher and Odeh (1976), and Bhattacharayya (1977). If the location parameters are unknown and possibly unequal, see Raghavachari (1965a), Puri (1968), and Nemenyi (1969) for modified tests. Further references may be found in an excellent review article by Duran (1976), or in a bibliography by Daniel (1979).

EXERCISES

1. A blood bank kept a record of the rate of heartbeats for several blood donors.

Men	Women
58	66
76	74
82	69
74	76
79	72
65	73
74	75
86	67
	68

   Is the variation among the men significantly greater than the variation among the women?

2. A particular watershed has been built up extensively in recent years, with housing developments, dams, and so forth. A random sample of stream flow rates (cubic feet per minute) for a stream in that watershed is compared with a sample of rates from earlier times to see if the variability has changed.

Present Rates	Past Rates
32	39
36	21
41	58
27	46
35	30
48	22
31	17
28	19

Is there a significant difference in variances?

3. Three different methods of instruction are compared by assigning fifth-grade students at random to three different classrooms. The grade-level attainment (as measured by a standardized exam) of each student is measured at the beginning of the year and again at the end of the year, and the increase for each student is noted.

Method of Instruction	Increase in Attainment
Structured Classes	0.7, 1.0, 2.0, 1.4, 0.5, 0.8, 1.0, 1.1, 1.9, 1.2, 1.5
Individual Studies	1.7, 2.1, −0.4, 0, 1.0, 1.1, 0.9, 2.3, 1.3, 0.4, 0.5
No Walls Classroom	0.9, 0.9, 1.0, 0, 0.1, −0.6, 2.2, −0.3, 0.6, 2.4, 2.5

Does there seem to be a difference in variance associated with the three methods of instruction? If so, which methods appear to differ in variability?

4. An investment class was divided into three groups of students. One group was instructed to invest in bonds, the second in blue chip stocks, and the third in speculative issues. Each student "invested" (on paper only) $10,000 and evaluated the hypothetical profit or loss at the end of 3 months with the following results.

Bonds	Blue Chip	Speculative
146	176	−540
180	110	1052
192	212	642
185	108	−281
153	196	67

Is the difference in variance significant? If so, which groups are significantly different?

PROBLEMS

1. Find the exact distribution of T as given by Equation 3 for $n = 3$ and $m = 4$ and compare it with the quantiles from Table A9.

2. Show that Equation 28 is equivalent to Equation 30.

3. Find the exact distribution for $n = 3$, $n = 4$, for the Mood statistic given by Equation 32 and the Hollander statistic given by Equation 33.

4. Show that the mean of the Mood statistic, given by Equation 32, is $n(N + 1)(N - 1)/12$.

5. Another test for equal variances was devised by Siegel and Tukey (1960). In the ordered combined sample of Xs and Ys, assign rank 1 to the smallest value, rank 2 to the largest value, rank 3 to the second largest value, rank 4 to the second smallest value, rank 5 to the third smallest value, and so on, alternately assigning ranks to the end values two at a time (after the first) and proceeding toward the middle. The test statistic is the sum of the ranks assigned to the sample of Xs.

   (a) Justify the use of the Table A7 for the statistic when both populations are identical.

   (b) Which tail (upper or lower) of the critical region should be used for the one-sided alternative Var $(X) >$ Var (Y)?

   (c) Use an extreme example to show that this test has little power if the two population means are far apart.

6. Show that if the F statistic, as defined by Equation 5.2.19, is computed on the scores $a(i)$ the result can be simplified to the form

   $$F = \frac{T_2/(k - 1)}{(N - 1 - T_2)/(N - k)}$$

   where T_2 is given in Equation 28 or 30. Note that this mathematical relationship holds for all types of scores.

5.4 MEASURES OF RANK CORRELATION

A measure of correlation is a random variable that is used in situations where the data consist of pairs of numbers, such as in bivariate data. Suppose a bivariate random sample of size n is represented by (X_1, Y_1), (X_2, Y_2), . . . , (X_n, Y_n). We will use (X, Y) when referring to the (X_i, Y_i) in general. That is, the (X_i, Y_i) for $i = 1, 2, . . . , n$ have identical bivariate distributions, the same bivariate distribution as (X, Y) has.

Examples of bivariate random variables include one where X_i represents the height of the ith man and Y_i represent his father's height, or where X_i represents a test score of the ith individual and Y_i represents her amount of training.

By tradition, a measure of correlation between X and Y should satisfy the following requirements in order to be acceptable.

1. The measure of correlation should assume only values between -1 and $+1$.

2. If the larger values of X tend to be paired with the larger values of Y, and hence the smaller values of X and Y tend to be paired together, then the measure of correlation should be positive, and close to $+1.0$ if the tendency is strong. Then we would speak of a positive correlation between X and Y.

3. If the larger values of X tend to be paired with the smaller values of Y, and vice versa, then the measure of correlation should be negative and close to

−1.0 if the tendency is strong. Then we say that X and Y are negatively correlated.

4. If the values of X seem to be randomly paired with the values of Y, the measure of correlation should be fairly close to zero. This should be the case when X and Y are independent, and possibly some cases where X and Y are not independent. We then say that X and Y are uncorrelated, or have no correlation, or have correlation zero.

The most commonly used measure of correlation is Pearson's product moment correlation coefficient, denoted by r and defined as

$$r = \frac{\sum_{i=1}^{n} (X_i - \overline{X})(Y_i - \overline{Y})}{\left[\sum_{i=1}^{n} (X_i - \overline{X})^2 \sum_{i=1}^{n} (Y_i - \overline{Y})^2 \right]^{\frac{1}{2}}} \tag{1}$$

where $\overline{X}$ and $\overline{Y}$ are the sample means as defined in Section 2.2. An easier form to use with a calculator is

$$r = \frac{\sum_{i=1}^{n} X_i Y_i - n\overline{X}\,\overline{Y}}{\left(\sum_{i=1}^{n} X_i^2 - n\overline{X}^2 \right)^{\frac{1}{2}} \left(\sum_{i=1}^{n} Y_i^2 - n\overline{Y}^2 \right)^{\frac{1}{2}}} \tag{2}$$

If the numerator and denominator in Equation 1 are divided by n, r becomes

$$r = \frac{\frac{1}{n} \sum_{i=1}^{n} (X_i - \overline{X})(Y_i - \overline{Y})}{\left[\frac{1}{n} \sum_{i=1}^{n} (X_i - \overline{X})^2 \right]^{\frac{1}{2}} \left[\frac{1}{n} \sum_{i=1}^{n} (Y_i - \overline{Y})^2 \right]^{\frac{1}{2}}} \tag{3}$$

which may be easily remembered as the sample covariance in the numerator, and the product of the two sample standard deviations in the denominator.

Pearson's r is a measure of the strength of the linear association between X and Y. This means that if a plot of Y versus X shows the points (X, Y) all lie on, or close to, a straight line, then r will equal, or be close to, 1.0 if the line is sloping upward, and −1.0 if the line is sloping downward.

This measure of correlation may be used with any data of a numeric nature without any requirements concerning the scale of measurement or the type of underlying distribution, although it is difficult to interpret unless the scale of measurement is at least interval. It meets the necessary requirements of an acceptable measure of correlation. However, r is a random variable and, as such, r has a distribution function. Unfortunately, the distribution function of r depends on the bivariate distribution function of (X, Y). Therefore r has no value as a test statistic in nonparametric tests or for forming confidence intervals unless, of course, the distribution of (X, Y) is known.

In addition to this widely accepted r, many other measures of correlation have been invented that satisfy the preceding requirements for acceptability. An excellent and readable survey article by Kruskal (1958) discusses many of these. Some measures of correlation possess distribution functions that do not depend on the bivariate distribution function of (X, Y) if X and Y are independent and, therefore, they may be used as test statistics in nonparametric tests of independence. The measures of correlation selected for presentation here are functions of only the ranks assigned to the observations. They possess distribution functions that are independent of the bivariate distribution function of (X, Y) if X and Y are independent and continuous. They may even be used as measures of correlation on certain types of nonnumeric data, if the data meet the ordinal scale of measurement. The first rank correlation coefficient we present is simply Pearson's r computed on the ranks of X and Y.

▶ **Spearman's Rho** _____

Data   The data may consist of a bivariate random sample of size n, (X_1, Y_1), $(X_2, Y_2), \ldots , (X_n, Y_n)$. Let $R(X_i)$ be the rank of X_i as compared with the other X values, for $i = 1, 2, \ldots , n$. That is, $R(X_i) = 1$ if X_i is the smallest of $X_1, X_2, \ldots , X_n$, $R(X_i) = 2$ if X_i is the second smallest, and so on, with rank n being assigned to the largest of the X_i. Similarly, let $R(Y_i)$ equal $1, 2, \ldots ,$ or n, depending on the relative magnitude of Y_i as compared with $Y_1, Y_2, \ldots , Y_n$, for each i.

Or the data may consist of nonnumeric observations occurring in n pairs if the observations are such that they can be ranked in the manner just described. The ranking may be based on the quality of the observations ("worst" observation to "best" observation) or according to the degree of preference attached to the observations, and so on.

In case of ties, assign to each tied value the average of the ranks that would have been assigned if there had been no ties, as was done in the Mann-Whitney and Kruskal-Wallis tests.

Measure of Correlation   The measure of correlation as given by Spearman (1904) is usually designated by ρ (rho) and is defined as

$$\rho = \frac{\sum\limits_{i=1}^{n} R(X_i)R(Y_i) - n\left(\dfrac{n+1}{2}\right)^2}{\left(\sum\limits_{i=1}^{n} R(X_i)^2 - n\left(\dfrac{n+1}{2}\right)^2\right)^{\frac{1}{2}}\left(\sum\limits_{i=1}^{n} R(Y_i)^2 - n\left(\dfrac{n+1}{2}\right)^2\right)^{\frac{1}{2}}} \tag{4}$$

which is simply Pearson's r computed on the ranks and average ranks.

If there are no ties, an equivalent but computationally easier form is given by

$$\rho = 1 - \frac{6\sum\limits_{i=1}^{n}[R(X_i) - R(Y_i)]^2}{n(n^2 - 1)} = 1 - \frac{6T}{n(n^2 - 1)} \tag{5}$$

where T represents the entire sum in the numerator. This form is equivalent only if there are no ties. If there are many ties use Equation 4. If a moderate number of ties is present in the data, Equation 5 is recommended for computational simplicity, since the difference between Equations 4 and 5 will be slight.

As we said, Spearman's ρ is merely what one obtains by replacing the observations by their ranks and then computing Pearson's r on the ranks. This may be seen as follows. If the data are replaced by their ranks, then $\overline{X}$ and $\overline{Y}$ correspond to

$$\overline{R(X)} = \frac{1}{n}\sum_{i=1}^{n} R(X_i) = \frac{1}{n}\sum_{i=1}^{n} i = \frac{1}{n}\frac{n(n+1)}{2}$$

$$= \frac{n+1}{2} \tag{6}$$

and

$$\overline{R(Y)} = \frac{n+1}{2} \tag{7}$$

so that Equation 2 becomes Equation 4.

EXAMPLE 1

Twelve MBA graduates are studied to measure the strength of the relationship between their score on the GMAT, which they took prior to entering graduate school, and their grade point average while they were in the MBA program. Their GMAT scores and their GPAs are given below, along with the ranks and some computations.

Student	GMAT (X)	GPA (Y)	R(X)	R(Y)	$[R(X) - R(Y)]^2$
1	710	4.0	12	11.5	0.25
2	610	4.0	9.5	11.5	4
3	640	3.9	11	10	1
4	580	3.8	8	9	1
5	545	3.7	3	8	25
6	560	3.6	5	7	4
7	610	3.5	9.5	5	20.25
8	530	3.5	1	5	16
9	560	3.5	5	5	0
10	540	3.3	2	3	1
11	570	3.2	7	1.5	30.25
12	560	3.2	5	1.5	12.25

Equation 4 should be used because there are ties involving about half of the observations.

$$\sum_{i=1}^{12} [R(X_i)]^2 = 647.5 \qquad \sum_{i=1}^{12} [R(Y_i)]^2 = 647$$

Without ties these would equal $n(n + 1)(2n + 1)/6 = 650$ from Lemma 2 in Section 1.4. Also,

$$\sum_{i=1}^{12} R(X_i)R(Y_i) = 589.75 \tag{8}$$

Substitution of the above results into Equation 4 gives

$$\rho = \frac{589.75 - 12\left(\frac{13}{2}\right)^2}{\left(647.5 - 12\left(\frac{13}{2}\right)^2\right)^{1/2}\left(647 - 12\left(\frac{13}{2}\right)^2\right)^{1/2}} = 0.5900 \tag{9}$$

For comparison, Equation (5) gives

$$\rho = 1 - \frac{6(115)}{12(12^2 - 1)} = 0.5979 \tag{10}$$

which is slightly larger than the more accurate $\rho = 0.5900$. The difference is due to the use of average ranks for ties.

As a point of interest, Pearson's r computed on the original data is $r = 0.6630$. In this case the linear relationship between X and Y appears stronger than the linear relationship between the rank of X and the rank of Y. ■

Hypothesis Test   The Spearman rank correlation coefficient is often used as a test statistic to test for independence between two random variables. See the Data section as given above for Spearman's ρ. The test statistic is given by Equation 4.

Null Distribution Exact quantiles of ρ when X and Y are independent are given in Table A10 for $n \leq 30$ and no ties. For larger n, or many ties, the pth quantile of ρ is given approximately by

$$w_p = \frac{z_p}{\sqrt{n-1}} \qquad (11)$$

where z_p is the standard normal quantile found in Table A1.

Hypotheses Spearman's ρ is insensitive to some types of dependence, so it is better to be specific as to what type of dependence may be detected. Therefore, the hypotheses take the following form.

A. (Two-Tailed Test)

H_0: The X_i and Y_i are mutually independent

H_1: Either (a) there is a tendency for the larger values of X to be paired with the larger values of Y, or (b) there is a tendency for the smaller values of X to be paired with the larger values of Y

Reject H_0 at the level α if the absolute value of ρ, $|\rho|$, is greater than its $1 - \alpha/2$ quantile obtained from Table A10 or Equation 11. The approximate two-tailed p-value is

$$p\text{-value} = 2 \cdot P(Z \geq |\rho| \sqrt{n-1}) \qquad (12)$$

using Table A1.

B. (Lower-Tailed Test for Negative Correlation)

H_0: The X_i and Y_i are mutually independent

H_1: There is a tendency for the smaller values of X to be paired with the larger values of Y, and vice versa

Reject H_0 at the level α if $\rho < -w_{1-\alpha}$ where $w_{1-\alpha}$ is found either in Table A10 or from Equation 11. The approximate lower-tailed p-value is

$$p\text{-value} = P(Z \leq \rho \sqrt{n-1}) \qquad (13)$$

using Table A1.

C. (Upper-Tailed Test for Positive Correlation)

H_0: The X_i and Y_i are mutually independent

H_1: There is a tendency for the larger values of X and Y to be paired together

Reject H_0 at the level α if $\rho > w_{1-\alpha}$ where $w_{1-\alpha}$ is found either in Table A10 or from Equation 11. The approximate upper-tailed p-value is

$$p\text{-value} = P(Z \geq \rho\sqrt{n-1}) \tag{14}$$

using Table A1.

Computer Assistance *Minitab, S-Plus, SAS,* and *StatXact* compute Spearman's ρ and perform a test for independence. These and other programs will convert the data to ranks and compute Pearson's r on the ranks, which automatically corrects for ties. —————————————————————————————— ◀

EXAMPLE 2

Let us continue with Example 1. Suppose the twelve MBA graduates in Example 1 are a random sample of all recent MBA graduates, and we want to know if there is a tendency for high GPAs to be associated with high GMAT scores. Then the null hypothesis is

H_0: GPAs are independent of GMAT scores

and the alternative hypothesis of interest is

H_1: High GPAs tend to be associated with high GMAT scores

which matches hypothesis C for an upper-tailed test. Therefore H_0 is rejected at $\alpha = 0.05$ if ρ exceeds its 0.95 quantile. Because of the extensive ties the quantile in Table A10 for $n = 12$, $w_{0.95} = 0.4965$, is only approximate. The normal approximation

$$w_{0.95} = \frac{1.6449}{\sqrt{11}} = 0.4960$$

is almost the same, but presumably more accurate.

The observed value of ρ is 0.5900 so we can safely assert that there is a positive correlation between GPA and GMAT scores for recent MBA graduates. The p-value is found approximately from Equation 14.

$$p\text{-value} = P(Z \geq 0.5900\sqrt{11}) = P(Z \geq 1.9568) = 0.025 \qquad ■$$

The next measure of correlation we are presenting resembles Spearman's ρ in that it is based on the order (ranks) of the observations rather than the numbers themselves, and the distribution of the measure does not depend on the distribution of X and Y if X and Y are independent and continuous. This measure, called Kendall's tau (τ), is usually considered to be more difficult to compute than Spearman's ρ. The chief advantage of Kendall's τ is that its distribution approaches the normal distribution quite rapidly so that the normal approximation is better for Kendall's τ than it is for Spearman's ρ, when the null hypothesis of independence between X and Y is true. Another advantage of Kendall's τ is its direct and simple interpretation in terms of probabilities of observing concordant and discordant pairs, as defined next.

▶ Kendall's Tau

Data The data may consist of a bivariate random sample of size n, (X_i, Y_i) for $i = 1, 2, \ldots, n$. Two observations, for example (1.3, 2.2) and (1.6, 2.7), are called *concordant* if both members of one observation are larger than their respective members of the other observation. Let N_c denote the number of concordant pairs of observations, out of the $\binom{n}{2}$ total possible pairs. A pair of observations, such as (1.3, 2.2) and (1.6, 1.1), are called *discordant* if the two numbers in one observation differ in opposite directions (one negative and one positive) from the respective members in the other observation. Let N_d be the total number of discordant pairs of observations. Pairs with ties between respective members are counted as discussed below under the Ties section. Because the n observations may be paired $\binom{n}{2} = n(n-1)/2$ different ways, the number of concordant pairs N_c plus the number of discordant pairs N_d plus the number of pairs with ties will add up to $n(n-1)/2$.

The data may also consist of nonnumeric observations occurring in n pairs if the observations are such that N_c and N_d just described may be computed.

Measure of Correlation The measure of correlation proposed by Kendall (1938) in the case of no ties is

$$\tau = \frac{N_c - N_d}{n(n-1)/2} \tag{15}$$

If all pairs are concordant, Kendall's τ equals 1.0. If all pairs are discordant, the value is -1.0. As a measure of correlation, Kendall's τ satisfies the requirements stated at the beginning of this section.

Ties In more precise terms a pair of bivariate observations (X_1, Y_1) and (X_2, Y_2) is considered concordant if $(Y_2 - Y_1)/(X_2 - X_1)$ is greater than zero, and discordant if it is less than zero. If $X_1 = X_2$ the denominator is zero so no comparison can be made. However, if $Y_1 = Y_2$ (and $X_1 \neq X_2$) the ratio $(Y_2 - Y_1)/(X_2 - X_1)$ is zero. In this case the pair should be counted as $\frac{1}{2}$ concordant and $\frac{1}{2}$ discordant. This makes no difference in the numerator of τ because the $\frac{1}{2}$ terms cancel when computing $N_c - N_d$. However, it makes a difference in the way τ should be computed when ties are present.

In the case of ties we can use

$$\tau = \frac{N_c - N_d}{N_c + N_d} \tag{16}$$

where all pairs (X_i, Y_i) and (X_j, Y_j) with $X_i \neq X_j$ are compared. This version of Kendall's τ has the advantage of achieving $+1$ or -1 even if ties are present. It was first discussed by Goodman and Kruskal (1963), and is sometimes called the *gamma coefficient*.

In summary,

if $\dfrac{Y_j - Y_i}{X_j - X_i} > 0$, add 1 to N_c (concordant)

if $\dfrac{Y_j - Y_i}{X_j - X_i} < 0$, add 1 to N_d (discordant)

if $\dfrac{Y_j - Y_i}{X_j - X_i} = 0$, add $\frac{1}{2}$ to N_c and $\frac{1}{2}$ to N_d

if $X_i = X_j$, no comparison is made.

The computation of τ is simplified if the observations (X_i, Y_i) are arranged in a column according to increasing values of X. Then each Y may be compared only with those below it, and the number of concordant and discordant comparisons is easily determined. Also, each pair of observations is considered only once. The procedure is illustrated in the following example.

EXAMPLE 3

Again we will use the data in Example 1 for purposes of illustration. Arrangement of the data (X_i, Y_i) according to increasing values of X gives the following.

	X_i, Y_i	Concordant Pairs Below (X_i, Y_i)	Discordant Pairs Below (X_i, Y_i)
	(530, 3.5)	7	4
	(540, 3.3)	8	2
	(545, 3.7)	4	5
tie	(560, 3.2)	5.5	0.5
	(560, 3.5)	4.5	1.5
	(560, 3.6)	4	2
	(570, 3.2)	5	0
	(580, 3.8)	3	1
tie	(610, 3.5)	2	0
	(610, 4.0)	0.5	1.5
	(640, 3.9)	1	0
	(740, 4.0)		
		$N_c = 44.5$	$N_d = 17.5$

Kendall's τ is given by

$$\tau = \frac{N_c - N_d}{N_c + N_d} = \frac{44.5 - 17.5}{44.5 + 17.5} = 0.4355$$

There is a positive rank correlation between the GMAT scores and the GPAs as measured by Kendall's τ. ∎

Hypothesis Test Kendall's τ may also be used as a test statistic to test the null hypothesis of independence between X and Y, with possible one-tailed or two-tailed alternatives as described with Spearman's ρ. Some arithmetic may be saved, however, by using $N_c - N_d$ as a test statistic, without dividing by $n(n-1)/2$ to obtain τ. Therefore we use T as the Kendall test statistic, where T is defined as

$$T = N_c - N_d \tag{17}$$

See the Data section as given above for Kendall's τ. The test statistic is given by Equation 17 in the case of no ties or few ties. If ties are extensive then τ, as given by Equation 16, should be used.

Null Distribution Exact upper quantiles for τ and T when X and Y are independent are given in Table A11 for $n \leq 60$ in the case of no ties. Lower quantiles are the negative of the upper quantiles given in the table. For larger n or many ties the pth quantile of τ is given approximately by

$$w_p = z_p \frac{\sqrt{2(2n+5)}}{3\sqrt{n(n-1)}} \tag{18}$$

where z_p is the pth quantile of a standard normal random variable, given in Table A1. The pth quantile of T is given approximately by

$$w_p = z_p \sqrt{n(n-1)(2n+5)/18} \tag{19}$$

Hypotheses

A. (Two-Tailed Test)

H_0: X and Y are independent

H_1: Pairs of observations either tend to be concordant, or tend to be discordant

Reject H_0 at the level α if T (or τ) is less than its $\alpha/2$ quantile or greater than its $1 - \alpha/2$ quantile in the null distribution (see Table A11).

The two-tailed p-value is twice the smaller of the one-tailed p-values, given approximately by

$$p(\text{lower-tailed}) = P\left(Z \le \frac{(T+1)\sqrt{18}}{\sqrt{n(n-1)(2n+5)}}\right) \tag{20}$$

and

$$p(\text{upper-tailed}) = P\left(Z \ge \frac{(T-1)\sqrt{18}}{\sqrt{n(n-1)(2n+5)}}\right) \tag{21}$$

where T is the observed value of $T_c - T_d$, the continuity correction is 1, and Z is a standard normal random variable whose probabilities are given in Table A1.

B. (Lower-Tailed Test)

H_0: X and Y are independent

H_1: Pairs of observations tend to be discordant

Reject H_0 at the level α if T (or τ) is less than its α quantile in the null distribution (see Table A11). The lower-tailed p-value is given approximately by Equation 20.

C. (Upper-Tailed Test)

H_0: X and Y are independent

H_1: Pairs of observations tend to be concordant

Reject H_0 at the level α if T (or τ) is greater than its $1 - \alpha$ quantile in the null distribution (see Table A11). The upper-tailed p-value is given approximately by Equation 21.

Computer Assistance   *Minitab, S-Plus, SAS,* and *StatXact* compute Kendall's τ and perform a test for independence. ──────────────────────◀

EXAMPLE 4

In Example 3 Kendall's τ was computed by first finding the value of

$$T = N_c - N_d = 44.5 - 17.5 = 27$$

If we are interested in using T to test the null hypothesis of independence between the student's GMAT score and his or her GPA, to see if higher GPAs tend to be associated with higher GMAT scores, then the null hypothesis is rejected at $\alpha = 0.05$ if T is greater than $w_{0.95} = 24$, found in Table A11. Because $T = 27$, the null hypothesis is rejected. The upper-tailed p-value is approximated from Equation 21.

$$p\text{-value} = P(T \geq 27)$$

$$\cong P\left(Z \geq \frac{(27-1)\sqrt{18}}{\sqrt{12 \cdot 11 \cdot 29}}\right)$$

$$= P(Z \geq 1.7829)$$

$$= 0.037$$

If we use τ as defined by Equation 16 as the test statistic, because of the presence of ties, the results are similar. ∎

The same data were used for both Spearman's ρ and Kendall's τ in order to compare the two statistics better. It was seen that Spearman's ρ ($\rho = 0.5900$) was a larger number than Kendall's τ ($\tau = 0.4355$). However, the two tests using the two statistics (or their equivalents) produced nearly identical results. Both of the preceding statements hold true in most, but not all, situations. Spearman's ρ tends to be larger than Kendall's τ, in absolute value. However, as a test of significance there is no strong reason to prefer one over the other, because both will produce nearly identical results in most cases.

▶ The Daniels Test for Trend

Daniels (1950) proposed the use of Spearman's ρ to test for trend by pairing measurements, called X_i, with the time (or order) at which the measurements were taken. The assumption is that the X_is are mutually independent, and the null hypothesis is that they are identically distributed. The alternative hypothesis is that the distribution of the X_is is related to time so that as time goes on, the X measurements tend to become larger (or smaller). The idea of trend was discussed more fully in Section 3.5, where the Cox and Stuart test for trend was presented. Tests of trend based on Spearman's ρ or Kendall's τ are generally considered to be more powerful than the Cox and Stuart test. It was mentioned in Section 3.5 that the A.R.E. of the Cox and Stuart test for trend, when applied to random variables known to be normally distributed, is about 0.78 with respect to the test based on the regression coefficient, while the A.R.E. of these tests using Spearman's ρ or Kendall's τ is about 0.98 under the same conditions, according to Stuart (1956). However, these tests are not as widely applicable as the Cox and Stuart test. For instance, these tests would be inappropriate in Example 3.5.3. These tests are appropriate in Example 3.5.2, and so we use that example to illustrate the use of Spearman's ρ as a test for trend. The procedure using Kendall's τ is similar. ◀

EXAMPLE 5

In Example 3.5.2, nineteen years of annual precipitation records are given. (See the following page.) The two-tailed test for trend involves rejection of the null

Year X_i	Precipitation Y_i (inches)	$R(X_i)$	$R(Y_i)$	$[R(X_i) - R(Y_i)]^2$
1950	45.25	1	12	121
1951	45.83	2	15	169
1952	41.77	3	11	64
1953	36.26	4	6	4
1954	45.27	5	13	64
1955	52.25	6	17	121
1956	35.37	7	2.5	20.25
1957	57.16	8	18	100
1958	35.37	9	2.5	42.25
1959	58.32	10	19	81
1960	41.05	11	9	4
1961	33.72	12	1	121
1962	45.73	13	14	1
1963	37.90	14	7	49
1964	41.72	15	10	25
1965	36.07	16	4	144
1966	49.83	17	16	1
1967	36.24	18	5	169
1968	39.90	19	8	121
				Total 1421.5

hypothesis of no trend if Spearman's ρ is too large or too small. The test statistic is given by Equation 5 because the number of ties is small.

$$T = \sum_{i=1}^{19} [R(X_i) - R(Y_i)]^2 = 1421.5$$

$$\rho = 1 - \frac{6T}{19(19^2 - 1)} = 1 - \frac{6(1421.5)}{6840} = -0.2469$$

and the quantiles of ρ, for $\alpha = 0.05$, are given in Table A10, for $n = 19$, as

$$w_{0.975} = 0.4579$$

and

$$w_{0.025} = -0.4579$$

As before, H_0 is readily accepted. The p-value is approximately given by Equation 12.

$$p\text{-value} \cong 2 \cdot P(Z \geq 0.2469 \sqrt{19 - 1}) = 2 \cdot P(Z \geq 1.0475)$$
$$= 2(0.147)$$
$$= 0.294$$

▶ **The Jonckheere-Terpstra Test** _____

Either Spearman's ρ or Kendall's τ can be used in the case of several independent samples to test the null hypothesis that all of the samples came from the same distribution

$$H_0: F_1(x) = F_2(x) = \cdots = F_k(x)$$

against the ordered alternative that the distributions differ in a specified direction

$$H_1: F_1(x) \geq F_2(x) \geq \cdots \geq F_k(x)$$

with at least one inequality. This alternative is sometimes written as

$$H_1: E(Y_1) \leq E(Y_2) \leq \cdots \leq E(Y_k)$$

where Y_i represents a random variable with the distribution function $F_i(x)$. Note that this is the same data setup, and the same null hypothesis, as in the Kruskal-Wallis test of Section 5.2. However, the Kruskal-Wallis test is sensitive against *any* differences in means, while this usage of Spearman's ρ or Kendall's τ is sensitive against only the ordering specified in the H_1 given above. When Kendall's τ is used, this test is equivalent to the Jonckheere-Terpstra test, which is found in the computer programs *SAS* and *StatXact*. We will illustrate the procedure in Example 6. _____ ◀

EXAMPLE 6

As the human eye ages, it loses its ability to focus on objects close to the eye. This is a well-recognized characteristic of people over 40 years old. In order to see if people in the 15- to 30-year-old range also exhibit this loss of ability to focus on nearby objects as they get older, eight people were selected from each of four age groups; about 15 years old, about 20, about 25, and about 30 years old. It was assumed that these people would behave as a random sample from their age group populations would, with regard to the characteristic being measured. Each person held a printed paper in front of his or her right eye, with the left eye covered. The paper was moved closer to the eye until the person declared that the print began to look fuzzy. The closest distance at which the print was still sharp was measured once for each person.

The null hypothesis was that the distance measured was identically distributed for all populations. The alternative hypothesis was that the older groups tended to furnish greater distances measured. The samples were numbered from 1 to 4 in order of age group.

$$H_0: F_1(x) = F_2(x) = F_3(x) = F_4(x) \quad \text{for all } x$$
$$H_1: F_i(x) > F_j(x) \quad \text{for some } x \text{ and some } i < j$$

We are assuming that ability to focus on close objects does not improve with age and, therefore, we are able to state the null hypothesis in this slightly simpler form.

The distances, measured in inches, are given next as Y values. The samples are ordered within themselves for convenience. The sample number is X.

15 years old		20 years old		25 years old		30 years old	
X	Y	X	Y	X	Y	X	Y
1	4.6	2	4.7	3	5.6	4	6.0
1	4.9	2	5.0	3	5.9	4	6.8
1	5.0	2	5.1	3	6.6	4	8.1
1	5.7	2	5.8	3	6.7	4	8.4
1	6.3	2	6.4	3	6.8	4	8.6
1	6.8	2	6.6	3	7.4	4	8.9
1	7.4	2	7.1	3	8.3	4	9.8
1	7.9	2	8.3	3	9.6	4	11.5

Notice that if the minimum focusing distance Y tends to get larger with age, then Y and X should show a positive correlation, using either Spearman's ρ or Kendall's τ. Thus we have an upper-tailed test with lots of ties among the Xs. Notice also that instead of $X = 1, 2, 3$, and 4 we could have used any increasing sequence of numbers for the different samples, such as $X = 15, 20, 25$, and 30 representing the age groups, and both ρ and τ will not be affected by the change in X values.

Spearman's ρ for these data (we omit the computational details) is $\rho = 0.5680$. The approximate 0.95 quantile for a 5% upper-tailed test is $1.6449/\sqrt{31} = 0.2954$, so the null hypothesis is easily rejected in favor of the ordered alternative. The upper-tailed p-value is less than 0.001.

Kendall's τ for these data, based on $N_c = 290.5$ and $N_d = 93.5$, is $\tau = 0.5130$ (again we omit the details). A comparison with $w_{0.95} = 0.2056$ shows again easy rejection at $\alpha = 0.05$. The upper-tailed p-value is again less than 0.001. This test using Kendall's τ is equivalent to a procedure introduced by Jonckheere (1954a) and Terpstra (1952), although Jonckheere's test statistic is simply N_c, the number of concordant pairs of observations. ∎

□*Theory*   The exact distributions of ρ and τ are quite simple to obtain in principle, although in practice the procedure is most tedious for even moderate-sized n. The exact distributions are found under the assumption that X_i and Y_i are independent and identically distributed. Then each of the $n!$ arrangements of the ranks of X_is paired with the ranks of the Y_is is equally likely. As in the previous sections of this chapter, the distribution functions are obtained simply by counting the number of arrangements that give a particular value of ρ or τ and dividing that number by $n!$ to get the probability of that value of ρ or τ.

A form of central limit theorem is applied to obtain the large sample approximate distributions, because both ρ and τ are based on the sum of random variables. Both ρ and τ have probability distributions symmetric about zero, so the means are zero for both. The variances are more difficult to obtain and will not be derived here. Division of ρ and τ by their respective variances thus results in a random variable that is approximately distributed as a standard normal random variable for large n. The approximation is considered quite good when used to find the quantiles of τ for $n \geq 8$, but not nearly as good when used to find the quantiles of ρ. $\square$

If (X_i, Y_i), $i = 1, \ldots, n$ are independent and identically distributed bivariate normal random variables, both ρ and τ have an asymptotic relative efficiency of $9/\pi^2 = 0.912$, relative to the parametric test for independence that uses Pearson's r as a test statistic (Stuart, 1954).

▶ Kendall's Partial Correlation Coefficient

The concept of partial correlation is not an easy one to grasp. However, in order to illustrate the manner in which Kendall's τ may be extended to partial correlation, a brief attempt to describe partial correlation will be made.

In a multivariate random variable $(X_1, X_2, \ldots, X_k)$ there may be correlation between X_1 and X_2, between X_2 and X_5, and so forth, and a measure of this correlation might be any of the measures already described. Those measures estimate the total influence (correlation) of one random variable on the other, including the indirect influence felt because the second random variable is correlated not only with the first random variable, but perhaps with a third random variable that is in turn correlated with the first random variable and hence acts as a carrier of indirect influence between the first and second random variables.

Sometimes it is desirable to measure the correlation between two random variables, under the condition that the indirect influence due to the other random variables is somehow eliminated. An estimate of this "partial" correlation between X_1 and X_2, say, while the indirect correlation due to X_3, X_4, $\ldots$, and X_n is eliminated, is denoted by $r_{12.34\ldots n}$ when using the extension of Pearson's r, or by $\tau_{12.34\ldots n}$ when using the extension of Kendall's τ.

In the simple case where $n = 3$, the partial correlation may be estimated by Pearson's partial correlation coefficient

$$r_{12.3} = \frac{r_{12} - r_{13}r_{23}}{\sqrt{(1 - r_{13}^2)(1 - r_{23}^2)}} \tag{22}$$

where r_{ij} is the ordinary Pearson r computed between X_i and X_j, and by Kendall's partial correlation coefficient

$$\tau_{12.3} = \frac{\tau_{12} - \tau_{13}\tau_{23}}{\sqrt{(1 - \tau_{13}^2)(1 - \tau_{23}^2)}} \tag{23}$$

where τ_{ij} is the ordinary Kendall's τ computed between X_i and X_j. *Minitab* can be used to compute this partial correlation coefficient. _____ ◀

The use of rank correlation methods to test for dependence in a series of measurements, using rank serial correlation coefficients, is discussed by Bartels (1982), Chan and Tran (1992), Cox (1966), Dufour (1981), Dufour and Roy (1985), Hallin and Melard (1988), Hallin et al. (1985), Hannan (1976), Harel and Puri (1990), Knoke (1977), Rao (1993), Sen (1981), and Tran (1990). Rank tests for dependence have much greater power than runs tests, a popular nonparametric alternative.

Spearman's ρ has also been extended to measure partial correlation in the same way as described for Kendall's τ. An advantage of using the extension of Spearman's ρ is that existing computer programs for finding Pearson's partial correlation coefficient may be used on the ranks instead of the data, and the rank partial correlation coefficients are obtained easily.

The distribution of $r_{12.3}$ depends on the multivariate distribution function of (X_1, X_2, X_3) and therefore may not be used as a test statistic in a nonparametric test. The distributions of $\tau_{12.3}$ and $\rho_{12.3}$ also depend on the multivariate distribution and therefore are not distribution-free except in the case where all three variables are mutually independent. For more on this subject see articles by Simon (1977a, 1977b), Agresti (1977), or Wolfe (1977b). Kendall (1942) presents a discussion of partial rank correlation.

Another measure of correlation proposed by Kendall for use in another situation is the *coefficient of concordance*. This may be used to measure total correlation when more than two variates are involved. However, the close relationship between Kendall's coefficient of concordance and a test statistic proposed by Friedman makes it advisable to present both of these statistics at the same time, which will be done in Section 5.8.

A comprehensive study of the rank correlation is contained in Kendall and Gibbons' (1980) book on the subject. See also Gibbons (1993), which contains *Minitab* and *SPSS* examples. Knight (1966) gives a computer method for calculating Kendall's τ. Best (1973, 1974) presents extended tables for Kendall's τ and even has tables for different cases when ties are present for $n \le 25$. Spearman's ρ for contingency tables is explained by Stuart (1963). More extensive tables for Spearman's ρ are given by Zar (1972), using some approximate methods that work quite well. Iman and Conover (1978) compare several approximations. A mechanical interpretation of Spearman's ρ is given by Evans (1973).

Usage of rank correlation methods in regression is discussed by Hotelling and Pabst (1936), Konijn (1961), Adichie (1967a, 1967b), and Sen (1968a) and is the topic of the next two sections of this chapter. Other papers on rank correlation and the concept of dependence are by Aitkin and Hume (1965), Lehmann (1966), Bell and Doksum (1967), Gokhale (1968), Ruymgaart et al. (1972), Ruymgaart (1973), Choi (1973), and Shirahata (1975, 1976). More references appear in a bibliography by Daniel (1980).

EXERCISES

1. A husband and wife who go bowling together kept their scores for 10 lines to see if there was a correlation between their scores. The scores were:

Line	Husband's Score	Wife's Score	Line	Husband's Score	Wife's Score
1	147	122	6	151	120
2	158	128	7	196	108
3	131	125	8	129	143
4	142	123	9	155	124
5	183	115	10	158	123

   (a) Compute ρ.

   (b) Compute τ.

   (c) Test the hypothesis of independence using a two-tailed test based on ρ.

   (d) Do the same as in part c for τ.

2. The following is an example of a situation in which τ and ρ yield widely varying estimates of correlation.

X_i	Y_i	X_i	Y_i	X_i	Y_i
-8.7	-0.6	-1.9	-4.7	2.2	3.8
-8.3	-0.8	-1.6	-5.5	4.0	3.5
-8.2	-1.3	-1.3	-5.6	5.6	3.1
-7.2	-1.9	-0.2	-6.0	5.9	2.6
-6.1	-2.0	0.7	4.6	6.2	2.0
-6.0	-2.1	1.3	4.4	6.6	1.2
-4.1	-4.0	1.6	4.2	6.7	0.6
-2.0	-4.6	2.1	3.9	8.1	0.4

   (a) Make a rough scatter diagram.

   (b) Compute τ.

   (c) Compute ρ.

   (d) Does either ρ or τ lead to rejection of the null hypothesis that X and Y are independent?

3. A new worker is assigned to a machine that manufacturers bolts. Each day a sample of bolts is examined and the percent defective is recorded. Do the following data indicate a significant improvement over time for that worker?

Day	Percent	Day	Percent	Day	Percent
1	6.1	6	6.1	10	4.6
2	7.5	7	5.3	11	3.0
3	7.7	8	4.5	12	4.0
4	5.9	9	4.9	13	3.7
5	5.2				

   (a) Use Spearman's ρ.

   (b) Use Kendall's τ.

4. Is there a significant correlation between the age at which a U.S. president was inaugurated for the first time and the age at which he died?

Name	Inaugurated	Died	Name	Inaugurated	Died
Washington	57	67	Hayes	54	70
J. Adams	61	90	Garfield	49	49
Jefferson	57	83	Arthur	50	56
Madison	57	85	Cleveland	47	71
Monroe	58	73	Harrison	55	67
J.Q. Adams	57	80	McKinley	54	58
Jackson	61	78	T. Roosevelt	42	60
Van Buren	54	79	Taft	51	72
Harrison	68	68	Wilson	56	67
Tyler	51	71	Harding	55	57
Polk	49	53	Coolidge	51	60
Taylor	64	65	Hoover	54	90
Fillmore	50	74	F. Roosevelt	51	63
Pierce	48	64	Truman	60	88
Buchanan	65	77	Eisenhower	62	78
Lincoln	52	56	Kennedy	43	46
A. Johnson	56	66	L. Johnson	55	64
Grant	46	63	Nixon	56	81

   (a) Use Spearman's ρ.

   (b) Use Kendall's τ.

   Note that these data do not represent a random sample, but one might assume that they behave as a random sample of all U.S. presidents, past, present, and future.

5. Five doctoral students took a test on current affairs. The ages of the doctoral students and their test scores are given below.

Age	Test Score
24	68
31	85
38	84
45	92
45	90

   Do older students tend to get higher test scores?

   (a) Use Spearman's ρ.

   (b) Use Kendall's τ.

6. In order to see whether a longer time lapse between the last day of class and the time of the final exam tends to improve student performance on the final exam, a class of 48 students was divided at random into four groups of 12 students each. Group 1 took the final exam 2 days after the last class period. Group 2 took the final exam 4 days after the last class period. Group 3 was given 6 days and Group 4, 8 days. All groups were given comparable exams under otherwise comparable conditions. The final exam scores are as follows.

Group 1			Group 2			Group 3			Group 4		
48	71	80	42	70	77	38	73	83	49	77	84
61	74	82	48	71	81	58	74	87	58	79	93
67	75	87	62	73	89	70	75	90	73	80	94
68	79	89	67	75	92	71	79	94	74	84	97

Does the increased time lapse tend to improve test performance?

7. Prior to the catastrophic explosion of the space shuttle *Challenger* in 1986 due to O-ring failure after a launch in subfreezing weather, engineers from the rocket manufacturer Thiokol recommended against launching on such a cold day because of the danger of O-ring failure due to cold weather. They presented the following data from the 24 previous launches. Does the number of O-ring incidents appear to increase as the temperature decreases?

O-ring Incidents	Temperature (Fahrenheit)								
None	66	67	67	67	68	68	70	70	
	72	73	75	76	76	78	79	80	81
One	57	58	63	70	70				
Two	75								
Three	53								

See Feynman (1988) for details.

PROBLEMS

1. Show that Equations 4 and 5 are equivalent expressions for ρ in the case of no ties.

2. For $n = 5$ what pairing of ranks results in

   (a) $\rho = 1$?

   (b) $\tau = 1$?

   (c) $\rho = -1$?

   (d) $\tau = -1$?

3. Generalize the result of Problem 2 to any value of n in general and show that ρ and τ do, in fact, assume the values indicated.

4. Suppose someone suggests using

$$R = 1 - \frac{\sum_{i=1}^{n} |R(X_i) - R(Y_i)|}{(1/4)n^2}$$

which is sometimes called "Spearman's footrule."

   (a) Under what conditions will $R = 1$?

   (b) Under what conditions will $R = -1$?

5. Find the exact distribution of ρ, τ, and R from Problem 4 for the case where $n = 3$ under the usual assumption of independence.

6. Compute R defined in Problem 4 for the data in Exercise 2. Does R seem to resemble ρ more than τ in its behavior?

5.5 NONPARAMETRIC LINEAR REGRESSION METHODS

This section is related closely to the previous section on rank correlation in that we are examining a random sample $(X_1, Y_1), \ldots, (X_n, Y_n)$ on the bivariate random variable (X, Y). Correlation methods emphasize estimating the degree of dependence between X and Y. Regression methods are used to inspect the relationship between X and Y more closely. One important objective of regression methods is to predict a value of Y in the pair (X, Y) where only the value for X is known, on the basis of information that we can obtain from previous observations (X_1, Y_1) through (X_n, Y_n). For example, if X represents the scores on a college entrance examination and Y represents the grade point average of that student four years later, observations on past students may help us to predict how well an incoming student will perform in the four years of college. Of course, Y is still a random variable, so we cannot expect to determine Y solely from knowing the associated value of X, but knowing X should help us make a better estimate concerning Y.

Regression methods also apply to controlled experiments where X may not be random at all, but may be set by the experimenter at various values to determine its effect on Y. For example, X may represent a measured amount of medication, such as medication intended to lower blood pressure in a patient. Several different levels of X may be selected in an experiment to determine the effect of the medication on Y, which is the patient's response such as the patient's reduction in blood pressure.

Formally, the regression of Y on X is merely the mean of Y for a given value of X, say x.

Definition 1   The *regression of Y on X* is $E(Y|X = x)$. The *regression equation* is $y = E(Y|X = x)$.

If the regression equation is known, we can represent the regression on a graph by plotting y as the ordinate and x as the abscissa. But the regression equation is seldom, if ever, known. It is estimated on the basis of past data. For example, if we would like to predict Y when $X = 6$, we could use $E(Y|X = 6)$ if we knew it; otherwise, we could use the sample mean or the sample median of several observed values of Y for which X is equal to 6 or close to 6. In this way point estimates and confidence intervals may be formed for $E(Y|X = 6)$ using the methods described in Sections 3.2 and 5.7. In order to have enough observations so that the regression of Y on X can be estimated for each value of X, many observations are needed. It is not unusual to have large data sets that contain hundreds or even thousands of observations, in which case the nonparametric methods just mentioned work very nicely.

A more difficult situation arises when we have only a few observations and wish to estimate the regression of Y on X. This is what we will examine in this section. It is helpful to know something about the relationship between $E(Y|X = x)$ and x and to be able to use this information when there are only a few observations. First, we will examine the case where $E(Y|X = x)$ is a linear function of x; in the next section we will consider a more general situation where $E(Y|X = x)$ is a monotonic (either increasing or decreasing) function of x.

The regression of Y on X is said to be linear if the graph of the regression equation is a straight line.

> **Definition 2**   The regression of Y on X is *linear regression* if the regression equation is of the form
>
> $$E(Y|X = x) = \alpha + \beta x \tag{1}$$
>
> for some constants α, called the *y-intercept*, and β, called the *slope*.

Usually the constants α and β are unknown and must be estimated from the data. If all of the observations of X and Y are used in estimating α and β, maximum usage is made of the data and a good estimate of $E(Y|X = x)$ for each x can be expected. A commonly accepted method for estimating α and β is called the *least squares method*.

> **Definition 3**   The *least squares method* for choosing estimates a and b of α and β in the regression equation $y = \alpha + \beta x$ is the method that minimizes the sum of squared deviations
>
> $$SS = \sum_{i=1}^{n} [Y_i - (a + bX_i)]^2 \tag{2}$$
>
> for the observations $(X_1, Y_1), \ldots , (X_n, Y_n)$.

The idea behind the least squares method is that an estimate of the regression line should be close to the observed values of X and Y because the true regression line is probably close to the observations. Therefore the estimate is selected so that the vertical distance D_i between Y_i and the estimated regression line, which equals $a + bX_i$ directly above or below Y_i, is small when all of the points are considered at once. We cannot merely make the sum of the Ds small because the sum of the Ds could be zero even though the estimated regression line is not at all close to the observations. That is, the absolute values of the distances D could be large, but the positive Ds could cancel the negative Ds, giving a sum of zero. To avoid this, we choose to minimize the sum of squares of the Ds:

$$SS = \sum_{i=1}^{n} D_i^2 \tag{3}$$

where

$$D_i = Y_i - (a + bX_i) \tag{4}$$

This usually produces a straight line that agrees well with the data and, therefore, is a reasonable estimate of the true regression line.

▶ **Nonparametric Methods for Linear Regression** _____

Data   The data consist of a random sample $(X_1, Y_1), (X_2, Y_2), \ldots, (X_n, Y_n)$ from some bivariate distribution.

Assumptions

1. The sample is a random sample. The methods of this section are valid if the values of X are nonrandom quantities as long as the Ys are independent with identical conditional distributions.

2. The regression of Y on X is linear. This implies an interval scale of measurement on both X and Y.

Least Squares Estimates   The method of least squares furnishes the estimate

$$y = a + bx \tag{5}$$

of the true regression line $y = \alpha + \beta x$, where a and b are computed from

$$b = \frac{n \sum_{i=1}^{n} X_i Y_i - \left(\sum_{i=1}^{n} X_i \right) \left(\sum_{i=1}^{n} Y_i \right)}{n \sum_{i=1}^{n} X_i^2 - \left(\sum_{i=1}^{n} X_i \right)^2} \tag{6}$$

and

$$a = \overline{Y} - b\overline{X} \tag{7}$$

where $\overline{X}$ and $\overline{Y}$ are the respective sample means.

Testing the Slope   To test the hypothesis concerning the slope, add the following assumption to Assumptions 1 and 2.

3. The "residual" $Y - E(Y|X)$ is independent of X.

Spearman's ρ may be adapted to test the following hypotheses concerning the slope. Let β_0 represent some specified number. For each pair (X_i, Y_i) compute $Y_i - \beta_0 X_i = U_i$ (say). Then find the Spearman rank correlation coefficient ρ on the pairs (X_i, U_i), $i = 1, \ldots, n$, as described in Section 5.4. Table A10 gives the quantiles of ρ when H_0 is true and there are no ties.

A. (Two-Tailed Test)

$$H_0: \beta = \beta_0$$
$$H_1: \beta \neq \beta_0$$

Reject H_0 if ρ exceeds the $1 - \alpha/2$ quantile, or is less than the $\alpha/2$ quantile as described in the two-tailed test for Spearman's ρ in Section 5.4.

B. (Lower-Tailed Test)

$$H_0: \beta = \beta_0$$
$$H_1: \beta < \beta_0$$

Reject H_0 if ρ is less than the α quantile, as described in the lower-tailed test for Spearman's ρ in Section 5.4.

C. (Upper-Tailed Test)

$$H_0: \beta = \beta_0$$
$$H_1: \beta > \beta_0$$

Reject H_0 if ρ exceeds the $1 - \alpha$ quantile as described in the upper-tailed test for Spearman's ρ in Section 5.4.

A Confidence Interval for the Slope   Assumptions 1, 2, and 3 are used in this procedure also. For each pair of points (X_i, Y_i) and (X_j, Y_j), such that $i < j$ and $X_i \neq X_j$, compute the "two-point slope,"

$$S_{ij} = \frac{Y_j - Y_i}{X_j - X_i} \tag{8}$$

Let N be the number of slopes computed. Order the slopes obtained and let

$$S^{(1)} \leq S^{(2)} \leq \cdots \leq S^{(N)}$$

denote the ordered slopes.

For a $1 - \alpha$ confidence interval, find $w_{1-\alpha/2}$, the $1 - \alpha/2$ quantile of $T = N_c - N_d$ from Table A11. Let r and s be given by

$$r = \tfrac{1}{2}(N - w_{1-\alpha/2}) \tag{9}$$

$$s = \tfrac{1}{2}(N + w_{1-\alpha/2}) + 1 = N + 1 - r \tag{10}$$

Round r downward and s upward to the next integer if they are not already integers. The $1 - \alpha$ confidence interval for β is given by the interval $(S^{(r)}, S^{(s)})$. That is,

$$P(S^{(r)} < \beta < S^{(s)}) \geq 1 - \alpha \tag{11}$$

Computer Assistance *Minitab* finds all pairwise slopes, making this confidence interval easier to obtain. ———————————————————————— ◄

Comment

The confidence interval for the slope is based on Kendall's τ, a completely different concept than the least squares concept; therefore it is possible, although unlikely, for the least squares estimator b for β to be outside the confidence interval for β. This could happen, for instance, when one value of Y is much larger or smaller than we would expect it to be, judging from the other observations. Such an outlying observation can "pull" the least squares line up to fit it more closely at the expense of the other observations. In such a case it makes more sense to choose an estimated regression line that passes through the point (sample median of X, sample median of Y), with slope equal to the median of the slopes S_{ij} defined by Equation 8. That is, we could choose our estimators to be

$$b_1 = \text{sample median of the } S_{ij}\text{s} = S_{(\text{median})} \tag{12}$$

and

$$a_1 = Y_{0.50} - b_1 X_{0.50} \tag{13}$$

where $X_{0.50}$ and $Y_{0.50}$ refer to the sample medians.

EXAMPLE I

Let us again use the data from the previous section. The GMAT score of each MBA graduate is denoted by X_i and that graduate's GPA is denoted by Y_i. The twelve observations (X, Y) are $(710, 4.0)$, $(610, 4.0)$, $(640, 3.9)$, $(580, 3.8)$, $(545, 3.7)$,

(560, 3.6), (610, 3.5), (530, 3.5), (560, 3.5), (540, 3.3), (570, 3.2), and (560, 3.2). These are plotted in Figure 1, along with the least squares regression line

$$y = 1.4287 + 0.003714x$$

which is obtained by substituting

$$\sum_{i=1}^{12} X_i = 7{,}015 \qquad \overline{X} = 584.58 \qquad \sum_{i=1}^{12} X_i^2 = 4{,}129{,}525$$

$$\sum_{i=1}^{12} Y_i = 43.2 \qquad \overline{Y} = 3.6 \qquad \sum_{i=1}^{12} X_i Y_i = 25{,}360.5$$

into Equations 6 and 7 to obtain $b = 0.003714$ and $a = 1.4287$. We may use the regression line as a description of the relationship between Y and X, or, more precisely, as an estimate of the conditional mean $E(Y|X)$ of Y given X. If a graduate student has a GMAT score of 550 we can predict that the student's graduating GPA will be about $1.4287 + 0.003714(550) = 3.47$. Individual students may have higher or lower GPAs because of other relevant factors, such as motivation, study habits, and competing obligations. This estimated regression line provides only a point estimate.

Suppose that a national study reports that "a 40 point increase in GMAT scores results in at least 0.4 increase in GPAs." Because slope is a change in Y divided by a change in X, this claim is equivalent to saying the slope in the regression of GPA onto GMAT score is at least $0.4/40 = 0.01$.

To see if our sample of 12 graduates is consistent with the national study we test

$$H_0: \beta \geq 0.01$$

versus

$$H_1: \beta < 0.01$$

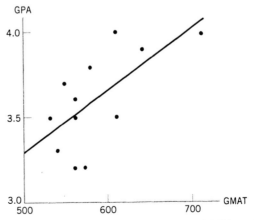

FIGURE I A scatterplot of GMAT scores versus GPAs for 12 MBA graduates, and the least squares regression line.

in a lower-tailed test. Sample slopes less than 0.01, such as ours, can simply be the result of chance fluctuations, and are not necessarily inconsistent with the national survey. Spearman's ρ is calculated between the GMAT scores X and the sample residuals $U = Y - (0.01)X$.

MBA Graduate i

	1	2	3	4	5	6
X_i	710	610	640	580	545	560
$U_i = Y - \beta_0 X_i$	−3.1	−2.1	−2.5	−2.0	−1.75	−2.0
$R(X_i)$	12	9.5	11	8	3	5
$R(U_i)$	1	7	3.5	9.5	12	9.5

	7	8	9	10	11	12
X_i	610	530	560	540	570	560
$U_i = Y - \beta_0 X_i$	−2.6	−1.8	−2.1	−2.1	−2.5	−2.4
$R(X_i)$	9.5	1	5	2	7	5
$R(U_i)$	2	11	7	7	3.5	5

Equation 5.4.4 is used to compute $\rho = -0.7273$, which is less than the 0.05 quantile of the null distribution from Table A10, so the null hypothesis is rejected at $\alpha = 0.05$. The p-value is approximately

$$P(Z \le -0.7273 \sqrt{11}) = P(Z \le -2.4121) = 0.008$$

from Equation 5.4.13. This sample of MBA graduates is not consistent with the national survey results.

To form a 95% confidence interval for the true slope in the population from which this sample of MBA graduates was obtained, all of the two-point slopes

$$S_{ij} = \frac{Y_j - Y_i}{X_j - X_i}$$

are computed for pairs of points where $X_i \ne X_j$. See Figure 2 for a convenient spreadsheet-like layout for computing the S_{ij}s. There are $N = 62$ pairs (X_i, Y_i) and (X_j, Y_j) with $X_i \ne X_j$, as seen in Figure 2.

From Table A11 for $n = 12$, the 0.975 quantile of T is found to be 28, so Equations 9 and 10 yield $r = 17$ and $s = 46$. The 17th ordered value of S_{ij} is found from Figure 2 to be

$$S^{(17)} = 0.00000$$

and the 46th ordered value of S_{ij} is

$$S^{(46)} = 0.00800$$

The 95% confidence interval for the true slope β is from 0.000 to 0.008.

If for some reason the least squares regression line is considered unsatisfactory, the median value of S_{ij}, which is the average of the 31st ordered value of S_{ij}, 0.1/30, and the 32nd ordered value, 0.4/110, is found. The median is 0.003485,

(530, 3.5)	.00278	.00364	.00625	0	.00600	−.00750	.00333	0	−.01000	.01333	−.02000
(540, 3.3)	.00412	.00600	.01000	.00286	.01250	−.00333	.01500	.01000	−.00500	.08000	(540, 3.3)
(545, 3.7)	.00182	.00211	.00462	−.00308	.00286	−.02000	−.00667	−.01333	−.03333	(545, 3.7)	
(560, 3.2)	.00533	.00875	.01600	.00600	.03000	0	NA	NA	(560, 3.2)		
(560, 3.5)	.00333	.00500	.01000	0	.01500	−.03000	NA	(560, 3.5)			
(560, 3.6)	.00267	.00375	.00800	−.00200	.01000	−.04000	(560, 3.6)				
(570, 3.2)	.00571	.01000	.02000	.00750	.06000	(570, 3.2)					
(580, 3.8)	.00154	.00167	.00667	−.01000	(580, 3.8)						
(610, 3.5)	.00500	.01333	NA	(610, 3.5)							
(610, 4.0)	0	−.00333	(610, 4.0)								
(640, 3.9)	.00143	(640, 3.9)									
(710, 4.0)											

FIGURE 2 A spreadsheet arrangement of points (X, Y), arranged by increasing Xs, to find the value of S_{ij}.

which provides an alternative estimate of β, as described in Equation 12. Then Equation 13 furnishes the estimate of α,

$$a_1 = Y_{0.50} - b_1 X_{0.50} = 3.55 - 0.003485(565)$$
$$= 1.581$$

The alternative estimate of the regression line is thus

$$y = 1.581 + 0.003485x$$

as discussed in the previous Comment section. ∎

Notice that the 95% confidence interval for the slope β in the previous example is from 0.000 to 0.008, and is consistent with the hypothesis test that rejected the null hypothesis $\beta = 0.01$. This will usually be the case; however, there are two reasons why there may be disagreement between the hypothesis test and the confidence interval. One reason is a two-sided confidence interval is the inversion of a two-sided hypothesis test, and this was a one-sided test. The other reason is that this confidence interval is the inversion of the hypothesis test based on Kendall's τ, while this hypothesis test is based on Spearman's ρ.

To explain further, the hypothesis test we presented was a nonparametric test for rank correlation between X and the residuals $U = Y - \beta_0 X$ under the null hypothesis. We used Spearman's ρ for this test, but we could have used Kendall's τ as a test statistic just as well, as suggested by Theil (1950). We chose Spearman's ρ because we find it easier to compute.

The confidence interval for β could have been found by inverting the hypothesis test for β as we presented it and finding all values of β_0 that are "acceptable" as a null hypothesis in a two-tailed test based on Spearman's ρ. In fact, this method was investigated by Taylor and Conover (1988) and found to result in more computations, and no advantages in efficiency, when compared to the method presented, which is based on Kendall's τ.

□ *Theory*   To derive a and b such that SS in Equation 2 is minimized, add and subtract the quantity $(\overline{Y} - b\overline{X})$ inside the brackets to get

$$SS = \sum_{i=1}^{n} [(Y_i - \overline{Y}) - b(X_i - \overline{X}) + (\overline{Y} - b\overline{X} - a)]^2 \tag{14}$$

Because of the algebraic identity

$$(c - d + e)^2 = c^2 + d^2 + e^2 - 2cd + 2ce - 2de \tag{15}$$

we can expand Equation 14 using $c = Y_i - \overline{Y}$ and so on, to get

$$\begin{aligned} SS = &\sum_{i=1}^{n} (Y_i - \overline{Y})^2 + b^2 \sum_{i=1}^{n} (X_i - \overline{X})^2 + \sum_{i=1}^{n} (\overline{Y} - b\overline{X} - a)^2 \\ &- 2b \sum_{i=1}^{n} (Y_i - \overline{Y})(X_i - \overline{X}) + 2(\overline{Y} - b\overline{X} - a) \sum_{i=1}^{n} (Y_i - \overline{Y}) \\ &- 2b(\overline{Y} - b\overline{X} - a) \sum_{i=1}^{n} (X_i - \overline{X}) \end{aligned} \tag{16}$$

Because $\sum (Y_i - \overline{Y}) = 0$ and $\sum (X_i - \overline{X}) = 0$ by the definition of $\overline{Y}$ and $\overline{X}$, the last two summations equal zero in Equation 16. The third summation is smallest (zero) when

$$a = \overline{Y} - b\overline{X} \tag{17}$$

which gives the least squares solution for a. We are left with the problem of finding the value of b that minimizes the sum of the second and fourth summations, that is, that minimizes

$$b^2 S_x - 2b S_{xy} \tag{18}$$

where

$$S_x = \sum_{i=1}^{n} (X_i - \overline{X})^2 \tag{19}$$

and

$$S_{xy} = \sum_{i=1}^{n} (X_i - \overline{X})(Y_i - \overline{Y}) \tag{20}$$

By adding and subtracting S_{xy}^2/S_x to Equation 18, the sum of the second and fourth summations becomes

$$S_x \left[b^2 - 2b \frac{S_{xy}}{S_x} + \left(\frac{S_{xy}}{S_x} \right)^2 \right] - \frac{S_{xy}^2}{S_x} = S_x \left(b - \frac{S_{xy}}{S_x} \right)^2 - \frac{S_{xy}^2}{S_x}$$

which is obviously a minimum when

$$b = \frac{S_{xy}}{S_x} \tag{21}$$

in agreement with Equation 6. Note that this reduces the second and fourth summation to $-S_{xy}^2/S_x$, so that the minimum sum of squares is

$$SS_{\min} = \sum_{i=1}^{n} (Y_i - \overline{Y})^2 - \frac{S_{xy}^2}{S_x}$$

$$= (1 - r^2) \sum_{i=1}^{n} (Y_i - \overline{Y})^2 \tag{22}$$

where r is the Pearson product moment correlation coefficient given by Equation 5.4.1. Also note that no assumptions regarding the distribution of (X, Y) were made, so the least squares method is distribution-free. In fact, the only purpose of assumptions 1 and 2 is to assure us that there is a regression line somewhere to be estimated.

Under assumption 3, the residuals

$$Y_i - E(Y_i|X_i) = Y_i - (\alpha + \beta X_i) \tag{23}$$

are independent of X_i, so the assumptions of Section 5.4 regarding Spearman's ρ are met. Note that the ranks of $(Y_i - \alpha - \beta X_i)$, $i = 1$ to n, are the same as the ranks of $U_i = (Y_i - \beta X_i)$, $i = 1$ to n, so we can test $H_0: \beta = \beta_0$ without knowing α. Just as Spearman's ρ is merely Pearson's r computed on ranks, this test is the rank analogue of computing r on the pairs (X_i, U_i), which is the usual parametric procedure for testing the same null hypothesis, valid with the additional assumption that (X, Y) has the bivariate normal distribution. Under that condition and the condition that the observations on X are equally spaced, the A.R.E. of this procedure is $(3/\pi)^{1/3} = 0.98$ according to Stuart (1954, 1956); for other distributions the A.R.E. is always greater than or equal to 0.95 (Lehmann, 1975).

To see the relationship between the slopes S_{ij} and Kendall's τ, note that for any hypothesized slope β_0 we have

$$
\begin{aligned}
S_{ij} &= \frac{Y_i - Y_j}{X_i - X_j} = \frac{U_i + \beta_0 X_i - U_j - \beta_0 X_j}{X_i - X_j} \\
&= \beta_0 + \frac{U_i - U_j}{X_i - X_j}
\end{aligned}
\tag{24}
$$

where $U_i = Y_i - \beta_0 X_i - \alpha$ is the residual of Y_i from the hypothesized regression line $y = \alpha + \beta_0 x$. The slope S_{ij} is greater than β_0 or less than β_0 according to whether the pair (X_i, U_i) and (X_j, U_j) is concordant or discordant in the sense described in Section 5.4 in the discussion of Kendall's τ. If we use the number of S_{ij}s less than β_0 as our test statistic for determining whether to accept $H_0: \beta = \beta_0$, we accept β_0 as long as the number of discordant pairs N_d is not too small or too large. Because N_d is related to the number of concordant pairs N_c by

$$
N_c + N_d = N
\tag{25}
$$

where N is the total number of pairs, and because the quantiles of $N_c - N_d$ are given in Table A11 if we have the true slope and Assumption 3 of independence, we can say N_d is too small if $N_c - N_d$ is greater than $w_{1-\alpha/2}$ from Table A11. This is equivalent to saying N_d is less than $r = (N - w_{1-\alpha/2})/2$. In other words, β_0 is acceptable if β_0 is greater than at least r of the S_{ij}s, or $\beta_0 > S^{(r)}$. The same argument gives an upper bound for β_0, and the confidence interval is obtained. This method, due to Theil (1950), was modified to handle ties by Sen (1968a). $\square$

For nonparametric tests applicable to several regression lines see Sen (1972), Adichie (1974, 1975), and Pothoff (1974). Alternative methods of estimating regression coefficients are given by Jureckova (1971, 1977), Huber (1973), and Hettmansperger and McKean (1977). Kalbfleish (1974) discusses ranks in nonlinear models. Further discussions of nonparametric regression appear in Puri (1985), Jaeckel (1972), Hollander and Wolfe (1973), Behnen (1976), and Stone (1977).

EXERCISES

1. A driver kept track of the number of miles she traveled and the number of gallons put in the tank each time she bought gasoline.

Miles	Gallons	Miles	Gallons
142	11.1	157	12.5
116	5.7	255	17.9
194	14.2	159	8.8
250	15.8	43	3.4
88	7.5	208	15.2

(a) Draw a diagram showing these points, using gallons as the x-axis.

(b) Estimate a and b using the method of least squares.

(c) Plot the least squares regression line on the diagram of part a.

(d) Suppose the EPA estimated this car's mileage at 18 miles per gallon. Test the null hypothesis that this figure applies to this particular car and driver. (Use the test for slope.)

(e) Find a 95% confidence interval for the mileage of this car and driver.

2. A random sample of American colleges and universities resulted in the following numbers of students and faculty (Spring 1973).

Name	Students	Faculty
American International	2546	129
Bethany Nazarene	1355	75
Carlow	1019	87
David Lipscomb	1858	99
Florida International University	4500	300
Heidelberg	1141	109
Lake Erie	784	77
Mary Hardin Baylor	1063	64
Mt. Angel	267	40
Newberry	753	61
Pacific Lutheran University	3164	190
St. Ambrose	1189	90
Smith	2755	240
Texas Women's University	5602	300
West Liberty State	2697	170
Wofford	988	73

(a) Draw a diagram showing these points using faculty as the x-axis.

(b) Estimate the regression line using the method of least squares.

(c) Plot the least squares regression line on the diagram of part a.

(d) Test the hypothesis that an increase of one faculty member is accompanied by an average increase of 15 students.

(e) Find a confidence interval for the slope.

3. A random sample of applicants for graduate school is examined. Test the null hypothesis that the regression of Verbal GRE Score (Y) on Math GRE Score (X) has a slope of 1.0, against the alternative that the slope is less than 1.0.

Student	Math	Verbal	Student	Math	Verbal
1	650	540	9	460	510
2	720	580	10	520	500
3	580	500	11	740	680
4	670	570	12	450	600
5	600	630	13	530	550
6	510	630	14	570	500
7	480	520	15	680	510
8	610	610	16	740	570

4. The volatility of a company's common stock is defined as the slope of its performance (Y) as compared with the performance of the S&P 500, an index of 500 stocks. For the last eight quarters these are the performance measures. Find a 95% confidence interval for the slope.

Quarter	Company X	S&P 500	Quarter	Company X	S&P 500
1	+4.5%	+2.6%	5	−4.6%	−2.5%
2	+5.1%	+2.7%	6	−0.6%	+0.1%
3	+8.0%	+3.1%	7	+10.3%	+4.9%
4	+2.2%	+0.8%	8	+2.2%	+1.0%

5.6 METHODS FOR MONOTONIC REGRESSION

In Section 5.5 nonparametric methods for linear regression were presented. These may be used in situations such as in Example 5.5.1, where the assumption of linear regression seems reasonable. In other situations it may be unreasonable to assume that the regression function is a straight line, but it may be reasonable to assume that $E(Y|X)$ increases (at least, it does not decrease) as X increases. In such a case we say the regression is *monotonically increasing*. If $E(Y|X)$ becomes smaller as X increases the regression is *monotonically decreasing*. Either case lends itself to the following method.

▶ **Nonparametric Methods for Monotonic Regression** _____

Data The data consist of a random sample $(X_1, Y_1), (X_2, Y_2), \ldots, (X_n, Y_n)$ from some bivariate distribution.

Assumptions

1. The sample is a random sample.
2. The regression of Y on X is monotonic.

An Estimate of $E(Y|X)$ at a Point To estimate the regression of Y on X at a particular value of $X = x_0$:

1. Obtain the ranks $R(X_i)$ of the Xs and $R(Y_i)$ of the Ys. Use average ranks in case of ties.
2. Find the least squares regression line on the ranks.

$$y = a_2 + b_2 x \tag{1}$$

where

$$b_2 = \frac{\sum\limits_{i=1}^{n} R(X_i)R(Y_i) - n(n+1)^2/4}{\sum\limits_{i=1}^{n} [R(X_i)]^2 - n(n+1)^2/4} \tag{2}$$

and

$$a_2 = (1 - b_2)(n+1)/2 \tag{3}$$

3. Obtain a rank $R(x_0)$ for x_0 as follows:
   (a) If x_0 equals one of the observed X_is, let $R(x_0)$ equal the rank of that X_i.
   (b) If x_0 lies between two adjacent values X_i and X_j where $X_i < x_0 < X_j$, interpolate between their respective ranks to get $R(x_0)$.

$$R(x_0) = R(X_i) + \frac{x_0 - X_i}{X_j - X_i}[R(X_j) - R(X_i)] \tag{4}$$

   This "rank" will not necessarily be an integer.
   (c) If x_0 is less than the smallest observed X or greater than the largest observed X, do not attempt to extrapolate. Information on the regression of Y on X is available only within the observed range of X.

4. Substitute $R(x_0)$ for x in Equation 1 to get an estimated rank $R(y_0)$ for the corresponding value of $E(Y|X = x_0)$.

$$R(y_0) = a_2 + b_2 R(x_0) \tag{5}$$

5. Convert $R(y_0)$ into $\hat{E}(Y|X = x_0)$, an estimate of $E(Y|X = x_0)$, by referring to the observed Y_is as follows.
   (a) If $R(y_0)$ equals the rank of one of the observations Y_i, let the estimate $\hat{E}(Y|X = x_0)$ equal that observation Y_i.
   (b) If $R(y_0)$ lies between the ranks of two adjacent values of Y, say Y_i and Y_j where $Y_i < Y_j$, so that $R(Y_i) < R(y_0) < R(Y_j)$, interpolate between Y_i and Y_j.

$$\hat{E}(Y|X = x_0) = Y_i + \frac{R(y_0) - R(Y_i)}{R(Y_j) - R(Y_i)}(Y_j - Y_i) \tag{6}$$

(c) If $R(y_0)$ is greater than the largest observed rank of Y, let $\hat{E}(Y|X = x_0)$ equal the largest observed Y. If $R(y_0)$ is less than the smallest observed rank of Y, let $\hat{E}(Y|X = x_0)$ equal the smallest observed Y.

An Estimate of the Regression of Y on X   To obtain the entire regression curve consisting of all points that can be obtained in the manner just described, the following procedure may be used.

1. For each X_i from $X^{(1)}$ to $X^{(n)}$ use the previously described procedure to estimate $E(Y|X)$.

2. For each rank of Y, $R(Y_i)$, find the estimated rank of X_i, $\hat{R}(X_i)$ from Equation 1.

$$\hat{R}(X_i) = [R(Y_i) - a_2]/b_2, \qquad i = 1, 2, \ldots, n \qquad (7)$$

3. Convert each $\hat{R}(X_i)$ to an estimate $\hat{X}_i$ in the manner of the preceding step 5. More specifically:
   (a) If $\hat{R}(X_i)$ equals the rank of some observation X_j, let $\hat{X}_i$ equal that observed value.
   (b) If $\hat{R}(X_i)$ falls between the ranks of two adjacent observations X_j and X_k, where $X_j < X_k$, then use interpolation,

$$\hat{X}_i = X_j + \frac{\hat{R}(X_i) - R(X_j)}{R(X_k) - R(X_j)}(X_k - X_j) \qquad (8)$$

   to get $\hat{X}_i$.
   (c) If $\hat{R}(X_i)$ is less than the smallest observed rank of X or greater than the largest observed rank, no estimate $\hat{X}_i$ is found.

4. Plot each of the points found in steps 1 and 3 on graph paper. That is, plot each $(X_i, \hat{Y}_i)$ and each $(\hat{X}_i, Y_i)$. All of these points should be monotonic, increasing if $b_2 > 0$ and decreasing if $b_2 < 0$.

5. Connect the adjacent points in step 4 with straight lines. This series of connected line segments is the estimate of the regression of Y on X. ____◄

EXAMPLE I

Seventeen jars of fresh grape juice were obtained to study how long it took for the grape juice to turn into wine as a function of how much sugar was added to the juice. Various amounts of sugar, ranging from none to about 10 pounds, were added to the jars, and each day the jars were checked to see if the transition to wine was complete. At the end of 30 days the experiment was terminated, with three jars still unfermented. An estimate of the regression curve of Y (number of days till fermentation) versus X (pounds of sugar) is desired.

The observations (X_i, Y_i), their ranks $R(X_i)$ and $R(Y_i)$, and the values $\hat{R}(Y_i)$, $\hat{Y}_i = \hat{E}(Y|X_i)$, $\hat{R}(X_i)$, and $\hat{X}_i$ computed from the preceding steps 1, 2, and 3 are

X_i	Y_i	$R(X_i)$	$R(Y_i)$	$\hat{R}(Y_i)$	$\hat{Y}_i$	$\hat{R}(X_i)$	$\hat{X}_i$
0	>30	1	16	16.47	>30	1.50	.25
.5	>30	2	16	15.54	29.54	1.50	.25
1.0	>30	3	16	14.60	28.60	1.50	.25
1.8	28	4	14	13.67	26.67	3.64	1.52
2.2	24	5	13	12.74	22.67	4.71	2.09
2.7	19	6	12	11.80	18.60	5.78	2.59
4.0	17	7.5	11	10.40	15.00	6.85	3.44
4.0	9	7.5	8	10.40	15.00	10.06	5.63
4.9	12	9	9.5	9.00	11.00	8.46	4.58
5.6	12	10	9.5	8.07	9.13	8.46	4.58
6.0	6	11	5	7.13	8.13	13.28	7.50
6.5	8	12	7	6.20	7.20	11.13	6.07
7.3	4	13	1.5	5.27	6.26	17.03	None
8.0	5	14	3	4.33	5.67	15.42	9.01
8.8	6	15	5	3.40	5.20	13.28	7.50
9.3	4	16	1.5	2.46	4.64	17.03	None
9.8	6	17	5	1.53	4.02	13.28	7.50

FIGURE 3 Calculations for finding the monotonic regression curve estimate.

given as indicated in Figure 3. Before obtaining $\hat{R}(Y_i)$, $\hat{Y}_i$, $\hat{R}(X_i)$, and $\hat{X}_i$, the least squares coefficients on the ranks are computed from Equations 2 and 3 and substituted into Equation 1 to get the least squares regression line on ranks

$$y = 17.4037 - 0.9337x \qquad (9)$$

The observations are plotted in Figure 4. The regression curve, consisting of line segments joining successive values of $(X_i, \hat{Y}_i)$ and $(\hat{X}_i, Y_i)$, is also plotted in Figure 4. An estimate $\hat{E}(Y|X = x_0)$ is obtained easily from Figure 4 by finding the ordinate that corresponds to the abscissa x_0. Note that the "censored" observations ">30" were used in the regression of ranks, but that portion of the regression curve of the data is not possible to plot using ordinary linear regression.

It is interesting to note how a set of observations, with a regression curve that is obviously nonlinear, is converted to ranks that have a regression curve that seems to be linear. The ranks are plotted in Figure 5 along with Equation 9. ∎

□ *Theory* The procedures for monotonic regression are based on the fact that if two variables have a monotonic relationship, their ranks will have a linear relationship. A scattering of the observations around the monotonic regression line should correspond to a scattering of the ranks around their linear regression line. The ranks serve as transformed variables, where the transformation seeks

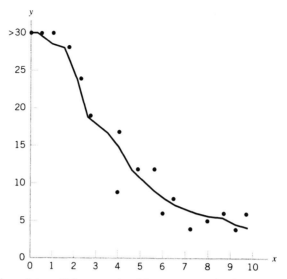

FIGURE 4   Number of days till fermentation (*y*) versus pounds of sugar (*x*), and the estimated monotonic regression curve.

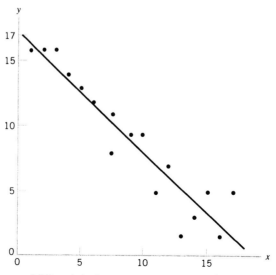

FIGURE 5 $R(Y_i)$ versus $R(X_i)$ and the least squares regression line.

to convert the monotonic regression function to a linear regression function. Interval estimates of $E(Y|X)$ can be found using the bootstrap method described in Section 2.2. □

Other methods of handling monotonic regression are compared and illustrated by Cryer, Robertson, Wright, and Casady (1972), Casady and Cryer (1976), Hogg (1975), and Iman and Conover (1979). This procedure is explained more fully in Iman and Conover (1979).

EXERCISES

1. Dose-response curves such as the following are widely used in biological studies and in the pharmaceutical industry. Suppose that a certain drug (X, measured in milliliters) is administered to guinea pigs to see whether a particular reaction (cancer, diabetes, etc.) occurs. Five guinea pigs are treated at each of several dosage levels of the drug, and the percent of the animals showing the reaction is recorded as the Y variable.

X (dosage)	0.5	1.0	1.5	2.0	2.5	3.0	3.5	4.0	4.5	5.0
Y (percent response)	0	0	20	0	40	60	40	80	100	100

   (a) Plot the points on a graph. Does the expected value of the response seem to be a linear function of the dosage? A monotonic function?

   (b) Estimate $E(Y|X)$ at $X = 3.0$ milliliters.

   (c) Estimate $E(Y|X)$ at $X = 3.3$ milliliters.

   (d) Estimate the regression of Y on X. Plot the estimated regression curve on the same graph used in part a.

2. Ten companies reported their percent increase in advertising expenses, X, and their percent increase in sales, Y, for last year as compared with the previous year.

					Company					
	1	2	3	4	5	6	7	8	9	10
X (advertising)	4	62	31	−11	47	88	16	−1	74	21
Y (sales)	10	33	39	−14	37	39	18	−8	45	33

   (a) Plot the points on a graph. Does expected value of the percent increase in sales seem to be a linear function of the percent increase in advertising? A monotonic function?

   (b) Estimate the expected percent increase in sales for a 25% increase in advertising.

   (c) Estimate the regression of Y on X. Plot the estimated regression curve on the same graph used in part a.

3. In a test to determine the probability of a land mine exploding, given a certain strength of stimulus, 17 land mines were tested by giving each one a different strength of shock stimulus and noting whether the mine exploded or not. Eight mines exploded and 9 did not. The respective strength of the shock stimulus for each is given as follows.

| *Exploded* | 10.7, 13.9, 15.8, 17.0, 18.1, 19.9, 20.7, 21.6 |
| *Did not explode* | 4.0, 4.4, 4.7, 5.1, 9.3, 11.2, 13.7, 15.0, 19.7 |

Use monotonic regression to estimate the probability of a land mine exploding given a shock stimulus of strength 20. (*Hint.* Let $Y = 0$ if the mine did not explode, and let $Y = 1$ if the mine exploded.)

PROBLEMS

1. Show that the estimates of $E(Y|X)$ can never be less than the smallest observed value of Y or greater than the largest observed value of Y. Discuss the advantages or disadvantages of this property relative to the situations described in Exercises 1 and 2.

2. Find the least squares regression line for the data in Exercise 1. Use this regression line to estimate the mean of Y given $X = 0.5$ milliliters. Does this estimate seem reasonable to you?

5.7 THE ONE-SAMPLE OR MATCHED-PAIRS CASE

The rank test of this section deals with the single random sample and the random sample of matched pairs that is reduced to a single sample by considering differences. A matched pair (X_i, Y_i) is actually a single observation on a bivariate random variable. The sign test of Section 3.4 analyzed matched pairs of data by reducing each pair to a plus, a minus, or a tie and applying the binomial test to the resultant single sample. The test of this section also reduces the matched pair (X_i, Y_i) to a single observation by considering the difference

$$D_i = Y_i - X_i \qquad \text{for } i = 1, 2, \ldots, n \tag{1}$$

The analysis is then performed on the D_is as a sample of single observations. Whereas the sign test merely noted whether D_i was positive, negative, or zero, the test of this section notes the sizes of the positive D_is relative to the negative D_is. The model of this section resembles the model used in the sign test. Also, the hypotheses resemble the hypotheses of the sign test. The important difference between the sign test and this test is an additional assumption of *symmetry* of the distribution of differences. Before we introduce the test, we should clarify the meaning of the adjective *symmetric* as it applies to a distribution and discuss the influence of symmetry on the scale of measurement.

Symmetry is easy to define if the distribution is discrete. A discrete distribution is symmetric if the left half of the graph of the probability function is the mirror image of the right half. For example, the binomial distribution is symmetric

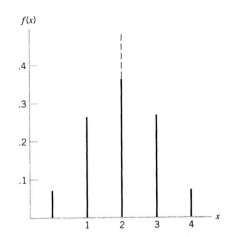

FIGURE 6 Symmetry in a binomial distribution.

if $p = 1/2$ (see Figure 6) and the discrete uniform distribution is always symmetric (see Figure 7). The dotted lines in the figures represent the lines about which the distributions are symmetric.

For other than discrete distributions we are not able to draw a graph of the probability function. Therefore a more abstract definition of symmetry is required, such as the following.

> **Definition 1** The distribution of a random variable X is *symmetric* about a line $x = c$, for some constant c, if the probability of $X \leq c - x$ equals the probability of $X \geq c + x$ for each possible value of x.

In Figure 6, $c = 2$ and the definition is easily verified for all real numbers x. In Figure 7, $c = 3.5$. Even though we may not know the exact distribution of a random variable, we are often able to say, "It is reasonable to assume that the distribution is symmetric." Such an assumption is not as strong as the assumption

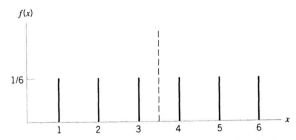

FIGURE 7 Symmetry in a discrete uniform distribution.

of a normal distribution; while all normal distributions are symmetric, not all symmetric distributions are normal.

If a distribution is symmetric, the mean (if it exists) coincides with the median because both are located exactly in the middle of the distribution, at the line of symmetry. One consequence of adding the assumption of symmetry to the model is that any inferences concerning the median are also valid statements for the mean.

A second consequence of adding the assumption of symmetry to the model is that the required scale of measurement is changed from ordinal to interval. With an ordinal scale of measurement, two observations of the random variable need only to be distinguished on the basis of which is larger and which is smaller. It is not necessary to know which one is farthest from the median, such as when the two observations are on opposite sides of the median. If the assumption of symmetry is a meaningful one, the distance from the median is a meaningful measurement and, therefore, the distance between two observations is a meaningful measurement. As a result, the scale of measurement is more than just ordinal, it is interval.

A test presented by Wilcoxon (1945) is designed to test whether a particular sample came from a population with a specified mean or median. It may also be used in situations where observations are paired, such as "before" and "after" observations on each of several subjects, to see if the second random variable in the pair has the same mean as the first. Note that in a symmetric distribution the mean equals the median, so the two terms can be used interchangeably.

▶ The Wilcoxon Signed Ranks Test

Data   The data consist of n' observations (x_1, y_1), (x_2, y_2), . . . , $(x_{n'}, y_{n'})$ on the respective bivariate random variables (X_1, Y_1), (X_2, Y_2), . . . , $(X_{n'}, Y_{n'})$. Find the n' differences $D_i = Y_i - X_i$. (In the one-sample problem, the Ds are the observations in the sample, as illustrated in Example 2.) The absolute differences (without regard to sign)

$$|D_i| = |Y_i - X_i| \qquad i = 1, 2, . . . , n' \tag{2}$$

are then computed for each of the n' pairs $(X_i - Y_i)$.

Omit from further consideration all pairs with a difference of zero (i.e., where $X_i = Y_i$, or $D_i = 0$). Let the number of pairs remaining be denoted by n, $n \leq n'$. Ranks from 1 to n are assigned to these n pairs according to the relative size of the absolute difference, as follows. The rank 1 is given to the pair (X_i, Y_i) with the smallest absolute difference $|D_i|$; the rank 2 is given to the pair with the second smallest absolute difference; and so on, with the rank n being assigned to the pair with the largest absolute difference.

If several pairs have absolute differences that are equal to each other, assign to each of these several pairs the *average* of the ranks that would have otherwise been assigned [i.e., if the ranks 3, 4, 5, and 6 belong to four pairs, but we do not know which rank to assign to which pair because all four absolute differences are exactly equal to each other, assign the average rank $\frac{1}{4}(3 + 4 + 5 + 6) = 4.5$ to each of the four pairs.]

Assumptions

1. The distribution of each D_i is symmetric.
2. The D_is are mutually independent.
3. The D_is all have the same mean.
4. The measurement scale of the D_is is at least interval.

Test Statistic Let R_i, called the signed rank, be defined for each pair (X_i, Y_i) as follows.

R_i = the rank assigned to (X_i, Y_i) if $D_i = Y_i - X_i$ is positive (i.e., $Y_i > X_i$)
R_i = the negative of the rank assigned to (X_i, Y_i) if D_i is negative (i.e., $Y_i < X_i$)

The test statistic is the sum of the positive signed ranks

$$T^+ = \Sigma \, (R_i \text{ where } D_i \text{ is positive}) \tag{3}$$

Null Distribution Lower quantiles of the exact distribution of T^+ when there are no ties and $n \leq 50$ are given in Table A12, under the null hypothesis that the D_is have mean 0. Upper quantiles are found from the relation

$$w_p = \frac{n(n + 1)}{2} - w_{1-p} \tag{4}$$

If there are many ties, or if $n > 50$, the normal approximation should be used. The normal approximation uses the sum of all of the signed ranks, with their + or − signs, and the statistic

$$T = \frac{\sum\limits_{i=1}^{n} R_i}{\sqrt{\sum\limits_{i=1}^{n} R_i^2}} \tag{5}$$

In case there are no ties, Equation 5 simplifies to

$$T = \frac{\sum\limits_{i=1}^{n} R_i}{\sqrt{n(n + 1)(2n + 1)/6}} \tag{6}$$

with the aid of Lemma 1.4.2. The null distribution of T is approximately standard normal, as in Table A1.

Hypotheses

(a) (Two-Tailed Test)

$$H_0: E(D) = 0 \qquad (\text{i.e., } E(Y_i) = E(X_i))$$
$$H_1: E(D) \neq 0$$

If the pairs (X_i, Y_i) have the same bivariate distribution, then H_1 can be written as $E(X) \neq E(Y)$. Reject H_0 at the level α if T^+ (or T) is less than its $\alpha/2$ quantile or greater than its $1 - \alpha/2$ quantile from Table A12 for T^+. Quantiles for T are obtained approximately from Table A1.

   The two-tailed p-value is twice the smaller of the one-tailed p-values, approximated from the normal distributions as either

$$\text{lower-tailed } p\text{-value} = P\left(Z \leq \frac{\sum\limits_{i=1}^{n} R_i + 1}{\sqrt{\sum\limits_{i=1}^{n} R_i^2}} \right) \tag{7}$$

or

$$\text{upper-tailed } p\text{-value} = P\left(Z \geq \frac{\sum\limits_{i=1}^{n} R_i - 1}{\sqrt{\sum\limits_{i=1}^{n} R_i^2}} \right) \tag{8}$$

(b) (Lower-Tailed Test)

$$H_0: E(D) \geq 0 \qquad (\text{i.e., } E(Y_i) \geq E(X_i))$$
$$H_1: E(D) < 0$$

If the pairs (X_i, Y_i) have the same bivariate distribution, H_1 can be written as $E(Y) < E(X)$. Reject H_0 at the level of α if T^+ (or T) is less than its α quantile from Table A12 (for T^+) or from Table A1 (for T). The lower-tailed p-value is given approximately by Equation 7.

(c) (Upper-Tailed Test)

$$H_0: E(D) \leq 0 \qquad (\text{i.e., } E(Y_i) \leq E(X_i))$$
$$H_1: E(D) > 0$$

If the pairs (X_i, Y_i) have the same bivariate distribution, H_1 can be written as $E(Y) > E(X)$. Reject H_0 at the level α if T^+ (or T) is greater than its α quantile given by Table A12 for T^+, or Table A1 for T. The upper-tailed p-value is given approximately by Equation 8.

Computer Assistance *Minitab, S-Plus, SAS,* and *StatXact* contain programs to perform the Wilcoxon signed ranks test. ──────────────────────────◀

EXAMPLE 1

Twelve sets of identical twins were given psychological tests to measure in some sense the amount of aggressiveness in each person's personality. We are interested in comparing the twins with each other to see if the firstborn twin tends to be more aggressive than the other. The results are as follows, where the higher score indicates more aggressiveness.

	Twin Set											
	1	*2*	*3*	*4*	*5*	*6*	*7*	*8*	*9*	*10*	*11*	*12*
Firstborn X_i	86	71	77	68	91	72	77	91	70	71	88	87
Second twin Y_i	88	77	76	64	96	72	65	90	65	80	81	72
Difference D_i	+2	+6	−1	−4	+5	0	−12	−1	−5	+9	−7	−15
Rank of $|D_i|$	3	7	1.5	4	5.5	—	10	1.5	5.5	9	8	11
R_i	3	7	−1.5	−4	5.5	—	−10	−1.5	−5.5	9	−8	−11

The hypotheses are:

H_0: The firstborn twin does not tend to be more aggressive than the other ($E(X_i) \le E(Y_i)$)

H_1: The firstborn twin tends to be more aggressive than the second twin ($E(X_i) > E(Y_i)$)

These correspond to hypotheses set B. We are assuming that the test scores are accurate measures of aggressiveness of the individuals. Because of the many ties, the test statistic is:

$$T = \frac{\sum R_i}{\sqrt{\sum R_i^2}} = \frac{-17}{\sqrt{505}} = -0.7565 \tag{9}$$

The critical region of size $\alpha = 0.05$ corresponds to values of T less than -1.6449 (from Table A1). Therefore H_0 is readily accepted. The p-value, from Equation 7, is 0.238.

If we had used T^+ and Table A12 we would have obtained $T^+ = 24.5$ and a critical region corresponding to values of T^+ less than 14. So the same conclusion would have been reached and a similar p-value would have been obtained by interpolation between $w_{0.20}$ and $w_{0.30}$ in Table A12. ∎

The Wilcoxon signed ranks test is equally appropriate as a median test, where the data consist of a single random sample of size n', $Y_1, Y_2, \ldots, Y_{n'}$. Let

Y be a random variable with the same distribution as the Y_i and let m be a specified constant. The hypotheses, corresponding to the preceding hypotheses sets A, B, and C, are as follows.

(a) (Two-Tailed Test)

H_0: The median of Y equals m
H_1: The median of Y is not m

(b) (Lower-Tailed Test)

H_0: The median of Y is $\geq m$
H_1: The median of Y is $< m$

(c) (Upper-Tailed Test)

H_0: The median of Y is $\leq m$
H_1: The median of Y is $> m$

The word "mean" may be substituted for "median" in these hypotheses because of the assumption of symmetry of the distribution of Y.

Pairs $(m, Y_1), (m, Y_2), \ldots, (m, Y_{n'})$ are formed, and the pairs are treated exactly the same as described in the Wilcoxon signed ranks test. The rest of the Wilcoxon test procedure remains unchanged. The following example illustrates the procedure.

EXAMPLE 2

Thirty observations on a random variable Y are obtained in order to test the hypothesis that $E(Y)$, the mean of Y, is no larger than 30 (hypotheses set C).

$$H_0: E(Y) \leq 30$$
$$H_1: E(Y) > 30$$

The observations, the differences $Y_i - m$, and the ranks of the pairs are as follows. (The random sample was ordered first, for convenience.)

Y_i	$D_i = Y_i - 30$	Rank of $|D_i|$	Y_i	$D_i = Y_i - 30$	Rank of $|D_i|$
23.8	−6.2	17	35.9	+5.9	15
26.0	−4.0	11	36.1	+6.1	16
26.9	−3.1	8	36.4	+6.4	18
27.4	−2.6	6	36.6	+6.6	19
28.0	−2.0	5	37.2	+7.2	20
30.3	+0.3	1	37.3	+7.3	21
30.7	+0.7	2	37.9	+7.9	22

31.2	+1.2	3	38.2	+8.2	23
31.3	+1.3	4	39.6	+9.6	24
32.8	+2.8	7	40.6	+10.6	25
33.2	+3.2	9	41.1	+11.1	26
33.9	+3.9	10	42.3	+12.3	27
34.3	+4.3	12	42.8	+12.8	28
34.9	+4.9	13	44.0	+14.0	29
35.0	+5.0	14	45.8	+15.8	30

The 0.05 quantile from Table A12 is 152 so the 0.95 quantile is $465 - 152 = 313$. Therefore the critical region of size ≤ 0.05 corresponds to values of the test statistic greater than 313.

The test statistic is defined by Equation 3. In this case T^+ equals the sum of the ranks associated with the positive D_i.

$$T^+ = 418 \tag{10}$$

The large value of T^+ results in rejection of H_0. We conclude that the mean of Y is greater than 30.

The approximate p-value is given by Equation 8.

$$P\left(Z \geq \frac{\sum R_i - 1}{\sqrt{n(n+1)(2n+1)/6}}\right) = P\left(Z \geq \frac{371 - 1}{\sqrt{9455}}\right) = P(Z \geq 3.8051)$$

Table A1 shows that the p-value is smaller than 0.0001. ∎

□ *Theory*   The model states that all of the differences D_i share a common median, say $d_{0.50}$, which equals zero when H_0 is true. By the definition of symmetry, the probability of each D_i being negative is the same as its probability of being positive, which equals 0.5 for continuous distributions, or for discrete distributions where values of D_i equal to zero are discarded. (Without symmetry it would be possible for the positive differences to tend to be much larger than the negative differences, or vice versa.)

The purpose of these considerations is to find the distribution of the test statistic T^+ when H_0 is true. First, we will consider the null hypothesis of the two-tailed test. The resulting distribution applies equally well in the one-tailed tests.

Consider n chips numbered from 1 to n, corresponding to the n ranks if there are no ties. Suppose each chip has its number written on one side and the negative of its number on the other side (like 6 and -6). Each chip is tossed into the air so that it is equally likely to land with either side showing, corresponding to the ranks of (X_i, Y_i), which are equally likely to correspond to a positive D_i, in which case the signed rank R_i equals the rank, or a negative D_i, in which case R_i is a negative rank. Let T^+ be the sum of the positive numbers showing after all n chips are tossed, corresponding to the definition of T^+ in Equation 3. The probability distribution of T^+ is the same in the game with the chips as it is when H_0 is true, but the game with the chips is easier to imagine.

The sample space in the game with the chips consists of points such as $(1, 2, 3, -4, -5, 6, 7, \ldots, n)$, simply a reordering of the R_i associated with a set of data like in Example 1. The tosses are independent of each other, so each of the 2^n points has probability $(1/2)^n$. The test statistic T^+ equals the sum of the positive numbers in the sample point. Therefore the probability that T^+ equals any number x is found by counting the points whose positive numbers add to x, then multiplying that count by the probability $(1/2)^n$.

For example, if $n = 8$, then T^+ can equal 0 one way (all the positive numbers landed face down), and so $P(T = 0) = (1/2)^8$. $T^+ = 1$ only one way, $T^+ = 2$ only one way, but $T^+ = 3$ two ways, points $(-1, -2, 3, -4, -5, -6, -7, -8)$ and $(1, 2, -3, -4, -5, -6, -7, -8)$. Also, $T^+ = 4$ two ways. That is,

$$P(T^+ = 0) = (1/2)^8 = 1/256 \qquad P(T^+ \le 0) = 0.0039$$
$$P(T^+ = 1) = 1/256 \qquad P(T^+ \le 1) = 0.0078$$
$$P(T^+ = 2) = 1/256 \qquad P(T^+ \le 2) = 0.0117$$
$$P(T^+ = 3) = 2/256 \qquad P(T^+ \le 3) = 0.0195$$
$$P(T^+ = 4) = 2/256 \qquad P(T^+ \le 4) = 0.0273$$
$$\text{etc.} \qquad\qquad\qquad \text{etc.}$$

The distribution function of T^+ is tabulated in Owen (1962) for $n \le 20$ and in Harter and Owen (1970) for $n \le 50$. A table of selected quantiles for $n \le 100$ is given by McCornack (1965). That table is more extensive than we need here, so the more useful quantiles were selected and are given in Table A12. The use of Table A12 will generally result in a slightly conservative test, because the probability of being less than the p quantile may be less than p. For example, if $n = 8$, as in the preceding paragraph, the 0.025 quantile of T^+ is given in Table A12 as 4, while the actual size of the critical region corresponding to values of T^+ less than 4 is 0.0195. Further results on the exact distribution of T^+ are given by Claypool (1970) and Chow and Hodges (1975).

For the one-tailed tests, the probability of getting a point in the critical region is a maximum when the median difference is 0, so this is the situation to be considered. Thus the preceding distribution of T^+ is equally valid when H_0 is true in the one-tailed tests.

To find the conditional distribution of T^+ when there are ties, only the initial step in the discussion is changed. That is, the numbers on the chips must agree with the ranks and average ranks assigned to the pairs (X_i, Y_i) in the particular set of data under consideration. Call these ranks and average ranks $a_1, a_2, \ldots, a_n$. In Example 1 we have $a_1 = 1.5, a_2 = 1.5, a_3 = 3$, and so on. For this set of numbers we can find the distribution T^+. Because there are 11 numbers in Example 1, there are $2^{11} = 2048$ points in the sample space. The smallest 5% of these, about 102 points, constitute the critical region. This is a large number of points to tabulate by hand, so the normal approximation is used.

To use the normal approximation, let S equal the sum of all the R_i. Then, to apply the central limit theorem from Section 1.5, we need the mean and variance of S when H_0 is true. Note that under H_0,

$$P(R_i = a_i) = 1/2 \quad \text{and} \quad P(R_i = -a_i) = 1/2$$

so that

$$E(R_i) = a_i(\tfrac{1}{2}) + (-a_i)(\tfrac{1}{2}) = 0 \tag{11}$$

and

$$\text{Var}(R_i) = a_i^2(\tfrac{1}{2}) + (-a_i)^2(\tfrac{1}{2}) = a_i^2 \tag{12}$$

Since the R_is are independent of each other (the tosses of the chips are independent), we can apply Theorems 1.4.1 and 1.4.3 to get

$$E(S) = \sum_{i=1}^{n} E(R_i) = 0 \tag{13}$$

and

$$\text{Var}(S) = \sum_{i=1}^{n} \text{Var}(R_i) = \sum_{i=1}^{n} a_i^2 \tag{14}$$

But since a_i^2 always equals R_i^2 (the sign always becomes $+$), we can say

$$\text{Var}(S) = \sum_{i=1}^{n} R_i^2 \tag{15}$$

and apply the central limit theorem to

$$T = \frac{\displaystyle\sum_{i=1}^{n} R_i}{\sqrt{\displaystyle\sum_{i=1}^{n} R_i^2}} \tag{16}$$

and use the normal distribution with a continuity correction as an approximation whenever exact tables are not available. Justification for the treatment of ties is given by Vorlickova (1970) and Conover (1973a).   □

The method presented here for handling zero differences was suggested by Wilcoxon (1949). Another method of handling zero differences is thoroughly discussed by Pratt (1959). It involves leaving the zero differences in, ranking the $|D_i|$ as described, but treating all the $D_i = 0$ values as a tie and assigning the average rank in the usual manner. Then R_i is defined as usual, except that $R_i = 0$ if $D_i = 0$. Then T is computed from Equation 5 and compared with Table A1. Table A12 is not used with Pratt's method when testing hypotheses, but some exact tables are given by Rahe (1974). A comparison by Conover (1973b) shows that each method of handling ties at zero is more powerful than the other in some situations, so there is little reason to prefer one over the other. Pratt's suggestion of retaining the zero differences is incorporated into the following method for finding a confidence interval for $d_{0.50}$, the common median of the D_is, which appears in Tukey (1949) and Walker and Lev (1953).

► **Confidence Interval for the Median Difference** —————————————

Data   The data consist of n observations (x_1, y_1), (x_2, y_2), . . . , (x_n, y_n) on the bivariate random variables (X_1, Y_1), (X_2, Y_2), . . . , (X_n, Y_n), respectively. Compute the difference

$$D_i = Y_i - X_i$$

for each pair and arrange them in order from the smallest (the most negative) to the largest (the most positive), denoted as follows.

$$D^{(1)} \le D^{(2)} \le \cdot \cdot \cdot \le D^{(n-1)} \le D^{(n)}$$

Or, in the one-sample situation, the data consist of a single sample $D_1, D_2,$ . . . , D_n, arranged in order as shown. We wish to find a confidence interval for the common median (or mean) of the D_is.

Assumptions

1. The distribution of each D_i is symmetric.
2. The D_is are mutually independent.
3. The D_is all have the same median.
4. The measurement scale of the D_is is at least interval.

Method   To obtain a $1 - \alpha$ confidence interval, obtain the $\alpha/2$ quantile $w_{\alpha/2}$ from Table A12. (If $w_{\alpha/2}$ = zero no confidence interval may be obtained for that value of α.) Then consider the $n(n + 1)/2$ possible averages $(D_i + D_j)/2$ for all i and j, including $i = j$, which is the average of D_i with itself, giving just D_i. The $w_{\alpha/2}$th largest of these averages and the $w_{\alpha/2}$th smallest of these averages constitute the upper and lower bounds for the $1 - \alpha$ confidence interval. It is not necessary to compute all $n(n + 1)/2$ averages; only the averages near the largest and the smallest need to be computed to obtain a confidence interval.

Computer Assistance Confidence intervals for the mean difference or median difference, also known as the Hodges-Lehmann estimates of shift, are given by *Minitab* and *StatXact*. ───────────────────────────────── ◄

EXAMPLE 1 (continued).

The 12 values of D_i, arranged in order, are

$$-15, -12, -7, -5, -4, -1, -1, 0, 2, 5, 6, 9$$

To find a 95% confidence interval for the median difference, Table A12 is entered with $n = 12$ to obtain $w_{0.025} = 14$. The 14 smallest averages, starting with $(-15 - 15)/2$, are

$$-15, -13.5, -12, -11, -10, -9.5, -9.5, -8.5, -8, -8, -8, -7.5, -7, -6.5$$

so the lower bound for the confidence interval is -6.5. The 14 largest averages are

$$9, 7.5, 7, 6, 5.5, 5.5, 5, 4.5, 4, 4, 4, 3.5, 3, 2.5$$

so the upper bound for the confidence interval is 2.5. The 95% confidence interval for the median difference of aggressiveness scores (firstborn twin minus second twin) is

$$P(-6.5 \leq d_{0.50} \leq 2.5) \geq 0.95 \tag{17}$$

A convenient method for finding the 14 smallest and largest averages is to form an upper-triangular matrix of averages, where the rows and columns of the matrix are the Ds.

	−15	−12	−7	−5	−4	−1	−1	0	2	5	6	9
−15	−15	−13.5	−11	−10	−9.5	−8	−8	−7.5	−6.5	−5	−4.5	−3
−12		−12	−9.5	−8.5	−8	−6.5	−6.5	−6	−5	−3.5	−3	−1.5
−7			−7	−6	−5.5	−4	−4	−3.5	−2.5	−1	−0.5	1
−5				−5	−4.5	−3	−3	−2.5	−1.5	0	0.5	2
−4					−4	−2.5	−2.5	−2	−1	0.5	1	2.5
−1						−1	−1	−0.05	0.5	2	2.5	4
−1							−1	−0.05	0.5	2	2.5	4
0								0	1	2.5	3	4.5
2									2	3.5	4	5.5
5										5	5.5	7
6											6	7.5
9												9

□ *Theory* To see the relationship between average differences $(D_i + D_j)/2$ and the rank of the difference, consider the following. The rank of any D_i, say $D_i = 6$ in the previous example, is equal to the number of D_js as close or closer to 0 than $D_i = 6$ is, assuming no ties. By counting the averages between 0 and 6, which involve D_i, we obtain the rank of D_i. (We must be careful to include the average of D_i with itself in the count.) By repeating this for all positive D_i we obtain, as a total count, the test statistic T^+.

A confidence interval for the median $d_{0.50}$ of D_i is found by using the Wilcoxon test to test

$$H_0: d_{0.50} = m$$

for various values of m. This procedure is equivalent to subtracting the value m from each D_i and testing to see if the median of the new D_is equals zero. But instead of subtracting m from each D_i and then reranking and recomputing T^+, it is easier to look at the averages of the original D_is, counting how many averages are above m (instead of zero, as we did before) and that equals T^+. By working backward, starting with the critical value for T^+ and finding those largest averages, the stopping point is the value of m that would have barely resulted in acceptance of H_0. Thus the bounds for the confidence interval are found. □

Noether (1967b) shows that if the continuity assumption is not true, the confidence interval with its end points (U and L) has a confidence coefficient of at least $1 - \alpha$, while the interval without its end points has a confidence coefficient of at most $1 - \alpha$. Therefore we recommend inclusion of the end points and a statement of the form

$$P(L \leq d_{0.50} \leq U) \geq 1 - \alpha$$

A discussion of this method of finding confidence intervals is given by Moses (1965). If the sampling is stratified rather than random, see the article by McCarthy (1965). For another type of dependency in the sample, see Høyland (1968). Confidence regions for the case of multivariate random variables are given by Puri and Sen (1968). Sequential sampling methods offer some advantages according to Geertsema (1970) and Srivastava and Sen (1973). Other methods of estimating the center of a distribution are discussed by Schuster and Navarte (1973), Noether (1973), Johns (1974), and Maritz, Wu, and Staudte (1977). See Seran (1977) for a theoretical discussion of robust location estimates.

Comparison with Other Procedures

When one encounters paired observations and wishes to test whether the mean difference is zero, and the scale of measurement is interval as in this section, the first test that usually comes to mind is the "paired t test," also called the "one-sample t test." This test uses the test statistic

$$t = \frac{\overline{D}}{\sqrt{\dfrac{1}{n(n-1)} \displaystyle\sum_{i=1}^{n} (D_i - \overline{D})^2}} \tag{18}$$

where $\overline{D}$ is the sample mean of the D_is, and compares it with quantiles of the t distribution from Table A21, in the row $k = n - 1$. In order for the quantiles from Table A21 to be accurate, an additional assumption of normality must be made. That is, add to the assumptions for the Wilcoxon test the assumption that the D_is are identically distributed normal random variables.

The assumptions of the Wilcoxon test are easier to justify than the assumption of normality. If the data are discrete, we know right away that the distribution is nonnormal because the normal distribution is continuous. If the data have an occasional very large or very small observation, called "outliers," the power of the t test drops considerably and should not be used.

If a computer program for the t test is available, it can be used for the Wilcoxon test also. Merely use the R_is instead of the D_is in computing t and compare the result with Table A21, as just described. This approximation is slightly more accurate than the normal approximation described earlier and works well with ties. See Iman (1974a) for more details.

The A.R.E. of the Wilcoxon signed ranks test relative to the paired t test is computed under the following restrictions.

1. The bivariate random variables $(X_1, Y_1), \ldots, (X_n, Y_n)$ constitute a random sample.

2. The distribution of X_i is identical with the distribution of Y_i, except for a possible difference in means.

Under these conditions the A.R.E. may range from $108/125 = 0.864$ up to infinity, but with the surprising insurance feature of never being less than 0.864. Therefore the Wilcoxon test never can be too bad, but it can be infinitely good as compared with the usual parametric test under these circumstances.

Under the further restriction that the differences D_i have the normal distribution, the A.R.E. is $3/\pi = 0.955$. If, instead, we assume that the differences D_i have the uniform distribution, the A.R.E. is 1.0. For a distribution known as the double exponential distribution, the A.R.E. is 1.5.

Under the preceding restrictions the sign test (Section 3.4) may be used to test the same hypotheses as the Wilcoxon test. Then the A.R.E. of the sign test relative to the Wilcoxon test is as follows.

Assumed Distribution	A.R.E.
Normal	$\frac{2}{3}$
Uniform	$\frac{1}{3}$
Double exponential	$\frac{4}{3}$

It may be surprising to some that the sign test may be more powerful than the Wilcoxon test under some circumstances. The Wilcoxon A.R.E., compared to the paired t test, is 3/2 for the double exponential distribution and, therefore, multiplication gives the A.R.E. of the sign test, relative to the paired t test, as

$$\left(\tfrac{4}{3}\right)\left(\tfrac{3}{2}\right) = 2$$

For the double exponential distribution the sign test has twice the asymptotic efficiency as the paired t test. However, in this case the A.R.E. is not a good indication of small sample power. See Conover, et al. (1978) for a discussion of the two-sample case. The power of the Wilcoxon signed ranks test is superior to both the sign test and the t test for small samples from the double exponential distribution. For other investigations of power and efficiency, see Klotz (1963, 1965), Arnold (1965), Noether (1967a), and Kraft and van Eeden (1972).

The Wilcoxon signed ranks test is sometimes called a test of symmetry. Schuster (1975) and Rao, Schuster, and Littell (1975) discuss estimation with symmetric distributions, while Rothman and Woodroofe (1972) present an alternative test for symmetry. Symmetry tests for bivariate random variables are discussed by Bell and Haller (1969), Hollander (1971), and Bhattacharyya, Johnson, and Neave (1971). Extensions to multivariate random variables are examined by Bennett (1965) and Sen and Puri (1967). Hollander (1970) adapts the Wilcoxon test to test for parallelism of two regression lines. Adaptations to sequential sampling are presented by Miller (1970), Weed, Bradley, and Govindarajulu (1974), Sen and Ghosh (1974), Reynolds (1975), and Spurrier and Hewett (1976). For other papers related to this section see Groeneveld (1972) and Bickel and Lehmann (1975). A test proposed by Walsh (1949) is identical to the Wilcoxon test for n less than 7, but not for n of 7 or more. An application of the Wilcoxon test to problems involving circular distributions, as presented by Batschelet (1965), is given by Schach (1969a).

EXERCISES

1. A random sample consisting of 20 people who drove automobiles was selected to see if alcohol affected reaction time. Each driver's reaction time was measured in a laboratory before and after drinking a specified amount of a beverage containing alcohol. The reaction times in seconds were as follows.

Subject	Before	After	Subject	Before	After
1	.68	.73	11	.65	.72
2	.64	.62	12	.59	.60
3	.68	.66	13	.78	.78
4	.82	.92	14	.67	.66
5	.58	.68	15	.65	.68
6	.80	.87	16	.76	.77
7	.72	.77	17	.61	.72
8	.65	.70	18	.86	.86
9	.84	.88	19	.74	.72
10	.73	.79	20	.88	.97

Does alcohol affect reaction time?

2. A grocer wishes to see whether the median number of items bought on each sale could be considered to be 10, so he observes 12 customers at the checkout counter.

Customer	Number of Items	Customer	Number of Items
1	22	7	15
2	9	8	26
3	4	9	47
4	5	10	8
5	1	11	31
6	16	12	7

Can the Wilcoxon test be used? Which assumptions of the model are violated in this problem?

3. Test the data of Example 3.5.3 to see if there is a tendency for the observations in the second year to be less than the observations in the first year.

4. Each member of a girls' basketball team was given a brief warm-up period and then told to shoot 25 free throws. The number X of goals was recorded. Then the team was given an extensive workout and, after a brief rest period, was told to shoot another 25 free throws each. The number Y of successful attempts was again recorded. Do the data indicate that the percentages tend to drop when the players are tired?

						Player						
	1	2	3	4	5	6	7	8	9	10	11	12
X_i (before)	18	12	7	21	19	14	8	11	19	16	8	11
Y_i (after)	16	10	8	23	13	10	8	13	9	8	8	5

5. A candidate for political office realizes that she will maximize her vote-getting ability if she adopts the median position of her constituents. Therefore she devises a questionnaire and distributes it to 15 voters (who hopefully resemble a random sample). The results of the questionnaire are scored from one extreme (0) to the other (10).

Voter	Score	Voter	Score	Voter	Score
1	6.7	6	9.3	11	8.8
2	4.2	7	8.9	12	5.4
3	4.1	8	7.4	13	6.1
4	2.3	9	7.4	14	6.0
5	6.1	10	9.3	15	4.9

Find a 90% confidence interval for the median score. Based on the procedure of this section, what is your point estimate of the median score?

6. An emergency rescue squad is responsible for accidents occurring in a long narrow lake. They wish to build a permanent station in the spot that will minimize the total distance they will have to travel going to accidents in the future. That spot should be at the median (by distance from some point of reference) spot at which accidents will occur. Assuming that the accidents occurred thus far resemble a random sample of all accidents yet to occur, the distances (from the dam) are measured.

Accident	Distance (miles)	Accident	Distance (miles)
1	7.1	8	6.1
2	4.4	9	2.2
3	3.9	10	6.7
4	2.2	11	4.9
5	4.2	12	7.3
6	3.4	13	0.3
7	1.1	14	7.6

What is a 95% confidence interval for the optimal distance from the dam for the station?

7. Seven married couples were selected at random, and each husband and each wife was asked how much money they spent on their spouse's Christmas present this year. The responses were as follows:

			Couple				
	1	2	3	4	5	6	7
Husband	25	21	38	64	52	16	26
Wife	16	42	56	41	19	26	24

(a) Find a 95% confidence interval for the median amount of what the husband spent in excess of what his wife spent.

(b) What is the exact level of confidence of the interval you found?

8. Four prospective graduate students took the GMAT twice, with the following scores.

Student	First Attempt	Second Attempt
1	470	510
2	530	550
3	610	600
4	440	490

(a) Find the exact distribution of the Wilcoxon signed ranks test statistic, the sum of the positive signed ranks, and draw a graph of its distribution function.

(b) Locate the observed value of the test statistic on the graph of part a and find the exact p-value in an upper-tailed test, or in a lower-tailed test, whichever is smaller.

(c) Find a nonparametric 80% (approximately) confidence interval for the mean *increase in score*.

(d) What is the exact confidence level of the interval you found?

PROBLEMS

1. Find the probability distribution of the test statistic T^+ of this section for $n = 5$. (Assume H_0 is true in the two-tailed test.)

2. In the Wilcoxon signed ranks test zero differences were discarded so the exact tables could be used. Why is it better to include the zero differences when finding a confidence interval for the median difference?

3. Suppose that the t statistic (Equation 18) is computed on the signed ranks R_i instead of the differences D_i. Show that this statistic is the following function of T,

$$t_R = \frac{T}{\left(\dfrac{n}{n-1} - \dfrac{1}{n-1}T^2\right)^{\frac{1}{2}}}$$

as stated in Equation 5.12.3. Also show that as T increases, t_R increases and, therefore, the test that rejects H_0 for large T is equivalent to the test that rejects H_0 for large t_R.

4. Compute the paired t test statistic on the data of Exercise 4 and compare the results with those of the Wilcoxon test. (Use Table A21 with row $k = 11$; zeros are not discarded in the paired t test.)

5.8. SEVERAL RELATED SAMPLES

In Section 5.2 we presented the Kruskal-Wallis rank test for several independent samples, which is an extension of the Mann-Whitney test for two independent samples introduced in Section 5.1. In this section we consider the problem of analyzing several related samples, which is an extension of the problem of matched pairs, or two related samples, examined in the previous section. First we will present the Friedman test, which is an extension of the sign test of Sections 3.4 and 3.5. Then we will present the Quade test, which is an extension of the Wilcoxon signed ranks test of the previous section. The Friedman test is the better-known test of the two and requires fewer assumptions, but it suffers from a lack of power when there are only three treatments, just as the sign test has less power than the Wilcoxon signed ranks test when there are only two treatments. When there are four or five treatments the Friedman test has about the same power as the Quade test, but when the number of treatments is six or more the Friedman test tends to have more power. See Iman et al. (1984) and Hora and Iman (1988) for power and A.R.E. comparisons.

The problem of several related samples arises in an experiment that is designed to detect differences in k possibly different treatments, $k \geq 2$. The observations are arranged in blocks, which are groups of k experimental units similar to each other in some important respects, such as k puppies that are littermates and therefore may tend to respond to a particular stimulus more similarly than would randomly selected puppies from various litters. The k experimental units within a block are matched randomly with the k treatments being scrutinized, so that each treatment is administered once and only once within each block. In this way the treatments may be compared with each other without an excess of unwanted effects confusing the results of the experiment. The total number of blocks used is denoted by $b, b > 1$.

The experimental arrangement described here is usually called a *randomized complete block design*. This design may be compared with the *incomplete* block design described in the next section, in which the blocks do not contain enough experimental units to enable all the treatments to be applied in all the blocks, and so each treatment appears in some blocks but not in others. Examples of randomized complete block designs are as follows.

1. *Psychology.* Five litters of mice, with four mice per litter, are used to examine the relationship between environment and aggression. Each litter is considered to be a block. Four different environments are designed. One mouse from each litter is placed in each environment, so that the four mice from each litter are in four different environments. After a suitable length of time, the mice are regrouped with their littermates and are ranked according to degree of aggressiveness.

2. *Home economics.* Six different types of bread dough are compared to see which bakes the fastest by forming three loaves with each type of dough. Three different ovens are used, and each oven bakes the six different types of bread at the same time. The ovens are the blocks and the doughs are the treatments.

3. *Environmental engineering.* One experimental unit may form a block if the different treatments may be applied to the same unit without leaving residual effects. Seven different men are used in a study of the effect of color schemes on work efficiency. Each man is considered to be a block and spends some time in each of three rooms, each with its own type of color scheme. While in the room, each man performs a work task and is measured for work efficiency. The three rooms are the treatments.

By now the reader should have some idea of the nature of a randomized complete block design. The usual parametric method of testing the null hypothesis of no treatment differences is called the two-way analysis of variance. The following nonparametric method depends only on the ranks of the observations within each block. Therefore it may be considered a two-way analysis of variance on ranks. This test is named after its inventor, the noted economist Milton Friedman.

▶ The Friedman Test

Data The data consist of b mutually independent k-variate random variables $(X_{i1}, X_{i2}, \ldots, X_{ik})$, called b blocks, $i = 1, 2, \ldots, b$. The random variable X_{ij} is in block i and is associated with treatment j. The b blocks are arranged as follows.

		Treatment		
Block	1	2	$\cdots$	k
1	X_{11}	X_{12}	$\cdots$	X_{1k}
2	X_{21}	X_{22}	$\cdots$	X_{2k}
3	X_{31}	X_{32}	$\cdots$	X_{3k}
$\cdots$	$\cdots$	$\cdots$	$\cdots$	$\cdots$
b	X_{b1}	X_{b2}	$\cdots$	X_{bk}

Let $R(X_{ij})$ be the rank, from 1 to k, assigned to X_{ij} within block (row) i. That is, for block i the random variables $X_{i1}, X_{i2}, \ldots, X_{ik}$ are compared with each other and the rank 1 is assigned to the smallest observed value, the rank 2 to the second smallest, and so on to the rank k, which is assigned to the largest observation in block i. Ranks are assigned in all of the b blocks. Use average ranks in case of ties.

Then sum the ranks for each treatment to obtain R_j where:

$$R_j = \sum_{i=1}^{b} R(X_{ij}) \tag{1}$$

for $j = 1, 2, \ldots, k$.

Assumptions

1. The b k-variate random variables are mutually independent. (The results within one block do not influence the results within the other blocks.)

2. Within each block the observations may be ranked according to some criterion of interest.

Test Statistic   Friedman suggested using the statistic

$$T_1 = \frac{12}{bk(k+1)} \sum_{j=1}^{k} \left(R_j - \frac{b(k+1)}{2} \right)^2 \tag{2}$$

If there are ties present an adjustment needs to be made. Let A_1 be the sum of the squares of the ranks and average ranks.

$$A_1 = \sum_{i=1}^{b} \sum_{j=1}^{k} [R(X_{ij})]^2 \tag{3}$$

Also compute the "correction factor" C_1 given by

$$C_1 = bk(k+1)^2/4 \tag{4}$$

Then the statistic T_1, adjusted for the presence of ties, becomes

$$T_1 = \frac{(k-1)\left[\sum_{j=1}^{k} R_j^2 - bC_1 \right]}{A_1 - C_1} = \frac{(k-1) \sum_{j=1}^{k} \left(R_j - \frac{b(k+1)}{2} \right)^2}{A_1 - C_1} \tag{5}$$

Current research indicates the preferred statistic, because of its more accurate approximate distribution, is the two-way analysis of variance statistic computed on the ranks $R(X_{ij})$, which simplifies to the following function of T_1 given above.

$$T_2 = \frac{(b-1)T_1}{b(k-1) - T_1} \tag{6}$$

See Iman and Davenport (1980) for more details on the closeness of these approximations.

Null Distribution   The exact distribution of T_1 (or T_2) is difficult to find and so an approximation is usually used. The approximate distribution of T_1 is the chi-squared distribution with $k - 1$ degrees of freedom. However, this approximation is sometimes rather poor, so it is recommended to use T_2 instead of T_1, which has the approximate quantiles given by the F distribution (Table A22) with $k_1 = k - 1$ and $k_2 = (b - 1)(k - 1)$, when the null hypothesis is true.

Hypotheses

H_0: Each ranking of the random variables within a block is equally likely (i.e., the treatments have identical effects)

H_1: At least one of the treatments tends to yield larger observed values than at least one other treatment

Reject H_0 at the approximate level α if T_2 exceeds the $1 - \alpha$ quantile of the F distribution given by Table A22 for $k_1 = k - 1$ and $k_2 = (b - 1)(k - 1)$. This approximation is fairly good and improves as b gets larger. The approximate p-value may be estimated from Table A22.

Multiple Comparisons The following method for comparing individual treatments may be used only if the Friedman test results in rejection of the null hypothesis. Treatments i and j are considered different if the following inequality is satisfied.

$$|R_j - R_i| > t_{1-\alpha/2} \left[\frac{2(bA_1 - \Sigma R_j^2)}{(b-1)(k-1)} \right]^{\frac{1}{2}} \tag{7}$$

where R_i, R_j, and A_1 are given previously and where $t_{1-\alpha/2}$ is the $1 - \alpha/2$ quantile of the t distribution given by Table A21 with $(b - 1)(k - 1)$ degrees of freedom. The value for α is the same one used in the Friedman test.

Alternatively, Equation 7 can be expressed as a function of T_1

$$|R_j - R_i| > t_{1-\alpha/2} \left[\frac{(A_1 - C_1)2b}{(b-1)(k-1)} \left(1 - \frac{T_1}{b(k-1)} \right) \right]^{\frac{1}{2}} \tag{8}$$

If there are no ties A_1 in Equation 7 simplifies to

$$A_1 = bk(k + 1)(2k + 1)/6$$

and $(A_1 - C_1)$ in Equation 8 simplifies to

$$A_1 - C_1 = bk(k + 1)(k - 1)/12$$

Computer Assistance The Friedman test appears in *Minitab, S-Plus, SAS,* and *StatXact.* ─────────────────────────────────◄

EXAMPLE 1

Twelve homeowners are randomly selected to participate in an experiment with a plant nursery. Each homeowner was asked to select four fairly identical areas in his yard and to plant four different types of grasses, one in each area. At the end of a specified length of time each homeowner was asked to rank the grass types in order of preference, weighing important criteria such as expense, maintenance and upkeep required, beauty, hardiness, wife's preference, and so on. The rank 1 was assigned to the least preferred grass and the rank 4 to the favorite. The null hypothesis was that there is no difference in preferences of the grass types, and the alternative was that some grass types tend to be preferred over others. Each of the 12 blocks consists of four fairly identical plots of land, each receiving care of approximately the same degree of skill because the four plots are presumably cared for by the same homeowner. The results of the experiment are as follows.

		Grass		
Homeowner	1	2	3	4
1	4	3	2	1
2	4	2	3	1
3	3	1.5	1.5	4
4	3	1	2	4
5	4	2	1	3
6	2	2	2	4
7	1	3	2	4
8	2	4	1	3
9	3.5	1	2	3.5
10	4	1	3	2
11	4	2	3	1
12	3.5	1	2	3.5
R_j (totals)	38	23.5	24.5	34

First A_1 is computed by squaring each $R(X_{ij})$ and summing to get $A_1 = 356.5$ as the total sum of squares. Equation 4 gives

$$C_1 = \frac{12(4)(25)}{4} = 300$$

and Equation 5 gives

$$T_1 = \frac{3[(38)^2 + (23.5)^2 + (24.5)^2 + (34)^2 - 12(300)]}{356.5 - 300}$$

$$= 8.097$$

Substitution of T_1 into Equation 6 gives

$$T_2 = \frac{11(8.097)}{12(3) - 8.097} = 3.19$$

The critical region of approximate size $\alpha = 0.05$ corresponds to all values of T_2 greater than 2.90, the 0.95 quantile of the F distribution with $k_1 = 3$, $k_2 = 33$, obtained from Table A22. Therefore the null hypothesis is rejected. We may conclude that there is a tendency for some types of grass to be preferred over others. The p-value is about 0.04, which is obtained by interpolation in Table A22. This means that the null hypothesis could have been rejected at a significance level as small as $\alpha = 0.04$. For multiple comparisons $t_{0.975}$ with $(11)(3) = 33$ degrees of freedom is found to be 2.036, by Table A21. From Equation 7 we have

$$t_{0.975} \left[\frac{2(bA_1 - \sum R_j^2)}{(b-1)(k-1)} \right]^{\frac{1}{2}} = 11.49$$

Any two grasses whose rank sums are more than 11.49 units apart may be regarded as unequal. Therefore grass 1 may be considered better than grasses 2 and 3. No other differences are significant. ■

The next test also tests the null hypothesis of equal treatment means in a randomized complete block design, but it uses information about the range of the blocks relative to each other and gives more weight to the blocks exhibiting the larger range. It also uses the rankings within the block as in the Friedman test. We will follow convention of naming the test after its inventor, Dana Quade (1972, 1979).

▶ **The Quade Test** _____

Data Find the ranks within blocks $R(X_{ij})$ as described in the previous test. The next step again uses the original observations X_{ij}. Ranks are assigned to the blocks themselves according to the size of the sample range in each block. The sample range within block i is the difference between the largest and the smallest observations within that block.

$$\text{Range in block } i = \underset{j}{\text{maximum}} \{X_{ij}\} - \underset{j}{\text{minimum}} \{X_{ij}\} \qquad (9)$$

There are b sample ranges, one for each block. Assign rank 1 to the block with the smallest range, rank 2 to the second smallest, and so on to the block with the largest range, which gets rank b. Use average ranks in case of ties. Let $Q_1, Q_2, \ldots, Q_b$ be the ranks assigned to blocks $1, 2, \ldots, b$, respectively.

Finally, the block rank Q_i is multiplied by the difference between the rank within block i, $R(X_{ij})$, and the average rank within blocks, $(k + 1)/2$, to get the product S_{ij}, where

$$S_{ij} = Q_i \left[R(X_{ij}) - \frac{k+1}{2} \right] \qquad (10)$$

is a statistic that represents the relative size of each observation within the block, adjusted to reflect the relative significance of the block in which it appears.

Let S_j denote the sum for each treatment:

$$S_j = \sum_{i=1}^{b} S_{ij} \tag{11}$$

for $j = 1, 2, \ldots, k$.

Assumptions   The first two assumptions are the same as the two assumptions of the previous test. A third assumption is needed because comparisons are made between blocks.

3. The sample range may be determined within each block so that the blocks may be ranked.

Test Statistic   For convenience, first calculate the term

$$A_2 = \sum_{i=1}^{b} \sum_{j=1}^{k} S_{ij}^2 \tag{12}$$

where S_{ij} is given by Equation 10. This is called the "total sum of squares." If there are no ties, A_2 simplifies to

$$A_2 = b(b + 1)(2b + 1)k(k + 1)(k - 1)/72 \tag{13}$$

Next calculate the term

$$B = \frac{1}{b} \sum_{j=1}^{k} S_j^2 \tag{14}$$

where S_j is given by Equation 11. This is called the "treatment sum of squares." The test statistic is

$$T_3 = \frac{(b - 1)B}{A_2 - B} \tag{15}$$

If $A_2 = B$, consider the point to be in the critical region and calculate the p-value as $(1/k!)^{b-1}$.

   Note that T_3 is the two-way analysis of variance test statistic computed on the scores S_{ij} given by Equation 10.

Null Distribution   The exact distribution of T_3 is difficult to find, so the F distribution, whose quantiles are given in Table A22, is used as an approximation, with $k_1 = k - 1$ and $k_2 = (b - 1)(k - 1)$ as before in the Friedman test.

Hypotheses The hypotheses are the same as in the Friedman test.

Reject the null hypothesis at the level α if T_3 exceeds the $1 - \alpha$ quantile of the F distribution as given in Table A22 with $k_1 = k - 1$ and $k_2 = (b - 1)(k - 1)$. Actually, the F distribution only approximates the exact distribution of T_3, but exact tables are not available at this time. As b becomes large, the F approximation comes closer to being exact.

Multiple Comparisons Only if the preceding procedure results in rejection of the null hypothesis are multiple comparisons made. Treatments i and j are considered different if the inequality

$$|S_i - S_j| > t_{1-\alpha/2} \left[\frac{2b(A_2 - B)}{(b - 1)(k - 1)} \right]^{1/2} \tag{16}$$

is satisfied, where S_i, S_j, A_2, and B are given previously, and where $t_{1-\alpha/2}$ is obtained from Table A21 with $(b - 1)(k - 1)$ degrees of freedom. This comparison is made for all pairs of treatments, using the same α used in the Quade test.

Computer Assistance The Quade test appears in *StatXact*. ——————◀

EXAMPLE 2

Seven stores are selected for a marketing survey. In each store five different brands of a new type of hand lotion are placed side by side. At the end of the week, the number of bottles of lotion sold for each brand is tabulated, with the following results.

Numbers of customers (rank within stores)

Store	A	B	C	D	E
			Brand		
1	5 (2)	4 (1)	7 (3)	10 (4)	12 (5)
2	1 (2.5)	3 (5)	1 (2.5)	0 (1)	2 (4)
3	16 (2)	12 (1)	22 (3.5)	22 (3.5)	35 (5)
4	5 (4.5)	4 (2.5)	3 (1)	5 (4.5)	4 (2.5)
5	10 (3.5)	9 (2)	7 (1)	13 (5)	10 (3.5)
6	19 (2)	18 (1)	28 (3)	37 (4)	58 (5)
7	10 (5)	7 (2.5)	6 (1)	8 (4)	7 (2.5)

The observations are ranked from 1 to 5 within each store, with average ranks assigned when there are ties. These ranks $R(X_{ij})$ appear in parentheses.

Next, the sample range within each store is computed by subtracting the smallest observation from the largest. In store 1 the sample range is 12–4 = 8. These sample ranges are listed next, along with the ranks Q_i of the sample ranges, and the products

$$S_{ij} = Q_i[R(X_{ij}) - (k + 1)/2]$$

$$S_{ij} = Q_i[R(X_{ij}) - 3]$$

Store Number	Sample Range	Rank Q_i	Brand A	B	C	D	E
1	8	5	−5	−10	0	+5	+10
2	3	2	−1	+4	−1	−4	+2
3	23	6	−6	−12	+3	+3	+12
4	2	1	+1.5	−0.5	−2	+1.5	−0.5
5	6	4	+2	−4	−8	+8	+2
6	40	7	−7	−14	0	+7	+14
7	4	3	+6	−1.5	−6	+3	−1.5
		$S_j =$	−9.5	−38	−14	+23.5	+38

From Equation 12,

$$A_2 = \sum_{i=1}^{7} \sum_{j=1}^{5} S_{ij}^2 = (-5)^2 + (-10)^2 + \cdots = 1366.5$$

which is slightly less than the more easily obtained value 1400, from Equation 13, applicable if there had been no ties. Equation 14 yields

$$B = \frac{1}{7} \sum_{j=1}^{5} S_j^2 = \frac{1}{7}[(-9.5)^2 + (-38)^2 + \cdots] = 532.4$$

which gives, when substituted into Equation 15, the test statistic

$$T_3 = \frac{6(532.4)}{1366.5 - 532.4} = 3.83$$

This value of T_3 is greater than 2.78, the 0.95 quantile of the F distribution with $k_1 = 4$ and $k_2 = 24$, obtained from Table A22; therefore the null hypothesis is rejected at $\alpha = 0.05$. In fact, perusal of Table A22 shows the p-value to be slightly less than 0.025. Some brands seem to be preferred over others by the store customers.

Because the null hypothesis is rejected, multiple comparisons are made. From Equation 16 two treatments are considered different if the difference between their sums $|S_i - S_j|$ exceeds

$$t_{1-\alpha/2}\left[\frac{2b(A_2 - B)}{(b - 1)(k - 1)}\right]^{1/2} = 2.064\left[\frac{14(834.1)}{24}\right]^{1/2} = 45.53$$

where $t_{1-\alpha/2} = t_{0.975}$ is obtained from Table A21 for $(b-1)(k-1) = 24$ degrees of freedom. Thus the brands that may be considered different from each other are brands A and E, brands B and D, brands B and E, and brands C and E.

Note that a summary of the multiple comparisons procedure may be presented by listing the treatments in order of increasing average scores, and underlining the groups of treatments that are not significantly different with a single underline, as follows.

B C A D E   ■

□ **Theory**   The exact distribution of T_1, T_2, and T_3 is found under the assumption that each ranking within a block is equally likely, which is the null hypothesis. There are $k!$ possible arrangements of ranks $R(X_{ij})$ within a block and, therefore, $(k!)^b$ possible arrangements of ranks in the entire array of b blocks. The preceding statements imply that each of these $(k!)^b$ arrangements is equally likely under the null hypothesis. Therefore the probability distributions of T_1, T_2, and T_3 may be found for a given number of samples k and blocks b, merely by listing all possible arrangements of ranks and by computing T_1, T_2, or T_3 for each arrangement.

For example, if $k = 2$ and $b = 3$, there are $(2!)^3 = 8$ equally likely arrangements of the ranks, which are listed next along with their associated values of T_1, and T_2. We will consider T_3 later.

Arrangements

Blocks	1	2	3	4	5	6	7	8
1	1, 2	1, 2	1, 2	2, 1	2, 1	2, 1	1, 2	2, 1
2	1, 2	1, 2	2, 1	1, 2	2, 1	1, 2	2, 1	2, 1
3	1, 2	2, 1	1, 2	1, 2	1, 2	2, 1	2, 1	2, 1
Probability	1/8	1/8	1/8	1/8	1/8	1/8	1/8	1/8
Value of T_2	∞	$\frac{1}{4}$	$\frac{1}{4}$	$\frac{1}{4}$	$\frac{1}{4}$	$\frac{1}{4}$	$\frac{1}{4}$	∞
Value of T_1	3	$\frac{1}{3}$	$\frac{1}{3}$	$\frac{1}{3}$	$\frac{1}{3}$	$\frac{1}{3}$	$\frac{1}{3}$	3

Therefore the probability distribution of T_1 is given by $P(T_1 = \frac{1}{3}) = 3/4$ and $P(T_1 = 3) = 1/4$ under H_0. The probability distribution of T_2 is given by $P(T_2 = \frac{1}{4}) = 3/4$ and $P(T_2 = \infty) = 1/4$.

To examine the behavior of T_3 under the null hypothesis we again start out with the eight equally likely arrangements of ranks $R(X_{ij})$, as just given. The average rank 1.5 is subtracted from each rank and, for the moment, we consider the case where the block ranks are given by $Q_1 = 1$, $Q_2 = 2$, $Q_3 = 3$. The resulting arrays of S_{ij} are given here.

	Arrangements			
Blocks	*1*	*2*	*3*	*4*
1	−0.5, +0.5	−0.5, +0.5	−0.5, +0.5	+0.5, −0.5
2	−1, +1	−1, +1	+1, −1	−1, +1
3	−1.5, +1.5	+1.5, −1.5	−1.5, +1.5	−1.5, +1.5
Conditional Probability	1/8	1/8	1/8	1/8
Value of T_3	12	0	$\frac{4}{19}$	$1\frac{3}{13}$

	Arrangements			
Blocks	*5*	*6*	*7*	*8*
1	+0.5, −0.5	+0.5, −0.5	−0.5, +0.5	+0.5, −0.5
2	+1, −1	−1, +1	+1, −1	+1, −1
3	−1.5, +1.5	+1.5, −1.5	+1.5, −1.5	+1.5, −1.5
Conditional Probability	1/8	1/8	1/8	1/8
Value of T_3	0	$\frac{4}{19}$	$1\frac{3}{13}$	12

The probability for each value of T_3 is:

$$\tfrac{1}{8} \cdot P(Q_1 = 1, Q_2 = 2, Q_3 = 3)$$

because 1/8 represents the conditional probability for that value of T_3, given the assignment of ranks Q_1, Q_2, and Q_3. Suppose a different assignment of ranks Q_1, Q_2, Q_3 is considered, say $Q_1 = 2$, $Q_2 = 1$, $Q_3 = 3$. Then the reader may easily verify, by listing the eight arrangements of values of S_{ij} as we just did, that again we observe the same eight values of T_3, and each of these eight values has probability

$$\tfrac{1}{8} \cdot P(Q_1 = 2, Q_2 = 1, Q_3 = 3)$$

By considering all six (3!) permutations of ranks for Q_1, Q_2, and Q_3, we arrive at the total probability for each value of T_3: 1/8. Thus, for purposes of calculating the null distribution of T_3, only the one case given here, $Q_i = i$, for $i = 1, 2, 3$, must be considered. The probability distribution of T_3 is obtained by collecting identical values of T_3 to get

$$P(T_3 = 0) = 1/4, \quad P(T_3 = \tfrac{4}{19}) = 1/4 \quad P(T_3 = 1\tfrac{3}{13}) = 1/4, \quad P(T_3 = 12) = 1/4$$

The approximation of the distributions of T_1, T_2, and T_3 that use the F or chi-squared distributions are justified using the central limit theorem. Some of the details are beyond the scope of this book, so the entire development of the asymptotic distributions is omitted. The reader is referred to Quade (1972, 1979) or Lawler (1978) for T_3, Iman and Davenport (1979) for T_2, and Friedman (1937) for T_1. □

The preceding distributions of T_1, T_2, and T_3 were obtained, as always, under the assumption that H_0 is true. If H_0 is false, the treatment sums S_j and R_j may be expected to vary widely from their mean values 0 and $b(k + 1)/2$, respectively. This causes all three statistics to tend to increase in size. Therefore the decision rule is to reject H_0 when T_1, T_2, or T_3 is large.

The parametric procedure for analyzing data from the randomized complete block design is valid only if the data are normally distributed with the same variance. The null hypothesis is that the random variables within the same block have the same mean. The test statistic is

$$F = \frac{(b - 1)SSB}{SST - SSB - SSR} \tag{17}$$

where

$$SSB = \frac{1}{b} \sum_{j=1}^{k} T_j^2 - \frac{T^2}{bk} \tag{18}$$

$$SSR = \frac{1}{k} \sum_{i=1}^{b} \left(\sum_{j=1}^{k} X_{ij} \right)^2 - \frac{T^2}{bk} \tag{19}$$

$$SST = \sum_{i=1}^{b} \sum_{j=1}^{k} X_{ij}^2 - \frac{T^2}{bk} \tag{20}$$

$$T_j = \sum_{i=1}^{b} X_{ij} \tag{21}$$

and

$$T = \sum_{i=1}^{b} \sum_{j=1}^{k} X_{ij} \tag{22}$$

The statistic in Equation 17 is compared with quantiles from the F distribution in Table A22, $k_1 = k - 1$, $k_2 = (b - 1)(k - 1)$.

If this same F statistic is computed on the ranks $R(X_{ij})$ instead of the data, the statistic is the same as T_2. If the F statistic is computed on the weighted ranks S_{ij}, the result is the statistic T_3. The adjustment for ties is automatically incorporated in the F statistic T_3. The method of making multiple comparisons described here is merely the parametric procedure, known as Fisher's least significant difference (LSD) method, but it is computed on the numbers S_{ij} or on the ranks $R(X_{ij})$ instead of on the data.

For two samples ($k = 2$) the A.R.E. of the Friedman test relative to the usual parametric t test is the same as that of the sign test, $2/\pi = 0.637$, in situations where the t test is the most powerful test. For k samples the A.R.E. of the Friedman test relative to the usual parametric F test is dependent on the number of samples k and equals $(0.955)k/(k + 1)$ if the populations are normal, $k/(k + 1)$ if the populations are uniform, and $3k/2(k + 1)$ if the populations are double exponential. In fact, the A.R.E. of the Friedman test relative to the popular F test never falls below $(0.864)k/(k + 1)$ under purely translation-type alternative hypotheses. The A.R.E. of the Friedman test is discussed more fully in Noether (1967a).

For $k = 2$ the A.R.E. of the Quade test relative to the usual parametric t test is the same as the Wilcoxon signed ranks test, $3/\pi = 0.955$, when the distributions are normal. As with the Wilcoxon test, the A.R.E. relative to the t test never falls below 0.864 but may be as high as infinity. The A.R.E. of this test for more than two samples has not yet been found but simulation studies by Iman et al. (1984) show that the power of the Quade test tends to drop below the power of the Friedman test when the number of treatments is five or more.

▶ **The Page Test for Ordered Alternatives** _____

In Section 5.4 we presented the Jonckheere-Terpstra test of k independent samples when the alternative of interest specifies an ordering of the treatment effects. It is equivalent to computing Kendall's τ between the observations and the ordering of the treatments specified in the alternative hypothesis. We mentioned that Spearman's ρ could have been used just as well.

In the randomized complete block design, Spearman's ρ is used to test for k related samples against the alternative hypothesis of a specified ordering of the treatment effects. The correlation between the Friedman within-block rankings and the ordering of the treatments as specified by H_1 is used in a test introduced by Page (1963).

Because of the many ties inherent in the data, Page uses a simpler statistic, which is a monotonic function of Spearman's ρ if there are no ties within blocks, namely

$$T_4 = \sum_{j=1}^{k} jR_j = R_1 + 2R_2 + \cdots + kR_k$$

where R_j is the treatment rank sum in the Friedman test, arranged in increasing order of the treatment effects as specified by H_1.

Although exact tables are given by Page (1963), we will consider only the large sample approximation, and reject H_0 when

$$T_5 = \frac{T_4 - bk(k + 1)^2/4}{[b(k^3 - k)^2/144(k - 1)]^{1/2}}$$

exceeds the $1 - \alpha$ quantile from a standard normal distribution, as given in Table A1, for an upper-tailed test of size α. *StatXact* finds exact p-values for Page's test. _____ ◀

EXAMPLE 3

Health researchers suspect that regular exercise has a tendency to lower the pulse rate of a resting individual. To test this theory eight healthy volunteers, who did not exercise on a regular basis, were enrolled in a supervised exercise program. Their resting pulse rate was measured at the beginning of the program, and again after each month for four months.

The null hypothesis is that there is no difference

$$H_0: \mu_1 = \mu_2 = \mu_3 = \mu_4 = \mu_5$$

against the ordered alternative

$$H_1: \mu_1 \le \mu_2 \le \mu_3 \le \mu_4 \le \mu_5$$

where μ_1 is the mean at the end of the fourth month, μ_5 is the initial mean, and there is at least one strict inequality in H_1.

The observed pulse rates are as follows, along with their Friedman within-blocks ranks.

Person	Initial	Month 1	Month 2	Month 3	Month 4
1	82(4)	84(5)	77(2)	76(1)	79(3)
2	80(4.5)	80(4.5)	76(1.5)	76(1.5)	78(3)
3	75(3)	78(5)	77(4)	74(2)	72(1)
4	65(1.5)	72(5)	68(4)	65(1.5)	66(3)
5	77(5)	74(2)	72(1)	75(3.5)	75(3.5)
6	68(4)	69(5)	65(2)	66(3)	64(1)
7	70(3.5)	74(5)	68(1.5)	70(3.5)	68(1.5)
8	77(4)	76(3)	78(5)	72(2)	70(1)
	$R_5 = 29.5$	$R_4 = 34.5$	$R_3 = 21$	$R_2 = 18$	$R_1 = 17$

Notice that R_1 is the rank sum predicted by H_1 to be the smallest, R_2 is predicted to be the second smallest, and so on.

$$T_4 = 17 + 2(18) + 3(21) + 4(34.5) + 5(29.5) = 401.5$$

$$T_5 = \frac{401.5 - 8(5)(36)/4}{\left[8(5^3 - 5)^2/144(4) \right]^{1/2}} = \frac{41.5}{\sqrt{200}} = 2.9345$$

A comparison of T_5 with Table A1 shows $p = 0.002$ and H_0 is easily rejected at $\alpha = 0.05$. Page's tables, exact only if there are no ties, show the same p-value.

Kendall's τ could have been used instead of Spearman's ρ (Jonckheere, 1954b). Other nonparametric tests for the same hypotheses are given by Shorak (1967) and Pirie and Hollander (1972). ∎

The following discussion shows that the Friedman test statistic is closely related to some other popular nonparametric statistics. This discussion is limited to the case when there are no ties merely for simplicity. A similar comparison can be made for the case when ties are present.

The Relationship with Kendall's Coefficient of Concordance

A statistic W called Kendall's coefficient of concordance was introduced independently by Kendall and Babington-Smith (1939) and Wallis (1939). It may be used in the same situation where Friedman's test statistic is applicable, although it was probably intended primarily as a measure of "agreement in rankings" in the b blocks rather than as a test statistic. Using the same notation as before, Kendall's W is defined as

$$W = \frac{12}{b^2 k(k+1)(k-1)} \sum_{j=1}^{k} \left[R_j - \frac{b(k+1)}{2} \right]^2 \tag{23}$$

If there is perfect agreement in the rankings in all b blocks, treatment 1 receives the same rank in all b blocks, treatment 2 receives the same rank in all b blocks, and so on, and the resulting value of W is 1.0. If there is "perfect disagreement" among rankings, the values of R_j will either be equal or very nearly equal to each other and their mean, and W will be 0 or very close to 0.

A comparison of Kendall's W with Friedman's T_1 of Equation 5 reveals the relationship

$$W = \frac{T_1}{b(k-1)} \tag{24}$$

Thus W is a simple modification of the Friedman test statistic, and any hypothesis test that uses W as a test statistic may be conducted by computing T_1 instead of W. If T_1 exceeds its $1 - \alpha$ quantile, W exceeds its own $1 - \alpha$ quantile. Kendall's W is computed in *Minitab*, *StatXact*, and *SPSS*.

The Relationship with Spearman's ρ

Spearman's ρ, defined by Equation 5.4.4, may be computed between any two blocks, say block i and block m, by considering the two blocks as two samples and the two ranks under each treatment as being a pair of related ranks. The average value of Spearman's ρ, averaged over all pairs of blocks, bears a direct relationship with Friedman's test statistic, as we will now demonstrate.

Let ρ_a denote the average value of Spearman's ρ. There are $b(b-1)$ values of ρ_{im} to be averaged, counting both ρ_{im} and ρ_{mi}, even though $\rho_{im} = \rho_{mi}$ by symmetry. To compute the average ρ, we will sum over all i and m and then subtract those ρ_{im} where i equals m; that is, we will subtract the values of ρ where a block is paired with itself. In those b cases ρ_{im} equals 1. Thus the average ρ may be expressed as

$$\rho_a = \frac{1}{b(b-1)} \left(\sum_{i=1}^{b} \sum_{m=1}^{b} \rho_{im} - b \right) \tag{25}$$

The random variable ρ_a equals 1 if there is "perfect agreement" among rankings, in the sense previously described, because then each ρ_{im} equals 1. If there is disagreement among rankings, ρ_a will be smaller than 1 and may even assume negative values. However, it is not possible for ρ_a to be as small as -1 except in the special case where there are only two blocks, $b = 2$.

Equation 25 and the definition of Spearman's ρ may be combined and simplified to reveal the relationship with Friedman's test statistic T_1 given by Equation 2.

$$\rho_a = \frac{T_1}{(b-1)(k-1)} - \frac{1}{b-1} \tag{26}$$

Thus the quantiles for the average Spearman's ρ may be easily obtained from the quantiles of the Friedman test statistic.

The preceding relationship between the Friedman test statistic and Spearman's ρ illustrates that the Friedman test may be used as a test for linear dependence in the two-sample case where Spearman's ρ was applicable. Spearman's ρ has the advantage of being tabulated for small samples, although the exact distribution of Friedman's test statistic could be easily obtained from the distribution of Spearman's ρ. Both tests would be equivalent, and therefore both would have an A.R.E. of 0.912 when compared with the usual parametric test using Pearson's r as a test statistic in the situation where Pearson's r is appropriate.

An Extension to the Case of Several Observations per Experimental Unit

If there are several (m) observations for each treatment in each block instead of only one observation per experimental unit as before, the null hypothesis of no differences among treatments may be tested by slightly modifying Friedman's procedure. The observations within each block are ranked as before, with the exception that the ranks range from 1 to mk. The sum of ranks R_j is defined, as before, as the sum of ranks assigned to all observations involving treatment j. Let the observations in block i using treatment j be denoted by $X_{ij1}, X_{ij2}, \ldots, X_{ijm}$. The mean of R_j becomes

$$E(R_j) = \sum_{i=1}^{b} \sum_{n=1}^{m} E[R(X_{ijn})] = \sum_{i=1}^{b} \sum_{n=1}^{m} \frac{mk+1}{2}$$
$$= \sum_{i=1}^{b} \frac{m(mk+1)}{2} = \frac{bm(mk+1)}{2} \tag{27}$$

The variance of R_j is found with the aid of Theorem 1.4.5, where n is replaced by m and N is replaced by mk.

$$\text{Var}(R_j) = \sum_{i=1}^{b} \text{Var}\left[\sum_{n=1}^{m} R(X_{ijn})\right] = \sum_{i=1}^{b} \frac{m(mk+1)(mk-m)}{12}$$
$$= \frac{bm^2(mk+1)(k-1)}{12} \tag{28}$$

If there are ties in the data, $\text{Var}(R_j)$ is given by

$$\text{Var}(R_j) = \frac{m(k-1)}{k(mk-1)}\left[\sum_{\substack{\text{all} \\ \text{ranks}}} R(X_{ijn})^2 - mkb(mk+1)^2/4\right] \tag{29}$$

which is the same for all values of j.

The test statistic

$$T_6 = \sum_{j=1}^{k} \frac{(k-1)}{k} \frac{[R_j - E(R_j)]^2}{\text{Var}(R_j)} \tag{30}$$

is used here. The mean and variance of R_j are given above. The chi-squared tables with $k-1$ degrees of freedom are used as before. Multiple comparisons use the inequality

$$|R_j - R_i| > t_{1-\alpha/2}\left\{\frac{2kb(mk-1)\,\text{Var}(R_j)}{(k-1)(mbk-k-b+1)}\left[1 - \frac{T_6}{b(mk-1)}\right]\right\}^{1/2} \tag{31}$$

where $t_{1-\alpha/2}$ is obtained from Table A21 with $mbk - k - b + 1$ degrees of freedom.

Interaction

As far as I know, there are no good, exact nonparametric tests for interaction. The exact tests either require that there are no differences in treatments, while the block effects are removed by a Friedman-type ranking within blocks (see Patel and Hoel, 1973, for example), or that they are "conditionally distribution-free," conditional on the observations. This latter class of tests requires an impractical amount of computation, and includes aligned ranks tests, where the treatment effects are "removed" by subtracting treatment means or medians from the observations, and the block effects are removed either by ranking only within blocks or by subtracting block means or medians from the observations. In terms of good power and ease of application, large-sample approximations to these aligned ranks tests, although not really nonparametric tests, appear to be the best alternatives to the parametric methods. They reduce some of the power problems caused by nonnormal distributions, and they are asymptotically distribution-free as the number of observations per cell goes to infinity. See Mansouri and Chang (1995) for a good comparison of some of the best aligned ranks tests for interaction.

Some references on rank sum tests for two-way layouts include Page (1963), Hollander (1967b), and Pirie (1974) if the alternative hypothesis specifies an ordering of treatment effects, and Dunn (1964) and McDonald and Thompson (1967) for multiple comparisons. Other methods of analysis are suggested by Doksum (1967), Puri and Sen (1967), Sen (1968b), and Lemmer, Stoker, and Reinach (1968). Asymptotic efficiency is studied by Mehra and Sarangi (1967) and Sen (1967a). Small-sample efficiency is studied by Gilbert (1972). Extensions to the multivariate case are considered by Gerig (1969, 1975). Koch (1970) discusses a split-plot variation with several observations per cell. Li and Schucany (1975) and Schucany and Beckett (1976) consider measuring the concordance between two sets of blocked rankings. For a complete presentation of an "aligned rank" procedure, which applies to the two-way layout and has a higher A.R.E. than the Friedman test, consult Lehmann (1975). Comparisons of the A.R.E. of the aligned ranks test and Friedman's test are given by Hora and Iman (1988). The interested reader should also see Section 5.12.

EXERCISES

1. A survey was taken of all seven hospitals in a particular city to obtain the number of babies born over a 12-month period. This time period was divided into the four seasons to test the hypothesis that the birth rate is constant over all four seasons.
   The results of the survey are as follows:

   Number of Births

Hospital	Winter	Spring	Summer	Fall
A	92	112	94	77
B	9	11	10	12
C	98	109	92	81
D	19	26	19	18
E	21	22	23	24
F	58	71	51	62
G	42	49	44	41

   (a) Analyze these data using the Friedman test.
   (b) Analyze these data using the Quade test
   (c) Can you account for the wide discrepancy in the results of the two tests?

2. Twelve randomly selected students are involved in a learning experiment. Four lists of words are made up by the experimenter. Each list contains 20 pairs of words, but different methods of pairing are used on the four lists. Each student is handed a list, given five

minutes to study it, and then examined on his or her ability to remember the words. This procedure is repeated for all four lists for each student, the order of the lists being rotated from one student to the next. The examination scores are as follows (20 is perfect).

List	Student											
	1	2	3	4	5	6	7	8	9	10	11	12
1	18	7	13	15	12	11	15	10	14	9	8	10
2	14	6	14	10	11	9	16	8	12	9	6	11
3	16	5	16	12	12	9	10	11	13	9	9	13
4	20	10	17	14	18	16	14	16	15	10	14	16

Are some lists easier to learn than others?

(a) Use the Friedman test.

(b) Use the Quade test.

3. Rework Example 2 using T_1 and compare p-values.

4. The emissions rate of nitrous oxide for used automobiles is measured on a dynamometer, where a driver follows set guidelines on rates of acceleration, deceleration, etc. for a 15-minute period. To see if there is a difference in measurements due to the different drivers, six automobiles are tested repeatedly by each of the three drivers at one station, with the following results.

Car	Driver		
	1	2	3
1	6.2	6.3	6.0
2	12.6	12.9	12.7
3	10.2	10.6	9.8
4	13.0	13.1	13.0
5	5.6	5.9	5.5
6	8.1	8.1	7.8

Do some drivers tend to get lower rates than others? Which drivers?

5. In an experiment attributed to Fox and Randall (1970) by Hollander and Wolfe (1973), increasing the amount of weight is expected to decrease the forearm tremor frequency of subjects. Six subjects are measured at each of five different weights with the following measurements of forearm tremor frequency. Do the data support the theory?

Subject	Weight (lb.)				
	0	1.25	2.5	5	7.5
1	3.01	2.85	2.62	2.63	2.58
2	3.47	3.43	3.15	2.83	2.70
3	3.35	3.14	3.02	2.71	2.78
4	3.10	2.86	2.58	2.49	2.36
5	3.41	3.32	3.08	2.96	2.67
6	3.07	3.06	2.85	2.50	2.43

6. The following is used by *StatXact* to illustrate Page's test. It is expected that decreasing the level of potash applied to a cotton plant tends to increase the strength of the fiber. Five different dosage levels were applied to plots in three different blocks. *StatXact* gives $p = 0.0025$. How does this compare with your analysis?

		Level of Potash (lb/acre)			
Block	144	108	72	54	36
1	7.46	7.17	7.76	8.14	7.63
2	7.68	7.57	7.73	8.15	8.00
3	7.21	7.80	7.74	7.87	7.93

PROBLEMS

1. Show that for $k = 2$ the statistic T_3 is a function of the Wilcoxon signed ranks statistic given by Equation 5.7.5 and, therefore, the two tests are equivalent. (*Hint.* First show that Q_i is equal to the absolute value of R_i.)

2. Show that for $k = 2$ the Friedman test is equivalent to the two-tailed sign test (large sample approximation).

5.9 THE BALANCED INCOMPLETE BLOCK DESIGN

In the randomized complete block design described at the beginning of Section 5.8 every treatment is applied in every block. However, it is sometimes impractical or impossible for all of the treatments to be applied to each block, especially when the number of treatments is large and the block size is limited. For example, if 20 different foods are to be tasted, each judge (block) may find it quite difficult to rank accurately all 20 foods in order of preference. But if each judge tastes only 5 foods and then four times as many judges are used (or each judge is used four times), the judging may be easier and more accurate. Those experimental designs in which not all treatments are applied to each block are called incomplete block designs. Furthermore, if the design is balanced so that (1) every block contains k experimental units, (2) every treatment appears in r blocks, and (3) every treatment appears with every other treatment an equal number of times, the design is called a *balanced* incomplete block design.

Durbin (1951) presented a rank test that may be used to test the null hypothesis of no differences among treatments in a balanced incomplete block design. Parametric methods of analyzing data obtained using a balanced incomplete block design exist and are based on certain normality assumptions that will not be explained here. The Durbin test may be preferred over the parametric test if the normality assumptions are not met, if an easier method of analysis is desired, or if the observations consist merely of ranks. The Durbin test reduces to the Friedman test if the number of treatments equals the number of experimental units per block. If the third condition just given is not completely satisfied, the Durbin test is still valid in most situations.

▶ **The Durbin Test** _____

Data   We will use the following notation.

t = the number of treatments to be examined.

k = the number of experimental units per block ($k < t$).

b = the total number of blocks.

r = the number of times each treatment appears ($r < b$).

λ = the number of blocks in which the ith treatment and the jth treatment appear together. (λ is the same for all pairs of treatments.)

The data are arrayed in a balanced incomplete block design, just defined. Let X_{ij} represent the result of treatment j in the ith block, if treatment j appears in the ith block.

Rank the X_{ij} within each block by assigning the rank 1 to the smallest observation in block i, the rank 2 to the second smallest, and so on to the rank k, which is assigned to the largest observation in block i, there being only k observations within each block. Let $R(X_{ij})$ denote the rank of X_{ij} where X_{ij} exists.

Compute the sum of the ranks assigned to the r observed values under the jth treatment and denote this sum by R_j. Then R_j may be written as

$$R_j = \sum_{i=1}^{b} R(X_{ij}) \tag{1}$$

where only r values of $R(X_{ij})$ exist under treatment j; therefore only r ranks are added to obtain R_j.

If the observations are nonnumeric but such that they are amenable to ordering and ranking within blocks according to some criterion of interest, the ranking of each observation is noted and the values R_j for $j = 1, 2, \ldots, t$ are computed as described.

If the ranks may be assigned in several different ways because of several observations being equal to each other, we recommend assigning the average of the disputed ranks to each of the tied observations. This procedure changes the null distribution of the test statistic, but the effect is negligible if the number of ties is not excessive.

Assumptions

1. The blocks are mutually independent of each other.
2. Within each block the observations have an ordinal scale of measurement. Ties cause no problem.

Test Statistic   Durbin (1951) suggested using the test statistic

$$T_1 = \frac{12(t-1)}{rt(k-1)(k+1)} \sum_{j=1}^{t} \left(R_j - \frac{r(k+1)}{2} \right)^2 \tag{2}$$

If there are ties within blocks, average ranks are used, and an adjustment needs to be made. Let A be the sum of the squares of the ranks and average ranks used.

$$A = \sum_{i=1}^{b} \sum_{j=1}^{t} [R(X_{ij})]^2 \tag{3}$$

Also compute the "correction factor" C given by

$$C = \frac{bk(k+1)^2}{4} \tag{4}$$

Then the statistic T_1, corrected for ties, becomes

$$T_1 = \frac{(t-1) \sum_{j=1}^{t} \left(R_j - \frac{r(k+1)}{2} \right)^2}{A - C} = \frac{(t-1) \left[\sum_{j=1}^{t} R_j^2 - rC \right]}{A - C} \tag{5}$$

An alternative procedure, equivalent to this one, is to use the ordinary analysis of variance procedure on the ranks and average ranks. This results in the following statistic T_2, which is merely a function of T_1. Current research indicates the approximate quantiles for T_2 are slightly more accurate than the approximate quantiles for T_1, making T_2 the preferred statistic.

$$T_2 = \frac{T_1/(t-1)}{(b(k-1) - T_1)/(bk - b - t + 1)} \tag{6}$$

In summary, use T_2 as the test statistic, but first compute T_1 using Equation 2 if there are no ties, or Equations 3, 4, and 5 if there are ties.

Null Distribution   The exact distribution of T_1 (or T_2) is difficult to find and so an approximation is usually used. The approximate distribution of T_1 is the chi-squared distribution with $t-1$ degrees of freedom. This approximation tends to be very conservative. The approximate distribution of T_2 is the F distribution

(Table A22) with $k_1 = t - 1$ and $k_2 = bk - b - t + 1$. This approximation tends to give inflated values of α, but closer than the values obtained from using T_1.

Hypotheses

H_0: Each ranking of the random variables within each block is equally likely (i.e., the treatments have identical effects)

H_1: At least one treatment tends to yield larger observed values than at least one other treatment

Reject H_0 at the approximate level α if T_2 exceeds the $(1 - \alpha)$ quantile of the F distribution given by Table A22, with $k_1 = t - 1$ and $k_2 = bk - b - t + 1$. The approximate p-value may be estimated using Table A22.

Multiple Comparisons   If the null hypothesis is rejected, then multiple comparisons between pairs of treatments may be made as follows. Consider treatments i and j to have different means if their rank sums R_i and R_j satisfy the inequality

$$|R_i - R_j| > t_{1-\alpha/2} \left[\frac{(A - C)2r}{bk - b - t + 1} \left(1 - \frac{T_1}{b(k - 1)} \right) \right]^{1/2} \tag{7}$$

where A, C, and T_1 are given by Equations 3, 4, and 5, respectively, and where $t_{1-\alpha/2}$ is the $1 - \alpha/2$ quantile of the t distribution given by Table A21 for $bk - b - t + 1$ degrees of freedom. If there are no ties Equation 7 simplifies to

$$|R_i - R_j| > t_{1-\alpha/2} \left[\frac{rk(k + 1)}{6(bk - b - t + 1)} (b(k - 1) - T_1) \right]^{1/2} \tag{8}$$

Computer Assistance   The balanced incomplete block design may be analyzed by converting the data to ranks within blocks, and then using a computer program, such as in *SAS* for the parametric balanced incomplete block design, on ranks, or using a general linear models program such as in *SPSS*, *Minitab*, or *SAS*, on the ranks. ◀

EXAMPLE I

Suppose an ice cream manufacturer wants to test the taste preferences of several people for her seven varieties of ice cream. She asks each person to taste three varieties and to rank them 1, 2, and 3, with the rank 1 being assigned to the favorite variety. In order to design the experiment so that each variety is compared with every other variety an equal number of times, a Youden square layout given by Federer (1963) is used. Seven people are each given three varieties to taste, and the resulting ranks are as follows.

				Variety			
Person	1	2	3	4	5	6	7
1	2	3		1			
2		3	1		2		
3			2	1		3	
4				1	2		3
5	3				1	2	
6		3				1	2
7	3		1				2
$R_j =$	8	9	4	3	5	6	7

In this experiment,

$t = 7 =$ total number of varieties.
$k = 3 =$ number of varieties compared at one time
$b = 7 =$ number of people (blocks)
$r = 3 =$ number of times each variety is tasted
$\lambda = 1 =$ number of times each variety is compared with each other variety

Therefore the design is a balanced incomplete block design, and the Durbin test may be used to test the null hypothesis that no variety of ice cream tends to be preferred over any other variety of ice cream.

The critical region of approximate size $\alpha = 0.05$ corresponds to all values of T_2 greater than 3.58, which is the 0.95 quantile of the F distribution with $k_1 = t - 1 = 6$ and $k_2 = bk - b - t + 1 = 8$, found in Table A22.

First T_1 is found using Equation 2 because there are no ties.

$$T_1 = \frac{12(t - 1)}{rt(k - 1)(k + 1)} \sum_{j=1}^{t} \left[R_j - \frac{r(k + 1)}{2} \right]^2$$

$$= \frac{(12)(6)}{(3)(7)(2)(4)} [(8 - 6)^2 + (9 - 6)^2 + \cdots + (7 - 6)^2]$$

$$= 12$$

Then T_2 is found using Equation 6.

$$T_2 = \frac{T_1 / (t - 1)}{(b(k - 1) - T_1) / (bk - b - t + 1)}$$

$$= \frac{12 / 6}{(14 - 12) / 8} = 8$$

The test statistic T_2 is in the critical region, so the null hypothesis is rejected. The approximate p-value is seen from Table A22 to be less than 0.01.

Multiple comparisons, using Equation 8, show variety 4 is preferred over varieties 6 and 7, and variety 3 is preferred over variety 7.

Notice that in this example there is perfect agreement among the seven people tasting ice cream, and so we can find the exact p-value without much difficulty. That is, everyone who tasted variety 4 preferred it over the others. Everyone who tasted variety 3 preferred it over the others except for variety 4. Everyone who tasted variety 5 preferred it over the others except for varieties 3 and 4, and so on. There is perfect agreement that the preference in varieties, from best to worst, is 4, 3, 5, 6, 7, 1, and 2. If the null hypothesis of no difference among varieties were true, each person would have 3! = 6 equally likely ways of ranking their three varieties, and the seven people would have $6^7 = 279{,}936$ equally likely combinations of rankings that could occur. Only one of these results is the ordering observed, with variety 4 being the best, variety 3 the second best, and so on. However, there are other perfect agreement cases that could occur, with any of the seven varieties being the best, any of the six remaining varieties being second best, and so on, giving 7! ways where perfect agreement could occur. Therefore the probability of getting perfect agreement, such as we have in this example, is the p-value, which is exactly

$$P(T_2 \geq 8) = P(T_2 = 8) = \frac{7!}{6^7} = 0.018$$

This exact p-value, 0.018, is slightly larger than the approximate p-value, <0.01, from the approximation in Table A22. However, it is closer than the approximate p-value obtained by comparing T_1 with the chi-squared distribution with 6 degrees of freedom, which is greater than 0.05. ∎

□ *Theory*   The theoretical development of the Durbin test is very similar to that of the Friedman test. That is, the exact distribution of the Durbin test statistic is found under the assumption that each arrangement of the k ranks within a block is equally likely because of no differences between treatments. There are $k!$ equally likely ways of arranging the ranks within each block, and there are b blocks. Therefore each arrangement of ranks over the entire array of b blocks is equally likely and has probability $1/(k!)^b$ associated with it, because there are $(k!)^b$ different arrays possible. The Durbin test statistic is calculated for each array and then the distribution function is determined, just as it was for the Friedman test statistic in the previous section.

The exact distribution is not practical to find in most cases, so the distribution of the Durbin test statistic T_1 is approximated by the chi-squared distribution with $t-1$ degrees of freedom, if the number of repetitions r of each treatment is large. The justification for this approximation is as follows.

If the number r of repetitions of each treatment is large, the sum of the ranks, R_j, under the jth treatment is approximately normal, according to the central limit theorem. Therefore the random variable

$$\frac{R_j - E(R_j)}{\sqrt{\text{Var}(R_j)}}$$

has approximately a standard normal distribution. As in the previous section, if the R_j were independent, the statistic

$$T' = \sum_{j=1}^{t} \frac{[R_j - E(R_j)]^2}{\text{Var}(R_j)} \tag{9}$$

could be considered as the sum of t independent, approximately chi-squared, random variables and the distribution of T' then could be approximated with a chi-squared distribution with t degrees of freedom. But the R_j are not independent. Their sum is fixed as

$$\sum_{j=1}^{t} R_j = \frac{bk(k+1)}{2} \tag{10}$$

so that the knowledge of $t - 1$ of the R_j enables us to state the value of the remaining R_j. Durbin (1951) shows that multiplication of T' by $(t - 1)/t$ results in a statistic that is approximately chi-squared with $t - 1$ degrees of freedom, with the form

$$T_1 = \frac{t-1}{t} T' = \frac{t-1}{t} \sum_{j=1}^{t} \frac{[R_j - E(R_j)]^2}{\text{Var}(R_j)} \tag{11}$$

It only remains to find the mean and variance of R_j in order to transform Equation 11 into the usual form given by Equation 2.

The sum of ranks R_j is the sum of independent random variables $R(X_{ij})$.

$$R_j = \sum_{i=1}^{b} R(X_{ij}) \tag{12}$$

Each $R(X_{ij})$, where it exists, is a randomly selected integer from 1 to k. Therefore the mean and variance of $R(X_{ij})$ are given by Theorem 1.4.5 as

$$E[R(X_{ij})] = \frac{k+1}{2} \tag{13}$$

and

$$\text{Var}[R(X_{ij})] = \frac{(k+1)(k-1)}{12} \tag{14}$$

Then the mean and variance of the R_j are easily found to be

$$E(R_j) = \sum_{i=1}^{b} E[R(X_{ij})] = \frac{r(k+1)}{2} \tag{15}$$

and

$$\text{Var}(R_j) = \sum_{i=1}^{b} \text{Var}[R(X_{ij})] = \frac{r(k+1)(k-1)}{12} \tag{16}$$

The mean and variance of R_j given here are substituted into the Durbin test statistic given by Equation 11 to obtain

$$T_1 = \frac{t-1}{t} \sum_{j=1}^{t} \frac{[R_j - r(k+1)/2]^2}{r(k+1)(k-1)/12}$$

$$= \frac{12(t-1)}{rt(k+1)(k-1)} \sum_{j=1}^{t} \left[R_j - \frac{r(k+1)}{2} \right]^2 \qquad (17)$$

which is in the same form as given in the explanation of the Durbin test.

The chi-squared approximation is based on the assumption that the number r of repetitions of each treatment is reasonably large. In practice the approximation is used even if r is as small as 3 or 2, out of sheer necessity. The stated α level is probably not very accurate in those circumstances. The stated α level is usually closer to the true α level if the analysis of variance F statistic is used on the ranks, which is the statistic T_2 given by Equation 6.

The Durbin test has been generalized to the case where some experimental units may contain several observations by Benard and van Elteren (1953). Noether (1967a) also discusses the Durbin test and its generalizations and shows that the A.R.E. of the Durbin test relative to its parametric counterpart is the same as that of the Friedman test relative to its parametric counterpart. See the preceding section for details. The case of paired comparisons ($k = 2$) is discussed in Puri and Sen (1969b). □

EXERCISES

1. Seven types of automobile tires are being tested for durability. It is felt that the best test is to see how the tires perform under actual driving conditions. However, only four tires may be compared at a time because only four-wheeled vehicles are available for testing. Therefore the experiment is designed using a balanced incomplete block design. Each of seven drivers is given four tires that are placed on each car in a random order and rotated regularly during the experiment. The tires are replaced when necessary, and ranks are assigned to the original tires according to the order of replacement.

| | | | | Tire Type | | | |
Driver	1	2	3	4	5	6	7
1			3		1	4	2
2	1			3		4	2
3	2	1			3		4
4	1	2	4			3	
5		1	4	3			2
6	2		4	1	3		
7		1		2	3	4	

Do the results indicate a significant difference in durability? (First examine the experiment to be sure it follows a balanced incomplete block design.) If there is a significant difference in durability, use the multiple comparisons procedure to determine which tire types are better than others.

2. An experiment is designed to determine which of five scents tends to be the most attractive to coyotes, for purposes of predator control. The experimenter has observed that the presence of more than three scents at a time tends to confuse the coyotes and produce inconsistent results. Therefore three scents at a time are placed in separate areas of a large pen. One coyote at a time is released into the pen, and the amount of time (seconds) the coyote spends at each scent is recorded.

   The scents are rotated according to a balanced incomplete block design, with the following results.

	Scent				
Coyote	1	2	3	4	5
1	12	23		14	
2		17	2		2
3	16		1	6	
4		42		10	0
5	8		6		1
6	22	31			0
7	28	16	4		
8	15			7	4
9		67	5	18	
10			6	16	1

Is this a balanced incomplete block design? Are there significant differences between scents? If so, which scents are better than others?

3. Students in a Finance class were divided into teams, where each team worked on a project and presented the results to the rest of the class. After all of the presentations were completed, each team rated the other teams from best (a score of 10) to worst (a score of 0). Here are their ratings of the other teams.

Team Doing the Rating	Team Being Rated				
	1	2	3	4	5
1		6.7	9.1	8.6	9.2
2	7.6		9.0	8.1	9.3
3	8.6	8.3		8.9	9.4
4	8.9	8.5	8.8		9.6
5	9.1	9.3	9.6	9.4	

Is there a significant difference in the ratings received by the teams? If so, which teams are significantly better than which other teams?

PROBLEMS

1. Show that $kb = rt$. (*Hint.* Count the number of observations in two different ways.)

2. Show that $\lambda = r(k-1)/(t-1)$. (*Hint.* First note that any particular treatment occurs in r blocks. Then count the number of units in which the treatment does not appear in those r blocks and count them in two different ways.)

5.10 TESTS WITH A.R.E. OF 1 OR MORE

The tests described in this section all share one property in common. The A.R.E. of each of these tests is 1 when compared with the usual parametric test in situations where the parametric test is appropriate. If the normality assumptions underlying the parametric test are not satisfied, under certain conditions that are easily met, the A.R.E. is always greater than 1 and may be as high as infinity. It seems to be a rather strong statement to say that the tests of this section are always at least as good as the usual parametric tests, such as the t test and the F test, as measured by their asymptotic relative efficiencies. However, it is true. Remember that A.R.E. is only one way to compare tests, although admittedly it is probably the most universally accepted method of comparison. Relative efficiency, without the word "asymptotic," is also a method of comparison. It compares the sample sizes required for two tests to have the same power under identical conditions where the sample sizes are finite. On the basis of relative efficiencies, the tests in this section may be better or worse than their usual parametric counterparts, depending on the circumstances. It is not possible to examine all of the possible circumstances, and that is why the A.R.E. is usually used to compare tests.

Unlike the previous sections of this chapter, no new experimental situations are introduced in this section. We have already introduced nonparametric methods, based on ranks, for handling the one-way layout in Section 5.2, correlation in Section 5.4, and the randomized complete block design in Section 5.8. Those methods are widely accepted, reasonably powerful, and not too difficult to administer. By comparison, the methods of this section are usually equally powerful, but they are slightly more difficult to administer. The assumptions behind these tests are practically identical to the assumptions underlying the earlier tests. Indeed, these tests are basically rank tests with a little dressing to improve the A.R.E. The user may decide whether to use these tests or the previously introduced tests; there is no solid statistical basis for preferring some to others.

The first tests we will describe are based on a simple idea suggested by van der Waerden (1952/1953). Instead of using the ranks of the observations as the basis for all of our computations, suppose we use other numbers that more nearly resemble observations from a normal distribution; in particular, instead of the rank k, suppose we use the $k/(N+1)$ quantile of a standard normal distribution, for $k = 1, 2, \ldots, N$, where N is the sample size. These quantiles, sometimes called *normal scores*, are readily available from Table A1. For example, suppose a random sample of five observations, arranged from smallest to largest, is given by 7.3, 7.7, 9.2, 12.0, and 26.4. Note that the three smallest observations are close together, the fourth observation is somewhat larger, and the largest observation is more than twice as large as any of the others. Replacement of these observations by their ranks 1, 2, 3, 4, and 5 amounts to a transformation of the nonsymmetrical original numbers into very symmetric, evenly spaced, somewhat "uniformly distributed" numbers. In the previous sections of this chapter we explained how the same kind of analysis customarily performed on the original observations can also be performed on the ranks. Now, following van der Waerden's suggestion, we transform the ranks into normal scores by replacing the rank k with the $k/(N + 1)$ quantile from the normal distribution given in Table A1. Thus rank 1 is replaced by $z_{1/6} = z_{0.167} = -0.9661$, rank 2 is replaced by $z_{2/6} = z_{0.333} = -0.4316$, and so on. Then, instead of performing the analysis on the ranks, we analyze the normal scores: -0.9661, -0.4316, 0.0000, 0.4316, and 0.9661. In general, these normal scores will be symmetrically distributed around zero and will be spread out much the same as the "perfect normal sample" would be although, of course, there is no such thing as a "perfect normal sample." The result of using normal scores is a nonparametric test that has the same asymptotic efficiency as the parametric test when the population is really normal and a larger asymptotic efficiency when the population is nonnormal.

We start out by showing how the normal scores may be used as a modification of the Kruskal-Wallis test of Section 5.2 for testing equality among k populations. The two-sample problem is a special case that relates to the Mann-Whitney test of Section 5.1.

▶ The van der Waerden (Normal Scores) Test for Several Independent Samples

Data   The data consist of k random samples of possibly unequal sample sizes. Denote the ith sample, of size n_i, by $X_{i1}, X_{i2}, \ldots, X_{in_i}$. Let N denote the total number of observations. Rank all N values from rank 1 to rank N, as explained in the Kruskal-Wallis test, using average ranks in cases of ties as usual. Let $R(X_{ij})$ denote the rank of X_{ij}.

Convert each rank R into the $R/(N + 1)$ quantile of a standard normal random variable obtained from Table A1. For brevity call these quantiles "normal scores" and denote them by A_{ij}.

$$A_{ij} = z_{R(X_{ij})/(N+1)} = \text{the } \frac{R(X_{ij})}{N + 1} \text{th quantile from Table A1} \qquad (1)$$

For convenience in obtaining the normal scores, round each value of $R(X_{ij})/(N + 1)$ off to three decimal places before consulting Table A1. Find the average score

$$\overline{A}_i = \frac{1}{n_i} \sum_{j=1}^{n_i} A_{ij} \qquad i = 1, 2, \ldots, k \qquad (2)$$

for each of the k samples and the variance

$$S^2 = \frac{1}{N-1} \sum_{\substack{\text{all} \\ \text{scores}}} A_{ij}^2 \tag{3}$$

Note that the overall mean equals zero if there are no ties and is essentially zero even if there are many ties, so the overall mean may be ignored when computing the variance.

Assumptions   The assumptions here are the same as in the Kruskal-Wallis test.

Test Statistic   The test statistic T_1 is defined as

$$T_1 = \frac{1}{S^2} \sum_{i=1}^{k} n_i(\overline{A}_i)^2 \tag{4}$$

where $\overline{A}_i$ and S^2 are given by Equations 2 and 3, respectively.

Null Distribution   The exact distribution of T_1 may be obtained by considering all permutations of the scores A_{ij}, just as the exact distribution of the Kruskal-Wallis statistic can be found by considering all permutations of the ranks $R(X_{ij})$. However, this is too difficult in most cases, so the chi-squared distribution with $k-1$ degrees of freedom is used as an approximation. This approximation is usually very good.

Hypotheses   As in the Kruskal-Wallis test, we have:

> H_0: All of the k population distribution functions are identical
>
> H_1: At least one of the populations tends to yield larger observations than at least one of the other populations

Reject H_0 at the level α if T_1 exceeds the $1-\alpha$ quantile of a chi-squared random variable with $k-1$ degrees of freedom, given in Table A2. Note that this is only an approximation, but it is good enough for most practical applications. The p-value is obtained by comparing T_1 with the quantiles in Table A2.

Multiple Comparisons   If the null hypothesis is rejected we can say populations i and j seem to be different if the inequality

$$|\overline{A}_i - \overline{A}_j| > t_{1-\alpha/2}\left(S^2 \frac{N-1-T_1}{N-k}\right)^{\frac{1}{2}}\left(\frac{1}{n_i} + \frac{1}{n_j}\right)^{\frac{1}{2}} \tag{5}$$

is satisfied, where $t_{1-\alpha/2}$ is the $1-\alpha/2$ quantile of the t distribution with $N-k$ degrees of freedom, obtained from Table A21, and the other terms are defined previously. This procedure is usually repeated for all pairs i and j. The same value for α as used before is also used for multiple comparisons.

Computer Assistance Normal scores tests are covered by *StatXact*, which calculates exact *p*-values. ——————————————————————————————————— ◄

EXAMPLE I

The same example that was used to illustrate the Kruskal-Wallis test in Section 5.2 and the median test in Section 4.3 will also be used here for ease in comparing these methods.

Four methods of growing corn resulted in the following observations and their ranks.

Method 1			Method 2			Method 3			Method 4		
Obser-vation	Rank	Normal Score	Obser-vation	Rank	Normal Score	Obser-vation	Rank	Normal Score	Obser-vation	Rank	Normal Score
83	11	−0.4845	91	23	0.4043	101	34	1.8957	78	2	−1.5805
91	23	0.4043	90	19.5	0.1434	100	33	1.5805	82	9	−0.6526
94	28.5	0.8927	81	6.5	−0.8927	91	23	0.4043	81	6.5	−0.8927
89	17	−0.0351	83	11	−0.4845	93	27	0.7421	77	1	−1.8957
89	17	−0.0351	84	13.5	−0.2898	96	31.5	1.2816	79	3	−1.3658
96	31.5	1.2816	83	11	−0.4845	95	30	1.0669	81	6.5	−0.8927
91	23	0.4043	88	15	−0.1789	94	28.5	0.8927	80	4	−1.2055
92	26	0.6526	91	23	0.4043				81	6.5	−0.8927
90	19.5	0.1434	89	17	−0.0351						
			84	13.5	−0.2898						
Average score $\bar{A}_i$:		0.3582			−0.1703			1.1234			−1.1723

The ranks are converted to normal scores in the following manner. The total sample size is $N = 34$, so each rank is divided by $N + 1 = 35$ and rounded off to three decimal places. The first observation has rank 11, and 11/35 equals .314. The .314 quantile from Table A1 is −0.4845, as noted.

The average score for each method of growing corn is also given above. The variance is computed using Equation 3 by squaring each of the 34 normal scores, summing them, and dividing by $N - 1 = 33$. The result is $S^2 = 0.8447$. In general, S^2 will always be slightly less than 1.0. The observed value of $T_1 = 25.1840$ is much greater than the 0.95 quantile of a chi-squared random variable with $k - 1 = 3$ degrees of freedom, which is 7.815 from Table A2. Therefore the null hypothesis is clearly rejected. The *p*-value is less than 0.001, as it was with the median test (Example 4.3.1, where $T = 17.6$) and the Kruskal-Wallis test (Example 5.2.1, where $T = 25.46$).

The multiple comparisons procedure uses the 0.975 quantile of the *t* distribution with $30(=34 - 4)$ degrees of freedom, which is given in Table A21 as 2.042. The results of the computations are as follows.

	$\lvert \bar{A}_i - \bar{A}_j \rvert$	$t_{0.975}\left(S^2\dfrac{N-1-T_1}{N-k}\right)^{\frac{1}{2}}\left(\dfrac{1}{n_i}+\dfrac{1}{n_j}\right)^{\frac{1}{2}}$
$i = 1, j = 2$	0.5286	0.4401
$i = 1, j = 3$	0.7652	0.4828
$i = 1, j = 4$	1.5305	0.4655
$i = 2, j = 3$	1.2937	0.4721
$i = 2, j = 4$	1.0020	0.4544
$i = 3, j = 4$	2.2957	0.4958

In each case the average scores are far enough apart to result in the conclusion that the two populations seem to be different. Note that these are the same conclusions that were reached using the Kruskal-Wallis test in Section 5.2. These two tests will agree in their conclusions quite often, but not always. To avoid ambiguous situations, either one test or the other, but not both, should normally be used. ∎

It should be clear by now that the normal scores are used in the same way the ranks were: as numbers used to replace the original observations. The analysis on the normal scores is analogous to the analysis on the ranks. Exact tables could be given, but we do not present them here. Instead we rely on the large sample approximation for all sample sizes, large or small.

The test of the van der Waerden type for the one-sample problem, as an analogue to the Wilcoxon signed ranks test, is mentioned by van Eeden (1963). Let R_i represent the signed rank of Section 5.7, defined prior to Equation 5.7.3. Instead of using the signed ranks R_i, use the $\frac{1}{2}[1 + R_i/(n + 1)]$th quantile from the normal distribution given in Table A1. Note that n is the number of nonzero differences obtained from the data. Call these signed normal scores, and denote them by A_i. Note that A_i will have the same sign as R_i. Then the test statistic

$$T_2 = \frac{\sum\limits_{i=1}^{n} A_i}{\sqrt{\sum\limits_{i=1}^{n} A_i^2}} \tag{6}$$

is compared with quantiles from the standard normal distribution as an approximate test of the same hypothesis tested by the Wilcoxon signed ranks test and under the same assumptions. Exact p-values are computed by *StatXact*.

EXAMPLE 2

For comparison purposes we will use the same data given in Example 5.7.1. to test

H_0: The firstborn twin does not tend to be more aggressive than the other

versus

H_1: The firstborn twin tends to be more aggressive than the second twin

The data are as follows.

Twin Set	Firstborn X_i	Secondborn Y_i	Difference D_i	Rank of $|D_i|$	Signed Rank R_i	Signed Normal Score A_i
1	86	88	+2	3	3	0.3186
2	71	77	+6	7	7	0.8134
3	77	76	−1	1.5	−1.5	−0.1560
4	68	64	−4	4	−4	−0.4316
5	91	96	+5	5.5	5.5	0.6098
6	72	72	0	—	—	—
7	77	65	−12	10	−10	−1.3852
8	91	90	−1	1.5	−1.5	−0.1560
9	70	65	−5	5.5	−5.5	−0.6098
10	71	80	+9	9	9	1.1503
11	88	81	−7	8	−8	−0.9661
12	87	72	−15	11	−11	−1.7279

The test statistic T_2, defined by Equation 6, equals

$$T_2 = \frac{\sum\limits_{i=1}^{n} A_i}{\sqrt{\sum\limits_{i=1}^{n} A_i^2}} = \frac{-2.5405}{\sqrt{8.9027}} = -0.8514 \tag{7}$$

which corresponds to a one-tailed p-value of 0.197, from Table A1, in reasonable agreement with the results of the Wilcoxon signed ranks test which had $T = -0.7565$ and p-value $= 0.238$. ■

The two-sample test for equal variances using normal scores was introduced by Klotz (1962). The test begins like the van der Waerden test for two samples but, in the end, the square of the normal scores is used rather than the normal scores themselves in the statistic

$$T_3 = \frac{\sum\limits_{i=1}^{n} A_i^2 - \frac{n}{N} \sum\limits_{i=1}^{N} A_i^2}{\left\{ \frac{nm}{N(N-1)} \left[\sum\limits_{i=1}^{N} A_i^4 - \frac{1}{N} \left(\sum\limits_{i=1}^{N} A_i^2 \right)^2 \right] \right\}^{\frac{1}{2}}} \tag{8}$$

where the normal scores are A_i, the sample sizes are m and n, and $N = n + m$ denotes the combined sample size. Then T_3 is compared with normal quantiles from Table A1. If the two samples come from populations with different means, the means (if they are known) or sample means are subtracted from the respective observations before the initial ranks are assigned. Exact p-values for the Klotz test are calculated in *StatXact*.

EXAMPLE 3

Refer to Example 5.3.1 for details of this example and a comparison with the squared ranks test. A new machine is being tested to see if it is less variable than the present machine. So the null hypothesis

H_0: Both machines have the same variability

is being tested against the one-sided alternative

H_1: The new machine has a smaller variance

We adjust the data by subtracting the sample means, because the population means are unknown. The result is an approximate test, just as in the squared ranks test.

X_i	$X_i - \overline{X}$	Ranks	$\dfrac{R_i}{N+1}$	Normal Score A_i	A_i^2
10.8	.06	8	.615	0.2924	.0855
11.1	.36	11	.846	1.0194	1.0392
10.4	−.34	2	.154	−1.0194	1.0392
10.1	−.64	1	.077	−1.4255	2.0321
11.3	.56	12	.923	1.4255	2.0321
Y_i	$Y_i - \overline{Y}$				
10.8	.01	6	.462	−0.0954	.0091
10.5	−.29	3	.231	−0.7356	.5411
11.0	.21	10	.769	0.7356	.5411
10.9	.11	9	.692	0.5015	.2515
10.8	.01	6	.462	−0.0954	.0091
10.7	−.09	4	.308	−0.5015	.2515
10.8	.01	6	.462	−0.0954	.0091

The sum of the A_i^2 from the first sample (from the present machine) is the basic measure of variability in the Klotz test. To check its level of significance, we subtract its mean and divide by its standard deviation under the null hypothesis to get

$$T_3 = \frac{6.2280 - 3.2669}{1.2629} = 2.3447 \qquad (9)$$

(see Equation 8), which is compared with Table A1 to get a one-sided p-value of about 0.01, similar to the results using the squared ranks test.   ■

To use the normal scores in regression and correlation, the normal scores replace the ranks of the X_is; then, in a separate step, the ranks of the Y_is are replaced by normal scores. If there are no ties the same set of normal scores is used for the X variable as is used for the Y variable, just as the same set of ranks 1 to n is used with each variable. The Pearson product moment correlation coefficient is computed on the normal scores (see Equation 5.4.2 for the coefficient). In the case of no ties it simplifies to

$$\rho = \frac{\sum\limits_{i=1}^{n} A_i B_i}{\sum\limits_{i=1}^{n} A_i^2} \tag{10}$$

(A_i and B_i represent the normal scores assigned to X_i and Y_i, respectively) because the mean scores are zero. Equation 10 may be used with ties unless the ties are quite extensive, in which case the safest procedure is to revert to computing Equation 5.4.2 on the actual normal scores used. Methods described in Sections 5.4 to 5.6 may be used on these scores, but we do not present the details here.

For the two-way layout, recall that the Friedman test of Section 5.8 used a ranking of the observations within each block. Normal scores may be substituted for these ranks in the usual manner. Let A_{ij} represent the normal score assigned to the variable X_{ij} in block i, treatment j, and let A_j be the sum of the scores in treatment j, analogous to $R(X_{ij})$ and R_j in the Friedman test. Then the test statistic

$$T_4 = \frac{k-1}{S^2}\left(\sum_{j=1}^{k} A_j^2\right) \tag{11}$$

where

$$S^2 = \sum_{\substack{all \\ scores}} A_{ij}^2 \tag{12}$$

is compared with the chi-squared distribution, $k-1$ degrees of freedom, given by Table A2, as with T_1 in Equation 5 of the Friedman test. The rest of the details are the same as in the discussion of the Friedman test, except that the chi-squared approximation is sufficiently good so the F approximation is not necessary. Multiple comparisons are made as described in Section 5.8, except that the preceding values of T_4 and S^2 are used instead of T_1 and $A_1 - C_1$ in Equation 5.8.8.

By now the pattern of using normal scores instead of ranks should be clear. The result is a slightly higher A.R.E. relative to the best parametric test. The A.R.E. relative to the rank tests presented in previous sections may be greater than 1 or less than 1, depending on the particular situation. Other scores may be used instead of normal scores to achieve identical A.R.E.s with the normal scores tests just described. Two of these types of scores are called "random normal deviates" and "expected normal scores." We will now describe them briefly.

Random Normal Deviates

One way to replace a random sample $X_1, \ldots, X_n$ from an arbitrary distribution, with numbers that seem to have come from a normal distribution, is to obtain somehow a group of n numbers that seem to have come from a normal population and to replace the smallest observation from the original sample with the smallest number from this pseudo normal sample, the second smallest with the second smallest, and so on. That is, the original observation of rank k is replaced by the number that has rank k in the pseudo normal sample. Note that only the ranks of the original observations need to be known in order to accomplish this replacement, so the resulting statistical procedures are rank tests. The pseudo normal sample may be obtained from tables of such numbers, such as in the book *A Million Random Digits with 100,000 Normal Deviates* by the Rand Corporation (1955), or by using computer programs specifically designed to produce such numbers. These numbers are called "random normal deviates," although in actuality they are not random in the true sense of the word. They are deliberately generated numbers that seem to resemble a random sample from a standard normal distribution.

For example, the random sample we used earlier to illustrate normal scores was 7.3, 7.7, 9.2, 12.0, 26.4. A group of five numbers from a table of normal deviates is .026, -1.388, 2.388, 1.066, $-.173$. The smallest of these, -1.388, replaces the 7.3. The next smallest, $-.173$, replaces 7.7, and so on. From this point these new numbers are used in much the way as the normal scores, or the ranks, were. Of course, someone else who analyzes the same data will most likely select a different set of five numbers to work with, and the resulting analysis will be slightly (or not so slightly, sometimes) different. This results in the unpleasant situation of two statisticians using the same test to analyze the same data but coming up with conflicting conclusions. For this reason these procedures are seldom, if ever, used in practice. However, they are very interesting to study from a theory point of view because their A.R.E. is the same as that of the normal scores tests, and their exact distributions are the same as in the parametric tests.

The principle of using random normal deviates is explained more fully by Bell and Doksum (1965). Earlier mention of the method appears in an article by Durbin (1961), the last problem in Fraser (1957), and in an article by Ehrenberg (1951).

Expected Normal Scores

One way of looking at the normal deviates procedures is to think of the actual order statistic $X^{(i)}$ as being replaced by order statistics $Z^{(i)}$ from a normal distribution. The next type of scores we consider uses the mean of the $Z^{(i)}$s, $E(Z^{(i)})$, instead of the order statistics themselves. These "expected normal scores" are well-defined numbers that are commonly available in tables such as Fisher and Yates (1957), Pearson and Hartley (1962), and Owen (1962). Therefore the unpleasant variability connected with using the $Z^{(i)}$s themselves, as random normal deviates, is eliminated. This type of procedure is still based only on the ranks of the observations and is therefore a rank test. Fisher and Yates (1957) suggest using these exact scores instead of the original data and then applying the usual parametric procedures to these expected normal scores as a nonparametric procedure. The A.R.E. of these methods is the same as that of the normal scores procedures and the random normal deviates procedures. A more complete presentation of this variation is given by Bradley (1968).

□ *Theory*   The same reasoning we used to find the exact distribution of the test statistics in previous sections is used here, with minor modifications. One obvious modification is that instead of working with ranks 1, 2, 3, . . . , we are working now with other numbers, which we will call $a(1)$, $a(2)$, $a(3)$, . . . These numbers $a(i)$ represent the normal scores or expected normal scores or any other set of numbers chosen independently of the data. Another modification to our previous tests is that these new numbers have different means and variances than the ranks have, and these means and variances need to be determined.

As one example of the method used to find the exact distribution, consider two independent samples $X_1, X_2, . . . , X_n$ of size n and $Y_1, Y_2, . . . , Y_m$ of size m, as in the Mann-Whitney test. Under the null hypothesis of identical distributions, the rank of X_i is equally likely to be any of the ranks from 1 to $n + m$. Therefore the score assigned to X_1 is equally likely to be any of the scores from $a(1)$ to $a(n + m)$. The same holds true for $X_2, X_3, . . . , Y_1, Y_2, Y_3$, and so on. Thus there are $\binom{n+m}{n}$ ways of selecting n ranks as belonging to the Xs and each of the $\binom{n+m}{n}$ ways is equally likely and has probability $1/\binom{n+m}{n}$. This also means that there are $\binom{n+m}{n}$ ways of selecting n of the scores $a(1)$ to $a(n + m)$, and each of these ways has probability $1/\binom{n+m}{n}$. This enables the null distribution of any statistic based on scores (or ranks) assigned to the Xs (or Ys) to be found by the same counting methods used earlier in this chapter.

To be more specific, let $n = 2$ and $m = 3$, and suppose the scores we are using are the normal scores.

$$a(1) = -0.9661$$
$$a(2) = -0.4316$$
$$a(3) = 0.0000$$
$$a(4) = 0.4316$$
$$a(5) = 0.9661$$

The possible ranks for X_1 and X_2, their associated scores, and the sum of scores are given as follows.

$(R(X_1), R(X_2))$	Scores (A_1, A_2)	Sum	Probability
(1, 2)	(−0.9661, −0.4316)	−1.3977	0.1
(1, 3)	(−0.9661, 0.0000)	−0.9661	0.1
(1, 4)	(−0.9661, 0.4316)	−0.5345	0.1
(1, 5)	(−0.9661, 0.9661)	0.0000	0.1
(2, 3)	(−0.4316, 0.0000)	−0.4316	0.1
(2, 4)	(−0.4316, 0.4316)	0.0000	0.1
(2, 5)	(−0.4316, 0.9661)	0.5345	0.1
(3, 4)	(0.0000, 0.4316)	0.4316	0.1
(3, 5)	(0.0000, 0.9661)	0.9661	0.1
(4, 5)	(0.4316, 0.9661)	1.3977	0.1

Thus the distribution function of the sum of scores can be found. In a similar manner, the distribution function of any of the statistics presented in this section can be found. However, we do not present tables of the exact distributions, but suggest using the approximate distributions instead for these tests. □

In order to find the mean and variance of the sum of ranks in the two sample situation, the same procedure shown in Section 5.3 is used. The result there, which is equally valid in this situation, is summarized in Equation 5.3.18 for the mean and Equation 5.3.24 for the variance. A more complete discussion of the use of scores instead of ranks is given by Hajek and Sidak (1967). There the method of choosing the best scores for each particular situation is described, and the complete theory is presented. The theory is well beyond the scope of this text, but Hajek and Sidak is strongly recommended reading for any statistician.

A discussion of random normal numbers may be found in Marsaglia (1968) or Lewis (1975), which are only two of the many references available on the subject. Jogdeo (1966) shows that the relative efficiency of the random normal deviates procedures is less than 1 for some fixed alternatives. Ramsey (1971) examines the small sample power of some of these two-sample tests, while large sample efficiencies are considered by Raghavachari (1965b), Thompson, Govindarajulu, and Doksum (1967), Bhattacharyya (1967), Stone (1968), and Gokhale (1968). Some variations of these tests are presented by Bradley, Patel, and Wackerly (1971) for the multivariate case, Johnson and Mehrotra (1972) for censored data, and Pirie and Hollander (1972) for ordered alternatives in the randomized block design. A broader perspective for these methods may be obtained by consulting Lehmann (1975) or Hogg (1976).

EXERCISES

1. Work Exercise 5.2.1 using normal scores instead of ranks and compare the results of the two methods.

2. Work Exercise 5.2.3 using normal scores instead of ranks and compare the results of the two methods.

3. Work Exericise 5.7.1 using normal scores instead of ranks and compare the results of the two methods.

4. Work Exercise 5.7.3 using normal scores instead of ranks and compare the results of the two methods.

5. Use the Klotz test on the data in Exercise 5.3.1 and compare the results of the two methods.

6. Use the Klotz test in Exercise 5.3.2.

7. Use normal scores in Exercise 5.4.1 to compute the correlation coefficient as given by Equation 10. How does this coefficient compare in size with Spearman's and Kendall's coefficients? Compare $\rho\sqrt{n-1}$ with the quantiles in Table A1 for significance, and test the hypothesis of independence. Compare with the results of Exercise 5.4.1.

8. Use normal scores in Exercise 5.4.3 to test for trend. Compare $\rho\sqrt{n-1}$ with the quantiles of Table A1 for significance. How do these results compare with the results of Exercise 5.4.3?

PROBLEMS

1. Find the exact distribution of the statistic given by Equation 6 for $n = 5$.

2. Find the exact distribution of the Klotz statistic given by Equation 8 for $n = 2$, $m = 3$.

3. Obtain 34 random normal deviates from a table of random numbers or from a computer program that generates random normal numbers. Rank them from smallest to largest and then use them in Example 1 instead of normal scores. How do the results of this Bell-Doksum procedure compare with the results of the van der Waerden and Kruskal-Wallis tests?

5.11 FISHER'S METHOD OF RANDOMIZATION

In the previous section we introduced a variety of ways of obtaining nonparametric tests. Each method consists of using a set of scores $a(1)$ to $a(N)$ in place of the ranks 1 through N. Some suggested scores included quantiles from a standard normal distribution, numbers that seem to be a random sample from a standard normal distribution, and expected values of order statistics from a standard normal distribution. We mentioned that any numbers whatsoever may be used as scores, but that some types of numbers resulted in more power against particular alternative hypotheses.

Suppose that in our search to find a "good" set of scores to use in place of the ranks from 1 to N we decide to use the numbers that actually occurred in the sample. These are convenient numbers to use, since there are exactly N of them and they are readily available. We use these numbers as scores just as we used the normal scores, for example, to obtain nonparametric tests in the manner described in the previous section. But is this choice of scores likely to have as much power as the normal scores? Apparently so, according to studies by Lehmann and Stein (1949), Hoeffding (1952), and many others, who find that these procedures have an A.R.E. of 1.0 when compared to the most powerful parametric tests in some situations. So these scores compare favorably not only with normal scores, but with any other type of test for those situations. Why, then, doesn't everybody use the data as scores in hypothesis tests?

The major disadvantage of this suggested procedure is that the test becomes very tedious to perform. This is because it is not possible to make tables of the critical regions or to present quantiles of the test statistics, since the scores are different in each test. Therefore each time a test of this type is performed, the critical region must be determined specifically for the set of data observed. Each different sample means a different set of scores and a different critical region. And even though the asymptotic distribution of the test statistic is, under easily met conditions, one of the standard distributions such as normal or chi-squared, the use of the asymptotic distribution as an approximation may not be accurate for some types of scores. At least when ranks, normal scores, or expected normal scores are used, we know what the scores are and the accuracy of each approximation may be determined. In those cases the asymptotic distribution works well as an approximation. But when the set of scores changes from one sample to the next, it is impossible to measure the accuracy of any asymptotic distribution. So, in short, exact p-values may be obtained for each case after some effort (considerable effort when sample sizes are not small). Methods for finding approximate p-values are available but may not be accurate. The computer package *StatXact* finds exact p-values and therefore essentially removes this handicap.

A second disadvantage of the randomization tests of this section is their relative lack of power. Unpublished simulation studies conducted by the author and R.L. Iman showed the power of the Fisher's randomization tests to be between the rank test and the parametric test for a variety of distributions. Overall, an ordinary rank test such as the Kruskal-Wallis test tends to have more power than Fisher's randomization test for nonnormal distributions with heavy tails, or for data with outliers.

One school of thought considers the situation where the sample or samples under investigation are not really random samples from some hypothetical populations, but are the populations themselves. Then the set of measurements is the complete population of interest, and the presence or absence of an effect on a subgroup of the set can and should be determined by a randomization test of the type discussed in this section. See Kempthorne and Doerfler (1969) for a more complete presentation of this philosophy.

The idea of using the data themselves as scores is credited to Fisher (1935), and the resulting tests are traditionally known as randomization tests. Although our presentation may lead one to think that randomization tests are a third generation of nonparametric tests, after rank tests and tests using other scores, randomization tests actually preceded these other tests in time. Randomization tests may be used in any of the situations for which we have described rank tests. We will describe in more detail the randomization tests for two independent samples and for matched pairs with examples of each, to clarify how these tests may be used. The first test is analogous to the Mann-Whitney test of Section 5.1.

▶ **The Randomization Test for Two Independent Samples** _____

Data The data consist of two random samples $X_1, X_2, \ldots X_n$ and $Y_1, Y_2, \ldots, Y_m$ of sizes n and m, respectively.

Assumptions

1. Both samples are random samples from their respective populations.
2. In addition to independence within each sample there is mutual independence between the two samples.
3. The measurement scale is at least interval.

Test Statistic The test statistic T_1 is the sum of the X observations:

$$T_1 = \sum_{i=1}^{n} X_i \tag{1}$$

Null Distribution The null distribution is found by considering all possible ways of selecting n of the numbers in the combined set of Xs and Ys, which are equally likely under the null hypothesis. Because the numbers on the Xs and Ys are different from one application to another, no tables are possible, and the validity of proposed approximate distributions is impossible to determine.

Hypotheses Only the two-tailed test is presented; the one-tailed tests may be surmised by direct analogy with the Mann-Whitney test in Section 5.1.

$$H_0: E(X) = E(Y)$$
$$H_1: E(X) \neq E(Y)$$

Reject H_0 at the level α if either $T_1 > w_{1-\alpha/2}$ or $T_1 < w_{\alpha/2}$, where the quantiles w_p are found as follows.

Consider the observed values of X_i and Y_j as merely a group of $n + m$ numbers, and consider the ways in which n of these numbers may be selected. There are $\binom{n+m}{n}$ such ways. To find the p quantile w_p, consider the $\binom{n+m}{n}(p)$ selections that yield the smallest sums, which sum we are calling T_1. The largest T_1 thus obtained is w_p.

As before, if $\binom{n+m}{n}(p)$ is not an integer, round upward to the next higher integer. If $\binom{n+m}{n}(p)$ is integer valued, w_p is the average of the largest T_1 thus obtained and the T_1 that would result from considering $\binom{n+m}{n}(p) + 1$ selections.

The p-value is obtained by counting the number of ways n of the $n + m$ observations may be selected so that their sum is smaller (or larger if the observed T_1 is in the upper tail) than, or equal to, the observed T_1 from the data. This number is doubled, because the test is two tailed, and divided by $\binom{n+m}{n}$ to get the p-value.

Computer Assistance   *StatXact* finds exact p-values for this and other tests of this type, by considering all permutations when the sample sizes are small, or by randomly selecting a large number of permutations to estimate the p-value when the total number of permutations is too large to handle. _____ ◄

> ## EXAMPLE I
>
> Suppose that a random sample yielded X_is of 0, 1, 1, 0, and -2 and an independent random sample of Y_js gave 6, 7, 7, 4, -3, 9, and 14. The null hypothesis
>
> $$H_0: E(X) = E(Y)$$
>
> is tested against
>
> $$H_1: E(X) \neq E(Y)$$
>
> at $\alpha = 0.05$, with the randomization test for two independent samples.
>   One sample is of size $n = 5$ and the other size $m = 7$, so there are $\binom{12}{5} = 792$ ways of forming a subset containing 5 of the 12 numbers. Because $(792)(0.025) = 19.8$, we need to find the 20 groups that yield the lowest values of T_1 to obtain $w_{0.025}$. These groups of numbers and the corresponding values of T_1 are as follows.

Combined Observations

Group	-3	-2	0	0	1	1	4	6	7	7	9	14	Observed T_1 ($T_1 = \Sigma X$)
1	X	X	X	X	X	Y	Y	Y	Y	Y	Y	Y	-4
2	X	X	X	X	Y	X	Y	Y	Y	Y	Y	Y	-4
3	X	X	X	Y	X	X	Y	Y	Y	Y	Y	Y	-3
4	X	X	Y	X	X	X	Y	Y	Y	Y	Y	Y	-3
5	X	X	X	X	Y	Y	X	Y	Y	Y	Y	Y	-1
6	X	Y	X	X	X	X	Y	Y	Y	Y	Y	Y	-1
7	Y	X	X	X	X	X	Y	Y	Y	Y	Y	Y	0
8	X	X	X	Y	X	Y	X	Y	Y	Y	Y	Y	0
9	X	X	X	Y	Y	X	X	Y	Y	Y	Y	Y	0
10	X	X	Y	X	X	Y	X	Y	Y	Y	Y	Y	0
11	X	X	Y	X	Y	X	X	Y	Y	Y	Y	Y	0
12	X	X	Y	Y	X	X	X	Y	Y	Y	Y	Y	1
13	X	X	X	X	Y	Y	Y	X	Y	Y	Y	Y	1
14	X	X	X	Y	X	Y	Y	X	Y	Y	Y	Y	2

15	X	X	X	Y	Y	X	Y	X	Y	Y	Y	Y	2
16	X	X	Y	X	X	Y	Y	X	Y	Y	Y	Y	2
17	X	X	Y	X	Y	X	Y	X	Y	Y	Y	Y	2
18	X	Y	X	X	X	Y	X	Y	Y	Y	Y	Y	2
19	X	Y	X	X	Y	X	X	Y	Y	Y	Y	Y	2
20	X	X	X	X	Y	Y	Y	Y	X	Y	Y	Y	2

The largest T_1 thus obtained is

$$w_{0.025} = 2$$

It is not necessary to find $w_{0.975}$ even though this is a two-tailed test, because the observed T_1 is in the lower tail. The observed value of T_1 from the data is

$$T_1 = \sum_{i=1}^{5} X_i = 0 + 1 + 1 + 0 - 2 = 0$$

which is less than $w_{0.025} = 2$, so H_0 is rejected. In fact, H_0 could have been rejected at the level

$$p\text{-value} = \frac{2(11)}{792} = 0.028$$

because there are 11 possible arrangements of the numbers that yield values less than or equal to zero. ∎

The previous randomization test is typical of randomization tests in general. Instead of the statistic T_1, the usual two-sample t statistic could have been used and computed for each of the 20 groups in each tail of the distribution, as was done in the example. This additional labor is unnecessary however, because the t statistic is a monotonic function of T_1, as mentioned in Section 5.1, so the 20 groups that provide the largest values of T_1 will be the same groups that provide the largest values of t. We mention this so that the extension of randomization tests to other cases, such as the one-way layout, two-way layout, or tests for correlation, becomes more obvious. The usual statistic for the situation, or any monotonic function of that statistic that is easier to compute, is used to determine the "most extreme" arrangements of the data and thus the critical region for the statistic being used. Because of the difficult calculations involved, the actual critical region is sometimes not found, but only the p-value is found, especially if the p-value is close to zero and only a few arrangements of the data need to be considered.

The randomization test for matched pairs follows a slightly different pattern than the other randomization tests, so we will now present it in detail. This test is analogous to the Wilcoxon signed ranks test for Section 5.7.

▶ **The Randomization Test for Matched Pairs** _____

Data The data consist of observations on n' bivariate random variables (X_1, Y_1), $(X_2, Y_2), \ldots, (X_{n'}, Y_{n'})$. Omit from further consideration all pairs (X_i, Y_i) whose difference $Y_i - X_i$ is zero and let the remaining number of pairs be denoted by n. Denote the nonzero differences $Y_i - X_i$ by $D_1, D_2, \ldots, D_n$.

Assumptions

1. The distribution of each D_i is symmetric.
2. The D_is are mutually independent.
3. The D_is all have the same mean.
4. The measurement scale of the D_is is at least interval.

Test Statistic The test statistic T_2 equals the sum of the positive differences.

$$T_2 = \Sigma\, D_i \qquad \text{only for those } D_i > 0 \tag{2}$$

Null Distribution The null distribution of T_2 is found by considering all possible assignments of plus and minus signs to the observed difference D_i. Each assignment is equally likely under the null hypothesis. Because the value of T_2 depends on the values of the Ds, which change from one application to another, it is not possible to obtain tables of quantiles in general, and impossible to verify the closeness of possible approximate distributions.

Hypotheses Only the two-tailed version of this test is presented, although the one-tailed version may be obtained by comparison with the Wilcoxon signed ranks test of Section 5.7.

$$H_0: E(D) = 0 \qquad (\text{i.e., } E(X) = E(Y))$$
$$H_0: E(D) \neq 0 \qquad (\text{i.e., } E(X) \neq E(Y))$$

Reject H_0 at the level α if $T_2 > w_{1-\alpha/2}$ or if $T_2 < w_{\alpha/2}$, where the quantiles $w_{1-\alpha/2}$ and $w_{\alpha/2}$ are found as follows.

Consider only the absolute values of the D_is, $|D_i|$, without regard for whether they were originally positive or negative. There are 2^n ways of assigning $+$ or $-$ signs to the set of absolute differences obtained, that is, we might assign $+$ signs to all n of the $|D_i|$, or we might assign a $+$ to $|D_1|$ but $-$ signs to $|D_2|$ to $|D_n|$, and so on. To find the p quantile w_p, $0 \leq p \leq 1$, first find the $(2^n)(p)$ assignments of signs that give the smallest values for T_2 the sum of the "positive" absolute differences. [If $(2^n)(p)$ is not an integer, use the next larger integer.] The largest value of T_2 thus obtained is the p quantile w_p of T_2 under the null hypothesis. [If $(2^n)(p)$ is an integer, use the average of the largest value of T_2 thus obtained, and largest T_2 possible if $(2^n)(p) + 1$ arrangements had been considered instead, according to our usual convention.]

The preceding method of finding w_p works for all values of p from 0 to 1, but in practice it should be used only for small values of p, such as $p = \alpha/2$. For large values of p such as $p = 1 - \alpha/2$ the relationship

$$w_{1-\alpha/2} = \sum_{i=1}^{n} |D_i| - w_{\alpha/2} \tag{3}$$

should be used. The relationship in Equation 3 is apparent if one considers that for every assignment of signs that results in a small value of T_2 a complete reversal of signs (pluses replaced by minuses, and vice versa) results in a large value of T_2. The latter value of T_2, the sum of the "positive" $|D_i|$s, plus the former value of T_2, the sum of the now "negative" $|D_i|$s, add up to the sum of all of the $|D_i|$s as indicated by Equation 3.

The p-value is obtained by counting the number of assignments of signs that result in a smaller (or larger, if the observed $T_2 > \frac{1}{2} \sum_{i=1}^{n} |D_i|$) value of T_2, or the same value for T_2, as the one obtained from the data. This number is doubled and divided by 2^n to get the p-value.

Computer Assistance   *StatXact* finds exact p-values for this version of Fisher's randomization test by considering all possible permutations whenever possible, or by randomly selecting a large number of permutations to estimate the p-value when the total number of permutations is too large to manage. ───────◄

EXAMPLE 2

Suppose that eight matched pairs resulted in the following differences: $-16, -4, -7, -3, 0, +5, +1, -10$. The zero is discarded, and we have

$$D_1 = -16, D_2 = -4, D_3 = -7, D_4 = -3, D_5 = +5, D_6 = +1, D_7 = -10$$

and $n = 7$. The null hypothesis

$$H_0: d_{0.50} = 0$$

is tested against the alternative

$$H_1: d_{0.50} \neq 0$$

using the randomization test at the level $\alpha = 0.05$.

The quantile $w_{0.025}$ is found by considering the 4 [because $(2^7)(0.025) = 3.2$] ways of assigning signs that result in the lowest sum of the "positive" absolute differences. These are given as follows.

Assignment of Signs	Σ *"positive"* $|D_i|$
$-16, -4, -7, -3, -5, -1, -10$	$T_2 = 0$
$-16, -4, -7, -3, -5, +1, -10$	$T_2 = 1$
$-16, -4, -7, +3, -5, -1, -10$	$T_2 = 3$
$-16, -4, -7, +3, -5, +1, -10$	$T_2 = 4$
(also $-16, +4, -7, -3, -5, -1, -10$ gives	$T_2 = 4$)

The largest of these T_2 values is 4, so

$$w_{0.025} = 4$$

From Equation 3 we have

$$w_{0.975} = \sum_{i=1}^{7} |D_i| - w_{0.025}$$
$$= 46 - 4 = 42$$

The value of the test statistic obtained from the data is

$$T_2 = \Sigma \text{ positive } D_i$$
$$= 5 + 1 = 6$$

which is neither less than 4 nor greater than 42, so H_0 is accepted.

The p-value is found by listing the assignments of signs that results in $T_2 \leq 6$, in addition to the five just listed.

Assignment of Signs	Σ *"positive"* $|D_i|$
$-16, +4, -7, -3, -5, +1, -10$	$T_2 = 5$
$-16, -4, -7, -3, +5, +1, -10$	$T_2 = 5$
$-16, -4, -7, -3, +5, +1, -10$	$T_2 = 6$

Thus there are eight arrangements that give values of T_2 less than or equal to the observed value of 6. Because this is a two-tailed test, this number is doubled, and the p-value is given by

$$p\text{-value} = \frac{2(8)}{2^7} = \frac{16}{128} = 0.125 \qquad \blacksquare$$

□ *Theory*   The theory behind the randomization tests is partially explained by the method of finding the critical region. In the test for two independent samples, for instance, it is obvious that we are considering each selection of n X observations to be equally likely, from the $n + m$ observations available. It just remains to explain why we may consider the selections to be equally likely and why we are working with the observations themselves as our "sample space," so to speak.

The selections may be considered to be equally likely because of the null hypothesis, which states (along with the assumptions) that the Xs and the Ys are all independent and identically distributed. Therefore the Xs should have no more of a tendency to be low than the Ys have, or to be high, or to be in the middle. Given any group of $m + n$ numbers, whether they be observations or not, each subgroup of n of those numbers is just as likely to be the n values of X as any other subgroup of n of those numbers, because the numbers that are not Xs have to be Ys and the overall probability attached to that group of numbers does not depend on which numbers are called Xs and which numbers are called Ys. Now, if the Xs are distributed differently than the Ys, it will matter which numbers are called Xs and which are called Ys but, for purposes of finding a critical region of size α, we restrict our consideration to identically distributed random variables. So that is the intuitive argument for considering each selection of n observations as Xs to be equally likely.

That also leads to the second question, "Why are we working with the observations themselves as our sample space?" We explained before that any set of $m + n$ numbers satisfies the "equally likely" criterion. But in a testing situation we need to identify the $m + n$ numbers used with the $m + n$ observations obtained. In a rank test the $m + n$ numbers used are the integers from 1 to $m + n$, and they are matched one for one with the observations by assigning ranks to the observations. In this case we are using the observations themselves as the numbers. This eliminates the problem of which numbers to assign to which observations, which occurs in the rank tests when ties confuse the ranking procedure. By using the observations themselves as the numbers, it is easy to identify one of the selections of n numbers as the one actually obtained in the data. Then, with the aid of the test statistic, all selections more extreme than the one obtained may also be identified, counted, and used to compute the p-value.

The critical region is thus determined for individual subsets of the sample space, such as for the subset of all outcomes that have the same numerical values as the observed values in the data. These subsets are mutually exclusive, cover the entire sample space (given *any* set of observations we can find the critical region for that subset of the sample space), and each subset has a critical region of size α relative to the size of the entire subset. So the overall size of all the critical regions combined is also α, which shows that the test is indeed a valid one.

The principal difference between the test for two independent samples and the test for matched pairs is that in the test for matched pairs the assumption of symmetry is used to justify the change of algebraic signs without changing the probability. If a difference D_i can be a $+6$, say, then it can be a -6 with the same probability when its distribution is symmetric about zero. Again, it does not matter which numbers are used. The Wilcoxon test uses ranks. The randomization test uses the observations themselves as numbers so that we may easily identify one of the assignments of signs as corresponding to the one actually obtained. □

The randomization test for matched pairs is discussed by Fisher (1935). The randomization test for two independent samples is presented by Pitman (1937/1938) along with a randomization test for correlation and an analysis of variance test.

A randomization test for multivariate data is presented by Chung and Fraser (1958). Further discussions of randomization tests may be found in articles by Welch (1937), Scheffé (1943), Moses (1952), Smith (1953), and Kempthorne (1955). A paper on multisample permutation tests is by Sen (1967b). Useful approximations to the distributions of the test statistics are discussed by Collier and Baker (1963, 1966), and Cleroux (1969). Other papers on Fisher's randomization tests are by Tsutakawa and Yang (1974), Oden and Wedel (1975), Boyett and Shuster (1977) and Soms (1977).

EXERCISES

1. A tire company did a follow-up study on ten customers, randomly selected from those who had purchased new tires from them three years earlier, and asked them how many times they had encountered tire failure from any cause, such as nails, valve leakage, etc. The study was restricted to two lines of long-life tires, called Brand A and Brand B. These were their results.

Customer	Brand A	Brand B
1	0	3
2	2	5
3	0	1
4	1	4
5	2	3

   Use Fisher's randomization method to get the exact p-value for testing the null hypothesis of equal likelihood for tire failure, against the one-sided alternative that Brand A tends to have fewer tire failures.

2. A random sample of eight adults were asked how old they were when they went on their first date. The three men responded with ages 15, 17, 16, while the five women answered 12, 14, 15, 10, and 12. Test the hypothesis that the average is the same for both sexes against the alternative that girls tend to be younger on the occasion of their first date.

3. The number of customers served by each of two salespersons is observed for each hour. The differences $Y_i - X_i$ are noted, where Y_i and X_i represent the number of customers served by each salesperson. Test whether the median difference $Y_i - X_i$ may be considered to be zero, where the observed differences are $+7$, $+3$, $+2$, $+8$, -2, $+3$, $+4$, and -1.

4. Two highway patrolmen kept track of the numbers of traffic tickets they wrote, Y_i and X_i, for seven days. The paired observations on (X_i, Y_i) are (17, 14), (15, 14), (12, 15), (9, 7), (17, 16), (18, 18), and (14, 10). Is the median of $Y_i - X_i$ zero?

PROBLEMS

1. Suppose someone suggests subtracting a constant from all of the observations in the randomization test for two independent samples to make the calculations easier, such as subtracting 10 from each observation in Exercise 2 before analyzing the data. Does this affect the results of the test? Explain. Would division of the observations by a constant affect the results?

2. Would the results of the randomization test for matched pairs be affected by subtracting a constant from all of the observations or by division of the data by a constant? Explain.

3. In the randomization test for correlation the critical region is determined, as in the rank correlation tests, by assuming that each pairing of Xs with Ys is equally likely, where the data consist of a bivariate sample (X_i, Y_i), $i = 1, 2, \ldots, n$. Explain how to find the p quantile w_p of the test statistic $T_3 = \sum_{i=1}^{n} X_i Y_i$ under the null hypothesis of independence between the Xs and the Ys.

5.12 SOME COMMENTS ON THE RANK TRANSFORMATION

Most of the nonparametric procedures presented in this chapter are examples of procedures that arise by applying the rank transformation to the data (i.e., replacing the data by their ranks) and then using the usual parametric procedure, but on the ranks instead of on the data. The most obvious use of the rank transformation is in Section 5.4, where the usual product moment correlation coefficient known as Pearson's r is applied to ranks and called Spearman's ρ. In Section 5.6 the usual least squares regression line is found for the ranks rather than for the original data.

Other nonparametric procedures, such as the Mann-Whitney test, the Wilcoxon signed ranks test, and the Kruskal-Wallis test, are less obvious applications of the rank transformation. Let us examine the Mann-Whitney test.

The Mann-Whitney test, as explained in Section 5.1, examines two random samples to see if the populations, from which the two samples were drawn, have equal means. If the populations are normal, the two-sample t test is the most powerful test. It compares the statistic t, given by Equation 5.1.17, with the student's t distribution, whose quantiles are given in Table A21, using $N - 2$ degrees of freedom.

The nonparametric Mann-Whitney test does not assume normality of the populations, and it compares the statistic T_1, given by Equation 5.1.2, with the normal distribution as an approximation to the exact distribution.

What would be the result if we compared the t statistic (Equation 5.1.17) on the ranks instead of on the data? It turns out this procedure is equivalent to the Mann-Whitney test, just using a different approximate distribution. In other words, let t_R be the t statistic of Equation 5.1.17 computed on the ranks of the observations instead of on the observations themselves. The two-sample t statistic, when computed on the ranks of the Xs and Ys, becomes

$$t_R = \frac{T_1}{\sqrt{\frac{N-1}{N-2} - \frac{1}{N-2}T_1^2}} \tag{1}$$

Note that as T_1 gets larger, t_R gets larger, and as T_1 gets smaller, t_R gets smaller. This means that a test that rejects the null hypothesis when T_1 is too large or too small achieves identical results as the test that rejects the null hypothesis when t_R is too large or too small. So the two tests, the Mann-Whitney test and the rank transformation procedure, are exactly equivalent. To find the exact 0.95 quantile of t_R merely substitute the exact 0.95 quantile of T_1 into Equation 1. The result will not be exactly the same as the 0.95 quantile from the t distribution (Table A21) with $N - 2$ degrees of freedom, but the latter value serves as a good approximation to the exact value when the exact value is not known. So tests based on T_1 and t_R are equivalent tests.

In Section 5.2 the Kruskal-Wallis test statistic T is given by Equation 5.2.3, and the F statistic used in the one-way analysis of variance is given by Equation 5.2.19. The F statistic computed on the ranks of the observation is a function of T:

$$F_R = \frac{T/(k-1)}{(N-1-T)/(N-k)} \tag{2}$$

and because F_R increases or decreases as T increases or decreases, the rank transformation procedure is equivalent to the Kruskal-Wallis test. (See Problem 5.2.5.)

Let us now look at the Wilcoxon signed ranks test. This test applies to a random sample of differences $D_1, D_2, \ldots, D_n$ to test the null hypothesis that the mean $E(D_i) = 0$. The usual parametric procedure assumes that the D_is are a random sample from a normal distribution and rejects the null hypothesis when the t statistic, given by Equation 5.7.18, is too large or too small. This is called the one-sample t test and was mentioned earlier in Section 5.7. For the Wilcoxon signed ranks test, the D_is are replaced by signed ranks (see Section 5.7 for further details) R_1 to R_n, and the null hypothesis is rejected when T, given by Equation 5.7.5, is too large or too small. The rank transformation procedure would suggest computing the statistic t on the signed ranks to obtain a new test. However, the test is not new, because the one-sample t statistic computed on the signed ranks, t_R say, is merely a function of the Wilcoxon signed ranks statistic T, as in the following equation. (See Problem 5.7.3.)

$$t_R = \frac{T}{\left(\frac{n}{n-1} - \frac{1}{n-1}T^2\right)^{1/2}} \tag{3}$$

Again, large values of T correspond to large values of t_R, and small values of T correspond to small values of t_R, so the two tests are equivalent.

These examples and others in this chapter, such as the test for slope in Section 5.5, the Friedman test, and the Durbin test, illustrate the idea of applying the usual parametric test statistic or its equivalent to the ranks of the observations to get a nonparametric procedure with high efficiency in most situations. Of course, the trick in each case is to rank the observations in such a way that all possible rankings are equally likely under the null hypothesis. This technique has been used successfully by Worsley (1977) in cluster analysis and by Shirley (1977) for contrasting increasing dose levels of a treatment.

Nonparametric tests that are equivalent to parametric tests computed on the ranks are easily computed using computer programs designed for parametric tests. Simply rank the data and use the parametric test on the ranks in situations where programs for the nonparametric tests are not readily available. The parametric test automatically corrects for ties, and usually the approximate p-value is as good or better than the usual normal or chi-squared approximation.

In situations where such a method of ranking is not possible, or quite difficult, the principle of the rank transformation may still be useful. Two areas of statistics in which the rank transformation is useful even though it does not result in nonparametric procedures are the areas of experimental design and multiple regression.

To analyze an experimental design using the rank transformation, first rank all of the observations together from smallest to largest and then apply the usual analysis of variance to the ranks. The result is a procedure that is only conditionally distribution-free.

That is, the exact distribution of the test statistic can be obtained as in a rank test, but the distribution is different for each different configuration of ranks. This means it is not practical to find the exact distribution, so the large sample approximation is, in many cases, the same F distribution that is used in the parametric test.

The rank transformation works well in a two-way layout without interaction (see Iman, Hora and Conover, 1984 and Hora and Iman, 1988) where it compares well with the Friedman test, the Quade test, and the parametric F test. However, attempts to apply the rank transformation procedure to test for interaction have met with mixed results, showing that it has good robustness and power by Iman (1974b), Conover and Iman (1976), and Pavur and Nath (1986), but poor robustness and power by Blair, Sawilowsky, and Higgins (1987). A theoretical study by Thompson (1991) shows a flaw in the rank transformation test for interaction showing conclusively that it is not a valid procedure and shouldn't be used. However, Mansouri and Chang (1995) used normal scores instead of ranks and found no problems with the normal scores transformation test for interaction, so the conversion to normal scores might correct the flaw found by Thompson.

The recommended procedure in experimental designs for which no nonparametric test exists is to use the usual analysis of variance on the data and then to use the same procedure on the rank transformed data. If the two procedures give nearly identical results the assumptions underlying the usual analysis of variance are likely to be reasonable and the regular parametric analysis valid. When the two procedures give substantially different results, the experimenter may want to take a closer look at the data and to look especially for outliers (observations that are unusually large compared with the bulk of the data) or very nonsymmetric distributions. These aberrations in the data affect the analysis of the data to a great extent by changing the level of significance and decreasing the power, but the analysis on the ranks is not affected nearly as much. The rank transformation is used in experimental designs by Crouse (1967), Lemmer and Stoker (1967), Crouse (1968), Macdonald (1971), Scheirer, Ray, and Hare (1976), and Hamilton (1976). See also more recent studies by Akritas (1990 and 1991).

In multiple regression each variable is ranked separately, as in the bivariate regression procedure of Section 5.5. Then the usual regression methods are applied to the ranks. The result is a robust regression method that is not sensitive to outliers or nonnormal distributions to the extent that the regular regression methods on the data are affected. As before, the recommended procedure is to analyze the data, analyze the ranks, and interpret the results in the light of both analyses. Prediction of the dependent variable may be accomplished as in Section 5.6 by predicting the rank from a regression equation and interpolating among the known values of the dependent variable. Examples of this procedure are given by Iman and Conover (1979).

Application of the rank transformation in discriminant analysis results in methods that are both simple to use and powerful in classifying observations. Briefly, each variable is ranked separately, and the popular linear discriminant function or quadratic discriminant function is computed on the ranks. A more complete discussion of this method and extensive Monte Carlo power comparisons are given by Conover and Iman (1978b and 1980).

Other areas of statistics are just as fertile for application of the rank transformation. These methods are usually not distribution-free. However, they seem to be more robust and often more powerful than the standard procedures when the assumptions behind the standard procedures are not reasonable. See Hettmansperger and McKean (1978) for a general discussion of the use of ranks. Other robust procedures, not necessarily based on ranks, are receiving a lot of attention currently. Some important references for these robust procedures include Huber (1972) and Hogg (1977). These methods are discussed by Labovitz (1970) and Allan (1976). The paper by Kim (1975) contains many references on the subject.

5.13   REVIEW PROBLEMS FOR CHAPTER 5

1. The state highway commission wants to buy a good grade of paint for painting lines on the highways. The choice has narrowed to two brands of paint. Twenty stripes are painted across a short stretch of highway. Ten of these stripes are painted with Brand A and the other 10 stripes with Brand B, in a random order. The stripes are inspected after 6 months and ranked according to wear. The results are as follows.

Ranks

Brand A 2, 3, 4, 6, 8, 9, 10, 12, 13, 14
Brand B 1, 5, 7, 11, 15, 16, 17, 18, 19, 20

Is there a significant difference between Brands A and B?

2. Gold rings in a jewelry store can be measured using two different scales. Scale A is electronic, and Scale B is a mechanical balance beam type. To see if Scale B tends to give a higher weight than Scale A, seven rings are measured using both scales, with the following results. Are the weights using Scale B significantly greater than the weights using Scale A?

Ring	Scale A	Scale B
1	22.6	22.9
2	13.8	14.3
3	19.0	19.1
4	26.5	26.4
5	24.9	25.2
6	16.0	16.4
7	23.3	23.4

(a) Use Fisher's randomization test.

(b) Use Wilcoxon's signed ranks test.

3. Ten golfers agreed to test a new type of golf ball in a tournament. Five golfers were selected at random from the 10 to try the new type of ball, and the other 5 were supplied with the old type of ball. The results after four rounds were as follows.

Scores with New Ball	295	301	288	290	289
Scores with Old Ball	302	306	292	306	314

(a) Do these results provide convincing evidence that the new type of ball tends to produce lower scores?

(b) What other statistical methods could you have used in part a? What are the main advantages and disadvantages of each, including the method you used?

4. While waiting for a customer, a caddy saw eight golfers finish their round of golf, pay their caddies, and leave. He estimated the age of each golfer and noted how much they paid their caddies.

				Golfer				
	1	2	3	4	5	6	7	8
Age (Estimated)	32	30	33	41	43	47	28	30
Amount Paid	10.00	11.50	9.00	12.00	16.00	17.00	8.75	10.50

(a) Does there seem to be a tendency for older golfers to pay their caddies more?

(b) What other statistical methods could you have used in part a? What are the main advantages and disadvantages of each, including the method you used?

5. Two racehorse trainers are comparing the results of their last five horses to see which trainer is best at teaching a horse to run faster. The first trainer gave the following times for running a quarter mile.

	Horse				
	1	**2**	**3**	**4**	**5**
Before Training	26.3	24.1	27.6	25.3	26.8
After Training	23.3	22.0	24.1	22.8	23.0

The second trainer gave the results for his last five horses.

	Horse				
	1	**2**	**3**	**4**	**5**
Before Training	25.4	26.2	24.0	26.0	27.7
After Training	23.6	23.9	21.8	23.6	25.7

Test the hypothesis that the two trainers are equally adept at training a horse to run faster.

6. To see if any further training was necessary, one particular horse was clocked on a quarter-mile distance each morning for 10 consecutive days with the following results.

Day	1	2	3	4	5	6	7	8	9	10
Speed (Seconds)	22.2	22.8	21.0	21.4	22.4	21.9	22.0	22.6	21.8	21.1

Do these results indicate that the horse is still improving?

7. Eighteen high school students, selected at random, were given a conduct rating X, where $X = 10$ represents a perfect score, and an achievement rating Y, where $Y = 20$ represents satisfactory achievement in each of 20 areas.

	Student								
	1	**2**	**3**	**4**	**5**	**6**	**7**	**8**	**9**
X	1.8	8.9	8.3	4.0	8.8	9.2	9.5	8.1	5.3
Y	11	17	16	10	16	17	20	16	11

	Student								
	10	**11**	**12**	**13**	**14**	**15**	**16**	**17**	**18**
X	7.3	7.7	6.8	7.9	8.8	9.9	9.0	9.3	9.2
Y	14	15	12	14	17	20	18	19	18

(a) Is there a significant positive correlation between X and Y?

(b) Find the least squares regression line.

(c) Find a 95% confidence interval for the slope of the least squares regression line.

(d) Estimate the regression curve $E(Y|X)$ using rank regression.

(e) Draw a graph, showing the data points, the least squares regression line, and the estimate of the monotonic regression curve using rank regression. Which estimate of the regression seems to agree better with the data?

8. A random sample of men and women resulted in the following measurement of height (inches):

Men	Women
$X_1 = 70.1$	$Y_1 = 62.2$
$X_2 = 67.8$	$Y_2 = 64.7$
$X_3 = 71.6$	$Y_3 = 65.3$

Test the null hypothesis that the heights of men and women are identically distributed against the alternative hypothesis that men tend to be taller than women. Let the test statistic T equal the *largest rank* assigned to the Y values, where the ranks 1 to 6 are assigned to the combined sample of both men and women in order of increasing height.

(a) Find the probability distribution T under the null hypothesis.

(b) Find, and sketch a graph of, the distribution function of T under the null hypothesis.

(c) Obtain a reasonable critical region and find the level of significance.

(d) Test the null hypothesis using the preceding test.

(e) Test the null hypothesis using any other nonparametric test you have learned or can invent.

9.   At the beginning of the year a first-grade class was randomly divided into two groups. One group was taught to read using a uniform method, where all students progressed from one stage to the next at the same time, following the teacher's direction. The second group was taught to read using an individual method, where each student progressed at his or her own rate according to a programmed workbook, under supervision of the teacher. At the end of the year each student was given a reading ability test, with the following results.

First Group				Second Group			
227	55	184	174	209	271	63	19
176	234	147	194	14	151	184	127
252	194	88	248	165	235	53	151
149	247	161	206	171	147	228	101
16	99	171	89	292	99	271	179

(a) Test the null hypothesis that there is no difference in the two teaching methods against the alternative that the two population means are different.

(b) Test the null hypothesis of equal variances against the alternative that the variance of the second population is greater than the variance of the population that used the uniform method of learning to read.

10.   A certain grade school has 121 students. One semester the number of students with their number of absences was summarized as follows.

Number of Absences	0	1	2	3	4	5	6	7	8	More Than 8
Number of Students	54	32	10	4	5	5	3	0	1	7
(Boys)	28	15	4	2	2	2	1	0	0	3
(Girls)	26	17	6	2	3	3	2	0	1	4

(a) Discuss the concepts of "target population" and "sampled population" as they apply to this problem.

(b) Are girls less likely than boys to have perfect attendance records?

(c) Do girls in general tend to have more absences than boys?

11.   Seven judges were asked to rank the five finalists in a local talent contest. The ranks went from 1 for the best to 5 for the worst. The results were as follows.

Performer

Judge	A	B	C	D	E
1	5	2	1	3	4
2	5	1	2	3	4
3	3	1	2	4	5
4	2	3	4	1	5
5	3	1	2	5	4
6	4	1	2	3	5
7	4	2	3	1	5

May the null hypothesis of random assignment of ranks be rejected?

12. A hairbrush marketing company wishes to choose one of five different styles of hairbrushes to market. As part of their analysis, they test the consumer preference of these brushes by selecting 10 girls from senior high school. Each of these 10 girls is given two styles of hairbrush to use for one month and is then asked to report her preference. For simplicity the different styles of hairbrushes are called A, B, C, D, and E. The results are as follows.

Alice prefers B over A	Fawn prefers B over D
Betty prefers A over C	Greta prefers E over B
Charlene prefers D over A	Heather prefers D over C
Donna prefers E over A	Inga prefers E over C
Ellen prefers B over C	Jean prefers E over D

Is there a significant difference in preferences? If so, which brushes are significantly different?

13. The rate of return on investment in several common stocks over a period of time is figured by taking the market price of each stock at the end of the time period plus any dividends that were paid during the time period and dividing the result by the price of the stock at the beginning of the time period. The rate of return is recorded here for several stocks over nine 3-month periods. Does there seem to be a significant difference in the rate of return for the different stocks?

Stock

Period	A	B	C	D	E
1	1.022	1.018	1.031	1.009	1.018
2	0.996	0.998	1.021	0.981	0.992
3	1.001	0.993	0.998	1.010	1.008
4	1.064	1.073	1.020	1.051	1.061
5	1.013	1.009	1.026	1.042	1.000
6	1.113	1.126	1.088	1.141	1.103
7	0.998	0.992	1.012	1.002	0.977
8	0.993	1.004	1.010	0.998	0.987
9	1.061	1.020	0.999	1.031	1.040

14. In another part of the same study as in Problem 13, the total rate of return over the 9-quarter period was calculated for 40 stocks. These 40 stocks were selected to represent four different types of industry, 10 stocks in each type of industry.

27-Month Rate of Return

Industry Type

Stock	A	B	C	D
1	1.062	1.060	1.101	1.003
2	1.021	1.001	.981	1.067
3	1.000	1.124	1.173	1.084
4	1.316	.961	1.126	1.049
5	1.177	1.054	1.002	1.056
6	1.289	1.048	.964	1.012
7	1.405	1.113	1.142	1.008
8	1.566	1.147	1.226	1.051
9	1.304	1.067	1.184	1.058
10	1.111	1.073	1.098	1.042

Does there seem to be a significant difference in the rate of return for stocks in the four types of industry?

15. A rural appraiser has kept a record of prices paid for all plots of land, 20 acres or more, sold in the vicinity of a certain town for the last year. She has reduced the data from each sale to two variables, X = the distance from the city limits and Y = the price per acre.

Parcel

	1	2	3	4	5	6	7
X (Miles)	12.1	4.8	13.9	1.6	17.4	7.5	19.9
Y (Dollars per Acre)	280	590	163	530	157	394	177

Parcel

	8	9	10	11	12	13	14	15
X (Miles)	21.8	2.4	5.8	2.3	12.8	25.6	8.8	7.3
Y (Dollars per Acre)	110	620	492	761	210	115	245	334

She is asked to suggest a fair market price for a parcel of land located 4.4 miles from the city limits. Taking into consideration only the preceding information, what should be the price per acre?

16. Six people were studied to see if their resting heartbeat rate was higher in the morning than it was in the evening. These are the results.

Person	Morning	Evening
1	78	73
2	86	81
3	64	64
4	74	73
5	74	69
6	72	71

(a) Use a normal scores type test to analyze these data.

(b) Find 90% confidence interval for slope, where Y = morning heartbeat and X = evening heartbeat.

17. A random sample of locally owned gas stations had the following numbers of employees:

$$X \quad 4, 5, 7, 12$$

A random sample of chain-owned gas stations had the following numbers of employees:

$$Y \quad 9, 11, 15$$

(a) Use the sum of the ranks of the Xs to test $H_0: \mu_x = \mu_y$, versus $H_1: \mu_x < \mu_y$ at $\alpha = 0.05$. Find the *exact* p-value (not from the tables).
(b) Use the van der Waerden test.
(c) Use Fisher's randomization test.

18. Four different contractors manufacture one type of chemical detection kit. All kits produced by all of the contractors are supposed to respond the same to toxic gases. A test is performed to see if this is the case. Ten kits are randomly selected from lots manufactured by each of the four contractors. The forty kits are put in a gas chamber under laboratory conditions for a specified length of time, and are then compared. The responses of the kits are different shades of color in a patch. The colors are ranked from pink to dark purple, as follows.

Contractors

A	B	C	D
	Ranks		
19	18	25	7
10	5	3	20.5
4	28	32	23
38	1	29	13
33	15	6	16
36	12	2	9
39	27	30	8
40	31	35	14
37	20.5	34	17
26	22	24	11

Is there a difference in kits manufactured by the various contractors?

19. Twelve battles are fought between the Red Army and the Blue Army, using a computer simulation model. The numbers of casualties recorded by each army are given below, for each battle. (Continued on next page.)

Battle	Red	Blue
1	41	38
2	8	14
3	65	41
4	28	31
5	11	8
6	15	18
7	73	48

8	54	32
9	7	7
10	50	37
11	59	42
12	24	28

(a) Make a scatterplot of the data.

(b) Use Spearman's ρ as a measure of the strength of monotonic dependence.

(c) Use Kendall's τ to measure the strength of concordance among the data pairs.

(d) Use monotonic regression to estimate the mean number of Red Army casualties, when the Blue Army has 40 casualties.

20. Eight volunteers were recruited to test the efficacy of using a telescopic sight on a rifle. It is believed that the use of the telescopic sight will raise the test scores on a shooting range. To prove this, the eight volunteers were asked to use a rifle on a shooting range, both with a telescopic sight on the rifle and without a telescopic sight on the rifle, using a random alternating pattern between the two. Here are their results.

	Volunteer							
	1	**2**	**3**	**4**	**5**	**6**	**7**	**8**
With Telescopic Sight	96	93	89	88	85	83	80	77
Without Telescopic Sight	92	92	89	96	82	79	80	78

Do the telescopic sights result in higher test scores?

(a) Use a test based on only ranks. Find the p-value three different ways:
   (1) Find the exact p-value.
   (2) Use the normal approximation without the continuity correction.
   (3) Use the normal approximation with the continuity correction.

(b) Use a test based on normal scores.

(c) Use a Fisher-type randomization test.

(d) Find a 90% confidence interval for the improvement in test score obtained by using a telescopic sight on the rifle.

CHAPTER 6

STATISTICS OF THE KOLMOGOROV-SMIRNOV TYPE

PRELIMINARY REMARKS

In Chapter 2 the empirical distribution function was introduced as a function based on a random sample that may be used to estimate the true distribution function of the population. If we want to see if two or more samples are governed by the same unknown distribution, it seems natural to compare the empirical distribution functions of those samples to see if they look somewhat similar. To be precise, however, some measure of disparity between or among these functions is needed. Kolmogorov and Smirnov developed statistical procedures that use the maximum vertical distance between these functions as a measure of how well the functions resemble each other. Their methods and other methods that use the same idea are presented in this chapter.

6.1 THE KOLMOGOROV GOODNESS-OF-FIT TEST

We will begin this chapter with a test for goodness of fit that was introduced by Kolmogorov (1933). This test is perhaps the most useful of the tests in this chapter, partly because it furnishes us with an alternative, designed for ordinal data, to the chi-squared test for goodness of fit introduced in Section 4.5, which was designed for nominal type data, and partly because the Kolmogorov test statistic enables us to form a "confidence band" for the unknown distribution function, as we will explain in this section.

A test for goodness of fit usually involves examining a random sample from some unknown distribution in order to test the null hypothesis that the unknown distribution function is in fact a known, specified function. That is, the null hypothesis specifies some distribution function $F^*(x)$, perhaps graphically as in Figure 1, or perhaps as a mathematical function that may be graphed. A random sample $X_1, X_2, \ldots, X_n$ is then drawn from some population and is compared with $F^*(x)$ in some way to see if it is reasonable to say that $F^*(x)$ is the true distribution function of the random sample.

One logical way of comparing the random sample with $F^*(x)$ is by means of the empirical distribution function $S(x)$, which was defined by Definition 2.2.1 as the fraction of X_is that are less than or equal to x for each x, $-\infty < x < +\infty$. We learned in Section 2.2 that the empirical distribution function $S(x)$ is useful as an estimator of $F(x)$, the unknown distribution function of the X_is. So we can compare the empirical distribution function $S(x)$ with the hypothesized distribution function $F^*(x)$ to see if there is good agreement. If there is not good agreement, then we may reject the null hypothesis and conclude that the true but unknown distribution function, $F(x)$, is in fact not given by the function $F^*(x)$ in the null hypothesis.

But what sort of test statistic can we use as a measure of the discrepancy between $S(x)$ and $F^*(x)$? One of the simplest measures imaginable is the largest distance between the two graphs $S(x)$ and $F^*(x)$, measured in a vertical direction. This is the statistic suggested by Kolmogorov (1933). That is, if $F^*(x)$ is given by Figure 1 and a random sample of size 5 is drawn from the population, the empirical distribution function $S(x)$ may be drawn on the same graph along with $F^*(x)$, as shown in Figure 2. If $F^*(x)$ and $S(x)$ are as given, the maximum vertical distance between the two graphs occurs just before the third step of $S(x)$. This distance is about 0.5 in Figure 2; therefore the Kolmogorov statistic T equals 0.5 in this case. Large values of T as determined by Table A13 lead to rejection of $F^*(x)$ as a reasonable approximation to the unknown true distribution function $F(x)$.

The Kolmogorov test may be preferred over the chi-squared test for goodness of fit if the sample size is small; the Kolmogorov test is exact even for small

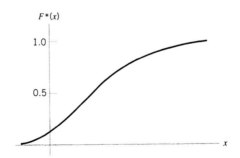

FIGURE I A hypothesized distribution function.

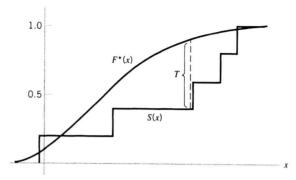

FIGURE 2   The hypothesized distribution function $F^*(x)$, the empirical distribution function $S(x)$, and Kolmogorov's statistic T.

samples, while the chi-squared test assumes that the number of observations is large enough so that the chi-squared distribution provides a good approximation as the distribution of the test statistic. There is controversy over which test is the more powerful, but the general feeling seems to be that the Kolmogorov test is probably more powerful than the chi-squared test in most situations involving ordinal data. For further comparisons see a paper by Slakter (1965).

The title of this chapter is "Statistics of the Kolmogorov-Smirnov Type." Statistics that are functions of the maximum vertical distance between $S(x)$ and $F^*(x)$ are considered to be Kolmogorov-type statistics. Statistics that are functions of the maximum vertical distance between two empirical distribution functions are of the Smirnov type. This chapter is concerned with statistics that are determined only by the vertical distances between distribution functions, either hypothesized or empirical distribution functions.

▶ **The Kolmogorov Goodness-of-Fit Test** _____

Data   The data consist of a random sample $X_1, X_2, \ldots, X_n$ of size n associated with some unknown distribution function, denoted by $F(x)$.

Assumption

  1. The sample is a random sample.

Test Statistic   Let $S(x)$ be the empirical distribution function based on the random sample $X_1, X_2, \ldots, X_n$. The test statistic is defined differently for the three different sets of hypotheses, A, B, and C. Let $F^*(x)$ be a completely specified hypothesized distribution function.

A. (Two-Sided Test) Let the test statistic T be the greatest (denoted by "sup" for supremum) vertical distance between $S(x)$ and $F^*(x)$. In symbols we say

$$T = \sup_x |F^*(x) - S(x)| \tag{1}$$

which is read "T equals the supremum, over all x, of the absolute value of the difference $F^*(x) - S(x)$."

B. (One-Sided Test) Denote this test statistic by T^+ and let it equal the greatest vertical distance attained by $F^*(x)$ above $S(x)$. That is,

$$T^+ = \sup_x [F^*(x) - S(x)] \tag{2}$$

which is similar to T except that we consider only the greatest difference where the function $F^*(x)$ is above the function $S(x)$.

C. (One-Sided Test) For this test use the test statistic T^-, defined as the greatest vertical distance attained by $S(x)$ above $F^*(x)$. Formally this becomes

$$T^- = \sup_x [S(x) - F^*(x)] \tag{3}$$

Null Distribution When $F(x)$ is continuous and the null hypothesis is true the exact distribution function of T^+ and T^- is given by

$$G(x) = 1 - x \sum_{j=0}^{[n(1-x)]} \binom{n}{j} \left(1 - x - \frac{j}{n}\right)^{n-j} \left(x + \frac{j}{n}\right)^{j-1} \tag{4}$$

where $[n(1 - x)]$ is the greatest integer less than or equal to $n(1 - x)$. This distribution is the same for T^+ and T^-. The asymptotic (as $n \to \infty$) distribution function of $\sqrt{n}T^+$ and $\sqrt{n}T^-$ is given by

$$H(x) = \lim_{n \to \infty} G\left(\frac{x}{\sqrt{n}}\right) = 1 - e^{-2x^2} \tag{5}$$

The approximate distribution function of T is

$$P(T \le x) \doteq [G(x)]^2 \tag{6}$$

because T is less than x only when both T^+ and T^- are less than x.

Exact quantiles for T in the two-sided test, and approximate quantiles for T^+ and T^- in the one-sided tests, are given in Table A13 for $n \le 40$. The asymptotic approximation is used for $n > 40$. Note that all of these tests are upper-tailed only. The designations "one-sided" and "two-sided" refer to the alternative hypothesis of interest, and the test statistics are redefined so that all three tests are upper-tailed.

Table A13 is exact only if $F(x)$ is continuous; otherwise these quantiles lead to a conservative test (Noether, 1967a). A method for finding the exact null distribution when $F(x)$ is discrete is described following Example 1.

Hypotheses

A. (Two-Sided Test)

$$H_0: F(x) = F^*(x) \quad \text{for all } x \text{ from } -\infty \text{ to } +\infty$$
$$H_1: F(x) \neq F^*(x) \quad \text{for at least one value of } x$$

Reject H_0 at the level of significance α if T exceeds the $1 - \alpha$ quantile as given by Table A13 for the two-tailed test. The approximate p-value can be found by interpolation in Table A13, or by using twice the one-tailed p-value given by

$$\text{one-tailed } p\text{-value} = t \sum_{j=0}^{[n(1-t)]} \binom{n}{j} \left(1 - t - \frac{j}{n}\right)^{n-j} \left(t + \frac{j}{n}\right)^{j-1} \tag{7}$$

where t is the observed value of the test statistic, and where $[n(1 - t)]$ is the greatest integer less than or equal to $n(1 - t)$.

B. (One-Sided Test)

$$H_0: F(x) \geq F^*(x) \quad \text{for all } x \text{ from } -\infty \text{ to } +\infty$$
$$H_1: F(x) < F^*(x) \quad \text{for at least one value of } x$$

Reject H_0 at the level of significance α if T^+ exceeds the $(1 - \alpha)$ quantile as given by Table A13 for the one-sided test. The approximate p-value can be found by interpolation in Table A13. The exact p-value can be found from Equation 7.

C. (One-Sided Test)

$$H_0: F(x) \leq F^*(x) \quad \text{for all } x \text{ from } -\infty \text{ to } +\infty$$
$$H_1: F(x) > F^*(x) \quad \text{for at least one value of } x$$

Reject H_0 at the level of significance α if T^- exceeds the $1 - \alpha$ quantile for a one-sided test as given by Table A13. The approximate p-value can be found by interpolation in Table A13. The exact p-value can be found from Equation 7.

Computer Assistance   *S-Plus* and *StatXact* calculate the Kolmogorov goodness-of-fit test. ◀

EXAMPLE 1

A random sample of size 10 is obtained: $X_1 = 0.621$, $X_2 = 0.503$, $X_3 = 0.203$, $X_4 = 0.477$, $X_5 = 0.710$, $X_6 = 0.581$, $X_7 = 0.329$, $X_8 = 0.480$, $X_9 = 0.554$, $X_{10} = 0.382$. The null hypothesis is that the distribution function is the uniform distribution function whose graph is given in Figure 3. The mathematical expression for the hypothesized distribution function is

$$
\begin{aligned}
F^*(x) &= 0 &&\text{if} &&x < 0 \\
&= x &&\text{if} &&0 \le x < 1 \\
&= 1 &&\text{if} &&1 \le x
\end{aligned}
\tag{8}
$$

Formally, the hypotheses are given by

$$
\begin{aligned}
H_0&: F(x) = F^*(x) &&\text{for all } x \\
H_1&: F(x) \ne F^*(x) &&\text{for at least one } x
\end{aligned}
$$

where $F(x)$ is the unknown distribution function common to the X_is and $F^*(x)$ is given by Equation 8.

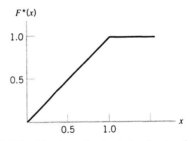

FIGURE 3   The hypothesized distribution function.

The two-sided Kolmogorov test for goodness of fit is used. The critical region of size $\alpha = 0.05$ corresponds to values of T greater than the 0.95 quantile 0.409, obtained from Table A13 for $n = 10$. The value of T is obtained by graphing the empirical distribution function $S(x)$ on top of the hypothesized distribution function $F^*(x)$, as shown in Figure 4. The largest vertical distance separating the two graphs in Figure 4 is 0.290, which occurs at $x = 0.710$ because $S(0.710) = 1.000$ and $F^*(0.710) = 0.710$. In other words,

$$
\begin{aligned}
T &= \sup_x |F^*(x) - S(x)| \\
&= |F^*(0.710) - S(0.710)| \\
&= 0.290
\end{aligned}
$$

Since $T = 0.290$ is less than 0.409, the null hypothesis is accepted. The p-value is seen, from Table A13, to be larger than 0.20.

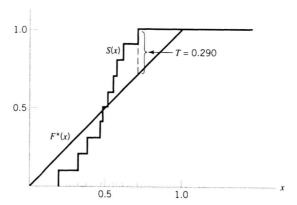

FIGURE 4   Graphs of $F^*(x)$ and $S(x)$, with T.

If we had wished to test the null hypothesis

$$H_0: F(x) \geq F^*(x) \qquad \text{for all } x$$

against the one-sided alternative

$$H_1: F(x) < F^*(x) \qquad \text{for some } x$$

the test statistic T^+ would have been used. The decision rule is to reject H_0 at $\alpha = 0.05$ if T^+ exceeds the 0.95 quantile for a one-sided test, 0.369, as given by Table A13 for $n = 10$. The value for T^+ in this case is computed just to the left of the second jump of $S(x)$.

$$T^+ = \sup_x [F^*(x) - S(x)]$$
$$= F^*(0.3289) - S(0.3289)$$
$$= 0.3289 - 0.100$$
$$= 0.2289$$

To be more precise, we should say that $T^+ = 0.228999. \ldots$, which is rounded off to 0.229. The end result is the same, with $p > 0.10$.

A one-sided test in the other direction would have resulted in

$$T^- = \sup_x [S(x) - F^*(x)]$$
$$= S(0.710) - F^*(0.710)$$
$$= 1.000 - 0.710$$
$$= 0.290$$

with $p > 0.10$.

The two-sided test is the appropriate test for this situation. The one-sided tests were presented merely to show how their test statistics are evaluated. In general, of course, the two-sided test statistic T always equals the larger of the two one-sided test statistics T^+ and T^-.

A more exact p-value in the two-sided test can be found by doubling the one-tailed p-value given by Equation 7.

$$p\text{-value} \doteq 2(0.29) \sum_{j=0}^{7} \binom{10}{j} \left(0.71 - \frac{j}{10}\right)^{10-j} \left(0.29 + \frac{j}{n}\right)^{j-1}$$

$$= 2(0.29)(0.112 + 0.117 + 0.101 + 0.081 + 0.061 + 0.040 + 0.017 + 0.000)$$

$$= 2(0.29)(0.530)$$

$$= 0.307 \qquad \blacksquare$$

A Method of Obtaining the Exact p-Value When $F^*(x)$ Is Discrete

If the hypothesized distribution function $F^*(x)$ is discrete and the conservative approximation for the p-value obtained from Table A13 is not satisfactory, the exact p-value may be obtained for a particular observed value of the test statistic. This computational procedure may be accomplished by hand for sample sizes of about 5 or less. A computer program such as *StatXact*, which uses the procedure described below, is recommended for larger sample sizes. The exact p-values for discrete distributions are often only about one-third as large as their approximations from Table A13. The following sections correspond to hypotheses A, B, and C given earlier.

A. (Two-Sided Test) Let t be the observed value of the test statistic T. Compute $P(T^+ \geq t)$ and $P(T^- \geq t)$ as described in parts B and C that follow, using t instead of t^+ and t^-. Then

$$P(T \geq t) \doteq P(T^+ \geq t) + P(T^- \geq t) \qquad (9)$$

is an approximation that is very close to the true p-value in most cases, unless t is small, in which case it may be greater than the true p-value.

B. (One-Sided Test) Let t^+ denote the observed value of T^+.

Step 1 Compute the probabilities f_j for $0 \leq j < n(1 - t^+)$ by drawing a horizontal line with ordinate $1 - t^+ - j/n$ directly on a graph of $F^*(x)$. Then $f_j = 1 - t^+ - j/n$ unless the horizontal line intersects $F^*(x)$ at a jump, in which case f_j equals the height of $F^*(x)$ at the bottom of the jump.

Step 2 Compute the constants $e_0, e_1, \ldots$, from the recursive relation $e_0 = 1$ and

$$e_k = 1 - \sum_{j=0}^{k-1} \binom{k}{j} f_j^{k-j} e_j \qquad k \geq 1 \tag{10}$$

for all k such that $f_k > 0$ in step 1. Note that these constants are of the form

$$e_0 = 1$$
$$e_1 = 1 - f_0$$
$$e_2 = 1 - f_0^2 - 2f_1 e_1$$
$$e_3 = 1 - f_0^3 - 3f_1^2 e_1 - 3f_2 e_2$$
$$e_4 = 1 - f_0^4 - 4f_1^3 e_1 - 6f_2^2 e_2 - 4f_3 e_3$$
$$e_5 = 1 - f_0^5 - 5f_1^4 e_1 - 10f_2^3 e_2 - 10f_3^2 e_3 - 5f_4 e_4$$

etc.

Step 3 Compute the exact one-sided p-value

$$P(T^+ \geq t^+) = \sum_{j=0}^{[n(1-t^+)]} \binom{n}{j} f_j^{n-j} e_j \tag{11}$$

from the f_j and e_j of steps 1 and 2.

C. (One-Sided Test) Let t^- denote the observed value of T^-.

Step 1 Compute the probability c_j for $0 \leq j < n(1 - t^-)$ as follows. Draw a horizontal line with the ordinate $t^- + j/n$ directly on a graph of $F^*(x)$. Then $c_j = 1 - t^- - j/n$ unless the horizontal line intersects $F^*(x)$ at a jump of $F^*(x)$. In that case $c_j = 1.0$ minus the height of $F^*(x)$ at the top of the jump.

Step 2 Compute the constants $b_0, b_1, \ldots$, from the recursive relationship $b_0 = 1$ and

$$b_k = 1 - \sum_{j=0}^{k-1} \binom{k}{j} c_j^{k-j} b_j \qquad k \geq 1 \tag{12}$$

for all k such that $c_k > 0$ in step 1. These constants follow the same pattern as the e_ks in part B, with the f_js replaced by c_js.

Step 3 Compute the exact one-sided p-value

$$P(T^- \geq t^-) = \sum_{j=0}^{[n(1-t^-)]} \binom{n}{j} c_j^{n-j} b_j \tag{13}$$

from the c_j and b_j of steps 1 and 2.

The following example illustrates the method of computing the exact p-value when $F^*(x)$ is discrete.

EXAMPLE 2

Let $F^*(x)$ be the discrete uniform distribution with equal probabilities $1/5$ at the five points $x = 1, 2, 3, 4, 5$. Suppose a random sample of size 10 with the (ordered) values 1, 1, 1, 2, 2, 2, 3, 3, 3, 3, is drawn from some population and the null hypothesis is that $F^*(x)$ is the population distribution function. The greatest distance between $F^*(x)$ and $S(x)$ occurs at $x = 3$ (see Figure 5), so the test statistic for the two-sided Kolmogorov test becomes

$$T = \sup_x |F^*(x) - S(x)| = 0.4 = t \tag{14}$$

To find the p-value associated with $t = 0.4$ first the probability $P(T^+ \geq 0.4)$ is computed.

Step 1 Because $n(1 - t) = 10(0.6) = 6$, the probabilities f_0 to f_5 need to be computed. The horizontal line with ordinate $1 - t = 0.6$ intersects $F^*(x)$ directly at the top of the jump at $x = 3$, so f_0 equals the ordinate of the horizontal line: $f_0 = 0.6$. For $j = 1$, the horizontal line $1 - t - 1/10 = 0.5$ intersects $F^*(x)$ at a jump, so f_1 equals the height of $F^*(x)$ at the bottom of the jump: $f_1 = 0.4$. Similarly, we find $f_2 = 0.4$, $f_3 = 0.2$, $f_4 = 0.2$, and $f_5 = 0$.

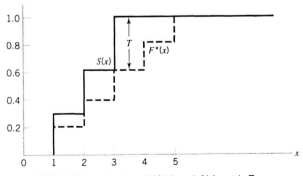

FIGURE 5 Graphs of $F^*(x)$ and $S(x)$, with T.

Step 2   The constants e_0 to e_4 are computed from Equation 10.

$e_0 = 1$
$e_1 = 1 - 0.6 = 0.4$
$e_2 = 1 - (0.6)^2 - 2(0.4)(0.4) = 0.32$
$e_3 = 1 - (0.6)^3 - 3(0.4)^2(0.4) - 3(0.4)(0.32) = 0.208$
$e_4 = 1 - (0.6)^4 - 4(0.4)^3(0.4) - 6(0.4)^2(0.32) - 4(0.2)(0.208) = 0.2944$

Step 3   The one-sided p-value $P(T^+ \geq t)$ is computed from Equation 11.

$$P(T^+ \geq t) = f_0^{10} + \binom{10}{1}f_1^9 e_1 + \binom{10}{2}f_2^8 e_2 + \binom{10}{3}f_3^7 e_3 + \binom{10}{4}f_4^6 e_4$$

$$= 0.021$$

Because $F^*(x)$ is symmetric, computation of the other one-sided p-value $P(T^- \geq 0.4)$ is identical with the preceding, so $P(T^- \geq 0.4) = 0.021$ and the p-value for the two-sided Kolmogorov test is approximately

$$P(T \geq 0.4) \doteq 2(0.021) = 0.042$$

It is interesting to note that this p-value shows that the correct decision is to reject the null hypothesis at $\alpha = 0.05$, while the use of Table A13 leads to the erroneous acceptance of $F^*(x)$ as the true distribution function at the same α level.   ∎

Comment

One of the most valuable features of the Kolmogorov two-sided test statistic is that its $1 - \alpha$ quantile $w_{1-\alpha}$ may be used to form a confidence band for the true unknown distribution function. Recall that in finding a confidence interval for some unknown parameter, we first drew a random sample and then, from that sample, computed an upper value U and a lower value L that contained the unknown parameter between them with a certain probability $1 - \alpha$, called the confidence coefficient. It would be convenient if we could do the same thing to obtain a "confidence band" within which the entire unknown distribution function would lie, with probability $1 - \alpha$. Then we could draw a random sample for some population whose distribution function is completely unknown, and we could place some bounds on a graph and make the statement that the unknown distribution function lies entirely within those bounds, with some probability $1 - \alpha$ that the statement is correct.

▶ Confidence Band for the Population Distribution Function _____

Data   The data consist of a random sample $X_1, X_2, \ldots , X_n$ of size n associated with some unknown distribution function, denoted by $F(x)$.

Assumptions

1. The sample is a random sample.
2. For the confidence coefficient to be exact, the random variables should be continuous. If the random variables are discrete, the confidence band is conservative; that is, the true but unknown confidence coefficient is greater than the stated one.

Method   Draw a graph of the empirical distribution function $S(x)$ based on the random sample. To form a confidence band with a confidence coefficient $1 - \alpha$, find the $1 - \alpha$ quantile of the Kolmogorov test statistic from Table A13 for the two-sided test (if a two-sided confidence band is desired) and for the appropriate sample size n. Let $w_{1-\alpha}$ denote this quantile. Draw a graph above $S(x)$ a distance $w_{1-\alpha}$ and call this graph $U(x)$. Draw a second graph a distance $w_{1-\alpha}$ below $S(x)$ and call this second graph $L(x)$. Then $U(x)$ and $L(x)$ form the upper and lower boundaries, respectively, of a $1 - \alpha$ confidence band that contains the unknown $F(x)$ completely within its boundaries.

There is no reason for $U(x)$ to be drawn above 1.0 even though $S(x) + w_{1-\alpha}$ might exceed 1.0, because we know that no distribution function ever exceeds 1.0. For the same reason $L(x)$ should not extend below the horizontal axis. The formal mathematical definitions of $U(x)$ and $L(x)$ are as follows.

$$
\begin{aligned}
U(x) &= S(x) + w_{1-\alpha} && \text{if} && S(x) + w_{1-\alpha} \leq 1 \\
U(x) &= 1.0 && \text{if} && S(x) + w_{1-\alpha} > 1 && (15) \\
L(x) &= S(x) - w_{1-\alpha} && \text{if} && S(x) - w_{1-\alpha} \geq 0 \\
L(x) &= 0 && \text{if} && S(x) - w_{1-\alpha} < 0 && (16)
\end{aligned}
$$

The resulting probability statement is

$$P[L(x) \leq F(x) \leq U(x), \text{ for all } x] \geq 1 - \alpha \qquad (17)$$

where the last inequality applies only when the random variables are discrete.

EXAMPLE 3

Suppose we wish to form a 90% confidence band for an unknown distribution function $F(x)$. A random sample of size 20 is obtained from the population with that distribution function. The results are ordered from smallest to largest for convenience.

16.7	17.4	18.1	18.2	18.8	19.3	22.4	22.4	24.0	24.7
25.7	27.0	35.1	35.8	36.5	37.6	39.8	42.1	43.2	46.2

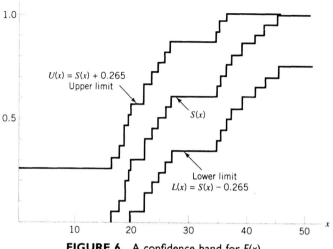

FIGURE 6   A confidence band for $F(x)$.

The 0.90 quantile is found from Table A13 to equal $w_{0.90} = 0.265$ for $n = 20$. The confidence band is $S(x) \pm 0.265$ as long as the band is between 0 and 1. Figure 6 shows $S(x)$, $U(x)$, and $L(x)$. The statement "$F(x)$ lies entirely between $U(x)$ and $L(x)$" is true with probability 0.90.   ∎

The derivation of the distribution of the Kolmogorov statistic is complicated and is not presented here. We will mention only some basic papers on the subject.

The asymptotic distribution of the two-sided statistic T was found by Kolmogorov (1933) and was tabulated by Smirnov (1948). The asymptotic distributions of the one-sided statistics T^+ and T^- were obtained by Smirnov (1939). The exact distribution of the test statistics for finite (small) sample sizes was studied by Wald and Wolfowitz (1939) and tabulated by Massey (1950a). The distribution function of T^- for finite sample sizes was derived by Birnbaum and Tingey (1951), and comparisons were made between the exact quantiles obtained from their distribution function and the asymptotic quantiles given by Smirnov (1939, 1948). It was found that use of the asymptotic quantiles leads to a conservative test.

The two-sided Kolmogorov test has the desirable property of being consistent against *all* differences between $F(x)$ and $F^*(x)$, the true and hypothesized distribution functions. However, it is biased for finite sample sizes (Massey, 1950b). A lower bound for the power of the two-sided test is given by Massey (1950b). The greatest lower bound for the power, under a certain class of alternative hypotheses, was obtained by Birnbaum (1953), and another greatest lower bound for the power, under a different class of alternative hypotheses, was obtained by Lee (1966).

Other power comparisons were made by van der Waerden (1953), Suzuki (1968), Shapiro, Wilk, and Chen (1968), and Knott (1970). Other papers on the Kolmogorov test and similar goodness-of-fit tests are by Finkelstein and Schafer

(1971), Maag and Dicaire (1971), Carnal and Riedwyl (1972), and Stephens (1974). Barr and Davidson (1973) and Pettitt and Stephens (1976) present modifications for censored data, while Barr and Shudde (1973) discuss a modification for observations on a circle. Govindarajulu and Klotz (1973) present a note on the asymptotic distribution. Estimating and testing symmetric distributions is the topic of papers by Schuster and Narvarte (1973), Schuster (1973), and Srinivasan and Godio (1974).

The modification for discrete distributions was developed independently by Conover (1972) and Coberly and Lewis (1973). Further analysis of this procedure appears in papers by Horn and Pyne (1976), Horn (1977), Bartels, Horn, Liebetrau, and Harris (1977), and Pettitt and Stephens (1977), who also present some tables. Maag, Streit, and Drouilly (1973) discuss goodness-of-fit-tests for grouped data. Wood and Altavela (1978) suggest using simulation techniques for large samples.

Another goodness-of-fit test is the Cramér-von Mises test, developed by Cramér (1928), von Mises (1931), and Smirnov (1936). Although it has more intuitive appeal than the Kolmogorov test to many people, there is not sufficient difference between the two tests to warrant its presentation here. The interested reader may find the asymptotic distribution of the Cramér-von Mises test given by Anderson and Darling (1952) and exact tables for finite sample sizes given by Stephens and Maag (1968). Earlier studies on this test and the Kolmogorov test are by Stephens (1964, 1965a), Tiku (1965), Suzuki (1967), Cronholm (1968), and Noé and Vandewiele (1968). The effect of discreteness (ties) on the two tests is discussed by Walsh (1960, 1963). Bias and power of the Cramér-von Mises test are examined by Thompson (1966). Relative efficiency of the Kolmogorov test is studied in Gelzer and Pyke (1965), Quade (1965), and Abrahamson (1967).

Goodness of fit for a sample density is discussed by Woodroofe (1967), for a circle by Stephens (1969), and in general by Riedwyl (1967). A different type of confidence interval for the distribution function is introduced by Durbin (1968).

EXERCISES

1. Five fourth-grade children were selected at random from the entire class and timed in a short race. The times in seconds were 6.3, 4.2, 4.7, 6.0, and 5.7. Give a 90% confidence band, either graphically or in tabled form, for the distribution function of times for all fourth-grade children in the class.

2. As a rural grocery store receives eggs from the neighboring farmers it "candles" the eggs to detect any eggs that are not fresh. Eight crates of eggs, 144 eggs per crate, were candled with the following numbers of eggs rejected from each crate: 4, 0, 2, 0, 2, 0, 2, 0. Present a 95% confidence band, either graphically or in tabled form, for the distribution function of the number of rejected eggs for the population of all crates received.

3. For the data in Exercise 1, test the hypothesis that the distribution of times is uniform on the interval from 4 to 8 seconds. Note that such a distribution is given by

$$F^*(x) = 0 \qquad \text{for} \quad x < 4$$
$$= (x-4)/4 \qquad \text{for} \quad 4 \le x < 8$$
$$= 1 \qquad \text{for} \quad 8 \le x$$

4. Previous records have indicated that the number of rejected eggs per crate follows the Poisson distribution with mean 1.5. For the data in Exercise 2 test the hypothesis that these eight crates came from the same distribution function. Note that the Poisson distribution with mean 1.5 has the following probabilities: $P(0) = 0.223$, $P(1) = 0.335$, $P(2) = 0.251$, $P(3) = 0.126$, $P(4) = 0.047$, $P(5) = 0.014$, and $P(6) = 0.004$.

5. An Olympic diver is rated on ten practice dives, with the following measurements: 1.7, 5.3, 7.6, 8.9, 9.0, 9.1, 9.3, 9.6, 9.9, and 9.9. Test the hypothesis that the distribution function of her scores is given by $F(x)$, where

$$F(x) = 0 \qquad \text{if } x < 0$$
$$F(x) = x^2/100 \qquad \text{if } 0 \le x \le 10$$
$$F(x) = 1 \qquad \text{if } 10 < x$$

6. The emissions of nitrous oxide from last year's model of automobile have been measured for thousands of cars and found to be approximately normal with mean 5.6 and standard deviation 1.2. Twelve of this year's model automobile have been tested with the results

$$4.8, 6.2, 6.0, 5.9, 6.6, 5.5, 5.8, 5.9, 6.3, 6.6, 6.2, 5.0$$

Does this year's model appear to have the same distribution as last year's model?

PROBLEM

1. Show that the confidence band given by Equation 17 is valid. That is, show that if $w_{1-\alpha}$ is a $1 - \alpha$ quantile of the Kolmogorov statistic, it follows that Equation 17 is also true.

6.2 GOODNESS-OF-FIT TESTS FOR FAMILIES OF DISTRIBUTIONS

The Kolmogorov goodness-of-fit test presented in Section 6.1 is a good test to use to see if a random sample agrees with some specified distribution function. The Kolmogorov test is intended for use only when the hypothesized distribution function is completely specified, that is, when there are no unknown parameters that must be estimated from the sample. Otherwise the test becomes conservative. The chi-squared goodness-of-fit test is flexible enough to allow for some parameters to be estimated from the data. One degree of freedom is simply subtracted for each parameter estimated in the "minimum chi-squared" manner described earlier. However, the chi-squared test requires that the data be grouped, and such a grouping of data is usually arbitrary. Also, the distribution of the test statistic is known only approximately, and sometimes the power of the chi-squared test is not very good. For these reasons other goodness-of-fit tests are sought, especially for frequently tested distributions.

The Kolmogorov test has been modified to allow it to be used in several situations where parameters are estimated from the data. Actually, the test statistic remains unchanged, but different tables of critical values are used. These tables are no longer the same for all distributions; they change from one hypothesized distribution to another. The test is still a nonparametric test because the validity of the test (the α level) does not depend on untested assumptions regarding the population distribution; instead, the population distributional form is the hypothesis being tested.

The first such modification is the Kolmogorov test as modified to test the composite hypothesis of normality. That is, the null hypothesis states that the population is one of the family of normal distributions without specifying the mean or the variance of the normal distribution. This test was first presented by Lilliefors (1967). One interesting feature of this test is that this is one of the earliest cases in which the computer was used to generate random numbers in order to obtain accurate estimates of the true quantiles of the exact distribution of the test statistic.

▶ **The Lilliefors Test for Normality** ────────────────────────

Data   The data consist of a random sample $X_1, X_2, \ldots, X_n$ of size n associated with some unknown distribution function, denoted by $F(x)$. Compute the sample mean

$$\overline{X} = \frac{1}{n} \sum_{i=1}^{n} X_i \tag{1}$$

for use as an estimate of μ and compute

$$s = \sqrt{\frac{1}{n-1} \sum_{i=1}^{n} (X_i - \overline{X})^2} \tag{2}$$

as an estimate of σ. Then compute the "normalized" sample values Z_i, defined by

$$Z_i = \frac{X_i - \overline{X}}{s} \qquad i = 1, 2, \ldots, n \tag{3}$$

The test statistic is computed from the Z_is instead of from the original random sample.

Assumption

1. The sample is a random sample.

Test Statistic   Ordinarily the test statistic is the usual two-sided Kolmogorov test statistic, defined as the maximum vertical distance between the empirical distribution function of the X_is and the normal distribution function with mean $\overline{X}$ and standard deviation s, as given by Equations 1 and 2. However, the following method of computing the test statistic is slightly easier and is equivalent to the method indicated.

Draw a graph of the standard normal distribution function, and call it $F^*(x)$. Actually, only the values of $F^*(x)$ at the observed Zs are needed. Table A1 may be of assistance. Also draw a graph of the empirical distribution function of the normalized sample, the Z_is defined by Equation 3, using the same set of coordinates as just used for $F^*(x)$. Find the maximum vertical distance between the two graphs, $F^*(x)$ and the empirical distribution function of the Z_is, which we will call $S(x)$. This distance is the test statistic. That is, the Lilliefors test statistic T_1 is defined by

$$T_1 = \sup_x |F^*(x) - S(x)| \tag{4}$$

The difference between T_1 and the Kolmogorov test statistic is that the empirical distribution function $S(x)$ in Equation 4 was obtained from the normalized sample, while $S(x)$ in the Kolmogorov test was based on the original unadjusted observations.

Null Distribution   The null distribution has been obtained approximately, by generating thousands of pseudo-random numbers on a computer, and estimating quantiles from the empirical distribution function of the thousands of subsequent values of the test statistic. These estimates of quantiles are given in Table A14. Exact quantiles, and the exact mathematical form of the null distribution, are unknown.

Hypotheses

H_0: The random sample comes from a population with the normal distribution, with unknown mean and standard deviation

H_1: The distribution function of the X_is is nonnormal

Reject H_0 at the approximate level of significance α if T_1 exceeds the $1 - \alpha$ quantile given in Table A14. The p-value can be approximated from the quantiles in Table A14.

Computer Assistance   The Lilliefors test for normality is computed in *Minitab, S-Plus,* and *StatXact.* ─────────────────────────────── ◀

EXAMPLE 1

The same data used to illustrate the chi-squared test for normality in Example 4.5.3 will be used to illustrate the Lilliefors test. Fifty two-digit numbers were drawn at random from a telephone book. Although the random variable sampled is clearly discrete, we may still justify testing for normality if we realize that acceptance of the null hypothesis of normality does not imply that the random variable has the normal distribution and is therefore continuous, but merely indicates that the difference between the normal distribution function and the true distribution function is sufficiently insignificant so as to remain undetected.

The numbers X_i are arranged from smallest to largest and converted to Z_i by subtracting $\overline{X} = 55.04$ and dividing by $s = 19.00$, as computed from Equations 1 and 2.

X_i	Z_i	X_i	Z_i	X_i	Z_i	X_i	Z_i	X_i	Z_i
23	−1.69	36	−1.00	54	−0.05	61	0.31	73	0.95
23	−1.69	37	−0.95	54	−0.05	61	0.31	73	0.95
24	−1.63	40	−0.79	56	0.05	62	0.37	74	1.00
27	−1.48	42	−0.69	57	0.10	63	0.42	75	1.05
29	−1.37	43	−0.63	57	0.10	64	0.47	77	1.16
31	−1.27	43	−0.63	58	0.16	65	0.52	81	1.37
32	−1.21	44	−0.58	58	0.16	66	0.58	87	1.68
33	−1.16	45	−0.53	58	0.16	68	0.68	89	1.79
33	−1.16	48	−0.37	58	0.16	68	0.68	93	2.00
35	−1.05	48	−0.37	59	0.21	70	0.79	97	2.21

The null hypothesis of normality is tested with the Lilliefors test statistic

$$T_1 = \sup_x |F^*(x) - S(x)|$$

where $F^*(x)$ is the standard normal distribution function and $S(x)$ is the empirical distribution function of the Z_is. Figure 7 presents the graphs of $F^*(x)$ and $S(x)$.

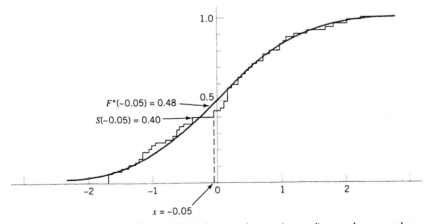

FIGURE 7 Graphs of $F^*(x)$ and $S(x)$ showing the maximum distance between them.

The maximum vertical distance between $F^*(x)$ and $S(x)$ is seen from Figure 7 to occur just to the left of $x = -0.05$, where $S(x) = 0.40$, $F^*(x) = 0.48$, and so $T = 0.08$. The vertical distance between the two curves equals 0.08 at other points, too, such as at $x = +0.05$ and $x = 0.10$. But at no point does the distance separating the two curves exceed 0.08.

The Lilliefors test calls for rejecting H_0 at $\alpha = 0.05$ if T_1 exceeds its 0.95 quantile, which is given by Table A14 as

$$w_{0.95} = \frac{0.886}{\sqrt{n}} = \frac{0.886}{\sqrt{50}} = 0.125$$

Because $T_1 = 0.08$ and is less than 0.125, the null hypothesis is accepted. In fact, the null hypothesis would still be accepted at $\alpha = 0.20$, because the 0.80 quantile is found to equal 0.104. Because Table A14 does not present smaller quantiles, we conclude that the p-value is some value greater than 0.20. Recall that the chi-squared test resulted in about the same conclusion.

Acceptance of the null hypothesis does not mean that the parent population is normal, but it does mean that the normal distribution does not seem to be an unreasonable approximation to the true unknown distribution; therefore either nonparametric methods or the parametric statistical procedures that assume a normal parent distribution may be appropriate for further testing with these data.

∎

□ *Theory*   One of the principal reasons for presenting the Lilliefors test is to show how the quantiles in Table A14 were obtained. The problem of finding the distribution of T_1 so that the Kolmogorov test could be used to test the composite hypothesis of normality with unspecified mean and variance had been too difficult to solve analytically. Therefore Lilliefors used a computer and random numbers to obtain an approximate solution. This same technique, described next, may be used to obtain an approximate solution to almost any problem in statistical inference.

Recall that to develop a statistical hypothesis test one must first invent a test statistic that acts as a reasonably sensitive indicator, indicating whether the null hypothesis is true or false. The statistic T_1 satisfies this requirement. Then one must select a certain region of values, corresponding to the critical region, that is unlikely to occur if H_0 is true but is more likely to occur if H_0 is false. Large values of T_1 meet this requirement. Then the difficulty comes in trying to find α, the probability of getting a point in the critical region when H_0 is true. To do this Lilliefors generated random normal deviates on a high-speed computer. Random normal deviates, as mentioned in Section 5.10, are numbers that seem to be observations on independent standard normal random variables.

These numbers were grouped into samples of size n for various values of n. For illustration let us say $n = 8$. A simulated sample of size 8 from a standard normal distribution, so that H_0 is true, was obtained from the computer. The sample mean $\overline{X}$ was computed and subtracted, and the result was divided by s as computed by Equation 2 for that sample to obtain the Z_i values. The empirical distribution function based on those Z_is was compared with the standard normal distribution function, and the maximum vertical difference T_1 was written down. The process was repeated with another set of eight computer-generated numbers to obtain another observed value of T_1. In all, over 1000 samples of size 8 were obtained and over 1000 values of T_1 were computed under the condition that H_0 is true. The empirical distribution function based on those 1000 or more values of T_1 was then used as an approximation to the true but unknown distribution function of T_1. From the empirical distribution function the selected quantiles given in Table A14 for $n = 8$ were obtained. Now a critical region of approximate size α may be specified.

The same procedure was repeated for other sample sizes ranging from $n = 4$ to $n = 30$. To obtain the approximation suggested by Lilliefors for n greater than 30, samples of size 40 were obtained, the quantiles were determined in the manner described, and the quantiles were multiplied by $\sqrt{40}$ and given in the table. This procedure is based on the unproved (at the time) conjecture that T_1 approaches its asymptotic distribution in much the same way that the Kolmogorov statistic is known to approach its limiting asymptotic distribution, as a function of $\sqrt{n}$.

Later simulation results by Mason and Bell (1986) used 20,000 simulations per sample size to obtain the more accurate quantile estimates reported in Table A14.

Lilliefors (1967) also compared the power of his test with the power of the chi-squared test in several nonnormal situations and found his test to be more powerful in the situations reported. □

A parametric confidence band for normal distributions was derived by Srinivasan and Wharton (1973). Other related papers are by Kanofsky and Srinivasan (1972) and Dyer (1974). A general discussion of simulation and presimulation days is found in Teichroew (1965).

A second modification of the Kolmogorov test was presented by Lilliefors in 1969. It tests the hypothesis that the parent distribution function is the exponential distribution $F(x) = 1 - e^{-x/t}$, $0 < x$, where t is an unspecified parameter that must be estimated from the data ($e = 2.718 \ldots$ is a well-known constant). Although Lilliefors obtained approximate critical values using random numbers as described, the exact distribution of the test statistic was subsequently obtained by Durbin (1975) and Margolin and Maurer (1976) using methods that are beyond the scope of this book and therefore not presented.

The exponential distribution is used to describe the length of time between consecutive "events," when the events occur randomly in time, according to a popular theory. Thus a test for the exponential distribution such as this one is actually used primarily as a test for randomness. Another application of this test is to see if the rate of failure of a product is constant over the life of the product, which is a property of the exponential distribution.

▶ **The Lilliefors Test for the Exponential Distribution** _____

Data   The data consist of a random sample $X_1, X_2, \ldots, X_n$ of size n associated with some unknown distribution function, denoted by $F(x)$. Compute the sample mean for use as an estimate of the unknown parameter. For each X_i, compute Z_i, defined by

$$Z_i = X_i / \overline{X} \tag{5}$$

for use in computing the test statistic.

Assumption

1. The sample is a random sample.

Test Statistic   First, the empirical distribution function $S(x)$ based on $Z_1, \ldots, Z_n$ is plotted on a graph. On the same graph the function $F^*(x) = 1 - e^{-x}$ is plotted for $x > 0$; actually, only values at n points need to be determined, the points being at $x = Z_1, x = Z_2$, and so on. Tables are available for evaluating e^{-x}; calculators that have this function may also be used. The maximum vertical distance between the two functions

$$T_2 = \sup_x |F^*(x) - S(X)| \tag{6}$$

is the test statistic.

   Although this is only the two-sided version of the test, one-sided versions are presented by Durbin (1975) along with tables.

Null Distribution   Although initially the quantiles of the null distribution were estimated by simulation on a computer, later research was successful in obtaining the exact distribution. Exact quantiles are given in Table A15.

Hypotheses

H_0: The random sample has the exponential distribution

$$F(x) = \begin{cases} 1 - e^{-x/t}, & x > 0 \\ 0, & x < 0 \end{cases} \tag{7}$$

where t is an unknown parameter

H_1: The distribution of X is not exponential

Reject H_0 at the level of significance α if T_2 exceeds the $1 - \alpha$ quantile as given in Table A15. The p-value can be obtained approximately by interpolation in Table A15.

Computer Assistance   *S-Plus* calculates this Lilliefors test for exponential distributions. ———————————————————————◄

EXAMPLE 2

The placement of long-distance telephone calls through a certain switchboard is believed to be a random process, with times between calls having an exponential distribution. The first 10 calls after 1 P.M. one Monday occurred at 1:06, 1:08, 1:16, 1:22, 1:23, 1:34, 1:44, 1:47, 1:51, and 1:57. The successive times between calls, counting the first time from 1:00 to 1:06, are (in minutes) 6, 2, 8, 6, 1, 11, 10, 3, 4, and 6, with sample mean $\overline{X} = 5.7$. The resulting values of Z_i, $1 - e^{-z_i}$, and the differences between $S(x)$ and $F^*(x)$ on both sides of each jump in $S(x)$ are given as follows. Note that the Xs are listed from smallest to largest for convenience.

i	X_i	$Z_i = X_i/\overline{X}$	$1 - e^{-z_i}$	$i/10 - 1 + e^{-z_i}$	$1 - e^{-z_i} - (i - 1)/10$
1	1	0.1754	0.1609	−0.0609	0.1609
2	2	0.3508	0.2959	−0.0959	0.1959
3	3	0.5263	0.4092	−0.1092	0.2092
4	4	0.7018	0.5043	−0.1043	0.2043
5	6	1.0526	0.6510	−0.1510	0.2510[b]
6	6	1.0526	0.6510	−0.0510	0.1510
7	6	1.0526	0.6510	0.0490	0.0510
8	8	1.4035	0.7543	0.0457	0.0543
9	10	1.7544	0.8270	0.0730	0.0270
10	11	1.9298	0.8548	0.1452[a]	−0.0452

[a] Largest difference $S(x) - F^*(x)$.
[b] Largest difference $F^*(x) - S(x)$.

The largest absolute deviation between $S(x)$ and $F^*(x)$ is seen to equal 0.2510. The null hypothesis of an exponential distribution may be rejected at $\alpha = 0.05$ only if T_2 exceeds 0.3244 (from Table A15, $n = 10$, $1 - \alpha = 0.95$). Since $T_2 = 0.2510$, the null hypothesis is accepted. The p-value, 0.25, is obtained by interpolation in Table A15. The times for the long-distance phone calls could be following a random process.   ■

The Kolmogorov test has been extended to the gamma distribution when parameters must be estimated, by Lilliefors (1973) and Schneider and Clickner (1976). A similar version of the Cramér-von Mises test is presented by Pettitt (1978). Other tests of a similar type are discussed by Green and Hegazy (1976).

We conclude this section by presenting a well-known goodness-of-fit test for normality that may be used instead of the Lilliefors test if desired. Some empirical studies indicate that this test has good power in many situations when compared with many other tests of the composite hypothesis of normality, including the Lilliefors test and the chi-squared test (Shapiro, Wilk, and Chen, 1968; La Brecque, 1977). Although this test is not a Kolmogorov-type test, it is included here because of its usefulness.

▶ **The Shapiro-Wilk Test for Normality** _____

Data   The data consist of a random sample $X_1, X_2, \ldots, X_n$ of size n associated with some unknown distribution function $F(x)$.

Assumption

1. The sample is a random sample.

Test Statistic   First compute the denominator D of the test statistic

$$D = \sum_{i=1}^{n} (X_i - \overline{X})^2 \tag{8}$$

where $\overline{X}$ is the sample mean. Then order the sample from smallest to largest,

$$X^{(1)} \le X^{(2)} \le \cdots \le X^{(n)}$$

and let $X^{(i)}$ denote the ith order statistic. From Table A16, for the observed sample size n, obtain the coefficients $a_1, a_2, \ldots, a_k$ where k is approximately $n/2$.
    The test statistic T_3 is given by

$$T_3 = \frac{1}{D} \left[\sum_{i=1}^{k} a_i (X^{(n-i+1)} - X^{(i)}) \right]^2 \tag{9}$$

Note that this test statistic is often denoted by W, and the test is often called the W test.

Null Distribution   The test statistic T_3 is basically the square of a correlation coefficient, where the Pearson correlation coefficient is computed between the order statistics $X^{(i)}$ in the sample and the scores a_i, which represent what the order statistics should look like if the population is normal. Thus if T_3 is close to 1.0 the sample behaves like a normal sample. If T_3 is too small, that is, too far below 1.0, the sample looks nonnormal. Quantiles of T_3 are given in Table A17.

Hypotheses

H_0: $F(x)$ is a normal distribution function with unspecified mean and variance
H_1: $F(x)$ is nonnormal

Reject H_0 at the level of significance α if T_3 is less than the α quantile as given by Table A17. If a more precise p-value for an observed value of T_3 is desired, the instructions in Table A18 allow T_3 to be converted to an approximately normal random variable, which may then be compared with the normal distribution in Table A1 to obtain an approximate p-value.

Computer Assistance The Shapiro-Wilk test is calculated in *Minitab, SAS,* and *StatXact.* ─── ◄

Comment

Although existing tables allow the Shapiro-Wilk test to be used only if $n \leq 50$, D'Agostino (1971) presents a test that may be used for n larger than 50, and Shapiro and Francia (1972) suggest an approximate test for n greater than 50 that is similar to the Shapiro-Wilk test.

EXAMPLE 3

The 50 two-digit numbers in Example 4.5.3 were drawn from a telephone book. The chi-squared goodness-of-fit test accepted the hypothesis of normality with a p-value well above 0.25. The Lilliefors test accepted the same hypothesis in Example 1 with a p-value greater than 0.20. The same data will be analyzed using the Shapiro-Wilk test.

The coefficients from Table A16 and the order statistics $X^{(n-i+1)} - X^{(i)}$ are given next.

i	a_i	$X^{(n-i+1)} - X^{(i)}$	i	a_i	$X^{(n-i+1)} - X^{(i)}$
1	0.3751	97–23	14	0.0846	66–42
2	0.2574	93–23	15	0.0764	65–43
3	0.2260	89–24	16	0.0685	64–43
4	0.2032	87–27	17	0.0608	63–44
5	0.1847	81–29	18	0.0532	62–45
6	0.1691	77–31	19	0.0459	61–48
7	0.1554	75–32	20	0.0386	61–48
8	0.1430	74–33	21	0.0314	59–54
9	0.1317	73–33	22	0.0244	58–54
10	0.1212	73–35	23	0.0174	58–56
11	0.1113	70–36	24	0.0104	58–57
12	0.1020	68–37	25	0.0035	58–57
13	0.0932	68–40			

The numerator of the test statistic becomes

$$\left[\sum_{i=1}^{k} a_i(X^{n-i+1} - X^{(i)})\right]^2 = [(0.3751)(97 - 23) + \cdots + (0.0035)(58 - 57)]^2$$

$$= [130.63]^2 = 17{,}064$$

and the denominator is given by

$$D = \sum_{i=1}^{n} (X_i - \overline{X})^2 = 17{,}698$$

so the test statistic becomes

$$T_3 = \frac{17{,}064}{17{,}698} = 0.9642$$

which lies somewhere between the 0.10 and the 0.50 quantiles of the distribution. Interpolation in Table A17 gives a p-value = 0.29 approximately.

In order to find a more precise value for the p-value, the coefficients from Table A18 are obtained for $n = 50$; $b_{50} = -7.677$, $c_{50} = 2.212$, and $d_{50} = 0.1436$. The observed value of T_3 is substituted into the formula

$$G = b_{50} + c_{50} \ln \left(\frac{T_3 - d_{50}}{1 - T_3}\right)$$

$$= -7.677 + (2.212) \ln \left(\frac{0.9642 - 0.1436}{1 - 0.9642}\right)$$

$$= -0.7488$$

which corresponds to $p = 0.227$ from Table A1. This is a more precise p-value than the one we obtained by interpolation earlier. ■

The theory behind the Shapiro-Wilk test is too lengthy to present here, but the interested reader is referred to the original papers by Shapiro and Wilk (1965, 1968). Basically, the test statistic is the square of Pearson's r computed between the order statistics and some constants a_i where the correlation should be close to 1.0 for samples from normal populations. Some efforts to extend existing tables (Stephens, 1975) apparently have not yet resulted in extended tables as far as we know. Other goodness-of-fit tests for the same composite hypothesis of normality have been offered by Hartley and Pfaffenberger (1972), Bowman and Shenton (1975), and Pearson, D'Agostino, and Bowman (1977).

One useful feature of the Shapiro-Wilk test is that several independent goodness-of-fit tests may be combined into one overall test of normality. This is convenient when several small samples from possibly different populations are insufficient by themselves to reject the hypothesis of normality, but their combined evidence is enough to disprove normality.

This technique of combining the results of several independent studies is called *meta-analysis*. One method consists of converting the test statistic in each study to a standard normal random variable, either by converting each p-value to a normal quantile, or by a direct conversion such as Table A18 allows in the Shapiro-Wilk test. Then the normal random variables are added and divided by the square root of the number of studies, to get one overall test statistic whose null distribution is the standard normal. See Wolf (1986) for elaboration of this technique. The technique is illustrated in the following example.

EXAMPLE 4

When an offshore lease is made available for bids, several oil companies usually submit bids for the right to drill for oil in that area. The distribution of these bids is often assumed to follow the "lognormal" distribution; that is, the logarithm of the bids is assumed to follow the normal distribution. However, the means and variances may vary from lease to lease. Also, the number of bids on any one lease is usually too small to be able to tell whether the normality assumption on the logarithms of the bids is reasonable or not.

To test the hypothesis

H_0: The bids are lognormally distributed

against the alternative that they are not lognormal, the bids on 16 different leases are observed. The Shapiro-Wilk test is conducted on the logarithms of the bids on each lease separately, with the result that the null hypothesis is rejected on 4 of the 16 leases at $\alpha = 0.05$. However, some of the leases show good agreement with the null hypothesis, with p-values well above 0.50. To combine the results from the 16 tests, the following steps are followed.

1. Each value of T_3 is converted to a value of G, as described in Table A18.
2. All $n = 16$ values of G are added together.
3. The result is divided by $\sqrt{n}$ to get Z, which is approximately standard normal under the null hypothesis.
4. If Z is less than the α quantile from Table A1, the null hypothesis is rejected at the level α.

For these leases the calculations are as follows.

Lease Number	Number of Bids	T_3	G
1	14	0.9243	−0.6550
2	14	0.9757	1.3559
3	14	0.9717	1.0939
4	14	0.8772	−1.5848
5	14	0.9537	0.2345
6	15	0.9135	−1.0093
7	15	0.8629[a]	−1.9321

8	15	0.8786[a]	−1.6806
9	15	0.8515[a]	−2.1011
10	15	0.9226	−0.7966
11	15	0.9581	0.3354
12	15	0.9625	0.5344
13	16	0.9178	−1.0151
14	16	0.8596[a]	−2.1011
15	15	0.9603	0.4323
16	16	0.9669	0.6795
		Total	−8.2099

[a] Significant at $\alpha = 0.05$.

$$Z = \frac{-8.2099}{\sqrt{16}} = -2.0525$$

This value of Z is smaller than -1.6449 from Table A1, so H_0 is rejected at $\alpha = 0.05$. The p-value is seen from Table A1 to equal 0.020. The assumption of lognormally distributed bids does not seem to be justified. ∎

We would be remiss if we did not point out that almost any goodness-of-fit test will result in rejection of the null hypothesis if the number of observations is very large. In other words, real data never really are distributed according to any known distribution. However, these known distributions are often "close enough" to the data for some reasonably accurate results to be obtained by assuming that the hypothesized distribution is the real one. A goodness-of-fit test is one way of ascertaining whether or not the agreement is close enough.

EXERCISES

1. The return on investment for 12 months on 20 randomly selected stocks is as follows.

9.1	5.0	7.3	7.4	5.5
8.6	7.0	4.3	4.7	8.0
4.0	8.5	6.4	6.1	5.8
9.5	5.2	6.7	8.3	9.2

Test the composite null hypothesis of normality using the Lilliefors test.

2. Fifteen entering freshmen had achievement scores as follows.

481	620	642	515	740
562	395	615	596	618
525	584	540	580	598

Test for normality using the Lilliefors test.

3. Test the hypothesis of Exercise 1 using the Shapiro-Wilk test.

4. Test the hypothesis of Exercise 2 using the Shapiro-Wilk test.

5. A store manager wanted to test the hypothesis that customers arrived randomly at her store, so she recorded the times between successive arrivals of customers one morning. These times (in minutes) were as follows.

3.6	6.2	12.7
14.2	38.0	3.8
10.8	6.1	10.1
22.1	4.2	4.6
1.4	3.3	8.2

Test the null hypothesis that these interarrival times follow an exponential distribution.

6. Twenty accidents occurred along a particular stretch of interstate highway one month. The nineteen distances between accidents, in miles, are as follows.

0.3	6.1	4.3	3.3	1.9
4.8	0.3	1.2	0.8	10.3
1.2	0.1	10.0	1.6	27.6
12.0	14.2	19.7	15.5	

Do the accidents appear to be distributed at random along the highway?

7. It is sometimes assumed that stream flow data (the amount of water flowing through a particular stream or river) are lognormally distributed. In order to test this assumption, data were collected on eight streams and rivers of various sizes. The data consisted of stream flow (cubic feet per second) measurements taken once a week for various numbers of weeks. The logarithms of the data were tested for normality using the Shapiro-Wilk test, with the following results.

Stream Number	Weeks of Record	Value of T_3
1	8	0.972
2	10	0.858
3	6	0.875
4	14	0.840
5	9	0.966
6	10	0.924
7	14	0.881
8	12	0.868

Do the combined results indicate that stream flow data tend to follow a lognormal distribution?

8. The total yearly rainfall is sometimes assumed to follow a normal distribution. Ten cities across the United States were selected to test this assumption. Annual rainfall records were analyzed using the Shapiro-Wilk test, with the following results. (Continued on next page.)

City	Years of Record	Value of T_3
1	18	0.875
2	34	0.874
3	26	0.948
4	43	0.980

5	40	0.937
6	29	0.915
7	35	0.915
8	38	0.890
9	42	0.963
10	47	0.941

Do the combined results indicate that annual rainfall follows a normal distribution?

6.3   TESTS ON TWO INDEPENDENT SAMPLES

The tests presented in this section are useful in situations where two samples are drawn, one from each of two possibly different populations, and the experimeter wishes to determine whether the two distribution functions associated with the two populations are identical or not. While other tests such as the median test, the Mann-Whitney test, or the parametric t test may also be appropriate, they are sensitive to differences between the two means or medians, but they may not detect differences of other types, such as differences in variances. One of the advantages of the two two-sided tests presented in this section is that both tests are consistent against all types of differences that may exist between the two distribution functions.

The first test presented is the Smirnov test (Smirnov, 1939). It is a two-sample version of the Kolmogorov test presented in Section 6.1 and is sometimes called the Kolmogorov-Smirnov two-sample test, while the Kolmogorov test is sometimes called the Kolmogorov-Smirnov one-sample test. The Smirnov test is presented in the one-sided and two-sided versions. Another two-sided test, the Cramér-von Mises test for two samples, is also presented. It is slightly more difficult to compute than the Smirnov test, but it appeals to some people because it seems to make more effective use of the data. Actually, there is probably little difference in power between the two tests.

▶ **The Smirnov Test** _____

Data   The data consist of two independent random samples, one of size n, X_1, $X_2, \ldots, X_n$, and the other of size m, $Y_1, Y_2, \ldots, Y_m$. Let $F(x)$ and $G(x)$ represent their respective, unknown, distribution functions.

Assumptions

1. The samples are random samples.
2. The two samples are mutually independent.

3. The measurement scale is at least ordinal.

4. For this test to be exact the random variables are assumed to be continuous.

If the random variables are discrete, the test is still valid but becomes conservative (Noether, 1967a).

Test Statistic Let $S_1(x)$ be the empirical distribution function based on the random sample $X_1, X_2, \ldots, X_n$, and let $S_2(x)$ be the empirical distribution function based on the other random sample $Y_1, Y_2, \ldots, Y_m$. The test statistic is defined differently for the three different sets of hypotheses, A, B, and C.

A. (Two-Sided Test) Define the test statistic T_1 as the greatest vertical distance between the two empirical distribution functions.

$$T_1 = \sup_x |S_1(x) - S_2(x)| \tag{1}$$

B. (One-Sided Test) Denote the test statistic by T_1^+ and let it equal the greatest vertical distance attained by $S_1(x)$ above $S_2(x)$.

$$T_1^+ = \sup_x [S_1(x) - S_2(x)] \tag{2}$$

C. (One-Sided Test) For the one-sided hypotheses in C below, let the test statistic, denoted by T_1^-, be the greatest vertical distance attained by $S_2(x)$ above $S_1(x)$.

$$T_1^- = \sup_x [S_2(x) - S_1(x)] \tag{3}$$

Null Distribution The exact null distribution of T_1, T_1^+ and T_1^- is found by considering all orderings of Xs and Ys to be equally likely under the null hypothesis, and computing T_1, T_1^+ or T_1^-, as appropriate, for each ordering, as with the Mann-Whitney test. Quantiles of the null distribution are given in Table A19 for equal sample sizes and in Table A20 for $m \neq n$.

In the case of equal sample sizes, $m = n$, the exact distribution of T_1^+ and T_1^- is given by

$$F(x) = 1 - \frac{\binom{2n}{n + c}}{\binom{2n}{n}} \tag{4}$$

where c is the greatest integer less than $x \cdot n$.

Hypotheses

A. (Two-Sided Test)

$$H_0: F(x) = G(x) \quad \text{for all } x \text{ from } -\infty \text{ to } +\infty$$
$$H_1: F(x) \neq G(x) \quad \text{for at least one value of } x$$

Reject H_0 at the level of significance α if T_1 is greater than the $1 - \alpha$ quantile for a two-sided test, given by Table A19 for $m = n$ or by Table A20 for $m \neq n$. Use the approximation at the end of the tables for larger sample sizes not covered by the tables. The approximate p-value can be found by interpolation in the appropriate table. If $m = n$ a more exact two-sided p-value can be found by taking twice the exact one-sided p-value

$$\text{one-sided } p\text{-value} = \frac{\dbinom{2n}{n + nt}}{\dbinom{2n}{n}} \tag{5}$$

where t is the observed value of the test statistic.

B. (One-Sided Test)

$$H_0: F(x) \leq G(x) \quad \text{for all } x \text{ from } -\infty \text{ to } +\infty$$
$$H_1: F(x) > G(x) \quad \text{for at least one value of } x$$

This alternative hypothesis is sometimes stated as, "The Xs tend to be *smaller* than the Ys," which is a more general form of location alternative than the statement that the Xs and Ys differ only by a location parameter (means or medians).

Reject H_0 at the level of significance α if $T_1{}^+$ is greater than the $1 - \alpha$ quantile for a one-sided test, given by Table A19 for $m = n$ or by Table A20 for $m \neq n$. Approximate p-values are found by interpolation in the appropriate table. An exact p-value can be found by using Equation 5 if $m = n$.

C. (One-Sided Test)

$$H_0: F(x) \geq G(x) \quad \text{for all } x \text{ from } -\infty \text{ to } +\infty$$
$$H_1: F(x) < G(x) \quad \text{for at least one value of } x$$

This is the one-sided test to use if it is suspected that the Xs might be shifted to the right (i.e., larger) of the Ys.

Reject H_0 at the level of significance α if $T_1{}^-$ is greater than the $1 - \alpha$ quantile for a one-sided test, given by Table A19 for $m = n$ or by Table A20 for $m \neq n$. Approximate p-values are found by interpolation in the appropriate table. An exact p-value can be found by using Equation 5 if $m = n$.

Computer Assistance *StatXact* performs this test, which it calls the Kolmogorov-Smirnov two-sample test, and finds exact *p*-values whenever possible. ____◀

EXAMPLE I

A random sample of size 9, $X_1, \ldots, X_9$ is obtained from one population, and a random sample of size 15, $Y_1, \ldots, Y_{15}$ is obtained from a second population. A graph of their empirical distribution functions is given in Figure 8.

The null hypothesis is that the two populations have identical distribution functions. If the respective distribution functions are denoted by $F(x)$ and $G(x)$, then the null hypothesis may be written as

$$H_0: F(x) = G(x) \quad \text{for all } x \text{ from } -\infty \text{ to } +\infty$$

The alternative hypothesis may be stated as

$$H_1: F(x) \neq G(x) \quad \text{for at least one value of } x$$

The two samples are ordered from smallest to largest for convenience, and their values, along with other pertinent information about their empirical distribution functions, are given next.

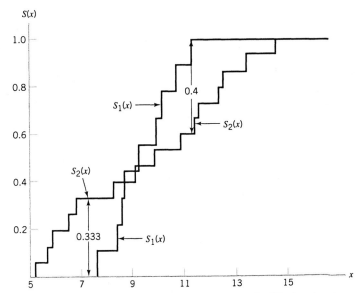

FIGURE 8 Graphs of $S_1(x)$, the e.d.f. of X, and $S_2(x)$, the e.d.f. of Y, showing the maximum distance between them.

X_i	Y_i	$S_1(x) - S_2(x)$	X_i	Y_i	$S_1(x) - S_2(x)$
	5.2	$0 - 1/15 = -1/15$		9.8	$5/9 - 8/15 = 1/45$
	5.7	$0 - 2/15 = -2/15$	9.9		$6/9 - 8/15 = 2/15$
	5.9	$0 - 3/15 = -1/5$	10.1		$7/9 - 8/15 = 11/45$
	6.5	$0 - 4/15 = -4/15$	10.6		$8/9 - 8/15 = 16/45$
	6.8	$0 - 5/15 = -1/3$		10.8	$8/9 - 9/15 = 13/45$
7.6		$1/9 - 5/15 = -2/9$	11.2		$1 - 9/15 = 2/5$
	8.2	$1/9 - 6/15 = -13/45$		11.3	$1 - 10/15 = 1/3$
8.4		$2/9 - 6/15 = -8/45$		11.5	$1 - 11/15 = 4/15$
8.6		$3/9 - 6/15 = -1/15$		12.3	$1 - 12/15 = 1/5$
8.7		$4/9 - 6/15 = 2/4$		12.5	$1 - 13/15 = 2/15$
	9.1	$4/9 - 7/15 = -1/15$		13.4	$1 - 14/15 = 1/15$
9.3		$5/9 - 7/15 = 4/45$		14.6	$1 - 1 = 0$

The test statistic for the two-sided test is given by Equation 1 as

$$T_1 = \sup_x |S_1(x) - S_2(x)|$$
$$= \tfrac{2}{5} = 0.400$$

the largest absolute difference between $S_1(x)$ and $S_2(x)$, which happens to occur between $x = 11.2$ and $x = 11.3$. The value of 0.400 for T_1 could also have been determined graphically in Figure 8. From the graph one can easily see that the difference $S_1(x)$ and $S_2(x)$ changes only at those observed values $x = X_i$ or $x = Y_i$, and that is why it is sufficient to compute $S_1(x) - S_2(x)$ only at the observed sample values, as done here.

From Table A20 we see that the 0.95 quantile of T_1, for the two-sided test and for $n = 9 = N_1$ and $m = 15 = N_2$, is given as $w_{0.95} = \tfrac{8}{15}$. For these data T_1 equals $\tfrac{2}{5}$ or $\tfrac{6}{15}$. Therefore H_0 is accepted at the 0.05 level. From the table, the p-value may be estimated as slightly larger than 0.20.

For the sake of comparison, the approximate 0.95 quantile based on the asymptotic distribution is found to be

$$w_{0.95} \cong 1.36\sqrt{\frac{m+n}{mn}} = 0.573$$

which is slightly larger than the exact value of $\tfrac{8}{15} = 0.533$. This illustrates the tendency of the asymptotic approximation to furnish a conservative test.

Note that many of the calculations performed in this example could have been eliminated because, either by an inspection of the data or a preliminary sketch of $S_1(x)$ and $S_2(x)$ as in Figure 8, many of the values of X_i and Y_j may be seen to be unlikely candidates for yielding the maximum value of $|S_1(x) - S_2(x)|$ and therefore may be ignored in favor of the more likely values of X_i and Y_j.

If a one-sided test had been appropriate instead of the two-sided test, the statistics

$$T_1^+ = \sup_x [S_1(x) - S_2(x)] = \tfrac{2}{5} = 0.400$$

for the set B of hypotheses, and

$$T_1^- = \sup_x [S_2(x) - S_1(x)] = \tfrac{1}{3} = 0.333$$

for the set C of hypotheses are easily determined from the preceding table of data. The p-values for both of the one-sided tests are seen from Table A20 to be greater than 0.10. ∎

□ **Theory** Although it may not be apparent at first, the statistics T_1, T_1^+, and T_1^- depend only on the order of the Xs and Ys in the ordered combined sample of Xs and Ys and do not require knowledge of the actual numerical values of the observations. To illustrate this, suppose there are three Xs and two Ys. There are $\binom{5}{2} = 10$ different ordered arrangements of the combined sample. These arrangements are given next, along with the values of T_1, T_1^+, and T_1^- for each ordered arrangement.

Arrangement	T_1	T_1^+	T_1^-	Arrangement	T_1	T_1^+	T_1^-
X < X < X < Y < Y	1	1	0	X < Y < X < Y < X	$\tfrac{1}{3}$	$\tfrac{1}{3}$	$\tfrac{1}{3}$
X < X < Y < X < Y	$\tfrac{2}{3}$	$\tfrac{2}{3}$	0	Y < X < X < Y < X	$\tfrac{1}{2}$	$\tfrac{1}{6}$	$\tfrac{1}{2}$
X < Y < X < X < Y	$\tfrac{1}{2}$	$\tfrac{1}{2}$	$\tfrac{1}{6}$	X < Y < Y < X < X	$\tfrac{2}{3}$	$\tfrac{1}{3}$	$\tfrac{2}{3}$
Y < X < X < X < Y	$\tfrac{1}{2}$	$\tfrac{1}{2}$	$\tfrac{1}{2}$	Y < X < Y < X < X	$\tfrac{2}{3}$	0	$\tfrac{2}{3}$
X < X < Y < Y < X	$\tfrac{2}{3}$	$\tfrac{2}{3}$	$\tfrac{1}{3}$	Y < Y < X < X < X	1	0	1

If the null hypothesis in the two-sided test is true, the two distribution functions are equal and each ordered arrangement is equally likely under the assumption of continuous random variables. This same point was discussed more thoroughly in connection with the Mann-Whitney test in Section 5.1. Therefore, in the two-sided test, the probability associated with each ordered arrangement is given by

$$\text{probability} = \frac{1}{\binom{m+n}{n}} = \frac{1}{\binom{5}{3}} = \frac{1}{10} \tag{6}$$

and from this the following probability distribution can be deduced.

$$P(T_1 = \tfrac{1}{3}) = 1/10 \qquad P(T_1^+ = 0) = 1/5 \qquad P(T_1^- = 0) = 1/5$$
$$P(T_1 = \tfrac{1}{2}) = 3/10 \qquad P(T_1^+ = \tfrac{1}{6}) = 1/10 \qquad P(T_1^- = \tfrac{1}{6}) = 1/10$$
$$P(T_1 = \tfrac{2}{3}) = 2/5 \qquad P(T_1^+ = \tfrac{1}{3}) = 1/5 \qquad P(T_1^- = \tfrac{1}{3}) = 1/5$$

$$P(T_1 = 1) = 1/5 \qquad P(T_1^+ = \tfrac{1}{2}) = 1/5 \qquad P(T_1^- = \tfrac{1}{2}) = 1/5$$
$$P(T_1^+ = \tfrac{2}{3}) = 1/5 \qquad P(T_1^- = \tfrac{2}{3}) = 1/5$$
$$P(T_1^+ = 1) = 1/10 \qquad P(T_1^- = 1) = 1/10$$

It is no coincidence that the distributions of T_1^+, and T_1^- are identical with each other for $n = 3$ and $m = 2$. They are identical with each other for all choices of n and m. However, the space-saving technique used in Tables A19 and A20 of stating that the $1 - \alpha$ quantile of T_1 in the two-sided test equals the $1 - \alpha/2$ quantile of T_1^+ in the one-sided test is a valid technique only if α is small. Notice, for example, in the preceding illustration that $P(T_1 \geq 1)$ equals twice $P(T_1^+ \geq 1)$ and $P(T_1 \geq \tfrac{2}{3})$ equals twice $P(T_1^+ \geq \tfrac{2}{3})$, but $P(T_1 \geq \tfrac{1}{2})$ does not equal twice $P(T_1^+ \geq \tfrac{1}{2})$.

The null distribution (i.e., the distribution when H_0 is true) in the one-sided tests is also found in the manner just described because, under the one-sided null hypotheses, the size of the critical region is a maximum when $F(x)$ is identical with $G(x)$. If the two samples are of equal size, it is not necessary to use this method to find the upper quantiles, because the distribution functions for T_1, T_1^+, and T_1^- were derived as a function of the sample size n by Gnedenko and Korolyuk (1951). The derivation of these distribution functions is interesting, and it is within the presumed mathematical grasp of the reader, but its length precludes its presentation here. The reader is referred to Fisz (1963) for a readable presentation of the derivation.

For samples of unequal size the method of finding quantiles is essentially as illustrated. However, many refinements using path-counting methods have simplified the bookkeeping enough so that extensive tables exist (Harter and Owen, 1970). See Steck (1969) for a general discussion of the Smirnov test. Kim (1969) gives some closer approximations to the exact quantiles when exact tables are not available. □

A modification of the Smirnov test was suggested by Tsao (1954) so that the test may be applied to truncated samples. That is, perhaps only the Xs and Ys less than $X^{(r)}$ are observed, as sometimes happens in life-testing experiments. The Smirnov test may then be applied to the truncated samples with the aid of tables derived recursively by Tsao (1954). The distribution functions of Tsao's statistics were derived analytically by Conover (1967a). Extensions of the Smirnov test to three or more samples are presented by Birnbaum and Hall (1960), who presented tables for three samples of equal sizes, and Conover (1965), who presented tables for k samples of equal size, $k \leq 10$ in Conover (1980). A one-sided multi-sample version of the Smirnov test was introduced by Conover (1967b) along with tables for up to 10 samples of equal size in Conover (1980).

The next test is the Cramér-von Mises two-sample test. This test is two-sided only and involves slightly more calculations than the Smirnov test does.

▶ **The Cramér-von Mises Two-Sample Test** _____

Data The data consist of two independent random samples, $X_1, \ldots, X_n$ and $Y_1, \ldots, Y_m$, with unknown distribution functions $F(x)$ and $G(x)$, respectively.

Assumptions

1. The samples are random samples, independent of each other.
2. The measurement scale is at least ordinal.
3. The random variables are continuous. If the random variables are actually discrete the test is likely to be conservative.

Test Statistic Let $S_1(x)$ and $S_2(x)$ be the empirical distribution functions of the two samples. The test statistic T_2 is defined as

$$T_2 = \frac{mn}{(m+n)^2} \sum_{\substack{x=X_i \\ x=Y_j}} [S_1(x) - S_2(x)]^2 \tag{7}$$

where the squared difference in the summation is computed at each X_i and at each Y_j. Perhaps it is clearer to write the test statistic as

$$T_2 = \frac{mn}{(m+n)^2} \left\{ \sum_{i=1}^{n} [S_1(X_i) - S_2(X_i)]^2 + \sum_{j=1}^{m} [S_1(Y_j) - S_2(Y_j)]^2 \right\} \tag{8}$$

Null Distribution The exact null distribution is found, as with the Smirnov test and the Mann-Whitney test, by considering all ordered arrangements of the combined sample to be equally likely, and by computing T_2 for each ordered arrangement. We will use the asymptotic distribution, as $n \to \infty$ and $m \to \infty$, as an approximate distribution for all sample sizes.

Hypotheses

$$H_0: F(x) = G(x) \qquad \text{for all } x \text{ from } -\infty \text{ to } +\infty$$
$$H_1: F(x) \neq G(x) \qquad \text{for at least one value of } x$$

Reject H_0 at the approximate level α if T_2 exceeds the $1 - \alpha$ quantile $w_{1-\alpha}$, as shown on the next page. These quantiles are approximations based on the asymptotic distribution, valid for large m and n, but they are considered fairly accurate even if the sample sizes are small (Burr, 1964).

$$w_{0.10} = 0.046 \qquad w_{0.50} = 0.119 \qquad w_{0.90} = 0.347$$
$$w_{0.20} = 0.062 \qquad w_{0.60} = 0.147 \qquad w_{0.95} = 0.461$$
$$w_{0.30} = 0.079 \qquad w_{0.70} = 0.184 \qquad w_{0.99} = 0.743$$
$$w_{0.40} = 0.097 \qquad w_{0.80} = 0.241 \qquad w_{0.999} = 1.168$$

These values were taken from Anderson and Darling (1952). Exact quantiles for $n + m \leq 17$ are given by Burr (1964). Approximate p-values are found by interpolation between the appropriate quantiles. ◀

EXAMPLE 2

From the data of Example 1 the test statistic T_2 may be computed by first finding

$$\sum_{i=1}^{9} [S_1(X_i) - S_2(X_i)]^2 = 0.459$$

and

$$\sum_{j=1}^{15} [S_1(Y_j) - S_2(Y_j)]^2 = 0.657$$

Then, from Equation 8, we have

$$T_2 = \frac{mn}{(m+n)^2} \left\{ \sum_{i=1}^{n} [S_1(X_i) - S_2(X_i)]^2 + \sum_{j=1}^{m} [S_1(Y_j) - S_2(Y_j)]^2 \right\}$$

$$= \frac{(15)(9)}{(24)^2} (0.459 + 0.657)$$

$$= 0.262$$

The null hypothesis of identical distribution functions is accepted at $\alpha = 0.05$, because $T_2 = 0.262$ is less than $w_{0.95} = 0.461$, as just given. The p-value is seen to be about 0.18. This is slightly smaller than the p-value for the Smirnov test on the same data. ■

□ *Theory*   The exact distribution of the Cramér-von Mises two-sample test statistic may be obtained in the same way as with the Smirnov test statistic. The different ordered arrangements of the combined sample are equally likely under the null hypothesis, and the statistic T_2 may be computed from the ordered combined sample. Exact quantiles of T_2 were obtained by Anderson (1962) and Burr (1963, 1964) for small samples in essentially the manner just described, with some computational shortcuts. □

The statistic T_2 was apparently introduced by Fisz (1960). He credits the statistic to Lehmann (1951) and the asymptotic distribution of the statistic to

Rosenblatt (1952), but the statistic studied by Lehmann and Rosenblatt is

$$T_3 = \frac{m}{2(m+n)} \sum_{i=1}^{n} \left[\frac{R(X^{(i)})}{m} - i\frac{m+n}{mn} \right]^2 + \frac{n}{2(m+n)} \sum_{j=1}^{m} \left[\frac{R(Y^{(j)})}{n} - j\frac{m+n}{mn} \right]^2 \qquad (9)$$

which differs from Fisz's statistic T_2 unless $m = n$. Fisz showed that T_2 has the same asymptotic distribution as T_3, which was shown by Rosenblatt to be the same as the asymptotic distribution of the Cramér-von Mises goodness-of-fit statistic. Therefore the asymptotic distribution of T_2 was actually obtained by Anderson and Darling (1952) in their paper on the Cramér-von Mises goodness-of-fit statistic. That is why T_2 is called the Cramér-von Mises two-sample test statistic, even though neither Cramér nor von Mises is credited with its invention.

For a two-sample test designed for points on a circle, see Stephens (1965b), Maag (1966), and Maag and Stephens (1968). A multivariate Smirnov test is described by Bickel (1969). Fine (1966) is concerned with the Cramér-von Mises statistic, while Csörgő (1965) and Percus and Percus (1970) work with variations of the Smirnov test. Papers on the asymptotic efficiency of the Smirnov test are by Capon (1965), Ramachandramurty (1966), Andel (1967), and Klotz (1967).

Adaptations of these tests to test for symmetry are presented by Rothman and Woodroofe (1972) and Rao, Schuster, and Littell (1975). Gail and Green (1976a) present more extensive tables for the one-sided test and discuss an interesting use of the test in another paper (1967b). More theoretical discussions are given by Takacs (1971) and Kalish and Mikulski (1971).

EXERCISES

1. Test the null hypothesis $F(x) \leq G(x)$, where the observations from $F(x)$ are 0.6, 0.8, 0.8, 1.2, and 1.4 and the observations from $G(x)$ are 1.3, 1.3, 1.8, 2.4, and 2.9.

2. A random sample of five sixth-grade boys in one section of town were given a literacy test with the following results; 82, 74, 87, 86, 75. A random sample of eight sixth-grade boys from a different section of town were given the same literacy test with these scores resulting: 88, 77, 91, 88, 94, 93, 83, 94. Is there a difference in literacy, as measured by this test, in the two populations of sixth-grade boys? (Use the Smirnov test).

3. Use the Cramér-von Mises test on the data in Exercise 2 and compare results with the Smirnov test.

4. A group of volunteer patients with high blood pressure are randomly assigned to Treatment A, which is a blood pressure medicine, or Treatment B, which involves a low salt diet and regular exercise. Is there any difference in the effectiveness of the two treatments, as measured by their change in blood pressure after six months?

Treatment	Change in Diastolic Blood Pressure
A	$-14, -62, -38, -19, -21, -28, -32, -40$
B	$-51, -31, +14, -12, -27, -38, -10, +6$

5.   Twenty home owners participated in a study of methods to reduce energy consumption. They were randomly assigned to Program A, an education program designed to instill energy-saving habits in their lifestyle, or Program B, where six inches of additional insulation was installed in their attic. Their energy savings for the following 12 months are given as follows.

Home Owner	Program	Savings	Home Owner	Program	Savings
1	A	$143	11	B	$175
2	A	106	12	B	142
3	B	182	13	B	111
4	B	158	14	A	82
5	B	161	15	A	12
6	A	108	16	A	58
7	B	131	17	A	42
8	A	138	18	B	96
9	A	101	19	B	90
10	A	83	20	B	144

Is there a difference in the effectiveness of the two programs?

6.   Cancer cells are injected into 15 laboratory mice to study the effectiveness of a proposed treatment. Six mice were given the cancer prevention treatment and the other nine were given a placebo (the control group). After several months the size of the tumors were measured in each animal.

Size of Tumors

Treatment	0.8, 0.0, 0.6, 1.1, 1.2, 0.5
Control	0.6, 1.6, 1.7, 1.3, 2.2, 1.5, 0.7, 0.7, 1.6

Is the treatment effective in reducing the eventual size of the tumor?

PROBLEMS

1.   Find the 0.80, 0.90, and 0.95 quantiles of T_1 for $n = 3$ and $m = 2$ from the exact distribution obtained in the text. Compare these with the quantiles in Table A20 and explain any differences.

2.   Find the exact distribution of T_1, T_1^+, T_1^-, and T_2 when $n = 3$ and $m = 3$.

3.   Compare the exact 0.95 quantiles of T_1 with the approximation based on the asymptotic distribution, for $n = m = 30$ and $n = m = 10$.

6.4 REVIEW PROBLEMS FOR CHAPTERS I THROUGH 6

1. A test is conducted to determine the breaking point for a particular type of rope. Ten pieces of rope were obtained. Their breaking points were recorded as follows (in pounds): 780, 620, 910, 900, 730, 700, 630, 690, 730, and 840.
   (a) Sketch the empirical distribution function.
   (b) Sketch a 90% confidence band for the population distribution function.
   (c) Find an approximate 90% confidence interval for the population median.

2. A certain town consists of five wards. Ten houses are selected at random from each ward and given a score from 0 to 100, depending on the level of deterioration of the house and yard (0 = no deterioration, 100 = no redeeming social value). These are the results.

House	Ward 1	Ward 2	Ward 3	Ward 4	Ward 5
1	08	74	92	03	37
2	45	42	79	09	28
3	43	77	99	22	42
4	64	09	38	06	44
5	03	32	31	26	01
6	85	66	83	20	32
7	74	16	27	56	65
8	48	45	76	20	02
9	19	15	82	04	80
10	57	24	37	29	93

   Make a list of all of the nonparametric tests that you could use to analyze these data to detect differences from ward to ward. Carefully state the advantages and disadvantages of each test. Select the test that you feel is the best one to use and test the hypothesis of no differences from ward to ward.

3. Gwen and Rich were teaching different sections of the same course. When the course was completed, they compared the grades they gave to see whether their distributions of grades were essentially the same.

	A	B	C	D & F
Gwen	14	28	17	3
Rich	6	22	23	7

   Use a chi-squared test to see whether the difference is significant.

4. Use a Kruskal-Wallis test for the data in Problem 3 and compare results. Under what circumstances would you recommend a chi-squared test, and when would you recommend the Kruskal-Wallis test?

5. (a) Answer true or false.
   (1) If two events are mutually exclusive, they are also independent.
   (2) The power of a consistent test approaches α as the sample size gets large.
   (3) The A.R.E. is a good approximation to use for the small sample relative efficiency.

(4) When sampling from the double exponential distribution, the sign test is more powerful than the Wilcoxon signed ranks test.

(5) The exact distribution of a rank statistic under H_0 can always be found by simple randomization methods.

(6) If extensive ties are present, the median test should be used rather than the Kruskal-Wallis test.

(b) Fill in the blanks with words.

(1) _____ is the size of the critical region.

(2) _____ is a subset of the sample space.

(3) _____ is the set of all possible outcomes of an experiment.

(4) If the experiment yields an outcome in the _____, the null hypothesis is rejected.

(5) The smallest level of significance at which the null hypothesis may be rejected is called the _____.

(6) The probability of rejecting a false null hypothesis is called _____.

6. A biased coin is tossed six times to test H_0: $P(H) = 1/3$ against H_1: $P(H) \neq 1/3$. If the outcome is "all tails" or "more than four heads" the null hypothesis is rejected.

(a) Is H_0 simple or composite?

(b) Is H_1 simple or composite?

(c) List the points in the critical region.

(d) What is the value of α?

(e) What is the equation of the power function?

7. An economist has been computing a monthly "prosperity index." For the last 24 months the values obtained are 123.6, 121.0, 124.1, 123.4, 125.7, 129.0, 126.8, 127.1, 127.3, 126.7, 124.8, 125.9, 124.7, 125.9, 125.6, 126.0, 125.7, 127.3, 127.7, 129.0, 128.2, 127.9, 127.8, and 127.1. Do these figures indicate a trend in the prosperity index?

8. The "maximum annual river stages" reported at a certain point each year for 16 years were (in feet): 7.4, 7.8, 6.9, 8.1, 8.0, 7.1, 7.4, 6.8, 6.9, 7.6, 7.6, 8.0, 8.3, 7.5, 7.8, and 7.1. Test the null hypothesis that the median maximum annual river stage is no greater than 8.0 feet. Find a 90% confidence interval for the median maximum annual river stage.

9. Sixty rolls of a die resulted in the following occurrences.

Number of Spots Showing	1	2	3	4	5	6
Number of Times Occurred	12	10	14	8	9	7

Test the hypothesis that the die is balanced, that is, that each number of spots has an equal probability of occurring.

10. Ninety graduating seniors, including 30 from Arts and Sciences, 30 from Engineering, and 30 from Agriculture, were selected at random from among those who had accepted salaried positions. Half of the 90 seniors accepted positions paying more than $2000 per month; the other half were to receive less than $2000 per month. Of those to receive more than $2000 per month, 9 were in Agriculture, 17 were in Arts and Sciences, and (therefore) 19 were in Engineering. Test the hypothesis that the median salary is the same for seniors in all three categories.

11. One hundred people were asked to taste the four new brands of cough syrup and state which new brands tasted better to them than the present formula and which brands

did not. As indicated in the following, 15 subjects preferred the new taste to the old for all four brands, 3 subjects preferred brands A, B, and C over the old brand but did not prefer brand D over the present formula, and so on. Test the null hypothesis that there is no significant difference in preferences among the four new brands of cough syrup.

A	B	C	D	Number of Subjects with this Response
1	1	1	1	15
1	1	1	0	3
1	1	0	1	3
1	0	1	1	6
0	1	1	1	21
1	1	0	0	1
1	0	1	0	1
0	1	1	0	1
1	0	0	1	2
0	1	0	1	2
0	0	1	1	19
1	0	0	0	3
0	1	0	0	3
0	0	1	0	2
0	0	0	1	13
0	0	0	0	5
				100

12.   The following data presents survival time in days of four samples of hypophysectomized rats without and with different dosages of adrenal cortical hormone.

A (None)	B	C	D
3	2	4	13
2	1	4	4
2	2	3	6
3	6	4	8
5	14	6	19
4	7	5	19
2	15	4	12
2	2	3	1
3	1	4	4
	4	5	4
			1
			12

Test the null hypothesis of no differences in treatment effects.

13.   These data represent the yields of four varieties of wheat grown in 13 different locations.

		Variety		
Location	A	B	C	D
1	43.60	24.05	19.47	19.41
2	40.40	21.76	16.61	23.84
3	18.08	14.19	16.69	16.08
4	19.57	18.61	17.78	18.29
5	45.20	29.33	20.19	30.08
6	25.87	25.60	23.31	27.04
7	55.20	38.77	21.15	39.95
8	55.32	34.19	18.56	25.12
9	19.79	21.65	23.31	22.45
10	46.24	31.52	22.48	29.28
11	14.88	15.68	19.79	22.56
12	7.52	4.69	20.53	22.08
13	41.17	32.59	29.25	43.95

Test the hypothesis of no difference in yields of the different varieties.

14. The following data represent days to death of 180 mice inoculated with three strains of typhoid organisms.

	Days to Death												
Strain	2	3	4	5	6	7	8	9	10	11	12	13	14
9D	10	8	18	16	3	4	1						
11C	1	3	3	6	6	14	11	4	6	2	3	1	
DSC1	1	2	1	3	8	11	10	7	7	3	4	2	1

For example, 10 mice inoculated with 9D died on the second day. Does there seem to be a significant difference in the reaction time to the various strains?

15. In Problem 14, is it reasonable to assume that the time till death for mice inoculated with strain 9D follows a normal distribution?

16. The pain threshold values for 12 males and 12 females were as follows.

Male	Females
8.5	6.4
7.9	7.8
6.7	7.1
7.4	8.0
7.5	6.6
8.6	7.3
8.0	8.1
8.1	7.4
7.2	8.3
8.0	8.9
7.8	7.8
7.8	7.7

Test the hypothesis of equal means. Test the hypothesis of equal variances.

17. The following data represent the 1951 and 1952 net earnings of common stocks in 20 representative corporations.

1951	1952	1951	1952
$1.68	$1.71	$4.64	$4.79
1.72	2.17	4.76	4.33
2.50	2.25	5.35	6.05
2.90	2.43	5.81	7.09
3.11	2.32	6.11	6.38
3.35	3.15	6.35	6.00
3.80	3.30	6.69	6.01
3.85	5.52	8.41	7.41
3.89	3.32	8.83	9.33
4.36	3.76	8.97	9.25

Is the increase in earnings from 1951 to 1952 statistically significant?

18. As an experiment in a sophomore statistics class, 10 students volunteered to take two tests. The first test was a basic mathematics test, and the second was a test on current events. The results were as follows.

Student	Math	Current Events
1	37	23
2	44	34
3	55	59
4	70	25
5	26	16
6	39	12
7	26	16
8	30	25
9	85	60
10	83	69

Does there seem to be a significant correlation between the scores on the two tests?

19. Four different types of automobile tires, 10 tires of each type, were tested under laboratory conditions for a given length of time. The average tread depth of each tire (in centimeters) was measured at the end of the experiment, with the following results.

Tire	Type 1	Type 2	Type 3	Type 4
1	0.34	0.18	0.40	0.33
2	0.31	0.31	0.21	0.29
3	0.08	0.16	0.27	0.13
4	0.26	0.00	0.38	0.24
5	0.29	0.07	0.00	0.10
6	0.00	0.12	0.08	0.45
7	0.09	0.00	0.19	0.37
8	0.14	0.00	0.36	0.19
9	0.26	0.04	0.34	0.53
10	0.19	0.09	0.44	0.56

The primary interest was in choosing the tire (or tires) that tended to show less wear. Which tire (or tires), if any, should be chosen?

20. Let X equal the number of automobiles in a household, and let Y equal the number of licensed drivers in the household. Suppose the population relative frequencies of values of X and Y are given as follows.

X	Y	Prob (X, Y)	X	Y	Prob (X, Y)
0	0	0.10	1	2	0.10
0	1	0.10	2	0	0.10
0	2	0.05	2	1	0.10
1	0	0.20	2	2	0.05
1	1	0.20			

(a) Find $E(X)$ and $E(Y)$.
(b) Find the median of X.
(c) Find the variance of Y.
(d) Are X and Y independent?
(e) Draw a graph of the distribution function of Y.
(f) If $f(x, y)$ is the probability function of (X, Y), find $f(1, 1)$.
(g) If $F(x, y)$ is the distribution function of (X, Y), find $F(1, 1)$.
(h) Find the correlation coefficient between X and Y.

21. A random sample of households resulted in the following data, where X and Y are the same as in Problem 20. That is, three households had no cars and no licensed drivers, etc.

X	Y	Frequency	X	Y	Frequency
0	0	3	1	2	1
0	1	2	2	0	2
0	2	0	2	1	0
1	0	3	2	2	2
1	1	4			

(a) Find the sample mean of X and Y.
(b) Find the sample median of X.
(c) Find the sample variance of Y (use n, not $n - 1$, as the divisor).
(d) Draw a graph of the empirical distribution function of Y.

22. Seven adults were enrolled in a speed-reading course designed to help them read faster. Their reading speeds were measured (words per minute) at the beginning of the course and at the end of the course with the following results.

Before	After
270	390
250	380
185	310
310	470
200	380
260	400
260	510

Find a 95% confidence interval for the average increase in reading speed a person can expect to attain from this course.

23. Two chimpanzees, Dick and Jane, are being compared to see which one appears to perform better at pressing the "correct" key. The number of times X that Dick pressed keys until the "correct" one was pressed is recorded for 30 trials. The number of times Y that Jane pressed keys until the "correct" one was pressed is recorded for 20 trials. Is one chimp significantly better than the other at finding the "correct" key?

	Dick		Jane
X	Frequency	Y	Frequency
1	12	1	12
2	8	2	7
3	7	3	1
4	3	4	0
Total	30	Total	20

24. A motorist conducts his own experiment to see if his car gets better gas mileage with Unleaded Plus gasoline than with Regular Unleaded. Each time he fills his car with gasoline he tosses a coin, and if it lands "heads" he fills his car with Unleaded Plus. If it lands "tails" he fills his car with Regular Unleaded. He runs the car until it is almost empty, and computes his gas mileage for each tankful of gas. By chance he filled up his car with Regular Unleaded three times and got mileages of 21.3, 21.2, and 21.6. He filled his car with Unleaded Plus five times and got mileages of 22.1, 22.7, 22.3, 21.5, and 21.8.
    (a) Does his car get more mileage with Unleaded Plus? Use a rank test.
    (b) Find the exact p-value using Fisher's randomization test to analyze the data.
    (c) Find the exact p-value using Smirnov's test to analyze the data.

25. The ages and blood pressures of 15 women are recorded as follows.

Age	Blood Pressure	Age	Blood Pressure
48	144	54	151
60	168	56	152
35	135	31	141
38	125	24	144
55	159	77	170
51	148	63	157
49	128	67	162
38	134		

Does there seem to be a significant monotonic relation between age and blood pressure?

26. Using the data in Problem 25 as a random sample, predict the mean blood pressure of 50-year-old women.

27. Twelve students took an exam and got the following grades:

62, 74, 82, 84, 86, 86, 89, 90, 94, 94, 95, 97

Use a goodness-of-fit test to test whether the data could have come from a normal population.

REFERENCES

Abrahamson, I.G. (1967). Exact Bahadur efficiencies for the Kolmogorov-Smirnov and Kuiper one- and two-sample statistics. *The Annals of Mathematical Statistics, 38,* 1475–1490 (6.1).

Adichie, J.N. (1967a). Asymptotic efficiency of a class of nonparametric tests for regression parameters. *The Annals of Mathematical Statistics, 38,* 884–893 (5.4).

Adichie, J.N. (1967b). Estimates of regression parameters based on rank tests. *The Annals of Mathematical Statistics, 38,* 894–904 (5.4).

Adichie, J.N. (1974). Rank score comparison of several regression parameters. *The Annals of Statistics, 2,* 396–402 (5.5).

Adichie, J.N. (1975). On the use of ranks for testing the coincidence of several regression lines. *The Annals of Statistics, 3,* 521–527 (5.5).

Agresti, A. (1977). Considerations in measuring partial association for ordinal categorical data. *Journal of the American Statistical Association, 72,* 37–45 (5.4).

Agresti, A. (1990). *Categorical Data Analysis.* Wiley, New York (4.7).

Aitchison, J., and Aitken, C.G.G. (1976). Multivariate binary discrimination by the kernel method. *Biometrika, 63,* 413–420 (2.4).

Aitkin, M.A., and Hume, M.W. (1965). Correlation in a singly truncated bivariate normal distribution. II. Rank correlation. *Biometrika, 52,* 639–643 (5.4).

Akritas, M.G. (1990). The rank transform method in some two-factor designs. *Journal of the American Statistical Association, 85,* 73–78 (5.12).

Akritas, M.G. (1991). Limitations of the rank transform procedures: A study of repeated measures designs, part I. *Journal of the American Statistical Association, 86,* 457–460 (5.12).

Allan, G.J.B. (1976). Ordinal-scaled variables and multivariate analysis: Comment on Hawkes. *American Journal of Sociology, 81,* 1498–1500 (5.12).

Alling, D.W. (1963). Early decision in the Wilcoxon two-sample test. *Journal of the American Statistical Association, 58,* 713–720 (5.1).

Altham, P.M.E. (1971). The analysis of matched proportions. *Biometrika, 58,* 561–566 (3.5).

Andel, J. (1967). Local asymptotic power and efficiency of tests of Kolmogorov-Smirnov type. *The Annals of Mathematical Statistics, 38,* 1705–1725 (6.3).

Anderson, T.W. (1962). On the distribution of the two-sample Cramer-von Mises criterion. *The Annals of Mathematical Statistics, 33,* 1148–1159 (6.3).

Anderson, T.W., and Burstein, H. (1967). Approximating the upper binomial confidence limit. *Journal of the American Statistical Association, 62,* 857–861 (3.1).

Anderson, T.W., and Burstein, H. (1968). Approximating the lower binomial confidence limit. *Journal of the American Statistical Association, 63*, 1413–1415 (corrections appear in Vol. 64, p. 669) (3.1).

Anderson, T.W., and Darling, D.A. (1952). Asymptotic theory of certain "goodness of fit" criteria based on stochastic processes. *The Annals of Mathematical Statistics, 23*, 193–212 (6.1, 6.3).

Ansari, A.R., and Bradley, R.A. (1960). Rank-sum tests for dispersion. *The Annals of Mathematical Statistics, 31*, 1174–1189 (5.3).

Arbuthnott, J. (1710). An argument for divine providence, taken from the constant regularity observed in the births of both sexes. *Philosophical Transactions, 27*, 186–190 (3.4).

Arnold, H.J. (1965). Small sample power for the one sample Wilcoxon test for non-normal shift alternatives. *The Annals of Mathematical Statistics, 36*, 1767–1778 (5.7).

Barlow, R.E., and Gupta, S.S. (1966). Distribution-free life test sampling plans. *Technometrics, 8*, 591–614 (3.2).

Barr, D.R., and Davidson, T. (1973). A Kolmogorov-Smirnov test for censored samples. *Technometrics, 15*, 739–757 (6.1).

Barr, D.R., and Shudde, R.H. (1973). A note on Kuiper's V_n statistic. *Biometrika, 60*, 663–664 (6.1).

Bartels, R.H. (1982). The rank version of von Neumann's ratio test for randomness. *Journal of the American Statistical Association, 77*, 40–46 (5.4).

Bartels, R.H., Horn, S.D., Liebetrau, A.M., and Harris, W.L. (1977). A computational investigation of Conover's Kolmogorov-Smirnov test for discrete distributions. Department of Mathematical Sciences, Johns Hopkins University, Technical Report No. 260 (6.1).

Basu, A.P. (1967a). On the large sample properties of a generalized Wilcoxon-Mann-Whitney statistic. *The Annals of Mathematical Statistics, 38*, 905–915 (5.1).

Basu, A.P. (1967b). On two k-sample rank tests for censored data. *The Annals of Mathematical Statistics, 38*, 1520–1535 (5.2).

Basu, A.P. (1968). On a generalized Savage statistic with applications to life testing. *The Annals of Mathematical Statistics, 39*, 1591–1604 (5.1).

Basu, A.P., and Woodworth, G. (1967). A note on nonparametric tests for scale. *The Annals of Mathematical Statistics, 38*, 274–277 (5.3).

Batschelet, E. (1965). *Statistical Methods for the Analysis of Problems in Animal Orientation and Certain Biological Rhythms.* The American Institute of Biological Sciences, Washington, D.C. (3.4, 5.1, 5.7).

Bauer, D.F. (1972). Constructing confidence sets using rank statistics. *Journal of the American Statistical Association, 67*, 687–690 (5.1, 5.3).

Behnen, K. (1976). Asymptotic comparison of rank tests for the regression problem when ties are present. *The Annals of Statistics, 4*, 157–174 (5.5).

Bell, C.B. (1964). Some basic theorems of distribution-free statistics. *The Annals of Mathematical Statistics, 35*, 150–156 (2.5).

Bell, C.B., and Doksum, K.A. (1965). Some new distribution-free statistics. *The Annals of Mathematical Statistics, 36*, 203–214 (5.10).

Bell, C.B., and Doksum, K.A. (1967). Distribution-free tests of independence. *The Annals of Mathematical Statistics, 38*, 429–446 (5.4).

Bell, C.B., and Haller, H.S. (1969). Bivariate symmetry tests: Parametric and nonparametric. *The Annals of Mathematical Statistics, 40*, 259–269 (5.7).

Benard, A., and van Elteren, P. (1953). A generalization of the method of m rankings.

Proceedings Koninklijke Nederlandse Akademie van Wetenschappen (A), 56 (Indagationes Mathematicae, 15), 358–369 (5.9).

Bennett, B.M. (1965). On multivariate signed rank tests. *Annals of the Institute of Statistical Mathematics, 17*, 55–61 (5.7).

Bennett, B.M., and Nakamura, E. (1963). Tables for testing significance in a 2×3 contingency table. *Technometrics, 5*, 501–511 (4.2).

Bennett, B.M., and Nakamura, E. (1964). The power function of the exact test for the 2×3 contingency table. *Technometrics, 6*, 439–458 (4.2).

Bennett, B.M., and Underwood, R.E. (1970). On McNemar's test for the 2×2 table and its power function. *Biometrics, 26*, 339–343 (3.5).

Beran, R.J. (1969). The derivation of nonparametric two-sample tests from tests for uniformity of a circular distribution. *Biometrika, 56*, 561–570 (5.1).

Beran, R.J. (1977). Robust location estimates. *The Annals of Statistics, 5*, 431–444 (5.7).

Berger, A., and Gold, R.Z. (1973). Note on Cochran's Q-test for the comparison of correlated proportions. *Journal of the American Statistical Association, 68*, 989–993 (4.6).

Berkson, J. (1980). Minimum chi-square, not maximum likelihood! *The Annals of Statistics, 8*, 457–487 (including comments by several discussants) (4.5).

Best, D.J. (1973). Extended tables for Kendall's tau. *Biometrika, 60*, 429–430 (5.4).

Best, D.J. (1974). Tables for Kendall's tau and an examination of the normal approximation. Division of Mathematical Statistics Technical Paper No. 39, Commonwealth Scientific and Industrial Research Organization, Australia (5.4, Appendix).

Bhapkar, V.P., and Deshpande, J.V. (1968). Some nonparametric tests for multisample problems. *Technometrics, 10*, 578–585 (5.2).

Bhapkar, V.P., and Koch, G.G. (1968). Hypotheses of "no interaction" in multidimensional contingency tables. *Technometrics, 10*, 107–124 (4.2).

Bhapkar, V.P., and Patterson, K.W. (1977). On some nonparametric tests for profile analysis of several multivariate samples. *Journal of Multivariate Analysis, 7*, 265–277 (2.5).

Bhapkar, V.P., and Somes, G.W. (1977). Distribution of Q when testing equality of matched proportions. *Journal of the American Statistical Association, 72*, 658–661 (4.6).

Bhattacharyya, G.K. (1967). Asymptotic efficiency of multivariate normal score test. *The Annals of Mathematical Statistics, 39*, 1731–1743 (5.1).

Bhattacharyya, G.K., and Johnson, R.A. (1968). Nonparametric tests for shift at unknown time point. *The Annals of Mathematical Statistics, 39*, 1731–1743 (5.1).

Bhattacharyya, G.K., Johnson, R.A., and Neave, H.R. (1971). A comparative power study of the bivariate rank sum test and T^2. *Technometrics, 13*, 191–198 (5.7).

Bhattacharyya, H.T. (1977). Nonparametric estimation of ratio of scale parameters. *Journal of the American Statistical Association, 72*, 459–463 (5.3).

Bickel, P.J. (1969). A distribution free version of the Smirnov two sample test in the p-variate case. *The Annals of Mathematical Statistics, 40*, 1–23 (6.3).

Bickel, P.J., and Lehmann, E.L. (1975). Descriptive statistics for nonparametric models. *The Annals of Statistics, 3*, 1038–1069 (5.1, 5.7).

Birnbaum, Z.W. (1953). On the power of a one-sided test of fit for continuous probability functions. *The Annals of Mathematical Statistics, 24*, 484–489 (6.1).

Birnbaum, Z.W. (1962). *Introduction to Probability and Mathematical Statistics.* Harper, New York (4.5).

Birnbaum, Z.W., and Hall, R.A. (1960). Small sample distributions for multisample statistics of the Smirnov type. *The Annals of Mathematical Statistics, 31*, 710–720 (6.4, Appendix).

Birnbaum, Z.W., and Tingey, F.H. (1951). One-sided confidence contours for probability distribution functions. *The Annals of Mathematical Statistics, 22,* 592–596 (6.1).

Birnbaum, Z.W., and Zuckerman, H.S. (1949). A graphical determination of sample size for Wilks' tolerance limits. *The Annals of Mathematical Statistics, 20,* 313–317 (3.3).

Bishop, Y.M.M. (1969). Full contingency tables, logits, and split contingency tables. *Biometrics, 25,* 383–400 (4.7).

Bishop, Y.M.M. (1971). Effects of collapsing multidimensional contingency tables. *Biometrics, 27,* 545–562 (4.7).

Bishop, Y.M.M., Fienberg, S.E., and Holland, P.W. (1975). *Discrete Multivariate Analysis: Theory and Practice.* The MIT Press, Cambridge, Mass. (4.7).

Blair, R.C., Sawilowsky, S.S., and Higgins, J.J. (1987). Limitations of the rank transform statistic in tests for interactions. *Communications in Statistics—Simulation, B16,* 1133–1145 (5.12).

Blomqvist, N. (1951). Some tests based on dichotomization. *The Annals of Mathematical Statistics, 22,* 362–371 (4.6).

Blum, J.R., and Fattu, N.A. (1954). Nonparametric methods. *Review of Educational Research, 24:5,* 467–487 (2.5).

Bohrer, R. (1968). A note on tolerance limits with type I censoring. *Technometrics, 10,* 392 (3.3).

Bowden, D.C. (1968). Query: Tolerance interval in regression. *Technometrics, 10,* 207–210 (3.3).

Bowman, K.O., and Shenton, L.R. (1975). Omnibus test contours for departures from normality based on $\sqrt{b_1}$ and b_2. *Biometrika, 62,* 243–250 (6.2).

Boyett, J.M., and Shuster, J.J. (1977). Nonparametric one-sided tests in multivariate analysis with medical applications. *Journal of the American Statistical Association, 72,* 665–668 (5.11).

Bradley, J.V. (1968). *Distribution-Free Statistical Tests.* Prentice-Hall, Englewood Cliffs, N.J. (5.10).

Bradley, R.A., Martin, D.C., and Wilcoxon, F. (1965). Sequential rank tests I. Monte Carlo studies of the two-sample procedure. *Technometrics, 7,* 463–483 (5.1).

Bradley, R.A., Merchant, S.D., and Wilcoxon, F. (1966). Sequential rank tests II. Modified two-sample procedures. *Technometrics, 8,* 615–624 (5.1).

Bradley, R.A., Patel, K.M., and Wackerly, D.D. (1971). Approximate small sample distributions for multivariate two-sample nonparametric tests. *Biometrics, 27,* 515–530 (5.10).

Breslow, N. (1970). A generalized Kruskal-Wallis test for comparing k samples and subject to unequal patterns of censorship. *Biometrika, 57,* 579–594 (5.2).

Breslow, N.E., and Liang, K.Y. (1982). The variance of the Mantel-Haenszel estimator. *Biometrics, 38,* 943–952 (4.1).

Broffitt, J.D., Randles, R.H., and Hogg, R.V. (1976). Distribution-free partial discriminant analysis. *Journal of the American Statistical Association, 71,* 934–939 (2.5).

Brundenk M.N. (1972). The analysis of non-independent 2×2 tables from $2 \times C$ tables using rank sums. *Biometrics, 28,* 603–607 (5.2).

Buckle, N., Kraft, C.H., and van Eeden, C. (1969). An approximation to the Wilcoxon-Mann-Whitney distribution. *Journal of the American Statistical Association, 64,* 591–599 (5.1).

Buringer, H. (1980). *Nonparametric Sequential Selection Procedures.* Birkhauser, Boston (2.5).

Burr, E.J. (1963). Distribution of the two-sample Cramér-von Mises criterion for small equal samples. *The Annals of Mathematical Statistics, 34,* 95–101 (6.3).

Burr, E.J. (1964). Small-sample distributions of the two-sample Cramér-von Mises' W^2 and Watson's U^2. *The Annals of Mathematical Statistics, 35,* 1091–1098 (6.3).

Capon, J. (1965). On the asymptotic efficiency of the Kolmogorov-Smirnov test. *Journal of the American Statistical Association, 60,* 843–853 (6.3).

Carnal, H., and Riedwyl, H. (1972). On a one-sample distribution-free test statistic V. *Biometrika, 59,* 465–467 (6.1).

Casady, R.J., and Cryer, J.D. (1976). Monotone percentile regression. *The Annals of Statistics, 4,* 532–541 (5.6).

Chacko, V.J. (1966). Modified chi-square test for ordered alternatives. *Sankhya (B), 28,* 185–190 (4.2).

Chan, N.H., and Tran, L.T. (1992). Nonparametric tests for serial dependence. *Journal of Time Series Analysis, 13,* 19–28 (5.4).

Chanda, K.C. (1963). On the efficiency of two-sample Mann-Whitney test for discrete populations. *The Annals of Mathematical Statistics, 34,* 612–617 (5.1).

Chapman, D.G., and Meng, R.C. (1966). The power of chi-square tests for contingency tables. *Journal of the American Statistical Association, 61,* 965–975 (4.2).

Chase, G.R. (1972). On the chi-square test when the parameters are estimated independently of the sample. *Journal of the American Statistical Association, 67,* 609–611 (4.5).

Chatterjee, S.K. (1966). A bivariate sign test for location. *The Annals of Mathematical Statistics, 37,* 1771–1782 (3.5).

Chen, T., and Fienberg, S.E. (1974). Two-dimensional contingency tables with both completely and partially cross-classified data. *Biometrics, 30,* 629–642 (4.2).

Chen, T., and Fienberg, S.E. (1976). The analysis of contingency tables with incompletely classified data. *Biometrics, 32,* 133–144 (4.7).

Chernoff, H. (1967). Query: Degrees of freedom for chi-square. *Technometrics, 9,* 489–490 (4.5).

Chernoff, H., and Lehmann, E.L. (1954). The use of maximum likelihood estimates in χ^2 tests for goodness of fit. *The Annals of Mathematical Statistics, 25,* 579–686 (4.5).

Chiacchierini, R.P., and Arnold, J.C. (1977). A two-sample test for independence in 2×2 contingency tables with both margins subject to misclassification. *Journal of the American Statistical Association, 72,* 170–174 (4.1).

Chmiel, J.J. (1976). Some properties of Spearman-type estimators of the variance and percentiles in bioassay. *Biometrika, 63,* 621–626 (2.5).

Choi, S.C. (1973). On nonparametric sequential tests for independence. *Technometrics, 15,* 625–629 (5.4).

Chow, W.K., and Hodges, J.L., Jr. (1975). An approximation for the distribution of the Wilcoxon one-sample statistic. *Journal of the American Statistical Association, 70,* 648–655 (5.7).

Chung, J.H., and Fraser, D.A.S. (1958). Randomization test for a multivariate two-sample problem. *Journal of the American Statistical Association, 53,* 729–735 (5.11).

Claringbold, P.J. (1961). The use of orthogonal polynomials in the partition of chi-square. *The Australian Journal of Statistics, 3,* 48–63 (4.2).

Claypool, P.L. (1970). Linear interpolation within McCormack's table of the Wilcoxon signed rank statistic. *Journal of the American Statistical Association, 65,* 974–975 (5.7).

Clayton, D.G. (1974). Some odds ratio statistics for the analysis of ordered categorical data. *Biometrika, 61,* 525–531 (4.2).

Cleroux, R. (1969). First and second moments of the randomization test in two-associate PBIB designs. *Journal of the American Statistical Association, 64,* 1424–1433 (5.11).

Clopper, C.J., and Pearson, E.S. (1934). The use of confidence or fiducial limits illustrated in the case of the binomial. *Biometrika, 26,* 404–413 (3.1).

Coberley, W.A., and Lewis, T.O. (1973). A note on a one-sided Kolmogorov-Smirnov test of fit for discrete distribution functions. *Annals of the Institute of Statistical Mathematics, 24,* 183–187 (6.1).

Cochran, W.G. (1937). The efficiencies of the binomial series tests of significance of a mean and of a correlation coefficient. *Journal of the Royal Statistical Society, 100,* 69–73 (3.5).

Cochran, W.G. (1950). The comparison of percentages in matched samples. *Biometrika, 37,* 256–266 (4.6).

Cochran, W.G. (1952). The χ^2 test of goodness of fit. *Annals of Mathematical Statistics, 23,* 315–345 (4.2, 4.5).

Cohen, A., and Sackrowitz, H.B. (1975). Unibiasedness of the chi-square, likelihood ratio, and other goodness of fit tests for the equal cell case. *The Annals of Statistics, 3,* 959–964 (4.5).

Collier, R.O. Jr., and Baker, F.B. (1963). The randomization distribution of F-ratios for the split-plot design—An empirical investigation. *Biometrika, 50,* 431–438 (5.11).

Collier, R.O. Jr., and Baker, F.B. (1966). Some Monte Carlo results on the power of the F-test under permutation in the simple randomized block design. *Biometrika, 53,* 199–203 (5.11).

Conover, W.J. (1965). Several k-sample Kolmogorov-Smirnov tests. *The Annals of Mathematical Statistics, 36,* 1019–1026 (6.4).

Conover, W.J. (1967a). The distribution functions of Tsao's truncated Smirnov statistics. *The Annals of Mathematical Statistics, 38,* 1208–1215 (6.3).

Conover, W.J. (1967b). A k-sample extension of the one-sided two-sample Smirnov test statistic. *The Annals of Mathematical Statistics, 38,* 1726–1730 (6.4).

Conover, W.J. (1972). A Kolmogorov goodness-of-fit test for discontinuous distributions. *Journal of the American Statistical Association, 67,* 591–596 (6.1).

Conover, W.J. (1973a). Rank tests for one sample, two samples and k samples without the assumption of a continuous distribution function. *The Annals of Statistics, 1,* 1105–1125 (5.1, 5.2, 5.7).

Conover, W.J. (1973b). On methods of handling ties in the Wilcoxon signed-rank test. *Journal of the American Statistical Association, 68,* 985–988 (5.7).

Conover, W.J. (1974). Some reasons for not using the Yates continuity correction on 2 × 2 contingency tables. *Journal of the American Statistical Association, 69,* 374–376 (4.1).

Conover, W.J. (1980). *Practical Nonparametric Statistics,* 2nd Ed. Wiley, New York (6.2).

Conover, W.J., and Iman, R.L. (1976). On some alternative procedures using ranks for the analysis of experimental designs. *Communications in Statistics—Theory and Methods, A5,* 1349–1368 (5.12).

Conover, W.J., and Iman, R.L. (1978a). Some exact tables for the squared ranks test. *Communications in Statistics, B7,* 491–513 (5.3).

Conover, W.J., and Iman, R.L. (1978b). The rank transformation as a method of discrimination with some examples. Technical Report SAND78-0583, Sandia Laboratories, Albuquerque (5.12).

Conover, W.J., and Iman, R.L. (1979). On multiple comparison procedures. Technical Report LA-7677-MS, Los Alamos Scientific Laboratory (5.2).

Conover, W.J., and Iman, R.L. (1980). The rank transformation as a method of discrimination with some examples. *Communications in Statistics, A9,* 465–487 (2.5, 5.12).

Conover, W.J., Johnson, M.E., and Johnson, M.M. (1981). A comparative study of tests for homogeneity of variances, with applications to the outer continental shelf bidding data. *Technometrics, 23,* 351–361 (5.3).

Conover, W.J., and Kemp, K.E. (1976). Comparisons of the asymptotic efficiencies of two sample tests for discrete distributions. *Communications in Statistics—Theory and Methods,* A5, 1–15 (5.1).

Conover, W.J., Wehmanen, O., and Ramsey, F.L. (1978). A note on the small sample power functions for nonparametric tests of location in the double exponential family. *Journal of the American Statistical Association, 73,* 188–190 (5.1, 5.7).

Cox, D.R. (1966). The null distribution of the first serial correlation coefficient. *Biometrika, 53,* 623–626 (5.4).

Cox, D.R., and Stuart, A. (1955). Some quick tests for trend in location and dispersion. *Biometrika, 42,* 80–95 (3.5).

Cramér, H. (1928). On the composition of elementary errors. *Skandinavisk Aktuarietidskrift, 11,* 13–74, 141–180 (6.1).

Cramér, H. (1946). *Mathematical Methods of Statistics.* Princeton University Press, Princeton, N.J. (4.1, 4.2, 4.4, 4.5).

Cronholm, J.N. (1968). Two tables connected with goodness-of-fit tests for equiprobable alternatives. *Biometrika, 55,* 441 (6.1).

Crouse, C.F. (1966). Distribution-free tests based on the sample distribution function. *Biometrika, 53,* 99–108 (5.2).

Crouse, C.F. (1967). A class of distribution-free analysis of variance tests. *South African Statistics Journal, 1,* 75–80 (5.12).

Crouse, C.F. (1968). A distribution-free method of analyzing a 2^m factorial experiment. *South African Statistics Journal, 2,* 101–108 (5.12).

Crowley, J., and Breslow N. (1975). Remarks on the conservatism of $\Sigma(0 - E)^2 / E$ in survival data. *Biometrics, 31,* 957–961 (4.2).

Cryer, J.D., Robertson, T., Wright, F.T., and Casady, R.J. (1972). Monotone median regression. *The Annals of Mathematical Statistics, 43,* 1459–1469 (5.6).

Csörgö, M. (1965). Some Smirnov type theorems of probability. *The Annals of Mathematical Statistics, 36,* 1113–1119 (6.3).

D'Agostino, R.B. (1971). An omnibus test of normality for moderate and large size samples. *Biometrika, 58,* 341–348 (6.2).

Dahiya, R.C. (1971). On the Pearson chi-squared goodness-of-fit test statistic. *Biometrika, 58,* 685–686 (4.5).

Dahiya, R.C., and Gurland, J. (1972). Pearson chi-squared test of fit with random intervals. *Biometrika, 59,* 147–153 (4.5).

Dahiya, R.C., and Gurland, J. (1973). How many classes in the Pearson chi-square test? *Journal of the American Statistical Association, 68,* 707–712 (4.5).

Daniel, W. W. (1979). *On Nonparametric and Robust Tests for Dispersion: A Selected Bibliography.* Public administration series—bibliography, P-382, Vance Bibliographies, Monticello, Ill. (5.3).

Daniel, W.W. (1980). *Nonparametric, Distribution-Free, and Robust Procedures in Regression Analysis: A Selected Bibliography.* Vance Publishers, Monticello, Ill. (5.4).

Daniels, H.E. (1950). Rank correlation and population models. *Journal of the Royal Statistical Society (B), 12,* 171–181 (5.4).

Danziger, L., and Davis, S.A. (1964). Tables of distribution-free tolerance limits. *The Annals of Mathematical Statistics, 35,* 1361–1365 (3.3).

Darroch, J.N. (1974). Multiplicative and additive interaction in contingency tables. *Biometrika, 61,* 207–214 (4.2).

Davis, J.A. (1967). A partial coefficient for Goodman and Kruskal's gamma. *Journal of the American Statistical Association, 62,* 189–193 (4.4).

Davison, A.C. and Hinkley, D.V. (1997). *Bootstrap Methods and their Applications.* Cambridge Univ. Press, Cambridge, England (2.2).

Deshpande, J.V. (1970). A class of multisample distribution-free tests. *The Annals of Mathematical Statistics, 41,* 227–236 (5.2).

Diamond, E.L. (1963). The limiting power of categorical data chi-square tests analogous to normal analysis of variance. *The Annals of Mathematical Statistics, 34,* 1432–1441 (4.2).

Dixon, W.J. (1953). Power functions of the sign test and power efficiency for normal alternatives. *The Annals of Mathematical Statistics, 24,* 467–473 (2.4, 3.4).

Doksum, K.A. (1967). Robust procedures for some linear models with one observation per cell. *The Annals of Mathematical Statistics, 38,* 878–883 (5.8).

Doksum, K.A., and Sievers, G.L. (1976). Plotting with confidence: Graphical comparisons of two populations. *Biometrika, 63,* 421–434 (5.1).

Dufour, J.M. (1981). Rank tests for serial dependence. *Journal of Time Series Analysis, 2,* 117–128 (5.4).

Dufour, J.M., and Roy, R. (1985). Some robust exact results on sample autocorrelations and tests for randomness. *Journal of Econometrics, 29,* 257–273 (5.4).

Dunn, O.J. (1964). Multiple comparisons using rank sums. *Technometrics, 6,* 241–252 (5.8).

Duran, B.S. (1976). A survey of nonparametric tests for scale. *Communications in Statistics—Theory and Methods, A5,* 1287–1312 (5.3).

Duran, B.S., and Mielke, P.W., Jr. (1968). Robustness of the sum of squared ranks test. *Journal of the American Statistical Association, 63,* 338–344 (5.3).

Duran, B.S., Tsai, W.S., and Lewis, T.O. (1976). A class of location-scale nonparametric tests. *Biometrika, 63,* 173–176 (5.3).

Durbin, J. (1951). Incomplete blocks in ranking experiments. *British Journal of Psychology (Statistical Section), 4,* 85–90 (5.9).

Durbin, J. (1961). Some methods of constructing exact tests. *Biometrika, 48,* 41–55 (5.10).

Durbin, J. (1968). The probability that the sample distribution function lies between two parallel straight lines. *The Annals of Mathematical Statistics, 39,* 398–411 (6.1).

Durbin, J. (1975). Kolmogorov-Smirnov tests when parameters are estimated with applications to tests of exponentiality and tests on spacings. *Biometrika, 62,* 5–22 (6.2, Appendix).

Dyer, A.R. (1974). Comparison of tests for normality with a cautionary note. *Biometrika, 61,* 185–189 (6.2).

Efron, B., and Morris, C. (1975). Data analysis using Stein's estimator and its generalizations. *Journal of the American Statistical Association, 70,* 311–319 (4.5).

Efron, B., and Tibshirani, R. (1986). Bootstrap methods for standard errors, confidence intervals, and other methods of statistical accuracy. *Statistical Science, 1* (1), 54–77 (2.2).

Ehrenberg, A.S.C. (1951). Note on normal transformations of ranks. *British Journal of Psychology (Statistical Section), 4,* 133–134 (5.10).

Elston, R.C. (1970). A new test of association for continuous variables. *Biometrics, 26,* 305–314 (4.2).

Evans, L.S. (1973). A mechanical interpretation of the coefficient of rank correlation and other analogies. *The American Statistician, 27* (2), 79–81 (5.4).

Federer, W. T. (1963). *Experimental Design.* Macmillan, New York (5.9).

Feller, W. (1968). *An Introduction to Probability Theory and Its Applications,* Vol. I, 3rd ed. J. Wiley, New York (1.1).

Festinger, L. (1946). The significance of difference between means without reference to the frequency distribution function. *Psychometrika, 11,* 97–105 (5.1).

Feynman, R.P. (1988). *What Do You Care What Other People Think?* W.W. Norton & Company, New York (5.4).

Fienberg, S.E. (1970a). An iterative procedure for estimation in contingency tables. *The Annals of Mathematical Statistics, 41,* 907–917 (correction appears in Vol. 42, p. 1778) (4.2).

Fienberg, S.E. (1970b). Quasi-independence and maximum likelihood estimation in incomplete contingency tables. *Journal of the American Statistical Association, 65,* 1610–1616 (4.7).

Fienberg, S.E. (1972). The analysis of incomplete multiway contingency tables. *Biometrics, 28,* 177–202 (4.7).

Fienberg, S.E. (1977). *The Analysis of Cross-Classified Categorical Data.* The MIT Press, Cambridge, Mass. (4.7).

Fienberg, S.E., and Gilbert, J.P. (1970). The geometry of a two by two contingency table. *Journal of the American Statistical Association, 65,* 694–701 (4.1).

Fienberg, S.E., and Larntz, K. (1976). Log linear representation for paired and multiple comparison models. *Biometrika, 63,* 245–254 (4.7).

Fine, T. (1966). On the Hodges and Lehmann shift estimator in the two sample problem. *The Annals of Mathematical Statistics, 37,* 1814–1818 (6.3).

Finkelstein, J.M., and Schafer, R.E. (1971). Improved goodness-of-fit tests. *Biometrika, 58,* 641–646 (6.1).

Finney, D.J. (1948). The Fisher-Yates test of significance in 2×2 contingency tables. *Biometrika, 35,* 145–156 (4.2).

Fisher, R.A. (1935). *The Design of Experiments.* Oliver & Boyd, Edinburgh-London (7th ed., 1960) (5.11).

Fisher, R.A., and Yates, F. (1957). *Statistical Tables for Biological, Agricultural, and Medical Research,* 5th ed. Oliver & Boyd, Edinburgh (5.10).

Fisz, M. (1960). On a result by M. Rosenblatt concerning the von Mises-Smirnov test. *The Annals of Mathematical Statistics, 31,* 427–429 (6.3).

Fisz, M. (1963). *Probability Theory and Mathematical Statistics,* 3rd ed. J. Wiley, New York (6.3).

Fleiss, J. L. (1965). A note on Cochran's Q test. *Biometrics, 21,* 1008–1010 (4.6).

Fleiss, J.L. (1973). *Statistical Methods for Rates and Proportions.* J. Wiley, New York (4.1).

Fligner, M.A., Hogg, R.V., and Killeen, T.J. (1976). Some distribution-free rank-like statistics having the Mann-Whitney-Wilcoxon null distribution. *Communications in Statistics— Theory and Methods, A5,* 373–376 (5.1).

Fligner, M.A., and Killeen, T.J. (1976). Distribution-free two-sample tests for scale. *Journal of the American Statistical Association, 71,* 210–213 (5.3).

Fox, J.R. and Randall, J.E. (1970). Relationship between forearm tremor and the biceps electromyogram. *Journal of Applied Physiology, 29,* 103–108 (5.8).

Fraser, D.A.S. (1957). *Nonparametric Methods in Statistics.* J. Wiley, New York (2.5, 5.10).

Freund, J.E. (1962). *Mathematical Statistics.* Prentice-Hall, Englewood Cliffs, N.J. (1.5).

Freund, J.E., and Ansari, A.R. (1957). *Two-way rank sum test for variances.* Technical Report No. 34, Virginia Polytechnic Institute, Blacksburg (5.1).

Friedman, M. (1937). The use of ranks to avoid the assumption of normality implicit in the analysis of variance. *Journal of the American Statistical Association, 32,* 675–701 (5.8).

Gabriel, K.R. (1966). Simultaneous test procedures for multiple comparisons on categorical data. *Journal of the American Statistical Association, 61,* 1081–1096 (4.3).

Gabriel, K.R., and Lachenbruch, P.A. (1969). Non-parametric ANOVA in small samples: A Monte Carlo study of the adequacy of the asymptotic approximation. *Biometrics, 25,* 593–596 (5.2).

Gail, M.H., and Gart, J.J. (1973). The determination of sample sizes for use with the exact conditional test in 2 × 2 comparative trials. *Biometrics, 29,* 441–448 (4.1).

Gail, M.H., and Green, S.B. (1976a). Critical values for the one-sided two-sample Kolmogorov-Smirnov statistic. *Journal of the American Statistical Association, 71,* 757–760 (6.3).

Gail, M.H., and Green, S.B. (1976b). A generalization of the one-sided two-sample Kolmogorov-Smirnov statistic for evaluating diagnostic tests. *Biometrics, 32,* 561–570 (6.3).

Garside, G.R., and Mack, C. (1976). Actual type 1 error probabilities for various tests in the homogeneity case of the 2 × 2 contingency table. *The American Statistician, 30* (1), 18–21 (4.1).

Gart, J.J. (1966). Alternative analyses of contingency tables. *Journal of the Royal Statistical Society (B), 28,* 164–179 (4.2).

Gart, J.J. (1972). Interaction tests for $2 \times s \times t$ contingency tables. *Biometrika, 59,* 309–316 (4.7).

Gastwirth, J.L. (1965a). Asymptotically most powerful rank tests for the two-sample problem with censored data. *The Annals of Mathematical Statistics, 36,* 1243–1248 (5.1).

Gastwirth, J.L. (1965b). Percentile modifications of two sample rank tests. *Journal of the American Statistical Association, 60,* 1127–1141 (5.1).

Geertsema, J.C. (1970). Sequential confidence intervals based on rank tests. *The Annals of Mathematical Statistics, 41,* 1016–1026 (5.7).

Gehan, E.A. (1965a). A generalized Wilcoxon test for comparing arbitrarily singly censored samples. *Biometrika, 52,* 203–224 (5.1).

Gehan, E.A. (1965b). A generalized two-sample Wilcoxon test for doubly censored data. *Biometrika, 52,* 650–653 (5.1).

Gehan, E.A., and Thomas, D.G. (1969). The performance of some two-sample tests in small samples with and without censoring. *Biometrika, 56,* 127–132 (5.1).

Gelzer, J., and Pyke, R. (1965). The asymptotic relative efficiency of goodness-of-fit tests against scalar alternatives. *Journal of the American Statistical Association, 60,* 410–419 (6.1).

Gerig, T.M. (1969). A multivariate extension of Friedman's χ_r^2-test. *Journal of the American Statistical Association, 64,* 1595–1608 (5.8).

Gerig, T.M. (1975). A multivariate extension of Friedman's χ_r^2-test with random covariates. *Journal of the American Statistical Association, 70,* 443–447 (5.8).

Gessaman, M.P., and Gessaman, P.H. (1972). A comparison of some multivariate discrimination procedures. *Journal of the American Statistical Association, 67* (338), 468–472 (2.5).

Ghosh, M., Grizzle, J.E., and Sen, P.K. (1973). Nonparametric methods in longitudinal studies. *Journal of the American Statistical Association, 68,* 29–36 (2.5).

Gibbons, J.D. (1964). Effect of non-normality on the power function of the sign test. *Journal of the American Statistical Association, 59,* 142–148 (3.4).

Gibbons, J.D. (1967). Correlation coefficients between nonparametric tests for location and scale. *Annals of the Institute of Statistical Mathematics, 19,* 519–526 (5.3).

Gibbons, J.D. (1993). *Nonparametric Measures of Association.* Sage University Paper series on Quantitative Applications in the Social Sciences, 07-091, Newbury Park, CA (5.4).

Gibbons, J.D., and Gastwirth, J.L. (1970). Properties of the percentile modified rank tests. *Annals of the Institute of Statistical Mathematics,* Supplement 6, 95–114 (5.1).

Gilbert, R.O. (1972). A Monte Carlo study of analysis of variance and competing rank tests for Scheffe's mixed model. *Journal of the American Statistical Association, 67,* 71–75 (5.8).

Glasser, G.J., and Winter, R.F. (1961). Critical values of the coefficient of rank correlation for testing the hypothesis of independence. *Biometrika, 48,* 444–448 (Appendix).

Gnedenko, B.V., and Korolyuk, V.S. (1951). On the maximum discrepancy between two empirical distributions (Russian). *Doklady Akademii Nauk SSSR (N.S.), 80,* 525–528. English translation in *IMS and American Mathematical Society* (1961) (6.3).

Gokhale, D.V. (1968). On asymptotic relative efficiencies of a class of rank tests for independence of two variables. *Annals of the Institute of Statistical Mathematics, 20,* 255–261 (5.4, 5.10).

Goodman, L.A. (1964). Simple methods for analyzing three-factor interaction in contingency tables. *Journal of the American Statistical Association, 59,* 319–352 (4.2).

Goodman, L.A. (1965). On simultaneous confidence intervals for multinomial proportions. *Technometrics, 7,* 247–254 (3.1).

Goodman, L.A. (1968). The analysis of cross-classified data: Independence, quasi-independence, and interactions in contingency tables with or without missing entries. *Journal of the American Statistical Association, 63,* 1091–1113 (4.2).

Goodman, L.A. (1970). The multivariate analysis of qualitative data: Interactions among multiple classifications. *Journal of the American Statistical Association, 65,* 226–256 (4.2).

Goodman, L.A. (1971). Partitioning of chi-square, analysis of marginal contingency tables, and estimation of expected frequencies in multidimensional contingency tables. *Journal of the American Statistical Association, 66,* 339–344 (4.2).

Goodman, L.A. (1971). The analysis of multidimensional contingency tables: Stepwise procedures and direct estimation methods for building models for multiple classifications. *Technometrics, 13,* 33–62 (4.7).

Goodman, L.A., and Kruskal, W.H. (1954). Measures of association for cross-classifications. *Journal of the American Statistical Association, 49,* 732–764 (correction appears in Vol. 52, p. 578) (4.4).

Goodman, L.A., and Kruskal, W.H. (1959). Measures of association for cross-classifications. II: Further discussion and references. *Journal of the American Statistical Association, 54,* 123–163 (4.4).

Goodman, L.A., and Kruskal, W.H. (1963). Measures of association for cross-classifications. III: Approximate sample theory. *Journal of the American Statistical Association, 58,* 310–364 (4.4).

Goodman, L.A., and Madansky, A. (1962). Parameter-free and nonparametric tolerance limits: The exponential case. *Technometrics, 4,* 75–95 (3.3).

Govindarajulu, Z. (1968). Distribution-free confidence bounds for $P(X < Y)$. *Annals of the Institute of Statistical Mathematics, 20,* 229–238 (5.1).

Govindarajulu, Z. (1976). A brief survey of nonparametric statistics. *Communications in Statistics—Theory and Methods, A5,* 429–453 (2.5).

Govindarajulu, Z., and Klotz, J.H. (1973). A note on the asymptotic distribution of the one-sample Kolmogorov-Smirnov statistic. *The American Statistician, 27* (4), 164–165 (6.1).

Govindarajulu, Z., and Leslie, R.T. (1972). Annotated bibliography on robustness studies on statistical procedures. Department of Health, Education, and Welfare Publication No. (HSM) 72-1051 (2.5).

Green, J.R., and Hegazy, Y.A.S. (1976). Powerful modified-EDF goodness-of-fit tests. *Journal of the American Statistical Association, 71,* 204–209 (6.2).

Gregory, G. (1961). Contingency tables with a dependent classification. *The Australian Journal of Statistics, 3,* 42–47 (4.2).

Grizzle, J.E. (1967). Continuity correction in the χ^2-test for 2×2 tables. *The American Statistician, 21* (4), 28–32 (4.1).

Grizzle, J.E., Starmer, C.F., and Koch, G.G. (1969). Analysis of categorical data by linear models. *Biometrics, 25,* 489–504 (4.7).

Grizzle, J.E., and Williams, O.D. (1972). Log linear models and tests of independence for contingency tables. *Biometrics, 28,* 137–156 (4.7).

Groeneveld, R.A. (1972). Asymptotically optimal group rank tests for location. *Journal of the American Statistical Association, 67,* 847–849 (5.7).

Haberman, S.J. (1973). Log-linear models for frequency data: Sufficient statistics and likelihood equations. *The Annals of Statistics, 1,* 617–632 (4.7).

Haga, T. (1960). A two-sample rank test on location. *Annals of the Institute of Statistical Mathematics, 11,* 211–219 (5.1).

Hajek, J., and Sidak, Z. (1967). *Theory of Rank Tests.* Academic Press, New York (5.1, 5.10).

Hallin, M., Ingenbleek, J.-Fr., and Puri, M.L. (1985). Linear serial rank tests for randomness against ARMA alternatives. *Annals of Statistics, 13,* 1156–1181 (5.4).

Hallin, M. and Melard, G. (1988). Rank-based tests for randomness against first-order serial dependence. *Journal of the American Statistical Association, 83,* 1117–1128 (5.4).

Halperin, M., Ware, J.H., Byar, D.P., Mantel, N., Brown, C.C., Koziol, J., Gail, M., and Greer, S.B. (1977). Testing for interaction in an $I \times J \times K$ contingency table. *Biometrika, 64,* 271–275 (4.2).

Hamilton, B.L. (1976). A Monte Carlo test of the robustness of parametric and nonparametric analysis of covariance against unequal regression slopes. *Journal of the American Statistical Association, 71,* 864–869 (5.12).

Hannan, E.J. (1976). The asymptotic distribution of serial covariances. *Annals of Statistics, 4,* 396–399 (5.4).

Hanson, D.L., and Owen, D.B. (1963). Distribution-free tolerance limits, elimination of the requirement that cumulative distribution functions be continuous. *Technometrics, 5,* 518–522 (3.3).

Harel, M. and Puri, M.L. (1990). Weak convergence of serial rank statistics under dependence with applications in time series and Markov processes. *Annals of Probability, 18,* 1361–1387 (5.4).

Harkness, W.L., and Katz, L. (1964). Comparison of the power functions for the test of independence in 2×2 contingency tables. *The Annals of Mathematical Statistics, 35,* 1115–1127 (4.1).

Harter, H.L., and Owen, D.B. (1970). *Selected Tables in Mathematical Statistics,* Vol. 1. Markham, Chicago (5.7, Appendix).

Hartley, H.O., and Pfaffenberger, R.C. (1972). Quadratic forms in order statistics used as goodness-of-fit criteria. *Biometrika, 59,* 605–611 (6.2).

Haynam, G.E., and Govindarajulu, Z. (1966). Exact power of the Mann-Whitney test for exponential and rectangular alternatives. *The Annals of Mathematical Statistics, 37,* 945–953 (5.1).

Haynam, G.E., and Leone, F.C. (1965). Analysis of categorical data. *Biometrika, 52,* 654–660 (4.2).

Healy, M.J.R. (1969). Exact tests of significance in contingency tables. *Technometrics, 11,* 393–395 (4.2).

Hemelrijk, J. (1952). A theorem on the sign test when ties are present. *Proceedings Koninklijke Nederlandse Akademie van Wetenschappen (A), 55,* 322–326 (3.4).

Henley, S. (1981). *Nonparametric Geostatistics.* Wiley, New York (2.5).

Hettmansperger, T.P. (1968). On the trimmed Mann-Whitney statistic. *The Annals of Mathematical Statistics, 39,* 1610–1614 (5.1).

Hettmansperger, T.P., and Malin, J.S. (1975). A modified Mood's test for location with no shape assumptions on the underlying distributions. *Biometrika, 62,* 527–529 (5.1).

Hettmansperger, T.P., and McKean, J.W. (1977). A robust alternative based on ranks to least squares in analyzing linear models. *Technometrics, 19,* 275–284 (5.5).

Hettmansperger, T.P., and McKean, J.W. (1978). Statistical inference based on ranks. *Psychometrika, 43,* 69–79 (5.12).

Hewett, J.E., and Tsutakawa, R.K. (1972). Two-stage chi-square goodness-of-fit test. *Journal of the American Statistical Association, 67,* 395–401 (4.5).

Hocking, R.R., and Oxspring, H.H. (1974). The analysis of partially categorized contingency data. *Biometrics, 30,* 469–484 (4.2).

Hodges, J.L., Jr., and Lehmann, E. (1956). The efficiency of some nonparametric competitors of the *t*-test. *The Annals of Mathematical Statistics, 27,* 324–335 (3.4, 5.1).

Hodges, J.L., Jr., and Lehmann, E.L. (1963). Estimates of location based on rank tests. *The Annals of Mathematical Statistics, 34,* 598–611 (5.1).

Hoeffding, W. (1952). The large-sample power of tests based on permutations of observations. *The Annals of Mathematical Statistics, 23,* 169–192 (5.11).

Hoeffding, W. (1965). Asymptotically optimal tests for multinomial distributions. *The Annals of Mathematical Statistics, 36,* 396–400 (4.2).

Hogg, R.V. (1974). Adaptive robust procedures. *Journal of the American Statistical Association, 69,* 909–923 (2.5).

Hogg, R.V. (1975). Estimates of percentile regression lines using salary data. *Journal of the American Statistical Association, 70,* 56–59 (5.6).

Hogg, R.V. (1976). A new dimension to nonparametric tests. *Communications in Statistics—Theory and Methods, A5,* 1313–1325 (5.10).

Hogg, R.V. (ed.) (1977). Robustness: A special issue of *Communications in Statistics—Theory and Methods, A6,* 789–894 (5.12).

Hollander, M. (1963). A nonparametric test for the two-sample problem. *Psychometrika, 28,* 395–403 (5.3).

Hollander, M. (1967a). Asymptotic efficiency of two nonparametric competitors of Wilcoxon's two sample test. *Journal of the American Statistical Association, 62,* 939–949 (5.1).

Hollander, M. (1967b). Rank tests for randomized blocks when the alternatives have an a priori ordering. *The Annals of Mathematical Statistics, 38,* 867–877 (5.8).

Hollander, M. (1968). Certain uncorrelated nonparametric test statistics. *Journal of the American Statistical Association, 63,* 707–714 (5.3).

Hollander, M. (1970). A distribution-free test for parallelism. *Journal of the American Statistical Association, 65,* 387–394 (5.7).

Hollander, M. (1971). A nonparametric test for bivariate symmetry. *Biometrika, 58,* 203–212 (5.7).

Hollander, M., Pledger, G., and Lin, P. (1974). Robustness of the Wilcoxon test to a certain dependency between samples. *The Annals of Statistics, 1,* 177–181 (5.1).

Hollander, M., and Wolfe, D.A. (1973). *Nonparametric Statistical Methods.* J. Wiley, New York (2.5, 5.5, 5.8).

Holst, Lars. (1972). Asymptotic normality and efficiency for certain goodness-of-fit tests. *Biometrika, 59,* 137–145 (4.5).

Hora, S.C., and Iman, R.L. (1988). Asymptotic relative efficiencies of the rank-transformation procedure in randomized complete blocks. *Journal of the American Statistical Association, 83,* 462–470 (5.8, 5.12).

Horn, S.D. (1977). Goodness-of-fit tests for discrete data: A review and an application to a health impairment scale. *Biometrics, 33,* 237–248 (4.5, 6.1).

Horn, S.D., and Pyne, D. (1976). Comparison of exact and approximate goodness-of-fit tests for discrete data. Department of Mathematical Sciences, Johns Hopkins University, Technical Report No. 257 (6.1).

Hotelling, H., and Pabst, M.R. (1936). Rank correlation and tests of significance involving no assumption of normality. *The Annals of Mathematical Statistics, 7,* 29–43 (5.4).

Høyland, A. (1965). Robustness of the Hodges-Lehmann estimates for shift. *The Annals of Mathematical Statistics, 36,* 174–197 (5.1).

Høyland, A. (1968). Robustness of the Wilcoxon estimate of location against a certain dependence. *The Annals of Mathematical Statistics, 39,* 1196–1201 (5.7).

Huber, P.J. (1972). Robust statistics: A review. *The Annals of Mathematical Statistics, 43,* 1041–1067 (5.12).

Huber, P.J. (1973). Robust regression: Asymptotics, conjectures and Monte Carlo. *The Annals of Statistics, 1,* 799–821 (5.5).

Hudimoto, H. (1959). On a two-sample nonparametric test in the case that ties are present. *Annals of the Institute of Statistical Mathematics, 11,* 113–120 (5.3).

Hwang, T.Y., and Klotz, J.H. (1975). Bahadur efficiency of linear rank statistics for scale alternatives. *The Annals of Statistics, 3,* 947–954 (5.3).

Iman, R.L. (1970). Use of summation operators for the derivation of common formulae. *Mathematics Teacher, 43* (4), 296–297 (1.4).

Iman, R.L. (1974a). Use of a *t*-statistic as an approximation to the exact distribution of the Wilcoxon signed ranks test statistic. *Communications in Statistics, 3,* 795–806 (5.7).

Iman, R.L. (1974b). A power study of a rank transform for the two-way classification model when interaction may be present. *The Canadian Journal of Statistics Section C: Applications, 2,* 227–239 (5.12).

Iman, R.L. (1976). An approximation to the exact distribution of the Wilcoxon-Mann-Whitney rank sum statistic. *Communications in Statistics—Theory and Methods, A5,* 587–598 (5.1).

Iman, R.L., and Conover, W.J. (1978). Approximations of the critical region for Spearman's rho with and without ties present. *Communications in Statistics—Series B, Computations and Simulation, 7,* 269–282 (5.4).

Iman, R.L., and Conover, W.J. (1979). The use of the rank transform in regression. *Technometrics, 21,* 499–509 (5.6).

Iman, R.L., and Davenport, J.M. (1976). New approximations to the exact distribution of the Kruskal-Wallis test statistic. *Communications in Statistics—Theory and Methods, A5,* 1335–1348 (5.2).

Iman, R.L., and Davenport, J.M. (1980). Approximations of the critical region of the Friedman statistic. *Communications in Statistics, A9* (6) (in press) (5.8).

Iman, R.L., Hora, S.C., and Conover, W.J. (1984). A comparison of asymptotically distribution-free procedures for the analysis of complete blocks. *Journal of American Statistical Association, 79,* 674–685 (5.8, 5.12).

Iman, R.L., Quade, D., and Alexander, D.A. (1975). Exact probability levels for the Kruskal-Wallis test. In Harter, H.L., and Owen, D.B., eds., *Selected Tables in Mathematical Statistics, 3,* 329–384 (5.2, Appendix).

Ireland, C.T., Ku, H.H., and Kullback, S. (1969). Symmetry, and marginal homogeneity of an $r \times r$ contingency table. *Journal of the American Statistical Association, 64,* 1323–1341 (4.2).

Ireland, C.T., and Kullback, S. (1968). Contingency tables with given marginals. *Biometrika, 55,* 179–188 (4.2).

Irwin, J.O. (1935). Tests of significance for differences between percentages based on small numbers. *Metron, 12,* 83–94 (4.1).

Ishii, G. (1960). Intraclass contingency tables. *Annals of the Institute of Statistical Mathematics, 12,* 161–207 (corrections appear in Vol. 12, p. 279) (4.2).

Ives, K.H., and Gibbons, J.D. (1967). A correlation measure for nominal data. *The American Statistician, 21* (5), 16–17 (4.4).

Jacobson, J.E. (1963). The Wilcoxon two-sample statistic: Tables and bibliography. *Journal of the American Statistical Association, 58,* 1086–1103 (5.1).

Jaeckel, L.A. (1972). Estimating regression coefficients by minimizing the dispersion of the residuals. *The Annals of Mathematical Statistics, 43,* 1449–1458 (5.5).

Jogdeo, K. (1966). On randomized rank score procedures of Bell and Doksum. *The Annals of Mathematical Statistics, 37,* 1697–1703 (5.10).

Johns, M.V., Jr. (1974). Nonparametric estimation of location. *Journal of the American Statistical Association, 69,* 453–460 (5.7).

Johnson, R.A., and Mehrotra, K.G. (1972). Locally most powerful rank tests for the two-sample problem with censored data. *The Annals of Mathematical Statistics, 43,* 823–831 (5.10).

Jonckheere, A.R. (1954a). A distribution-free k-sample test against ordered alternatives. *Biometrika, 41,* 133–145 (5.4).

Jonckheere, A.R. (1954b). A test of significance for the relation between m rankings and k ranked categories. *British Journal of Statistical Psychology, 7,* 93–100 (5.8).

Jureckova, J. (1971). Nonparametric estimate of regression coefficients. *The Annals of Mathematical Statistics, 42,* 1328–1338 (5.5).

Jureckova, J. (1977). Asymptotic relations of M-estimates and R-estimates in linear regression model. *The Annals of Statistics, 5,* 464–472 (5.5).

Kalbfleish, J.D. (1974). Some efficiency calculations for survival distributions. *Biometrika, 61,* 31–38 (5.5).

Kalish, G., and Mikulski, P.W. (1971). The asymptotic behavior of the Smirnov test compared to standard "optimal procedures." *The Annals of Mathematical Statistics, 42,* 1742–1747 (6.3).

Kanofsky, P., and Srinivasan, R. (1972). An approach to the construction of parametric confidence bands on cumulative distribution functions. *Biometrika, 59,* 623–631 (6.2).

Kaplan, E.L., and Meier, P. (1958). Nonparametric estimation from incomplete observations. *Journal of the American Statistical Association, 53,* 457–481 (2.2).

Kempthorne, O. (1955). The randomization theory of experimental inference. *Journal of the American Statistical Association, 50,* 946–967 (5.11).

Kempthorne, O., and Doerfler, T.E. (1969). The behavior of some significance tests under experimental randomization. *Biometrika, 56,* 231–248 (5.11).

Kendall, M.G. (1938). A new measure of rank correlation. *Biometrika, 30,* 81–93 (5.4).

Kendall, M.G. (1942). Partial rank correlation. *Biometrika, 32,* 277–283 (5.4).

Kendall, M.G., and Babington-Smith, B. (1939). The problem of m rankings. *The Annals of Mathematical Statistics, 10,* 275–287 (5.8).

Kendall, M.G. and Gibbons, J.D. (1980). *Rank Correlation Methods,* 5th ed. Edward Arnold, London (5.4).

Kendall, M.G., and Sundrum, R.M. (1953). Distribution-free methods and order properties. *Review of the International Statistical Institute, 21* (3), 124–134 (2.5).

Kim, J-O. (1975). Multivariate analysis of ordinal variables. *American Journal of Sociology, 81,* 261–298 (5.12).

Kim, P.J. (1969). On the exact and approximate sampling distribution of the two sample

Kolmogorov-Smirnov criterion $D_{mn}m \le n$. *Journal of the American Statistical Association*, 64, 1625–1635 (6.3).

Klotz, J. (1962). Nonparametric tests for scale. *The Annals of Mathematical Statistics*, 33, 498–512 (5.10).

Klotz, J. (1963). Small sample power and efficiency for the one sample Wilcoxon and normal scores tests. *The Annals of Mathematical Statistics*, 34, 624–632 (5.7).

Klotz, J. (1965). Alternative efficiencies for signed rank tests. *The Annals of Mathematical Statistics*, 36, 1759–1766 (5.7).

Klotz, J. (1966). The Wilcoxon, ties, and the computer. *Journal of the American Statistical Association*, 61, 772–787 (5.1).

Klotz, J. (1967). Asymptotic efficiency of the two sample Kolmogorov-Smirnov test. *Journal of the American Statistical Association*, 62, 932–938 (6.3).

Klotz, J., and Teng, J. (1977). One-way layout for counts and the exact enumeration of the Kruskal-Wallis H distribution with ties. *Journal of the American Statistical Association*, 72, 165–169 (5.2).

Knight, W.R. (1966). A computer method for calculating Kendall's tau with ungrouped data. *Journal of the American Statistical Association*, 61, 436–439 (5.4).

Knoke, J.D. (1976). Multiple comparisons with dichotomous data. *Journal of the American Statistical Association*, 71, 849–853 (4.3).

Knoke, J.D. (1977). Testing for randomness against autocorrelation: Alternative tests. *Biometrika*, 64, 523–529 (5.4).

Knott, M. (1970). The small sample power of one-sided Kolmogorov tests for a shift in location of the normal distribution. *Journal of the American Statistical Association*, 65, 1384–1391 (6.1).

Koch, G.G. (1970). The use of non-parametric methods in the statistical analysis of a complex split plot experiment. *Biometrics*, 26, 105–128 (5.8).

Koch, G.G., Imrey, P.B., and Reinfurt, D.W. (1972). Linear model analysis of categorical data with incomplete response vectors. *Biometrics*, 28, 663–692 (4.7).

Koch, G.G., Johnson, W.D., and Tolley, H.D. (1972). A linear models approach to the analysis of survival and extent of disease in multidimensional contingency tables. *Journal of the American Statistical Association*, 67, 783–796 (4.2).

Koch, G.G., and Reinfurt, D.W. (1971). The analysis of categorical data from mixed models. *Biometrics*, 27, 157–175 (4.7).

Koehler, K.J. and Larntz, K. (1980). An empirical investigation of goodness-of-fit statistics for sparse multinomials. *Journal of the American Statistical Association*, 75, 336–344 (4.5).

Kolmogorov, A.N. (1933). Sulla determinazione empirica di una legge di distribuzione. *Giornale dell' Istituto Italiano degli Attuari*, 4, 83–91 (6.1).

Konijn, H.S. (1961). Non-parametric, robust and short-cut methods in regression and structural analysis. *The Australian Journal of Statistics*, 3, 77–86 (5.4).

Kraft, C.H., and van Eeden, C. (1972). Asymptotic efficiencies of quick methods of computing efficient estimates based on ranks. *Journal of the American Statistical Association*, 67, 199–202 (5.1, 5.7).

Krewski, D. (1976). Distribution-free confidence intervals for quantile intervals. *Journal of the American Statistical Association*, 71, 420–422 (3.2).

Kruskal, W.H. (1952). A nonparametric test for the several sample problem. *The Annals of Mathematical Statistics*, 23, 525–540 (5.2).

Kruskal, W.H. (1958). Ordinal measures of association. *Journal of the American Statistical Association*, 53, 814–861 (5.4).

Kruskal, W.H., and Wallis, W.A. (1952). Use of ranks on one-criterion variance analysis. *Journal of the American Statistical Association, 47,* 583–621 (corrections appear in Vol. 48, pp. 907–911) (5.2).

Ku, H.H. (1963). A note on contingency tables involving zero frequencies and the 2I test. *Technometrics, 5,* 398–400 (4.2).

Ku, H.H., and Kullback, S. (1974). Loglinear models in contingency table analysis. *The American Statistician, 28* (4), 115–122 (4.7).

Ku, H.H., Varner, R.N., and Kullback, S. (1971). On the analysis of multidimensional contingency tables. *Journal of the American Statistical Association, 66,* 55–64 (4.2).

Kullback, S. (1971). Marginal homogeneity of multidimensional contingency tables. *The Annals of Mathematical Statistics, 42,* 594–606 (4.2).

Kullback, S., Kupperman, M., and Ku, H.H. (1962). Tests for contingency tables and Markov chains. *Technometrics, 4,* 573–608 (4.2).

Labovitz, S. (1970). The assignment of numbers to rank order categories. *American Sociological Review, 35,* 515–524 (5.12).

La Brecque, J. (1977). Goodness-of-fit tests based on non-linearity in probability plots. *Technometrics, 19,* 293–306 (6.2).

Laubscher, N.F., and Odeh, R.E. (1976). A confidence interval for the scale parameter based on Sukhatme's two-sample statistic. *Communications in Statistics—Theory and Methods, A5,* 1393–1407 (5.3).

Laubscher, N.F., Steffens, F.E., and DeLange, E.M. (1968). Exact critical values for Mood's distribution-free test statistic for dispersion and its normal approximation. *Technometrics, 10,* 497–508 (5.3).

Lawler, K.R. (1978). A distribution-free generalization of the Wilcoxon signed-ranks test for the two-way layout. Master's Thesis, Texas Tech University, Lubbock.

Lee, S.K. (1978). An example for teaching some basic concepts in multidimensional contingency table analysis. *The American Statistician, 32* (2), 69–70 (4.7).

Lee, S.W. (1966). The power of the one-sided and one-sample Kolmogorov-Smirnov test. Unpublished master's report, Kansas State University (6.1).

Lehmann, E.L. (1951). Consistency and unbiasedness of certain nonparametric tests. *The Annals of Mathematical Statistics, 22,* 165–179 (6.3).

Lehmann, E.L. (1959). *Testing Statistical Hypotheses.* J. Wiley, New York (2.4).

Lehmann, E.L. (1966). Some concepts of dependence. *The Annals of Mathematical Statistics, 37,* 1137–1153 (5.4).

Lehmann, E.L. (1975). *Nonparametrics: Statistical Methods Based on Ranks.* Holden-Day, San Francisco (5.5, 5.8, 5.10).

Lehmann, E.L., and Stein, C. (1949). On the theory of some nonparametric hypotheses. *The Annals of Mathematical Statistics, 20,* 28–45 (5.11).

Lemmer, H.H., and Stoker, D.J. (1967). A distribution-free analysis of variance for the two-way classification. *South African Statistics Journal, 1,* 67–74 (5.12).

Lemmer, H.H., Stoker, D.J., and Reinach, S.G. (1968). A distribution-free analysis of variance technique for block designs. *South African Journal of Statistics, 2,* 9–32 (5.8).

Lepage, Y. (1971). A combination of Wilcoxon's and Ansari-Bradley's statistics. *Biometrika, 58,* 213–217 (5.3).

Lepage, Y. (1973). A table for a combined Wilcoxon Ansari-Bradley statistic. *Biometrika, 60,* 113–116 (5.3).

Lepage, Y. (1977). A class of nonparametric tests for location and scale parameters. *Communications in Statistics—Theory and Methods, A6,* 649–659 (5.3).

Lewis, T.G. (1975). *Distribution Sampling for Computer Simulation*. D.C. Heath, Lexington, Mass. (5.10).

Li, L., and Schucany, W.R. (1975). Some properties of a test for concordance of two groups of rankings. *Biometrika, 62*, 417–423 (5.8).

Li, S.-H., Simon, R.M., and Gart, J.J. (1979). Small sample properties of the Mantel-Haenszel test. *Biometrika, 66*, 181–183 (4.1).

Light, R.J., and Margolin, B.H. (1971). An analysis of variance for categorical data. *Journal of the American Statistical Association, 66*, 534–544 (4.2).

Lilliefors, H.W. (1967). On the Kolmogorov–Smirnov test for normality with mean and variance unknown. *Journal of the American Statistical Association, 62*, 399–402 (6.2, Appendix).

Lilliefors, H.W. (1969). On the Kolmogorov–Smirnov test for the exponential distribution with mean unknown. *Journal of the American Statistical Association, 64*, 387–389 (6.2).

Lilliefors, H.W. (1973). The Kolmogorov–Smirnov and other distance tests for the gamma distribution and for the extreme-value distribution when parameters must be estimated. Department of Statistics, George Washington University, unpublished manuscript (6.2).

Maag, U.R. (1966). A k-sample analogue of Watson's U^2 statistic. *Biometrika, 53*, 579–584 (6.3).

Maag, U.R., and Dicaire, G. (1971). On Kolmogorov–Smirnov type one-sample statistics. *Biometrika, 58*, 653–656 (6.1).

Maag, U.R., and Stephens, M.A. (1968). The V_{NM} two-sample test. *The Annals of Mathematical Statistics, 39*, 923–935 (6.3).

Maag, U.R., Streit, F., and Drouilly, P.A. (1973). Goodness-of-fit tests for grouped data. *Journal of the American Statistical Association, 68*, 462–465 (6.1).

Macdonald, P. (1971). The analysis of a 2^n experiment by means of ranks. *Journal of the Royal Statistical Society (Series C) Applied Statistics, 20*, 259–275 (5.12).

Mack, C. (1969). Query: Tolerance limits for a binomial distribution. *Technometrics, 11*, 201 (3.3).

MacKinnon, W.J. (1964). Table for both the sign test and distribution-free confidence intervals of the median for sample sizes to 1,000. *Journal of the American Statistical Association, 59*, 935–959 (3.4).

Mann, H.B., and Whitney, D.R. (1947). On a test of whether one of two random variables is stochastically larger than the other. *The Annals of Mathematical Statistics, 18*, 50–60 (5.1).

Manoukin, B. (1986). *Mathematical Nonparametric Statistics*. Gordon and Breach Science Publishers, New York (2.5).

Mansfield, E. (1962). Power functions for Cox's test of randomness against trend. *Technometrics, 4*, 430–432 (3.5).

Mansouri, H. and Chang, G-H. (1995). A comparative study of some rank tests for interaction. *Computational Statistics & Data Analysis, 19*, 85–96 (5.8, 5.12).

Mantel, N., and Fleiss, J.L. (1975). The equivalence of the generalized McNemar tests for marginal homogeneity in 2^3 and 3^2 tables. *Biometrics, 31*, 727–729 (3.5).

Mantel, N., and Greenhouse, S.W. (1968). What is the continuity correction? *The American Statistician, 22* (5), 27–30 (4.1).

Mantel, N., and Haenszel, W. (1959). Statistical aspects of the analysis of data from retrospective studies of disease. *Journal of the National Cancer Institute, 22*, 719–748 (4.1, 4.2).

Mardia, K.V. (1967a). A non-parametric test for the bivariate two-sample location problem. *Journal of the Royal Statistical Society (B), 29*, 320–342 (5.1).

Mardia, K.V. (1967b). Some contributions to contingency-type bivariate distributions. *Biometrika, 54*, 235–249 (4.2).

Mardia, K.V. (1968). Small sample power of a nonparametric test for the bivariate two-sample location problem in the normal case. *Journal of the Royal Statistical Society (B), 30,* 83–92 (5.1).

Margolin, B.H., and Light, R.J. (1974). An analysis of variance for categorical data, II: Small sample comparisons with chi-square and other competitors. *Journal of the American Statistical Association, 69,* 755–764 (4.2).

Margolin, B.H., and Maurer, W. (1976). Tests of the Kolmogorov–Smirnov type for exponential data with unknown scale, and related problems. *Biometrika, 63,* 149–160 (6.2).

Maritz, J.S., Wu, M, and Staudte, R.G., Jr. (1977). A location estimator based on a U-statistic. *The Annals of Statistics, 5,* 779–786 (5.7).

Marsaglia, G. (1968). Query: Pseudo random normal numbers. *Technometrics, 10,* 401–402 (5.10).

Mason, A.L., and Bell, C.B. (1986). New Lilliefors and Srinivasan tables with applications. *Communications in Statistics—Simulation, 15,* 451–467 (6.2, Appendix).

Massey, F.J. (1950a). A note on the estimation of a distribution function by confidence limits. *The Annals of Mathematical Statistics, 21,* 116–119 (6.1).

Massey, F.J. (1950b). A note on the power of a nonparametric test. *The Annals of Mathematical Statistics, 21,* 440–443 (Corrections appear in Vol. 23, pp. 637–638) (6.1).

Massey, F.J. (1952). Distribution table for the deviation between two sample cumulatives. *The Annals of Mathematical Statistics, 23,* 435–441. Corrections appear in Davis, L.S. (1958), *Mathematical Tables and Other Aids to Computation, 12,* 262–263 (Appendix).

Matthes, T.K., and Truax, D.R. (1965). Optimal invariant rank tests for the k-sample problem. *The Annals of Mathematical Statistics, 36,* 1207–1222 (5.2).

Maxwell, E.A. (1961). *Analyzing Qualitative Data.* Wiley, New York (4.2).

Maxwell, E.A. (1976). Analysis of contingency tables and further reasons for not using Yates correction in 2 × 2 tables. *The Canadian Journal of Statistics Section C: Applications, 4,* 277–290 (4.1).

McCarthy, P.J. (1965). Stratified sampling and distribution-free confidence intervals for a median. *Journal of the American Statistical Association, 60,* 772–783 (5.7).

McCornack, R.L. (1965). Extended tables of the Wilcoxon matched pairs signed rank statistics. *Journal of the American Statistical Association, 60,* 864–871 (5.7).

McDonald, B.J., and Thompson, W.A. (1967). Rank sum multiple comparisons in one- and two-way classifications, *Biometrika, 54,* 487–498 (5.2, 5.8).

McDonald, L.L., Davis, B.M., and Milliken, G.A. (1977). A nonrandomized unconditional test for comparing two proportions in 2 × 2 contingency tables. *Technometrics, 19,* 145–158 (4.1).

McKean, J.W., and Ryan T.A., Jr. (1977). An algorithm for obtaining confidence intervals and point estimates based on ranks in the two-sample location problem. *AMC Transactions on Mathematical Software, 3,* 183–185 (5.1).

McKinlay, S.M. (1975). A note on the chi-square test for pair-matched samples. *Biometrics, 31,* 731–735 (3.5).

McNeil, D.R. (1967). Efficiency loss due to grouping in distribution-free tests. *Journal of the American Statistical Association, 62,* 954–965 (5.1).

McNeil, D.R., and Tukey, J.W. (1975). Higher order diagnosis of two-way tables, illustrated on two sets of demographic empirical distributions. *Biometrics, 31,* 487–510 (4.2).

McNemar, Q. (1962). *Psychological Statistics,* 3rd ed. Wiley, New York (4.4).

Meeker, W.Q. (1978). Sequential tests of independence for 2 × 2 contingency tables. *Biometrika, 65* (1), 85–90 (4.1).

Mehra, K.L., and Sarangi, J. (1967). Asymptotic efficiency of certain rank tests for comparative experiments. *The Annals of Mathematical Statistics, 38,* 90–107 (5.8).

Mielke, P.W., Jr. (1967). Note on some squared rank tests with existing ties. *Technometrics, 9,* 312–314 (5.3).

Mielke, P.W., Jr. (1972). Asymptotic behavior of two-sample tests based on powers of ranks for detecting scale and location alternatives. *Journal of the American Statistical Association, 67,* 850–854 (5.1, 5.3).

Mielke, P.W., Jr., and Siddiqui, M.M. (1965). A combinatorial test for independence of dichotomous responses. *Journal of the American Statistical Association, 60,* 437–441 (4.2).

Mikulski, P.W. (1963). On the efficiency of optimal nonparametric procedures in the two sample case. *The Annals of Mathematical Statistics, 34,* 22–32 (5.1).

Miller, L.H. (1956). Table of percentage points of Kolmogorov statistics. *Journal of the American Statistical Association, 51,* 111–121 (Appendix).

Miller, R.G., Jr. (1970). A sequential signed rank test. *Journal of the American Statistical Association, 65,* 1554–1561 (5.7).

Miller, R.G. (1973). Nonparametric estimators of the mean tolerance in bioassay. *Biometrika, 60,* 535–542 (2.5).

Milton, R.C. (1964). An extended table of critical values for the Mann–Whitney (Wilcoxon) two-sample statistic. *Journal of the American Statistical Association, 59,* 925–934 (5.1).

Molinari, L. (1977). Distribution of the chi-square test in nonstandard situations. *Biometrika, 64,* 115–121 (4.5).

Mood, A.M. (1954). On the asymptotic efficiency of certain nonparametric two-sample tests. *The Annals of Mathematical Statistics, 25,* 514–522 (5.3).

Moses, L.E. (1952). Nonparametric statistics for psychological research. *Psychological Bulletin, 49,* 122–143 (5.11).

Moses, L.E. (1963). Rank tests of dispersion, *The Annals of Mathematical Statistics, 34,* 973–983 (5.3).

Moses, L.E. (1965). Query: Confidence limits from rank tests, *Technometrics, 7,* 257–260 (5.1, 5.7).

Mosteller, F. (1968). Association and estimation in contingency tables. *Journal of the American Statistical Association, 63,* 1–29 (4.2).

Mote, V.L., and Anderson, R.L. (1965). An investigation of the effect of misclassification on the properties of χ^2-tests in the analysis of categorical data. *Biometrika, 52,* 95–110 (4.2).

Murphy, R.B. (1948). Nonparametric tolerance limits. *The Annals of Mathematical Statistics, 19,* 581–588 (3.3).

Nelson, L.S. (1966). Query: Combining values of observed χ^2's. *Technometrics, 8,* 709 (1.5, 4.1).

Nemenyi, P. (1969). Variances: An elementary proof and a nearly distribution-free test. *The American Statistician, 23* (5), 35–37 (5.3).

Noé, M., and Vandewiele, G. (1968). The calculation of distributions of Kolmogorov–Smirnov type statistics including a table of significance points for a particular case. *The Annals of Mathematical Statistics, 39,* 233–241 (6.1).

Noether, G.E. (1963). Efficiency of the Wilcoxon two-sample statistic for randomized blocks. *Journal of the American Statistical Association, 58,* 894–898 (5.1).

Noether, G.E. (1967a). *Elements of Nonparametric Statistics.* Wiley, New York (2.4, 2.5, 3.3, 5.1, 5.7, 5.8, 5.9).

Noether, G.E. (1967b). Wilcoxon confidence intervals for location parameters in the discrete case. *Journal of the American Statistical Association, 62,* 184–188 (5.7).

Noether, G.E. (1973). Some simple distribution-free confidence intervals for the center of a symmetric distribution. *Journal of the American Statistical Association, 68,* 716–719 (5.7).

Odeh, R.E. (1967). The distribution of the maximum use of ranks. *Technometrics, 9,* 271–278 (5.2).

Odeh, R.E. (1971). On Jonckheere's k-sample test against ordered alternatives. *Technometrics, 13,* 912–918 (5.2).

Odeh, R.E. (1972). On the power of Jonckheere's k-sample test against ordered alternatives. *Biometrika, 59,* 467–471 (5.2).

Oden, A., and Wedel, H. (1975). Arguments for Fisher's permutation test. *The Annals of Statistics, 3,* 518–520 (5.11).

Odoroff, C.L. (1970). A comparison of minimum logit chi-square estimation and maximum likelihood estimation in $2 \times 2 \times 2$ and $3 \times 2 \times 2$ contingency tables: Tests for interaction. *Journal of the American Statistical Association, 65,* 1617–1631 (4.7).

Olshen, R.A. (1967). Sign and Wilcoxon tests for linearity. *The Annals of Mathematical Statistics, 38,* 1759–1769 (3.5).

Ott, R.L., and Free, S.M. (1969). A short-cut rule for a one-sided test of hypothesis for qualitative data. *Technometrics, 11,* 197–200 (4.1).

Owen, D.B. (1962). *Handbook of Statistical Tables.* Addison-Wesley, Reading, Mass. (3.3, 5.7, 5.8, 5.10).

Page, E.B. (1963). Ordered hypotheses for multiple treatments: A significance test for linear ranks. *Journal of the American Statistical Association, 58,* 216–230 (5.8).

Pahl, P.J. (1969). On testing for goodness-of-fit of the negative binominal distribution when expectations are small. *Biometrics, 25,* 143–151 (4.5).

Patel, K.D. (1975). Cochran's Q test: Exact distribution. *Journal of the American Statistical Association, 70,* 186–189 (4.6).

Patel, K.M., and Hoel, David G. (1973). A nonparametric test for interaction in factorial experiments. *Journal of the American Statistical Association, 68,* 615–620 (5.8).

Pavur, R.J. and Nath, R. (1986). Parametric versus rank transform procedures in the two-way factorial experiment: A comparative study. *Journal of Statistical Computer Simulation, 23,* 231–240 (5.12).

Pearson, E.S. (1947). The choice of statistical test illustrated on the interpretation of data classed in a 2×2 table. *Biometrika, 34,* 139–167 (4.1).

Pearson, E.S., D'Agostino, R.B., and Bowman, K.O. (1977). Tests for departure from normality: Comparison of powers. *Biometrika, 64,* 231–246 (6.2).

Pearson, E.S., and Hartley, H.O. (1962). *Biometrika Tables for Statisticians,* Vol. I, 2nd ed. Cambridge University Press, Cambridge, England (5.10).

Pearson, E.S., and Hartley, H.O. (1970). *Biometrika Tables for Statisticians,* Vol. I. 3rd ed. Reprinted with corrections 1976. Cambridge University Press, Cambridge, England (Appendix).

Pearson, E.S., and Hartley, H.O. (1972). *Biometrika Tables for Statisticians,* Vol. II. Reprinted with corrections 1976. Cambridge University Press, Cambridge, England (Appendix).

Pearson, E.S., and Please, N.W. (1975). Relation between the shape of population distribution and the robustness of four simple test statistics. *Biometrika, 62,* 223–241 (2.5).

Pearson, K. (1900). On the criterion that a given system of deviations from the probable in the case of a correlated system of variables is such that it can reasonably be supposed to have arisen from random sampling. *Philosophical Magazine, 50* (5), 157–175. (Reprinted 1948 in *Karl Person's Early Statistical Papers,* ed. by E.S. Pearson, Cambridge: Cambridge University Press.) (4.5, 4.7)

Pearson, K. (1922). On the χ^2 test of goodness of fit. *Biometrika, 14,* 186–191 (4.7).

Percus, O.E., and Percus, J.K. (1970). Extended criterion for comparison of empirical distributions. *Journal of Applied Probability, 7,* 1–20 (6.3).

Pettitt, A.N. (1976). A two-sample Anderson–Darling rank statistic. *Biometrika, 63,* 161–168 (5.1).

Pettitt, A.N. (1978). Generalized Cramér-von Mises statistics for the gamma distribution. *Biometrika, 65* (1), 232–235 (6.2).

Pettitt, A.N., and Stephens, M.A. (1976). Modified Cramér-von Mises statistics for censored data. *Biometrika, 63,* 291–298 (6.1).

Pettitt, A.N., and Stephens, M.A. (1977). The Kolmogorov–Smirnov goodness-of-fit statistic with discrete and grouped data. *Technometrics, 19,* 205–210 (6.1).

Pierce, A. (1970). *Fundamentals of Nonparametric Statistics.* Dickenson Pub. Co., Belmont, Calif. (2.5)

Pirie, W.R. (1974). Comparing rank tests for ordered alternatives in randomized blocks. *The Annals of Statistics, 2,* 374–382 (correction appears in Vol. 3, p. 796) (5.8).

Pirie, W.R., and Hamdan, M.A. (1972). Some revised continuity corrections for discrete distributions. *Biometrics, 28,* 693–701 (4.1).

Pirie, W.R., and Hollander, M. (1972). A distribution-free normal scores test for ordered alternatives in the randomized block design. *Journal of the American Statistical Association, 67,* 855–857 (5.9).

Pitman, E.J.G. (1937/1938). Significance tests which may be applied to samples from any populations, I. *Supplement to the Journal of the Royal Statistical Society, 4,* (1937), 119–130, II, The correlation coefficient test. *Supplement to the Journal of the Royal Statistical Society, 4,* (1937), 225–232. III. The analysis of variance test. *Biometrika, 29* (1938), 322–335 (5.11).

Plackett, R.L. (1964). The continuity correction in 2×2 tables. *Biometrika, 51,* 327–338 (4.1).

Plackett, R.L. (1965). A class of bivariate distributions. *Journal of the American Statistical Association, 60,* 516–522 (4.2).

Plackett, R.L. (1977). The marginal totals of a 2×2 table. *Biometrika, 64,* 37–42 (4.1).

Policello, G.E., II, and Hettmansperger, T.P. (1976). Adaptive robust procedures for the one-sample location problem. *Journal of the American Statistical Association, 71,* 624–633 (2.5).

Potthoff, R.F. (1963). Use of the Wilcoxon statistic for a generalized Behrens–Fisher problem. *The Annals of Mathematical Statistics,* 1596–1599 (5.1).

Potthoff, R.F. (1974). A non-parametric test of whether two simple regression lines are parallel. *The Annals of Statistics, 2,* 295–310 (5.5).

Pratt, J.W. (1959). Remarks on zeros and ties in the Wilcoxon signed rank procedures. *Journal of the American Statistical Association, 54,* 655–667 (5.7).

Pratt, J.W. (1981). *Concepts of Nonparametric Theory.* Springer-Verlag, New York (2.5).

Puri, M.L. (1965). On some tests of homogeneity of variances. *Annals of the Institute of Statistical Mathematics, 17,* 323–330 (5.3).

Puri, M.L. (1968). Multi-sample problem: Unknown location parameters. *Annals of the Institute of Statistical Mathematics, 20,* 99–106 (5.3).

Puri, M.L. (1985). *Nonparametric Methods in General Linear Models.* Wiley, New York (5.5).

Puri, M.L., and Puri, P.S. (1969). Multiple decision procedures based on ranks for certain problems in analysis of variance. *The Annals of Mathematical Statistics, 40,* 619–632 (5.2).

Puri, M.L., and Sen, P.K. (1967). On some optimum nonparametric procedures in two-way layouts. *Journal of the American Statistical Association, 62,* 1214–1229 (5.8).

Puri, M.L., and Sen, P.K. (1968). Nonparametric confidence regions for some multivariate location problems. *Journal of the American Statistical Association, 63,* 1373–1378 (5.7).

Puri, M.L., and Sen, P.K. (1969a). Analysis of covariance based on general rank scores. *The Annals of Mathematical Statistics, 40,* 610–618 (5.2).

Puri, M.L., and Sen, P.K. (1969b). On the asymptotic theory of rank order tests for experiments involving paired comparisons. *Annals of the Institute of Statistical Mathematics, 21,* 163–173 (5.9).

Puri, M.L., and Sen, P.K. (1971). *Nonparametric Methods in Multivariate Analysis.* Wiley, New York (2.5).

Putter, J. (1964). The χ^2 goodness-of-fit test for a class of dependent observations. *Biometrika, 51,* 250–252 (4.5).

Quade, D. (1965). On the asymptotic power of the one-sample Kolmogorov–Smirnov tests. *The Annals of Mathematical Statistics, 36,* 1000–1018 (6.1).

Quade, D. (1966). On analysis of variance for the k-sample problem. *The Annals of Mathematical Statistics, 37,* 1747–1785 (5.2).

Quade, D. (1967). Rank analysis of covariance. *Journal of the American Statistical Association, 62,* 1187–1200 (5.2).

Quade, D. (1972). Analyzing randomized blocks by weighted rankings. Technical Report SW 18/72, Stichting Mathematisch Centrum, Amsterdam (5.8).

Quade, D. (1979). Using weighted rankings in the analysis of complete blocks with additive block effects. *Journal of the American Statistical Association, 74,* 680–683 (5.8).

Quade, D., and Salama, I.A. (1975). A note on minimum chi-square statistics in contingency tables. *Biometrics, 31,* 953–956 (4.2).

Quesenberry, C.P., and Gessaman, M.P. (1968). Nonparametric discrimination using tolerance regions. *The Annals of Mathematical Statistics, 39,* 664–673 (3.3).

Quesenberry, C.P., and Hurst, D.C. (1964). Large sample simultaneous confidence intervals for multinominal proportions. *Technometrics, 6,* 191–195 (3.1).

Radhakrishna, S. (1965). Combination of results from several 2×2 contingency tables. *Biometrics, 21,* 86–99 (1.5, 4.1).

Raghavachari, M. (1965a). The two-sample scale problem when locations are unknown. *The Annals of Mathematical Statistics, 36,* 1236–1242 (5.3).

Raghavachari, M. (1965b). On the efficiency of the normal scores test relative to the F-test. *The Annals of Mathematical Statistics, 36,* 1306–1307 (5.10).

Rahe, A.J. (1974). Tables of critical values for the Pratt matched pair signed rank statistics. *Journal of the American Statistical Association, 69,* 368–373 (5.7).

Ramachandramurty, P.V. (1966). On the Pitman efficiency of one-sided Kolmogorov and Smirnov tests for normal alternatives. *The Annals of Mathematical Statistics, 37,* 940–944 (6.3).

Ramsey, F.L. (1971). Small sample power functions for nonparametric tests of location in the double exponential family. *Journal of the American Statistical Association, 66,* 149–151 (correction appears in Vol. 72, p. 703) (5.10).

Rand Corporation (1955). *A Million Random Digits with 100,000 Normal Deviates.* Free Press, Glencoe, Ill. (5.10).

Randles, R.H. (1979). *Introduction to the Theory of Nonparametric Statistics.* Wiley, New York (2.5).

Randles, R.H., Broffitt, J.D., Ramberg, J.S., and Hogg, R.V. (1978). Discriminant analysis based on ranks. *Journal of the American Statistical Association, 73* (362), 379–384 (2.5).

Rao, P.V., Schuster, Eugene F., and Littell, R.C. (1975). Estimation of shift and center of symmetry based on Kolmogorov–Smirnov statistics. *The Annals of Statistics, 3,* 862–873 (5.1, 5.6, 6.3).

Rao, T.S. (1968). A note on the asymptotic relative efficiencies of Cox and Stuart's tests for testing trend in dispersion of a p-dependent time series. *Biometrika, 55,* 381–386 (3.5).

Rao, T.S., Ed. (1993). *Developments in Time Series Analysis, In honour of Maurice B. Priestley.* Chapman & Hall, London (5.4).

Ray, R.M. (1976). A new $C(\alpha)$ test for 2×2 tables. *Communications in Statistics—Theory and Methods, A5,* 545–563 (4.1).

Read, C.B. (1977). Partitioning chi-square in contingency tables: A teaching approach. *Communications in Statistics—Theory and Methods, A6,* 553–562 (4.7).

Reiss, R.D., and Rüschendorf, L. (1976). On Wilks' distribution-free confidence intervals for quantile intervals. *Journal of the American Statistical Association, 71,* 940–944 (3.2).

Reynolds, Marion R., Jr. (1975). A sequential signed-rank test for symmetry. *The Annals of Statistics, 3,* 382–400 (5.7).

Rhyne, A.L., and Steel, R.G.D. (1965). Tables for a treatments versus control multiple comparisons sign test. *Technometrics, 7,* 293–306 (3.5).

Riedwyl, H. (1967). Goodness of fit. *Journal of the American Statistical Association, 62,* 390–398 (6.1).

Rizvi, M.H., and Sobel, M. (1967). Nonparametric procedures for selecting a subset containing the population with the largest α-quantile. *The Annals of Mathematical Statistics, 36,* 1788–1803 (5.2).

Rizvi, M.H., Sobel, M., and Woodworth, G.G. (1968). Nonparametric ranking procedures for comparison with a control. *The Annals of Mathematical Statistics, 39,* 2075–2093 (5.2).

Robertson, W.H. (1960). Programming Fisher's exact method of comparing two percentages. *Technometrics, 2,* 103–107 (4.2).

Roscoe, J.T., and Byars, J.A. (1971). Sample size restraints commonly imposed on the use of the chi-square statistic. *Journal of the American Statistical Association, 66,* 755–759 (4.2).

Rosenblatt, M. (1952). Limit theorems associated with variants of the von Mises statistic. *The Annals of Mathematical Statistics, 23,* 617–623 (6.3).

Rothman, E.D., and Woodroofe, M. (1972). A Cramér-von Mises type statistic for testing symmetry. *The Annals of Mathematical Statistics, 43,* 2035–2038 (5.7, 6.3).

Ruist, E. (1955). Comparison of tests for nonparametric hypotheses. *Arkiv För Matematik, 3* (2), 131–163 (2.4).

Ruymgaart, F.H. (1973). *Asymptotic Theory of Rank Tests of Independence.* Mathematical Centre Tracts, 43, Mathematisch Centrum, Amsterdam (5.4).

Ruymgaart, F.H., Shorack, G.R. and van Zwet, W.R. (1972). Asymptotic normality of nonparametric tests for independence. *Annals of Mathematical Statistics, 43,* 1122–1135 (5.4).

Savage, I.R. (1962). *Bibliography of Nonparametric Statistics.* Harvard University, Cambridge, Mass. (Introduction).

Savage, I.R. (1969). Nonparametric statistics: A personal review. *Sankhya: The Indian Journal of Statistics. Series A, 31,* 107–144 (2.5).

Saw, J.G. (1966). A nonparametric comparison of two samples, one of which is censored. *Biometrika, 53,* 599–603 (5.1).

Schaafsma, W. (1973). Paired comparisons with order-effects. *The Annals of Statistics, 1,* 1027–1045 (3.5).

Schach, S. (1969a). Nonparametric symmetry tests of circular distributions. *Biometrika, 56,* 571–577 (5.7).

Schach, S. (1969b). On a class of nonparametric two-sample tests for circular distributions. *The Annals of Mathematical Statistics, 40,* 1791–1800 (5.1).

Scheffé, H. (1943). Statistical inference in the nonparametric case. *The Annals of Mathematical Statistics, 14,* 305–332 (5.11).

Scheffé, H., and Tukey, J.W. (1944). A formula for sample sizes for population tolerance limits. *The Annals of Mathematical Statistics, 15,* 217 (3.3).

Scheirer, C.J., Ray, W.S., and Hare, N. (1976). The analysis of ranked data derived from completely randomized factorial designs. *Biometrics, 32,* 429–434 (5.12).

Schneider, B.E., and Clickner, R.P. (1976). On the distribution of the Kolmogorov–Smirnov test statistic for the gamma distribution with unknown parameters. Department of Statistics, Temple University, Mimeo Series No. 36 (6.2).

Schucany, W.R., and Beckett, J., III (1976). Analysis of multiple sets of incomplete rankings. *Communications in Statistics—Theory and Methods, A5,* 1327–1334 (5.8).

Schuster, E.F. (1973). On the goodness-of-fit problem for continuous symmetric distributions. *Journal of the American Statistical Association, 68,* 713–715 (6.1).

Schuster, E.F. (1975). Estimating the distribution function of a symmetric distribution. *Biometrika, 62,* 631–635 (Correction appears in Vol. 63, p. 412) (5.7).

Schuster E.F., and Narvarte, J.A. (1973). A new nonparametric estimator of the center of a symmetric distribution. *The Annals of Statistics, 1,* 1096–1104 (5.7, 6.1).

Sen, P.K. (1962). On studentized non-parametric multi-sample location tests. *Annals of the Institute of Statistical Mathematics, 14,* 119–131 (5.2).

Sen, P.K. (1963). On weighted rank-sum tests for dispersion. *Annals of the Institute of Statistical Mathematics, 15,* 117–135 (5.3).

Sen, P.K. (1966). On nonparametric simultaneous confidence regions and tests for the one criterion analysis of variance problem. *Annals of the Institute of Statistical Mathematics, 18,* 319–336 (5.2).

Sen, P.K. (1967a). A note on the asymptotic efficiency of Friedman's χ_r^2-test. *Biometrika, 54,* 677–679 (5.8).

Sen, P.K. (1967b). On some multisample permutation tests based on a class of U-statistics. *Journal of the American Statistical Association, 62,* 1201–1213 (5.11).

Sen, P.K. (1968a). Estimates of the regression coefficient based on Kendall's tau. *Journal of the American Statistical Association, 63,* 1379–1389 (5.4, 5.5).

Sen, P.K. (1968b). On a class of aligned rank order tests in two-way layouts. *The Annals of Mathematical Statistics, 39,* 1115–1124 (5.8).

Sen, P.K. (1972). On a class of aligned rank order tests for the identity of the intercepts of several regression lines. *The Annals of Mathematical Statistics, 43,* 2004–2012 (5.5).

Sen, P.K. (1981). *Sequential Nonparametrics: Invariance Principles and Statistical Inference.* Wiley, New York (5.4).

Sen, P.K., and Ghosh, M. (1974). Sequential rank tests for location. *The Annals of Statistics, 2,* 540–552 (5.1, 5.7).

Sen, P.K., and Govindarajulu, Z. (1966). On a class of c-sample weighted rank-sum tests for location and scale. *Annals of the Institute of Statistical Mathematics, 18,* 87–105 (5.2).

Sen, P.K., and Puri, M.L. (1967). On the theory of rank order tests for location in the multivariate one sample problem. *The Annals of Mathematical Statistics, 38,* 1216–1228 (5.7).

Serfling, R.J. (1968). The Wilcoxon two-sample statistic on strongly mixing processes. *The Annals of Mathematical Statistics, 39,* 1202–1209 (5.1).

Shapiro, S.S., and Francia, R.S. (1972). An approximate analysis of variance test for normality. *Journal of the American Statistical Association, 67,* 215–216 (6.2).

Shapiro, S.S., and Wilk, M.B. (1965). An analysis of variance test for normality (complete samples). *Biometrika, 52,* 591–611 (6.2).

Shapiro, S.S., and Wilk, M.B. (1968). Approximations for the null distribution of the *W* statistic. *Technometrics, 10,* 861–866 (6.2).

Shapiro, S.S., Wilk, M.B., and Chen, Mrs. H.J. (1968). A comparative study of various tests for normality. *Journal of the American Statistical Association, 63,* 1343–1372 (6.1, 6.2).

Sherman, E. (1965). A note on multiple comparisons using rank sums. *Technometrics, 7,* 255–256 (5.2).

Shirahata, S. (1975). Locally most powerful rank tests for independence with censored data. *The Annals of Statistics, 3,* 241–245 (5.4).

Shirahata, S. (1976). Test for independence in infinite contingency tables. *The Annals of Statistics, 4,* 542–553 (5.4).

Shirley, E. (1977). A non-parametric equivalent of Williams' test for contrasting increasing dose levels of a treatment. *Biometrics, 33,* 386–389 (5.12).

Shorack, G.R. (1967). Testing against ordered alternatives in model I analysis of variance: Normal theory and nonparametric. *The Annals of Mathematical Statistics, 38,* 1740–1752 (5.2, 5.8).

Shorack, G.R. (1969). Testing and estimating ratios of scale parameters. *Journal of the American Statistical Association, 64,* 999–1013 (5.3).

Shorack, R.A. (1967). Tables of the distribution of the Mann–Whitney–Wilcoxon *U*-statistic under Lehman alternatives. *Technometrics, 9,* 666–678 (5.1).

Shorack, R.A. (1968). Recursive generation of the distribution of several nonparametric test statistics under censoring. *Journal of the American Statistical Association, 63,* 353–366 (5.1).

Shuster, J.J., and Downing, D.J. (1976). Two-way contingency tables for complex sampling schemes. *Biometrika, 63,* 271–276 (4.2).

Siegel, S. (1956). *Nonparametric Statistics for the Behavioral Sciences.* McGraw-Hill, New York (4.4).

Siegel, S., and Tukey, J.W. (1960). A nonparametric sum of ranks procedure for relative spread in unpaired samples. *Journal of the American Statistical Association, 55,* 429–444 (corrections appear in Vol. 56, p. 1005) (5.1, 5.3).

Simon, G. (1974). Alternative analyses for the singly-ordered contingency table. *Journal of the American Statistical Association, 69,* 971–976 (4.2).

Simon, G. (1977a). A nonparametric test of total independence based on Kendall's tau. *Biometrika, 64,* 277–282 (5.4).

Simon, G. (1977b). Multivariate generalization of Kendall's tau with application to data reduction. *Journal of the American Statistical Association, 72,* 367–376 (5.4).

Slakter, M.J. (1965). A comparison of the Pearson chi-square and Kolmogorov goodness-of-fit tests with respect to validity. *Journal of the American Statistical Association, 60,* 854–858 (6.1).

Slakter, M.J. (1966). Comparative validity of the chi-square and two modified chi-square goodness-of-fit tests for small but equal expected frequencies. *Biometrika, 53,* 619–623 (4.5).

Slakter, M.J. (1968). Accuracy of an approximation to the power of the chi-square goodness of fit test with small but equal expected frequencies. *Journal of the American Statistical Association, 63,* 912–918 (4.5).

Slakter, M.J. (1973). Large values for the number of groups with the Pearson chi-squared goodness-of-fit test. *Biometrika, 60,* 420–421 (4.5).

Smirnov, N.V. (1936). Sui la distribution de w^2 (Criterium de M.R.v. Mises). *Comptes Rendus (Paris), 202,* 449–452 (6.1).

Smirnov, N.V. (1939). Estimate of deviation between empirical distribution functions in two independent samples. (Russian) *Bulletin Moscow University, 2* (2), 3–16 (6.1, 6.3).

Smirnov, N.V. (1948). Table for estimating goodness of fit of empirical distributions. *The Annals of Mathematical Statistics, 19,* 279–281 (6.1).

Smith, K. (1953). Distribution-free statistical methods and the concept of power efficiency. In L. Festinger and D. Katz (eds.) *Research Methods in the Behavioral Sciences.* Dryden, New York, 536–577 (5.11).

Sobel, M. (1967). Nonparametric procedures for selecting the t populations with the largest α-quantile. *The Annals of Mathematical Statistics, 38,* 1804–1816 (5.2).

Soms, A.P. (1977). An algorithm for the discrete Fisher's permutation test. *Journal of the American Statistical Association, 72,* 662–664 (5.11).

Spearman, C. (1904). The proof and measurement of association between two things. *American Journal of Psychology, 15,* 72–101 (5.4).

Spurrier, J.D., and Hewett, J.E. (1976). Two-stage Wilcoxon tests of hypotheses. *Journal of the American Statistical Association, 71,* 982–987 (5.1, 5.7).

Srinivasan, R., and Godio, L.B. (1974). A Cramér-von Mises type statistic for testing symmetry. *Biometrika, 61,* 196–198 (6.1).

Srinivasan, R., and Wharton, R.M. (1973). The limit distribution of a random variable used in the construction of confidence bands. *Biometrika, 60,* 431–433 (6.2).

Srivastava, M.S. and Sen, A.K. (1973). On sequential confidence intervals based on Wilcoxon type estimates. *The Annals of Statistics, 1,* 1200–1202 (5.7).

Steck, G.P. (1968). A note on contingency-type bivariate distributions. *Biometrika, 55,* 262–264 (4.2).

Steck, G.P. (1969). The Smirnov two sample tests as rank tests. *The Annals of Mathematical Statistics, 40,* 1449–1466 (6.3).

Steel, R.G.D. (1960). A rank sum test for comparing all pairs of treatments. *Technometrics, 2,* 197–207 (5.2).

Stephens, M.A. (1964). The distribution of the goodness-of-fit statistic, U_N^2. II. *Biometrika, 51,* 393–398 (6.1).

Stephens, M.A. (1965a). The goodness-of-fit statistic, V_N: Distribution and significant points. *Biometrika, 52,* 309–322 (6.1).

Stephens, M.A. (1965b). Significance points for the two-sample statistic $U_{M,N}^2$. *Biometrika, 52,* 661–663 (6.3).

Stephens, M.A. (1969). A goodness-of-fit statistic for the circle, with some comparisons. *Biometrika, 56,* 161–168 (6.1).

Stephens, M.A. (1974). EDF statistics for goodness-of-fit and some comparisons. *Journal of the American Statistical Association, 69,* 730–737 (6.1).

Stephens, M.A. (1975). Asymptotic properties for covariance matrices of order statistics. *Biometrika, 62,* 23–28 (6.2).

Stephens, M.A., and Maag, U.R. (1968). Further percentage points for W_N^2. *Biometrika, 55,* 428–430 (6.1).

Stevens, S.S. (1946). On the theory of scales of measurement. *Science, 103,* 677–680 (2.1).

Stone, C.J. (1977). Consistent nonparametric regression. *The Annals of Statistics, 5,* 595–645 (5.5).

Stone, M. (1967). Extreme tail probabilities for the null distribution of the two-sample Wilcoxon statistic. *Biometrika, 54,* 629–640 (5.1).

Stone, M. (1968). Extreme tail probabilities for sampling without replacement and exact Bahadur efficiency of the two-sample normal scores test. *Biometrika, 55,* 371–376 (5.10).

Stuart, A. (1953). The estimation and comparisons of strengths of association in contingency tables. *Biometrika, 40,* 105–110 (4.4).

Stuart, A. (1954). Asymptotic relative efficiency of tests and the derivatives of their power functions. *Skandinavisk Aktaurietidskrift,* Parts 3–4, pp. 163–169 (2.4, 5.4, 5.5).

Stuart, A. (1956). The efficiencies of test of randomness against normal regression. *Journal of the American Statistical Association, 51,* 285–287 (3.5, 5.4, 5.5).

Stuart, A. (1963). Calculation of Spearman's rho for ordered two-way classifications. *The American Statistician, 17* (4), 23–24 (5.4).

Sugiura, N., and Otake, M. (1968). Numerical comparison of improved methods of testing in contingency tables with small frequencies. *Annals of the Institute of Statistical Mathematics, 20,* 505–517 (4.2).

Susarla, V., and Van Ryzin, J. (1976). Nonparametric bayesian estimation of survival curves from incomplete observations. *Journal of the American Statistical Association, 71,* 897–902 (2.5).

Suzuki, G. (1967). On exact probabilities of some generalized Kolmogorov's *D*-statistics. *Annals of the Institute of Statistical Mathematics, 19,* 373–388 (6.1).

Suzuki, G. (1968). Kolmogorov–Smirnov tests of fit based on some general bounds. *Journal of the American Statistical Association, 63,* 919–924 (6.1).

Switzer, P. (1976). Confidence procedures for two-sample problems. *Biometrika, 63,* 13–25 (5.1).

Takacs, L. (1971). On the comparison of two empirical distribution functions. *The Annals of Mathematical Statistics, 42,* 1157–1166 (6.3).

Talwar, P.P., and Gentle, J.E. (1977). A robust test for the homogeneity of scales. *Communications in Statistics—Theory and Methods, A6,* 363–369 (5.3).

Tamura, R. (1963). On a modification of certain rank tests. *The Annals of Mathematical Statistics, 34,* 1101–1103 (5.1).

Tapia, R.A. (1978). *Nonparametric Probability Density Estimation.* Johns Hopkins series in the mathematical sciences; no. 1. Johns Hopkins University Press, Baltimore. (2.5).

Tarone, R.E., and Ware, J. (1977). On distribution-free tests for equality of survival distributions. *Biometrika, 64,* 156–160 (2.5).

Tate, M.W. (1957). *Nonparametric and Shortcut Statistics in the Social, Biological, and Medical Sciences.* Interstate Printers and Publishers, Danville, Ill. (2.5).

Tate, M.W., and Brown, S.M. (1970). Note on the Cochran *Q* test. *Journal of the American Statistical Association, 65,* 155–160 (4.6).

Taylor, G. and Conover, W.J. (1988). A nonparametric confidence interval for slope based on Spearman's rho. *Communications in Statistics—Simulation, 17* (3), 905–916 (5.5).

Teichroew, D. (1965). A history of distribution sampling prior to the era of the computer and its relevance to simulation. *Journal of the American Statistical Association, 60,* 27–49 (6.2).

Terpstra, T.J. (1952). The asymptotic normality and consistency of Kendall's test against trend, when ties are present in one ranking. *Indagationes Mathematicae, 14,* 327–333 (5.4).

Theil, H. (1950). A rank-invariant method of linear and polynomial regression analysis. *Indagationes Mathematicae, 12,* 85–91 (5.5).

Thompson, G.L. (1991). A note on the rank transform for interactions. *Biometrika, 78,* 697–701 (5.12).

Thompson, R. (1966). Bias of the one-sample Cramér-von Mises test. *Journal of the American Statistical Association, 61,* 246–247 (6.1).

Thompson, R., Govindarajulu, Z., and Doksum, K.A. (1967). Distribution and power of the absolute normal scores test. *Journal of the American Statistical Association, 62,* 966–975 (5.10).

Tiku, M.L. (1965). Chi-square approximations for the distribution of goodness-of-fit statistics, U_N^2 and W_N^2. *Biometrika, 52,* 630–633 (6.1).

Tobach, E., Smith, M., Rose, G., and Richter, D. (1967). A table for rank sum multiple paired comparisons. *Technometrics, 9,* 561–568 (5.2).

Tran, L.T. (1990). Rank statistics for serial dependence. *Journal of Statistical Planning and Inference, 24,* 215–232 (5.4).

Tryon, P.V. and Hettmansperger, T.P. (1973). A class of nonparametric tests for homogeneity against ordered alternatives. *The Annals of Statistics, 1,* 1061–1070 (5.2).

Tsai, W.S., Duran, B.S., and Lewis, T.O. (1975). Small sample behavior of some multisample nonparametric tests for scale. *Journal of the American Statistical Association, 70,* 791–796 (5.3).

Tsao, C.K. (1954). An extension of Massey's distribution of the maximum deviation between two sample cumulative step functions. *The Annals of Mathematical Statistics, 25,* 587–592 (6.3).

Tsutakawa, R.K., and Yang, S.L. (1974). Permutation tests applied to antibiotic drug resistance. *Journal of the American Statistical Association, 69,* 87–92 (5.11).

Tukey, J.W. (1949). The simplest signed-rank tests. Mimeographed Report No. 17, Statistical Research Group, Princeton University (5.7).

Tukey, J.W. (1977). *Exploratory Data Analysis.* Addison-Wesley, Reading, Mass. (5.0).

Upton, G.J.G., and Lee, R.D. (1976). The importance of the patient horizon in the sequential analysis of binomial clinical trials. *Biometrika, 63,* 335–342 (4.1).

Ury, H.K. (1966). Large-sample sign tests for trend in dispersion. *Biometrika, 53,* 289–291 (3.5).

Ury, H.K. (1972). On distribution-free confidence bounds for $Pr(Y < X)$. *Technometrics, 14,* 577–582 (5.1).

Ury, H.K. (1975). Efficiency of case-control studies with multiple controls per case: Continuous or dichotomous data. *Biometrics, 3,* 643–650 (3.5).

van der Parren, J.L. (1970). Tables for distribution-free confidence limits for the median. *Biometrika, 57,* 613–618 (3.2).

van der Parren, J.L. (1973). Tables of distribution-free one-sided and two-sided confidence limits for quantiles. *Biometrika, 73,* 433–434 (3.2).

van der Reyden, D. (1952). A simple statistical significance test. *Rhodesia Agricultural Journal, 49,* 96–104 (5.1).

van der Waerden, B.L. (1952/1953). Order tests for the two-sample problem and their power. *Proceedings Koninklijke Nederlandse Akademie van Wetenschappen (A), 55 (Indagationes Mathematicae, 14),* 453–458, and *56 (Indagationes Mathematicae, 15),* 303–316 (corrections appear in Vol. 56, p. 80) (5.10).

van der Waerden, B.L. (1953). Testing a distribution function. *Proceedings Koninklijke Nederlandse Akademie van Wetenschappen (A), 56 (Indagationes Mathematicae, 15),* 201–207 (6.1).

van Eeden, C. (1963). The relation between Pitman's asymptotic relative efficiency of two tests and the correlation coefficient between their test statistics. *Annals of Mathematical Statistics, 34,* 1442–1451 (5.10).

van Eeden, C. (1964). Note on the consistency of some distribution-free tests for dispersion. *Journal of the American Statistical Association, 59,* 105–119 (5.3).

Verdooren, L.R. (1963). Extended tables of critical values for Wilcoxon's test statistic. *Biometrika, 50,* 177–186 (5.1).

von Mises, R. (1931). *Wahrscheinlichkeitsrechnung und Ihre Anwendung in der Statistik und Theoretischen Physik.* F. Deuticke, Leipzig (6.1).

Vorlickova, D. (1970). Asymptotic properties of rank tests under discrete distributions. *Zeitschrift fur Wahrscheinlichkeitstheorie, 14,* 275–289 (5.7).

Wagner, S.S. (1970). The maximum likelihood estimate for contingency tables with zero diagonal. *Journal of the American Statistical Association, 65,* 1362–1383 (4.7).

Wald, A., and Wolfowitz, J. (1939). Confidence limits for continuous distribution functions. *The Annals of Mathematical Statistics, 10,* 105–118 (6.1).

Walker, H.M., and Lev, J. (1953). *Statistical Inference.* Holt, New York (5.1, 5.7).

Wallis, W.A. (1939). The correlation ratio for ranked data. *Journal of the American Statistical Association, 34,* 533–538 (5.8).

Walsh, J.E. (1949). Some significance tests for the median which are valid under very general conditions. *The Annals of Mathematical Statistics, 20,* 64–81 (5.7).

Walsh, J.E. (1951). Some bounded significance level properties of the equal-tail sign test. *Annals of Mathematical Statistics, 22,* 408–417 (3.4).

Walsh, J.E. (1960). Probabilities for Cramér-von Mises–Smirnov test using grouped data. *Annals of the Institute of Statistical Mathematics, 12,* 143–145 (6.1).

Walsh, J.E. (1962). *Handbook of Nonparametric Statistics.* Van Nostrand, Princeton, N.J. (2.5).

Walsh, J.E. (1963). Bounded probability properties of Kolmogorov–Smirnov and similar statistics for discrete area. *Annals of the Institute of Statistical Mathematics, 15,* 153–158 (6.1).

Weed, H.D., Jr., Bradley, R.A., and Govindarajulu, Z. (1974). Stopping times of some one-sample sequential rank tests. *The Annals of Statistics, 2,* 1314–1322 (5.6).

Welch, B.L. (1937). On the z-test in randomized blocks and Latin squares. *Biometrika, 29,* 21–52 (5.11).

Wheeler, S., and Watson, G.S. (1964). A distribution-free two-sample test on a circle. *Biometrika, 51,* 256–257 (5.1).

White, C. (1952). The use of ranks in a test of significance for comparing two treatments. *Biometrika, 8,* 33–41 (5.1).

Wilcoxon, F. (1945). Individual comparisons by ranking methods. *Biometrika, 1,* 80–83 (5.1, 5.7).

Wilcoxon, F. (1949). *Some Rapid Approximate Statistical Procedures.* American Cyanamid Co., Stamford Research Laboratories (5.7).

Wilks, S.S. (1935). The likelihood test of independence in contingency tables. *The Annals of Mathematical Statistics, 6,* 190–196 (4.7).

Wilks, S.S. (1938). The large-sample distribution of the likelihood ratio for testing composite hypotheses. *The Annals of Mathematical Statistics, 9,* 60–62 (4.7).

Wilks, S.S. (1962). *Mathematical Statistics.* Wiley, New York (3.3).

Williams, O.S. and Grizzle, J.E. (1972). Analysis of contingency tables having ordered response categories. *Journal of the American Statistical Association, 67,* 55–63 (4.2).

Woinsky, M.N., and Kurz, L. (1969). Sequential nonparametric two-way classification with prescribed maximum asymptotic error probability. *The Annals of Mathematical Statistics, 40,* 445–455 (5.1).

Wolf, F.M. (1986). *Meta-analysis: Quantitative Methods for Research Synthesis.* Sage University Paper series on Quantitative Applications in the Social Sciences 07-059, Newbury Park, CA (6.2).

Wolfe, D.A. (1977a). Two-stage two-sample median test. *Technometrics, 19,* 495–501 (4.3).

Wolfe, D.A. (1977b). A distribution-free test for related correlation coefficients. *Technometrics, 19,* 507–509 (5.4).

Wood, C.L., and Altavela, M.M. (1978). Large sample results for Kolmogorov–Smirnov statistics for discrete distributions. *Biometrika, 65* (1), 235–239 (6.1).

Woodbury, M.A., Manton, K.G., and Woodbury, L.A. (1977). An extension of the sign test for replicated measurements. *Biometrics, 33,* 453–461 (3.5).

Woodroofe, M. (1967). On the maximum deviation of the sample density. *The Annals of Mathematical Statistics, 38,* 475–481 (6.1).

Worsley, K.J. (1977). A nonparametric extension of a cluster analysis method by Scott and Knott. *Biometrics, 33,* 532–535 (5.12).

Yarnold, J.K. (1970). The minimum expectations in χ^2 goodness of fit tests and the accuracy of approximations for the null distribution. *Journal of the American Statistical Association, 65,* 864–886 (4.5).

Yates, F. (1934). Contingency tables involving small numbers and the χ^2 test. *Journal of the Royal Statistical Society, 1,* 217–235 (4.1).

Yule, G.U., and Kendall, M.G. (1950). *An Introduction to the Theory of Statistics,* 14th ed. Hafner, New York (4.4, 4.5).

Zahn, D.A., and Roberts, G.C. (1971). Exact χ^2 criterion tables with cell expectations one: An application to Coleman's measure of consensus. *Journal of the American Statistical Association, 66,* 145–148 (4.5).

Zar, J.H. (1972). Significance testing of the Spearman rank correlation coefficient. *Journal of the American Statistical Association, 67,* 578–580 (5.4).

Zaremba, S.K. (1965). Note on the Wilcoxon–Mann–Whitney statistic. *The Annals of Mathematical Statistics, 36,* 1058–1060 (5.1).

Zelen, M. (1971). The analysis of several 2×2 contingency tables. *Biometrika, 58,* 129–137 (4.1).

APPENDIX

TABLES

TABLE A1 Normal Distribution[a]

		Selected values	$z_{0.0001} = -3.7190$ $z_{0.9999} = 3.7190$		$z_{0.0005} = -3.2905$ $z_{0.9995} = 3.2905$		$z_{0.025} = -1.9600$ $z_{0.975} = 1.9600$	$z_{0.05} = -1.6449$ $z_{0.95} = 1.6449$		
p	0.000	0.001	0.002	0.003	0.004	0.005	0.006	0.007	0.008	0.009
0.00		-3.0902	-2.8782	-2.7478	-2.6521	-2.5758	-2.5121	-2.4573	-2.4089	-2.3656
0.01	-2.3263	-2.2904	-2.2571	-2.2262	-2.1973	-2.1701	-2.1444	-2.1201	-2.0969	-2.0749
0.02	-2.0537	-2.0335	-2.0141	-1.9954	-1.9774	-1.9600	-1.9431	-1.9268	-1.9110	-1.8957
0.03	-1.8808	-1.8663	-1.8522	-1.8384	-1.8250	-1.8119	-1.7991	-1.7866	-1.7744	-1.7624
0.04	-1.7507	-1.7392	-1.7279	-1.7169	-1.7060	-1.6954	-1.6849	-1.6747	-1.6646	-1.6546
0.05	-1.6449	-1.6352	-1.6258	-1.6164	-1.6072	-1.5982	-1.5893	-1.5805	-1.5718	-1.5632
0.06	-1.5548	-1.5464	-1.5382	-1.5301	-1.5220	-1.5141	-1.5063	-1.4985	-1.4909	-1.4833
0.07	-1.4758	-1.4684	-1.4611	-1.4538	-1.4466	-1.4395	-1.4325	-1.4255	-1.4187	-1.4118
0.08	-1.4051	-1.3984	-1.3917	-1.3852	-1.3787	-1.3722	-1.3658	-1.3595	-1.3532	-1.3469
0.09	-1.3408	-1.3346	-1.3285	-1.3225	-1.3165	-1.3106	-1.3047	-1.2988	-1.2930	-1.2873
0.10	-1.2816	-1.2759	-1.2702	-1.2646	-1.2591	-1.2536	-1.2481	-1.2426	-1.2372	-1.2319
0.11	-1.2265	-1.2212	-1.2160	-1.2107	-1.2055	-1.2004	-1.1952	-1.1901	-1.1850	-1.1800
0.12	-1.1750	-1.1700	-1.1650	-1.1601	-1.1552	-1.1503	-1.1455	-1.1407	-1.1359	-1.1311
0.13	-1.1264	-1.1217	-1.1170	-1.1123	-1.1077	-1.1031	-1.0985	-1.0939	-1.0893	-1.0848
0.14	-1.0803	-1.0758	-1.0714	-1.0669	-1.0625	-1.0581	-1.0537	-1.0494	-1.0450	-1.0407
0.15	-1.0364	-1.0322	-1.0279	-1.0237	-1.0194	-1.0152	-1.0110	-1.0069	-1.0027	-0.9986
0.16	-0.9945	-0.9904	-0.9863	-0.9822	-0.9782	-0.9741	-0.9701	-0.9661	-0.9621	-0.9581
0.17	-0.9542	-0.9502	-0.9463	-0.9424	-0.9385	-0.9346	-0.9307	-0.9269	-0.9230	-0.9192
0.18	-0.9154	-0.9116	-0.9078	-0.9040	-0.9002	-0.8965	-0.8927	-0.8890	-0.8853	-0.8816
0.19	-0.8779	-0.8742	-0.8705	-0.8669	-0.8633	-0.8596	-0.8560	-0.8524	-0.8488	-0.8452
0.20	-0.8416	-0.8381	-0.8345	-0.8310	-0.8274	-0.8239	-0.8204	-0.8169	-0.8134	-0.8099
0.21	-0.8064	-0.8030	-0.7995	-0.7961	-0.7926	-0.7892	-0.7858	-0.7824	-0.7790	-0.7756
0.22	-0.7722	-0.7688	-0.7655	-0.7621	-0.7588	-0.7554	-0.7521	-0.7488	-0.7454	-0.7421
0.23	-0.7388	-0.7356	-0.7323	-0.7290	-0.7257	-0.7225	-0.7192	-0.7160	-0.7128	-0.7095
0.24	-0.7063	-0.7031	-0.6999	-0.6967	-0.6935	-0.6903	-0.6871	-0.6840	-0.6808	-0.6776

TABLE A1 (Continued)

p	0.000	0.001	0.002	0.003	0.004	0.005	0.006	0.007	0.008	0.009
0.25	−0.6745	−0.6713	−0.6682	−0.6651	−0.6620	−0.6588	−0.6557	−0.6526	−0.6495	−0.6464
0.26	−0.6433	−0.6403	−0.6372	−0.6341	−0.6311	−0.6280	−0.6250	−0.6219	−0.6189	−0.6158
0.27	−0.6128	−0.6098	−0.6068	−0.6038	−0.6008	−0.5978	−0.5948	−0.5918	−0.5888	−0.5858
0.28	−0.5828	−0.5799	−0.5769	−0.5740	−0.5710	−0.5681	−0.5651	−0.5622	−0.5592	−0.5563
0.29	−0.5534	−0.5505	−0.5476	−0.5446	−0.5417	−0.5388	−0.5359	−0.5330	−0.5302	−0.5273
0.30	−0.5244	−0.5215	−0.5187	−0.5158	−0.5129	−0.5101	−0.5072	−0.5044	−0.5015	−0.4987
0.31	−0.4959	−0.4930	−0.4902	−0.4874	−0.4845	−0.4817	−0.4789	−0.4761	−0.4733	−0.4705
0.32	−0.4677	−0.4649	−0.4621	−0.4593	−0.4565	−0.4538	−0.4510	−0.4482	−0.4454	−0.4427
0.33	−0.4399	−0.4372	−0.4344	−0.4316	−0.4289	−0.4261	−0.4234	−0.4207	−0.4179	−0.4152
0.34	−0.4125	−0.4097	−0.4070	−0.4043	−0.4016	−0.3989	−0.3961	−0.3934	−0.3907	−0.3880
0.35	−0.3853	−0.3826	−0.3799	−0.3772	−0.3745	−0.3719	−0.3692	−0.3665	−0.3638	−0.3611
0.36	−0.3585	−0.3558	−0.3531	−0.3505	−0.3478	−0.3451	−0.3425	−0.3398	−0.3372	−0.3345
0.37	−0.3319	−0.3292	−0.3266	−0.3239	−0.3213	−0.3186	−0.3160	−0.3134	−0.3107	−0.3081
0.38	−0.3055	−0.3029	−0.3002	−0.2976	−0.2950	−0.2924	−0.2898	−0.2871	−0.2845	−0.2819
0.39	−0.2793	−0.2767	−0.2741	−0.2715	−0.2689	−0.2663	−0.2637	−0.2611	−0.2585	−0.2559
0.40	−0.2533	−0.2508	−0.2482	−0.2456	−0.2430	−0.2404	−0.2378	−0.2353	−0.2327	−0.2301
0.41	−0.2275	−0.2250	−0.2224	−0.2198	−0.2173	−0.2147	−0.2121	−0.2096	−0.2070	−0.2045
0.42	−0.2019	−0.1993	−0.1968	−0.1942	−0.1917	−0.1891	−0.1866	−0.1840	−0.1815	−0.1789
0.43	−0.1764	−0.1738	−0.1713	−0.1687	−0.1662	−0.1637	−0.1611	−0.1586	−0.1560	−0.1535
0.44	−0.1510	−0.1484	−0.1459	−0.1434	−0.1408	−0.1383	−0.1358	−0.1332	−0.1307	−0.1282
0.45	−0.1257	−0.1231	−0.1206	−0.1181	−0.1156	−0.1130	−0.1105	−0.1080	−0.1055	−0.1030
0.46	−0.1004	−0.0979	−0.0954	−0.0929	−0.0904	−0.0878	−0.0853	−0.0828	−0.0803	−0.0778
0.47	−0.0753	−0.0728	−0.0702	−0.0677	−0.0652	−0.0627	−0.0602	−0.0577	−0.0552	−0.0527
0.48	−0.0502	−0.0476	−0.0451	−0.0426	−0.0401	−0.0376	−0.0351	−0.0326	−0.0301	−0.0276
0.49	−0.0251	−0.0226	−0.0201	−0.0175	−0.0150	−0.0125	−0.0100	−0.0075	−0.0050	−0.0025
0.50	0.0000	0.0025	0.0050	0.0075	0.0100	0.0125	0.0150	0.0175	0.0201	0.0226
0.51	0.0251	0.0276	0.0301	0.0326	0.0351	0.0376	0.0401	0.0426	0.0451	0.0476
0.52	0.0502	0.0527	0.0552	0.0577	0.0602	0.0627	0.0652	0.0677	0.0702	0.0728
0.53	0.0753	0.0778	0.0803	0.0828	0.0853	0.0878	0.0904	0.0929	0.0954	0.0979
0.54	0.1004	0.1030	0.1055	0.1080	0.1105	0.1130	0.1156	0.1181	0.1206	0.1231

Table A1 (Continued)

p	0.000	0.001	0.002	0.003	0.004	0.005	0.006	0.007	0.008	0.009
0.55	0.1257	0.1282	0.1307	0.1332	0.1358	0.1383	0.1408	0.1434	0.1459	0.1484
0.56	0.1510	0.1535	0.1560	0.1586	0.1611	0.1637	0.1662	0.1687	0.1713	0.1738
0.57	0.1764	0.1789	0.1815	0.1840	0.1866	0.1891	0.1917	0.1942	0.1968	0.1993
0.58	0.2019	0.2045	0.2070	0.2096	0.2121	0.2147	0.2173	0.2198	0.2224	0.2250
0.59	0.2275	0.2301	0.2327	0.2353	0.2378	0.2404	0.2430	0.2456	0.2482	0.2508
0.60	0.2533	0.2559	0.2585	0.2611	0.2637	0.2663	0.2689	0.2715	0.2741	0.2767
0.61	0.2793	0.2819	0.2845	0.2871	0.2898	0.2924	0.2950	0.2976	0.3002	0.3029
0.62	0.3055	0.3081	0.3107	0.3134	0.3160	0.3186	0.3213	0.3239	0.3266	0.3292
0.63	0.3319	0.3345	0.3372	0.3398	0.3425	0.3451	0.3478	0.3505	0.3531	0.3558
0.64	0.3585	0.3611	0.3638	0.3665	0.3692	0.3719	0.3745	0.3772	0.3799	0.3826
0.65	0.3853	0.3880	0.3907	0.3934	0.3961	0.3989	0.4016	0.4043	0.4070	0.4097
0.66	0.4125	0.4152	0.4179	0.4207	0.4234	0.4261	0.4289	0.4316	0.4344	0.4372
0.67	0.4399	0.4427	0.4454	0.4482	0.4510	0.4538	0.4565	0.4593	0.4621	0.4649
0.68	0.4677	0.4705	0.4733	0.4761	0.4789	0.4817	0.4845	0.4874	0.4902	0.4930
0.69	0.4959	0.4987	0.5015	0.5044	0.5072	0.5101	0.5129	0.5158	0.5187	0.5215
0.70	0.5244	0.5273	0.5302	0.5330	0.5359	0.5388	0.5417	0.5446	0.5476	0.5505
0.71	0.5534	0.5563	0.5592	0.5622	0.5651	0.5681	0.5710	0.5740	0.5769	0.5799
0.72	0.5828	0.5858	0.5888	0.5918	0.5948	0.5978	0.6008	0.6038	0.6068	0.6098
0.73	0.6128	0.6158	0.6189	0.6219	0.6250	0.6280	0.6311	0.6341	0.6372	0.6403
0.74	0.6433	0.6464	0.6495	0.6526	0.6557	0.6588	0.6620	0.6651	0.6682	0.6713
0.75	0.6745	0.6776	0.6808	0.6840	0.6871	0.6903	0.6935	0.6967	0.6999	0.7031
0.76	0.7063	0.7095	0.7128	0.7160	0.7192	0.7225	0.7257	0.7290	0.7323	0.7356
0.77	0.7388	0.7421	0.7454	0.7488	0.7521	0.7554	0.7588	0.7621	0.7655	0.7688
0.78	0.7722	0.7756	0.7790	0.7824	0.7858	0.7892	0.7926	0.7961	0.7995	0.8030
0.79	0.8064	0.8099	0.8134	0.8169	0.8204	0.8239	0.8274	0.8310	0.8345	0.8381
0.80	0.8416	0.8452	0.8488	0.8524	0.8560	0.8596	0.8633	0.8669	0.8705	0.8742
0.81	0.8779	0.8816	0.8853	0.8890	0.8927	0.8965	0.9002	0.9040	0.9078	0.9116
0.82	0.9154	0.9192	0.9230	0.9269	0.9307	0.9346	0.9385	0.9424	0.9463	0.9502

Table A1 (Continued)

p	0.000	0.001	0.002	0.003	0.004	0.005	0.006	0.007	0.008	0.009
0.83	0.9542	0.9581	0.9621	0.9661	0.9701	0.9741	0.9782	0.9822	0.9863	0.9904
0.84	0.9945	0.9986	1.0027	1.0069	1.0110	1.0152	1.0194	1.0237	1.0279	1.0322
0.85	1.0364	1.0407	1.0450	1.0494	1.0537	1.0581	1.0625	1.0669	1.0714	1.0758
0.86	1.0803	1.0848	1.0893	1.0939	1.0985	1.1031	1.1077	1.1123	1.1170	1.1217
0.87	1.1264	1.1311	1.1359	1.1407	1.1455	1.1503	1.1552	1.1601	1.1650	1.1700
0.88	1.1750	1.1800	1.1850	1.1901	1.1952	1.2004	1.2055	1.2107	1.2160	1.2212
0.89	1.2265	1.2319	1.2372	1.2426	1.2481	1.2536	1.2591	1.2646	1.2702	1.2759
0.90	1.2816	1.2873	1.2930	1.2988	1.3047	1.3106	1.3165	1.3225	1.3285	1.3346
0.91	1.3408	1.3469	1.3532	1.3595	1.3658	1.3722	1.3787	1.3852	1.3917	1.3984
0.92	1.4051	1.4118	1.4187	1.4255	1.4325	1.4395	1.4466	1.4538	1.4611	1.4684
0.93	1.4758	1.4833	1.4909	1.4985	1.5063	1.5141	1.5220	1.5301	1.5382	1.5464
0.94	1.5548	1.5632	1.5718	1.5805	1.5893	1.5982	1.6072	1.6164	1.6258	1.6352
0.95	1.6449	1.6546	1.6646	1.6747	1.6849	1.6954	1.7060	1.7169	1.7279	1.7392
0.96	1.7507	1.7624	1.7744	1.7866	1.7991	1.8119	1.8250	1.8384	1.8522	1.8663
0.97	1.8808	1.8957	1.9110	1.9268	1.9431	1.9600	1.9774	1.9954	2.0141	2.0335
0.98	2.0537	2.0749	2.0969	2.1201	2.1444	2.1701	2.1973	2.2262	2.2571	2.2904
0.99	2.3263	2.3656	2.4089	2.4573	2.5121	2.5758	2.6521	2.7478	2.8782	3.0902

SOURCE. Generated by R. L. Iman. Used with permission.

[a]The entries in this table are quantiles z_p of the standard normal random variable Z selected so $P(Z \leq z_p) = p$ and $P(Z > z_p) = 1 - p$. Note that the value of p to two decimal places determines which row to use; the third decimal place of p determines which column to use to find z_p.

TABLE A2   Chi-Squared Distribution[a]

	$p = 0.750$	0.900	0.950	0.975	0.990	0.995	0.999
$k = 1$	1.323	2.706	3.841	5.024	6.635	7.879	10.83
2	2.773	4.605	5.991	7.378	9.210	10.60	13.82
3	4.108	6.251	7.815	9.348	11.34	12.84	16.27
4	5.385	7.779	9.488	11.14	13.28	14.86	18.47
5	6.626	9.236	11.07	12.83	15.09	16.75	20.51
6	7.841	10.64	12.59	14.45	16.81	18.55	22.46
7	9.037	12.02	14.07	16.01	18.48	20.28	24.32
8	10.22	13.36	15.51	17.53	20.09	21.96	26.13
9	11.39	14.68	16.92	19.02	21.67	23.59	27.88
10	12.55	15.99	18.31	20.48	23.21	25.19	29.59
11	13.70	17.28	19.68	21.92	24.73	26.76	31.26
12	14.85	18.55	21.03	23.34	26.22	28.30	32.91
13	15.98	19.81	22.36	24.74	27.69	29.82	34.53
14	17.12	21.06	23.68	26.12	29.14	31.32	36.12
15	18.25	22.31	25.00	27.49	30.58	32.80	37.70
16	19.37	23.54	26.30	28.85	32.00	34.27	39.25
17	20.49	24.77	27.59	30.19	33.41	35.72	40.79
18	21.60	25.99	28.87	31.53	34.81	37.16	42.31
19	22.72	27.20	30.14	32.85	36.19	38.58	43.82
20	23.83	28.41	31.41	34.17	37.57	40.00	45.32
21	24.93	29.62	32.67	35.48	38.93	41.40	46.80
22	26.04	30.81	33.92	36.78	40.29	42.80	48.27
23	27.14	32.01	35.17	38.08	41.64	44.18	49.73
24	28.24	33.20	36.42	39.37	42.98	45.56	51.18
25	29.34	34.38	37.65	40.65	44.31	46.93	52.62
26	30.43	35.56	38.89	41.92	45.64	48.29	54.05
27	31.53	36.74	40.11	43.19	46.96	49.64	55.48
28	32.62	37.92	41.34	44.46	48.28	50.99	56.89
29	33.71	39.09	42.56	45.72	49.59	52.34	58.30
30	34.80	40.26	43.77	46.98	50.89	53.67	59.70
40	45.62	51.81	55.76	59.34	63.69	66.77	73.40
50	56.33	63.17	67.50	71.42	76.15	79.49	86.66
60	66.98	74.40	79.08	83.30	88.38	91.95	99.61
70	77.58	85.53	90.53	95.02	100.4	104.2	112.3
80	88.13	96.58	101.9	106.6	112.3	116.3	124.8
90	98.65	107.6	113.1	118.1	124.1	128.3	137.2
100	109.1	118.5	124.3	129.6	135.8	140.2	149.4
z_p	0.675	1.282	1.645	1.960	2.326	2.576	3.090

For $k > 100$ use the approximation $w_p = (\frac{1}{2})(z_p + \sqrt{2k - 1})^2$, or the more accurate $w_p =$ $k\left(1 - \frac{2}{9k} + z_p\sqrt{\frac{2}{9k}}\right)^3$, where z_p is the value from the standardized normal distribution shown in the bottom of the table.

SOURCE: Abridged from Table 8, Vol. 1 of Pearson and Hartley (1976), with permission from the *Biometrika*, Trustees.

[a] The entries in this table are quantiles w_p of a chi-squared random variable W with k degrees of freedom, selected so $P(W \leq w_p) = p$ and $P(W > w_p) = 1 - p$.

TABLE A3 Binomial Distribution[a]

n	y	p = 0.05	0.10	0.15	0.20	0.25	0.30	0.35	0.40	0.45
1	0	0.9500	0.9000	0.8500	0.8000	0.7500	0.7000	0.6500	0.6000	0.5500
	1	1.0000	1.0000	1.0000	1.0000	1.0000	1.0000	1.0000	1.0000	1.0000
2	0	0.9025	0.8100	0.7225	0.6400	0.5625	0.4900	0.4225	0.3600	0.3025
	1	0.9975	0.9900	0.9775	0.9600	0.9375	0.9100	0.8775	0.8400	0.7975
	2	1.0000	1.0000	1.0000	1.0000	1.0000	1.0000	1.0000	1.0000	1.0000
3	0	0.8574	0.7290	0.6141	0.5120	0.4219	0.3430	0.2746	0.2160	0.1664
	1	0.9928	0.9720	0.9392	0.8960	0.8438	0.7840	0.7182	0.6480	0.5748
	2	0.9999	0.9990	0.9966	0.9920	0.9844	0.9730	0.9571	0.9360	0.9089
	3	1.0000	1.0000	1.0000	1.0000	1.0000	1.0000	1.0000	1.0000	1.0000
4	0	0.8145	0.6561	0.5220	0.4096	0.3164	0.2401	0.1785	0.1296	0.0915
	1	0.9860	0.9477	0.8905	0.8192	0.7383	0.6517	0.5630	0.4752	0.3910
	2	0.9995	0.9963	0.9880	0.9728	0.9492	0.9163	0.8735	0.8208	0.7585
	3	1.0000	0.9999	0.9995	0.9984	0.9961	0.9919	0.9850	0.9744	0.9590
	4	1.0000	1.0000	1.0000	1.0000	1.0000	1.0000	1.0000	1.0000	1.0000
5	0	0.7738	0.5905	0.4437	0.3277	0.2373	0.1681	0.1160	0.0778	0.0503
	1	0.9774	0.9185	0.8352	0.7373	0.6328	0.5282	0.4284	0.3370	0.2562
	2	0.9988	0.9914	0.9734	0.9421	0.8965	0.8369	0.7648	0.6826	0.5931
	3	1.0000	0.9995	0.9978	0.9933	0.9844	0.9692	0.9460	0.9130	0.8688
	4	1.0000	1.0000	0.9999	0.9997	0.9990	0.9976	0.9947	0.9898	0.9815
	5	1.0000	1.0000	1.0000	1.0000	1.0000	1.0000	1.0000	1.0000	1.0000
6	0	0.7351	0.5314	0.3771	0.2621	0.1780	0.1176	0.0754	0.0467	0.0277
	1	0.9672	0.8857	0.7765	0.6554	0.5339	0.4202	0.3191	0.2333	0.1636
	2	0.9978	0.9842	0.9527	0.9011	0.8306	0.7443	0.6471	0.5443	0.4415
	3	0.9999	0.9987	0.9941	0.9830	0.9624	0.9295	0.8826	0.8208	0.7447
	4	1.0000	0.9999	0.9996	0.9984	0.9954	0.9891	0.9777	0.9590	0.9308
	5	1.0000	1.0000	1.0000	0.9999	0.9998	0.9993	0.9982	0.9959	0.9917
	6	1.0000	1.0000	1.0000	1.0000	1.0000	1.0000	1.0000	1.0000	1.0000
7	0	0.6983	0.4783	0.3206	0.2097	0.1335	0.0824	0.0490	0.0280	0.0152
	1	0.9556	0.8503	0.7166	0.5767	0.4449	0.3294	0.2338	0.1586	0.1024
	2	0.9962	0.9743	0.9262	0.8520	0.7564	0.6471	0.5323	0.4199	0.3164
	3	0.9998	0.9973	0.9879	0.9667	0.9294	0.8740	0.8002	0.7102	0.6083
	4	1.0000	0.9998	0.9988	0.9953	0.9871	0.9712	0.9444	0.9037	0.8471
	5	1.0000	1.0000	0.9999	0.9996	0.9987	0.9962	0.9910	0.9812	0.9643
	6	1.0000	1.0000	1.0000	1.0000	0.9999	0.9998	0.9994	0.9984	0.9963
	7	1.0000	1.0000	1.0000	1.0000	1.0000	1.0000	1.0000	1.0000	1.0000

TABLE A3   (Continued)

n	y	p = 0.50	0.55	0.60	0.65	0.70	0.75	0.80	0.85	0.90	0.95
1	0	0.5000	0.4500	0.4000	0.3500	0.3000	0.2500	0.2000	0.1500	0.1000	0.0500
	1	1.0000	1.0000	1.0000	1.0000	1.0000	1.0000	1.0000	1.0000	1.0000	1.0000
2	0	0.2500	0.2025	0.1600	0.1225	0.0900	0.0625	0.0400	0.0225	0.0100	0.0025
	1	0.7500	0.6975	0.6400	0.5775	0.5100	0.4375	0.3600	0.2775	0.1900	0.0975
	2	1.0000	1.0000	1.0000	1.0000	1.0000	1.0000	1.0000	1.0000	1.0000	1.0000
3	0	0.1250	0.0911	0.0640	0.0429	0.0270	0.0156	0.0080	0.0034	0.0010	0.0001
	1	0.5000	0.4252	0.3520	0.2818	0.2160	0.1562	0.1040	0.0608	0.0280	0.0072
	2	0.8750	0.8336	0.7840	0.7254	0.6570	0.5781	0.4880	0.3859	0.2710	0.1426
	3	1.0000	1.0000	1.0000	1.0000	1.0000	1.0000	1.0000	1.0000	1.0000	1.0000
4	0	0.0625	0.0410	0.0256	0.0150	0.0081	0.0039	0.0016	0.0005	0.0001	0.0000
	1	0.3125	0.2415	0.1792	0.1265	0.0837	0.0508	0.0272	0.0120	0.0037	0.0005
	2	0.6875	0.6090	0.5248	0.4370	0.3483	0.2617	0.1808	0.1095	0.0523	0.0140
	3	0.9375	0.9085	0.8704	0.8215	0.7599	0.6836	0.5904	0.4780	0.3439	0.1855
	4	1.0000	1.0000	1.0000	1.0000	1.0000	1.0000	1.0000	1.0000	1.0000	1.0000
5	0	0.0312	0.0185	0.0102	0.0053	0.0024	0.0010	0.0003	0.0001	0.0000	0.0000
	1	0.1875	0.1312	0.0870	0.0540	0.0308	0.0156	0.0067	0.0022	0.0005	0.0000
	2	0.5000	0.4069	0.3174	0.2352	0.1631	0.1035	0.0579	0.0266	0.0086	0.0012
	3	0.8125	0.7438	0.6630	0.5716	0.4718	0.3672	0.2627	0.1648	0.0815	0.0226
	4	0.9688	0.9497	0.9222	0.8840	0.8319	0.7627	0.6723	0.5563	0.4095	0.2262
	5	1.0000	1.0000	1.0000	1.0000	1.0000	1.0000	1.0000	1.0000	1.0000	1.0000
6	0	0.0156	0.0083	0.0041	0.0018	0.0007	0.0002	0.0001	0.0000	0.0000	0.0000
	1	0.1094	0.0692	0.0410	0.0223	0.0109	0.0046	0.0016	0.0004	0.0001	0.0000
	2	0.3438	0.2553	0.1792	0.1174	0.0705	0.0376	0.0170	0.0059	0.0013	0.0001
	3	0.6562	0.5585	0.4557	0.3529	0.2557	0.1694	0.0989	0.0473	0.0158	0.0022
	4	0.8906	0.8364	0.7667	0.6809	0.5789	0.4661	0.3446	0.2235	0.1143	0.0328
	5	0.9844	0.9723	0.9533	0.9246	0.8824	0.8220	0.7379	0.6229	0.4686	0.2649
	6	1.0000	1.0000	1.0000	1.0000	1.0000	1.0000	1.0000	1.0000	1.0000	1.0000
7	0	0.0078	0.0037	0.0016	0.0006	0.0002	0.0001	0.0000	0.0000	0.0000	0.0000
	1	0.0625	0.0357	0.0188	0.0090	0.0038	0.0013	0.0004	0.0001	0.0000	0.0000
	2	0.2266	0.1529	0.0963	0.0556	0.0288	0.0129	0.0047	0.0012	0.0002	0.0000
	3	0.5000	0.3917	0.2898	0.1998	0.1260	0.0706	0.0333	0.0121	0.0027	0.0002
	4	0.7734	0.6836	0.5801	0.4677	0.3529	0.2436	0.1480	0.0738	0.0257	0.0038
	5	0.9375	0.8976	0.8414	0.7662	0.6706	0.5551	0.4233	0.2834	0.1497	0.0444
	6	0.9922	0.9848	0.9720	0.9510	0.9176	0.8665	0.7903	0.6794	0.5217	0.3017
	7	1.0000	1.0000	1.0000	1.0000	1.0000	1.0000	1.0000	1.0000	1.0000	1.0000

TABLE A3 (Continued)

n	y	p = 0.05	0.10	0.15	0.20	0.25	0.30	0.35	0.40	0.45
8	0	0.6634	0.4305	0.2725	0.1678	0.1001	0.0576	0.0319	0.0168	0.0084
	1	0.9428	0.8131	0.6572	0.5033	0.3671	0.2553	0.1691	0.1064	0.0632
	2	0.9942	0.9619	0.8948	0.7969	0.6785	0.5518	0.4278	0.3154	0.2201
	3	0.9996	0.9950	0.9786	0.9437	0.8862	0.8059	0.7064	0.5941	0.4770
	4	1.0000	0.9996	0.9971	0.9896	0.9727	0.9420	0.8939	0.8263	0.7396
	5	1.0000	1.0000	0.9998	0.9988	0.9958	0.9887	0.9747	0.9502	0.9115
	6	1.0000	1.0000	1.0000	0.9999	0.9996	0.9987	0.9964	0.9915	0.9819
	7	1.0000	1.0000	1.0000	1.0000	1.0000	0.9999	0.9998	0.9993	0.9983
	8	1.0000	1.0000	1.0000	1.0000	1.0000	1.0000	1.0000	1.0000	1.0000
9	0	0.6302	0.3874	0.2316	0.1342	0.0751	0.0404	0.0207	0.0101	0.0046
	1	0.9288	0.7748	0.5995	0.4362	0.3003	0.1960	0.1211	0.0705	0.0385
	2	0.9916	0.9470	0.8591	0.7382	0.6007	0.4628	0.3373	0.2318	0.1495
	3	0.9994	0.9917	0.9661	0.9144	0.8343	0.7297	0.6089	0.4826	0.3614
	4	1.0000	0.9991	0.9944	0.9804	0.9511	0.9012	0.8283	0.7334	0.6214
	5	1.0000	0.9999	0.9994	0.9969	0.9900	0.9747	0.9464	0.9006	0.8342
	6	1.0000	1.0000	1.0000	0.9997	0.9987	0.9957	0.9888	0.9750	0.9502
	7	1.0000	1.0000	1.0000	1.0000	0.9999	0.9996	0.9986	0.9962	0.9909
	8	1.0000	1.0000	1.0000	1.0000	1.0000	1.0000	0.9999	0.9997	0.9992
	9	1.0000	1.0000	1.0000	1.0000	1.0000	1.0000	1.0000	1.0000	1.0000
10	0	0.5987	0.3487	0.1969	0.1074	0.0563	0.0282	0.0135	0.0060	0.0025
	1	0.9139	0.7361	0.5443	0.3758	0.2440	0.1493	0.0860	0.0464	0.0233
	2	0.9885	0.9298	0.8202	0.6778	0.5256	0.3828	0.2616	0.1673	0.0996
	3	0.9990	0.9872	0.9500	0.8791	0.7759	0.6496	0.5138	0.3823	0.2660
	4	0.9999	0.9984	0.9901	0.9672	0.9219	0.8497	0.7515	0.6331	0.5044
	5	1.0000	0.9999	0.9986	0.9936	0.9803	0.9527	0.9051	0.8338	0.7384
	6	1.0000	1.0000	0.9999	0.9991	0.9965	0.9894	0.9740	0.9452	0.8980
	7	1.0000	1.0000	1.0000	0.9999	0.9996	0.9984	0.9952	0.9877	0.9726
	8	1.0000	1.0000	1.0000	1.0000	1.0000	0.9999	0.9995	0.9983	0.9955
	9	1.0000	1.0000	1.0000	1.0000	1.0000	1.0000	1.0000	0.9999	0.9997
	10	1.0000	1.0000	1.0000	1.0000	1.0000	1.0000	1.0000	1.0000	1.0000
11	0	0.5688	0.3138	0.1673	0.0859	0.0422	0.0198	0.0088	0.0036	0.0014
	1	0.8981	0.6974	0.4922	0.3221	0.1971	0.1130	0.0606	0.0302	0.0139
	2	0.9848	0.9104	0.7788	0.6174	0.4552	0.3127	0.2001	0.1189	0.0652
	3	0.9984	0.9815	0.9306	0.8389	0.7133	0.5696	0.4256	0.2963	0.1911
	4	0.9999	0.9972	0.9841	0.9496	0.8854	0.7897	0.6683	0.5328	0.3971
	5	1.0000	0.9997	0.9973	0.9883	0.9657	0.9218	0.8513	0.7535	0.6331
	6	1.0000	1.0000	0.9997	0.9980	0.9924	0.9784	0.9499	0.9006	0.8262
	7	1.0000	1.0000	1.0000	0.9998	0.9988	0.9957	0.9878	0.9707	0.9390
	8	1.0000	1.0000	1.0000	1.0000	0.9999	0.9994	0.9980	0.9941	0.9852
	9	1.0000	1.0000	1.0000	1.0000	1.0000	1.0000	0.9998	0.9993	0.9978
	10	1.0000	1.0000	1.0000	1.0000	1.0000	1.0000	1.0000	1.0000	0.9998
	11	1.0000	1.0000	1.0000	1.0000	1.0000	1.0000	1.0000	1.0000	1.0000

TABLE A3   (Continued)

n	y	p = 0.50	0.55	0.60	0.65	0.70	0.75	0.80	0.85	0.90	0.95
8	0	0.0039	0.0017	0.0007	0.0002	0.0001	0.0000	0.0000	0.0000	0.0000	0.0000
	1	0.0352	0.0181	0.0085	0.0036	0.0013	0.0004	0.0001	0.0000	0.0000	0.0000
	2	0.1445	0.0885	0.0498	0.0253	0.0113	0.0042	0.0012	0.0002	0.0000	0.0000
	3	0.3633	0.2604	0.1737	0.1061	0.0580	0.0273	0.0104	0.0029	0.0004	0.0000
	4	0.6367	0.5230	0.4059	0.2936	0.1941	0.1138	0.0563	0.0214	0.0050	0.0004
	5	0.8555	0.7799	0.6846	0.5722	0.4482	0.3215	0.2031	0.1052	0.0381	0.0058
	6	0.9648	0.9368	0.8936	0.8309	0.7447	0.6329	0.4967	0.3428	0.1869	0.0572
	7	0.9961	0.9916	0.9832	0.9681	0.9424	0.8999	0.8322	0.7275	0.5695	0.3366
	8	1.0000	1.0000	1.0000	1.0000	1.0000	1.0000	1.0000	1.0000	1.0000	1.0000
9	0	0.0020	0.0008	0.0003	0.0001	0.0000	0.0000	0.0000	0.0000	0.0000	0.0000
	1	0.0195	0.0091	0.0038	0.0014	0.0004	0.0001	0.0000	0.0000	0.0000	0.0000
	2	0.0898	0.0498	0.0250	0.0112	0.0043	0.0013	0.0003	0.0000	0.0000	0.0000
	3	0.2539	0.1658	0.0994	0.0536	0.0253	0.0100	0.0031	0.0006	0.0001	0.0000
	4	0.5000	0.3786	0.2666	0.1717	0.0988	0.0489	0.0196	0.0056	0.0009	0.0000
	5	0.7461	0.6386	0.5174	0.3911	0.2703	0.1657	0.0856	0.0339	0.0083	0.0006
	6	0.9102	0.8505	0.7682	0.6627	0.5372	0.3993	0.2618	0.1409	0.0530	0.0084
	7	0.9805	0.9615	0.9295	0.8789	0.8040	0.6997	0.5638	0.4005	0.2252	0.0712
	8	0.9980	0.9954	0.9899	0.9793	0.9596	0.9249	0.8658	0.7684	0.6126	0.3698
	9	1.0000	1.0000	1.0000	1.0000	1.0000	1.0000	1.0000	1.0000	1.0000	1.0000
10	0	0.0010	0.0003	0.0001	0.0000	0.0000	0.0000	0.0000	0.0000	0.0000	0.0000
	1	0.0107	0.0045	0.0017	0.0005	0.0001	0.0000	0.0000	0.0000	0.0000	0.0000
	2	0.0547	0.0274	0.0123	0.0048	0.0016	0.0004	0.0001	0.0000	0.0000	0.0000
	3	0.1719	0.1020	0.0548	0.0260	0.0106	0.0035	0.0009	0.0001	0.0000	0.0000
	4	0.3770	0.2616	0.1662	0.0949	0.0473	0.0197	0.0064	0.0014	0.0001	0.0000
	5	0.6230	0.4956	0.3669	0.2485	0.1503	0.0781	0.0328	0.0099	0.0016	0.0001
	6	0.8281	0.7340	0.6177	0.4862	0.3504	0.2241	0.1209	0.0500	0.0128	0.0010
	7	0.9453	0.9004	0.8327	0.7384	0.6172	0.4744	0.3222	0.1798	0.0702	0.0115
	8	0.9893	0.9767	0.9536	0.9140	0.8507	0.7560	0.6242	0.4557	0.2639	0.0861
	9	0.9990	0.9975	0.9940	0.9865	0.9718	0.9437	0.8926	0.8031	0.6513	0.4013
	10	1.0000	1.0000	1.0000	1.0000	1.0000	1.0000	1.0000	1.0000	1.0000	1.0000
11	0	0.0005	0.0002	0.0000	0.0000	0.0000	0.0000	0.0000	0.0000	0.0000	0.0000
	1	0.0059	0.0022	0.0007	0.0002	0.0000	0.0000	0.0000	0.0000	0.0000	0.0000
	2	0.0327	0.0148	0.0059	0.0020	0.0006	0.0001	0.0000	0.0000	0.0000	0.0000
	3	0.1133	0.0610	0.0293	0.0122	0.0043	0.0012	0.0002	0.0000	0.0000	0.0000
	4	0.2744	0.1738	0.0994	0.0501	0.0216	0.0076	0.0020	0.0003	0.0000	0.0000
	5	0.5000	0.3669	0.2465	0.1487	0.0782	0.0343	0.0117	0.0027	0.0003	0.0000
	6	0.7256	0.6029	0.4672	0.3317	0.2103	0.1146	0.0504	0.0159	0.0028	0.0001
	7	0.8867	0.8089	0.7037	0.5744	0.4304	0.2867	0.1611	0.0694	0.0185	0.0016
	8	0.9673	0.9348	0.8811	0.7999	0.6873	0.5448	0.3826	0.2212	0.0896	0.0152
	9	0.9941	0.9861	0.9698	0.9394	0.8870	0.8029	0.6779	0.5078	0.3026	0.1019
	10	0.9995	0.9986	0.9964	0.9912	0.9802	0.9578	0.9141	0.8327	0.6862	0.4312
	11	1.0000	1.0000	1.0000	1.0000	1.0000	1.0000	1.0000	1.0000	1.0000	1.0000

TABLE A3 (Continued)

n	y	p = 0.05	0.10	0.15	0.20	0.25	0.30	0.35	0.40	0.45
12	0	0.5404	0.2824	0.1422	0.0687	0.0317	0.0138	0.0057	0.0022	0.0008
	1	0.8816	0.6590	0.4435	0.2749	0.1584	0.0850	0.0424	0.0196	0.0083
	2	0.9804	0.8891	0.7358	0.5583	0.3907	0.2528	0.1513	0.0834	0.0421
	3	0.9978	0.9744	0.9078	0.7946	0.6488	0.4925	0.3467	0.2253	0.1345
	4	0.9998	0.9957	0.9761	0.9274	0.8424	0.7237	0.5833	0.4382	0.3044
	5	1.0000	0.9995	0.9954	0.9806	0.9456	0.8822	0.7873	0.6652	0.5269
	6	1.0000	0.9999	0.9993	0.9961	0.9857	0.9614	0.9154	0.8418	0.7393
	7	1.0000	1.0000	0.9999	0.9994	0.9972	0.9905	0.9745	0.9427	0.8883
	8	1.0000	1.0000	1.0000	0.9999	0.9996	0.9983	0.9944	0.9847	0.9644
	9	1.0000	1.0000	1.0000	1.0000	1.0000	0.9998	0.9992	0.9972	0.9921
	10	1.0000	1.0000	1.0000	1.0000	1.0000	1.0000	0.9999	0.9997	0.9989
	11	1.0000	1.0000	1.0000	1.0000	1.0000	1.0000	1.0000	1.0000	0.9999
	12	1.0000	1.0000	1.0000	1.0000	1.0000	1.0000	1.0000	1.0000	1.0000
13	0	0.5133	0.2542	0.1209	0.0550	0.0238	0.0097	0.0037	0.0013	0.0004
	1	0.8646	0.6213	0.3983	0.2336	0.1267	0.0637	0.0296	0.0126	0.0049
	2	0.9755	0.8661	0.6920	0.5017	0.3326	0.2025	0.1132	0.0579	0.0269
	3	0.9969	0.9658	0.8820	0.7473	0.5843	0.4206	0.2783	0.1686	0.0929
	4	0.9997	0.9935	0.9658	0.9009	0.7940	0.6543	0.5005	0.3530	0.2279
	5	1.0000	0.9991	0.9925	0.9700	0.9198	0.8346	0.7159	0.5744	0.4268
	6	1.0000	0.9999	0.9987	0.9930	0.9757	0.9376	0.8705	0.7712	0.6437
	7	1.0000	1.0000	0.9998	0.9988	0.9944	0.9818	0.9538	0.9023	0.8212
	8	1.0000	1.0000	1.0000	0.9998	0.9990	0.9960	0.9874	0.9679	0.9302
	9	1.0000	1.0000	1.0000	1.0000	0.9999	0.9993	0.9975	0.9922	0.9797
	10	1.0000	1.0000	1.0000	1.0000	1.0000	0.9999	0.9997	0.9987	0.9959
	11	1.0000	1.0000	1.0000	1.0000	1.0000	1.0000	1.0000	0.9999	0.9995
	12	1.0000	1.0000	1.0000	1.0000	1.0000	1.0000	1.0000	1.0000	1.0000
	13	1.0000	1.0000	1.0000	1.0000	1.0000	1.0000	1.0000	1.0000	1.0000
14	0	0.4877	0.2288	0.1028	0.0440	0.0178	0.0068	0.0024	0.0008	0.0002
	1	0.8470	0.5846	0.3567	0.1979	0.1010	0.0475	0.0205	0.0081	0.0029
	2	0.9699	0.8416	0.6479	0.4481	0.2811	0.1608	0.0839	0.0398	0.0170
	3	0.9958	0.9559	0.8535	0.6982	0.5213	0.3552	0.2205	0.1243	0.0632
	4	0.9996	0.9908	0.9533	0.8702	0.7415	0.5842	0.4227	0.2793	0.1672
	5	1.0000	0.9985	0.9885	0.9561	0.8883	0.7805	0.6405	0.4859	0.3373
	6	1.0000	0.9998	0.9978	0.9884	0.9617	0.9067	0.8164	0.6925	0.5461
	7	1.0000	1.0000	0.9997	0.9976	0.9897	0.9685	0.9247	0.8499	0.7414
	8	1.0000	1.0000	1.0000	0.9996	0.9978	0.9917	0.9757	0.9417	0.8811
	9	1.0000	1.0000	1.0000	1.0000	0.9997	0.9983	0.9940	0.9825	0.9574
	10	1.0000	1.0000	1.0000	1.0000	1.0000	0.9998	0.9989	0.9961	0.9886
	11	1.0000	1.0000	1.0000	1.0000	1.0000	1.0000	0.9999	0.9994	0.9978
	12	1.0000	1.0000	1.0000	1.0000	1.0000	1.0000	1.0000	0.9999	0.9997
	13	1.0000	1.0000	1.0000	1.0000	1.0000	1.0000	1.0000	1.0000	1.0000
	14	1.0000	1.0000	1.0000	1.0000	1.0000	1.0000	1.0000	1.0000	1.0000

TABLE A3 (Continued)

n	y	p = 0.50	0.55	0.60	0.65	0.70	0.75	0.80	0.85	0.90	0.95
12	0	0.0002	0.0001	0.0000	0.0000	0.0000	0.0000	0.0000	0.0000	0.0000	0.0000
	1	0.0032	0.0011	0.0003	0.0001	0.0000	0.0000	0.0000	0.0000	0.0000	0.0000
	2	0.0193	0.0079	0.0028	0.0008	0.0002	0.0000	0.0000	0.0000	0.0000	0.0000
	3	0.0730	0.0356	0.0153	0.0056	0.0017	0.0004	0.0001	0.0000	0.0000	0.0000
	4	0.1938	0.1117	0.0573	0.0255	0.0095	0.0028	0.0006	0.0001	0.0000	0.0000
	5	0.3872	0.2607	0.1582	0.0846	0.0386	0.0143	0.0039	0.0007	0.0001	0.0000
	6	0.6128	0.4731	0.3348	0.2127	0.1178	0.0544	0.0194	0.0046	0.0005	0.0000
	7	0.8062	0.6956	0.5618	0.4167	0.2763	0.1576	0.0726	0.0239	0.0043	0.0002
	8	0.9270	0.8655	0.7747	0.6533	0.5075	0.3512	0.2054	0.0922	0.0256	0.0022
	9	0.9807	0.9579	0.9166	0.8487	0.7472	0.6093	0.4417	0.2642	0.1109	00.196
	10	0.9968	0.9917	0.9804	0.9576	0.9150	0.8416	0.7251	0.5565	0.3410	0.1184
	11	0.9998	0.9992	0.9978	0.9943	0.9862	0.9683	0.9313	0.8578	0.7176	0.4596
	12	1.0000	1.0000	1.0000	1.0000	1.0000	1.0000	1.0000	1.0000	1.0000	1.0000
13	0	0.0001	0.0000	0.0000	0.0000	0.0000	0.0000	0.0000	0.0000	0.0000	0.0000
	1	0.0017	0.0005	0.0001	0.0000	0.0000	0.0000	0.0000	0.0000	0.0000	0.0000
	2	0.0112	0.0041	0.0013	0.0003	0.0001	0.0000	0.0000	0.0000	0.0000	0.0000
	3	0.0461	0.0203	0.0078	0.0025	0.0007	0.0001	0.0000	0.0000	0.0000	0.0000
	4	0.1334	0.0698	0.0321	0.0126	0.0040	0.0010	0.0002	0.0000	0.0000	0.0000
	5	0.2905	0.1788	0.0977	0.0462	0.0182	0.0056	0.0012	0.0002	0.0000	0.0000
	6	0.5000	0.3563	0.2288	0.1295	0.0624	0.0243	0.0070	0.0013	0.0001	0.0000
	7	0.7095	0.5732	0.4256	0.2841	0.1654	0.0802	0.0300	0.0075	0.0009	0.0000
	8	0.8666	0.7721	0.6470	0.4995	0.3457	0.2060	0.0991	0.0342	0.0065	0.0003
	9	0.9539	0.9071	0.8314	0.7217	0.5794	0.4157	0.2527	0.1180	0.0342	0.0031
	10	0.9888	0.9731	0.9421	0.8868	0.7975	0.6674	0.4983	0.3080	0.1339	0.0245
	11	0.9983	0.9951	0.9874	0.9704	0.9363	0.8733	0.7664	0.6017	0.3787	0.1354
	12	0.9999	0.9996	0.9987	0.9963	0.9903	0.9762	0.9450	0.8791	0.7458	0.4867
	13	1.0000	1.0000	1.0000	1.0000	1.0000	1.0000	1.0000	1.0000	1.0000	1.0000
14	0	0.0001	0.0000	0.0000	0.0000	0.0000	0.0000	0.0000	0.0000	0.0000	0.0000
	1	0.0009	0.0003	0.0001	0.0000	0.0000	0.0000	0.0000	0.0000	0.0000	0.0000
	2	0.0065	0.0022	0.0006	0.0001	0.0000	0.0000	0.0000	0.0000	0.0000	0.0000
	3	0.0287	0.0114	0.0039	0.0011	0.0002	0.0000	0.0000	0.0000	0.0000	0.0000
	4	0.0898	0.0426	0.0175	0.0060	0.0017	0.0003	0.0000	0.0000	0.0000	0.0000
	5	0.2120	0.1189	0.0583	0.0243	0.0083	0.0022	0.0004	0.0000	0.0000	0.0000
	6	0.3953	0.2586	0.1501	0.0753	0.0315	0.0103	0.0024	0.0003	0.0000	0.0000
	7	0.6047	0.4539	0.3075	0.1836	0.0933	0.0383	0.0116	0.0022	0.0002	0.0000
	8	0.7880	0.6627	0.5141	0.3595	0.2195	0.1117	0.0439	0.0115	0.0015	0.0000
	9	0.9102	0.8328	0.7207	0.5773	0.4158	0.2585	0.1298	0.0467	0.0092	0.0004
	10	0.9713	0.9368	0.8757	0.7795	0.6448	0.4787	0.3018	0.1465	0.0441	0.0042
	11	0.9935	0.9830	0.9602	0.9161	0.8392	0.7189	0.5519	0.3521	0.1584	0.0301
	12	0.9991	0.9971	0.9919	0.9795	0.9525	0.8990	0.8021	0.6433	0.4154	0.1530
	13	0.9999	0.9998	0.9992	0.9976	0.9932	0.9822	0.9560	0.8972	0.7712	0.5123
	14	1.0000	1.0000	1.0000	1.0000	1.0000	1.0000	1.0000	1.0000	1.0000	1.0000

TABLE A3   (Continued)

n	y	p = 0.05	0.10	0.15	0.20	0.25	0.30	0.35	0.40	0.45
15	0	0.4633	0.2059	0.0874	0.0352	0.0134	0.0047	0.0016	0.0005	0.0001
	1	0.8290	0.5490	0.3186	0.1671	0.0802	0.0353	0.0142	0.0052	0.0017
	2	0.9638	0.8159	0.6042	0.3980	0.2361	0.1268	0.0617	0.0271	0.0107
	3	0.9945	0.9444	0.8227	0.6482	0.4613	0.2969	0.1727	0.0905	0.0424
	4	0.9994	0.9873	0.9383	0.8358	0.6865	0.5155	0.3519	0.2173	0.1204
	5	0.9999	0.9978	0.9832	0.9389	0.8516	0.7216	0.5643	0.4032	0.2608
	6	1.0000	0.9997	0.9964	0.9819	0.9434	0.8689	0.7548	0.6098	0.4522
	7	1.0000	1.0000	0.9994	0.9958	0.9827	0.9500	0.8868	0.7869	0.6535
	8	1.0000	1.0000	0.9999	0.9992	0.9958	0.9848	0.9578	0.9050	0.8182
	9	1.0000	1.0000	1.0000	0.9999	0.9992	0.9963	0.9876	0.9662	0.9231
	10	1.0000	1.0000	1.0000	1.0000	0.9999	0.9993	0.9972	0.9907	0.9745
	11	1.0000	1.0000	1.0000	1.0000	1.0000	0.9999	0.9995	0.9981	0.9937
	12	1.0000	1.0000	1.0000	1.0000	1.0000	1.0000	0.9999	0.9997	0.9989
	13	1.0000	1.0000	1.0000	1.0000	1.0000	1.0000	1.0000	1.0000	0.9999
	14	1.0000	1.0000	1.0000	1.0000	1.0000	1.0000	1.0000	1.0000	1.0000
	15	1.0000	1.0000	1.0000	1.0000	1.0000	1.0000	1.0000	1.0000	1.0000
16	0	0.4401	0.1853	0.0743	0.0281	0.0100	0.0033	0.0010	0.0003	0.0001
	1	0.8108	0.5147	0.2839	0.1407	0.0635	0.0261	0.0098	0.0033	0.0010
	2	0.9571	0.7892	0.5614	0.3518	0.1971	0.0994	0.0451	0.0183	0.0066
	3	0.9930	0.9316	0.7899	0.5981	0.4050	0.2459	0.1339	0.0651	0.0281
	4	0.9991	0.9830	0.9209	0.7982	0.6302	0.4499	0.2892	0.1666	0.0853
	5	0.9999	0.9967	0.9765	0.9183	0.8103	0.6598	0.4900	0.3288	0.1976
	6	1.0000	0.9995	0.9944	0.9733	0.9204	0.8247	0.6881	0.5272	0.3660
	7	1.0000	0.9999	0.9989	0.9930	0.9729	0.9256	0.8406	0.7161	0.5629
	8	1.0000	1.0000	0.9998	0.9985	0.9925	0.9743	0.9329	0.8577	0.7441
	9	1.0000	1.0000	1.0000	0.9998	0.9984	0.9929	0.9771	0.9417	0.8759
	10	1.0000	1.0000	1.0000	1.0000	0.9997	0.9984	0.9938	0.9809	0.9514
	11	1.0000	1.0000	1.0000	1.0000	1.0000	0.9997	0.9987	0.9951	0.9851
	12	1.0000	1.0000	1.0000	1.0000	1.0000	1.0000	0.9998	0.9991	0.9965
	13	1.0000	1.0000	1.0000	1.0000	1.0000	1.0000	1.0000	0.9999	0.9994
	14	1.0000	1.0000	1.0000	1.0000	1.0000	1.0000	1.0000	1.0000	0.9999
	15	1.0000	1.0000	1.0000	1.0000	1.0000	1.0000	1.0000	1.0000	1.0000
	16	1.0000	1.0000	1.0000	1.0000	1.0000	1.0000	1.0000	1.0000	1.0000

TABLE A3   (Continued)

n	y	p = 0.50	0.55	0.60	0.65	0.70	0.75	0.80	0.85	0.90	0.95
15	0	0.0000	0.0000	0.0000	0.0000	0.0000	0.0000	0.0000	0.0000	0.0000	0.0000
	1	0.0005	0.0001	0.0000	0.0000	0.0000	0.0000	0.0000	0.0000	0.0000	0.0000
	2	0.0037	0.0011	0.0003	0.0001	0.0000	0.0000	0.0000	0.0000	0.0000	0.0000
	3	0.0176	0.0063	0.0019	0.0005	0.0001	0.0000	0.0000	0.0000	0.0000	0.0000
	4	0.0592	0.0255	0.0093	0.0028	0.0007	0.0001	0.0000	0.0000	0.0000	0.0000
	5	0.1509	0.0769	0.0338	0.0124	0.0037	0.0008	0.0001	0.0000	0.0000	0.0000
	6	0.3036	0.1818	0.0950	0.0422	0.0152	0.0042	0.0008	0.0001	0.0000	0.0000
	7	0.5000	0.3465	0.2131	0.1132	0.0500	0.0173	0.0042	0.0006	0.0000	0.0000
	8	0.6964	0.5478	0.3902	0.2452	0.1311	0.0566	0.0181	0.0036	0.0003	0.0000
	9	0.8491	0.7392	0.5968	0.4357	0.2784	0.1484	0.0611	0.0168	0.0022	0.0001
	10	0.9408	0.8796	0.7827	0.6481	0.4845	0.3135	0.1642	0.0617	0.0127	0.0006
	11	0.9824	0.9576	0.9095	0.8273	0.7031	0.5387	0.3518	0.1773	0.0556	0.0055
	12	0.9963	0.9893	0.9729	0.9383	0.8732	0.7639	0.6020	0.3958	0.1841	0.0362
	13	0.9995	0.9983	0.9948	0.9858	0.9647	0.9198	0.8329	0.6814	0.4510	0.1710
	14	1.0000	0.9999	0.9995	0.9984	0.9953	0.9866	0.9648	0.9126	0.7941	0.5367
	15	1.0000	1.0000	1.0000	1.0000	1.0000	1.0000	1.0000	1.0000	1.0000	1.0000
16	0	0.0000	0.0000	0.0000	0.0000	0.0000	0.0000	0.0000	0.0000	0.0000	0.0000
	1	0.0003	0.0001	0.0000	0.0000	0.0000	0.0000	0.0000	0.0000	0.0000	0.0000
	2	0.0021	0.0006	0.0001	0.0000	0.0000	0.0000	0.0000	0.0000	0.0000	0.0000
	3	0.0106	0.0035	0.0009	0.0002	0.0000	0.0000	0.0000	0.0000	0.0000	0.0000
	4	0.0384	0.0149	0.0049	0.0013	0.0003	0.0000	0.0000	0.0000	0.0000	0.0000
	5	0.1051	0.0486	0.0191	0.0062	0.0016	0.0003	0.0000	0.0000	0.0000	0.0000
	6	0.2272	0.1241	0.0583	0.0229	0.0071	0.0016	0.0002	0.0000	0.0000	0.0000
	7	0.4018	0.2559	0.1423	0.0671	0.0257	0.0075	0.0015	0.0002	0.0000	0.0000
	8	0.5982	0.4371	0.2839	0.1594	0.0744	0.0271	0.0070	0.0011	0.0001	0.0000
	9	0.7728	0.6340	0.4728	0.3119	0.1753	0.0796	0.0267	0.0056	0.0005	0.0000
	10	0.8949	0.8024	0.6712	0.5100	0.3402	0.1897	0.0817	0.0235	0.0033	0.0001
	11	0.9616	0.9147	0.8334	0.7108	0.5501	0.3698	0.2018	0.0791	0.0170	0.0009
	12	0.9894	0.9719	0.9349	0.8661	0.7541	0.5950	0.4019	0.2101	0.0684	0.0070
	13	0.9979	0.9934	0.9817	0.9549	0.9006	0.8029	0.6482	0.4386	0.2108	0.0429
	14	0.9997	0.9990	0.9967	0.9902	0.9739	0.9365	0.8593	0.7161	0.4853	0.1892
	15	1.0000	0.9999	0.9997	0.9990	0.9967	0.9900	0.9719	0.9257	0.8147	0.5599
	16	1.0000	1.0000	1.0000	1.0000	1.0000	1.0000	1.0000	1.0000	1.0000	1.0000

TABLE A3   (Continued)

n	y	p = 0.05	0.10	0.15	0.20	0.25	0.30	0.35	0.40	0.45
17	0	0.4181	0.1668	0.0631	0.0225	0.0075	0.0023	0.0007	0.0002	0.0000
	1	0.7922	0.4818	0.2525	0.1182	0.0501	0.0193	0.0067	0.0021	0.0006
	2	0.9497	0.7618	0.5198	0.3096	0.1637	0.0774	0.0327	0.0123	0.0041
	3	0.9912	0.9174	0.7556	0.5489	0.3530	0.2019	0.1028	0.0464	0.0184
	4	0.9988	0.9779	0.9013	0.7582	0.5739	0.3887	0.2348	0.1260	0.0596
	5	0.9999	0.9953	0.9681	0.8943	0.7653	0.5968	0.4197	0.2639	0.1471
	6	1.0000	0.9992	0.9917	0.9623	0.8929	0.7752	0.6188	0.4478	0.2902
	7	1.0000	0.9999	0.9983	0.9891	0.9598	0.8954	0.7872	0.6405	0.4743
	8	1.0000	1.0000	0.9997	0.9974	0.9876	0.9597	0.9006	0.8011	0.6626
	9	1.0000	1.0000	1.0000	0.9995	0.9969	0.9873	0.9617	0.9081	0.8166
	10	1.0000	1.0000	1.0000	0.9999	0.9994	0.9968	0.9880	0.9652	0.9174
	11	1.0000	1.0000	1.0000	1.0000	0.9999	0.9993	0.9970	0.9894	0.9699
	12	1.0000	1.0000	1.0000	1.0000	1.0000	0.9999	0.9994	0.9975	0.9914
	13	1.0000	1.0000	1.0000	1.0000	1.0000	1.0000	0.9999	0.9995	0.9981
	14	1.0000	1.0000	1.0000	1.0000	1.0000	1.0000	1.0000	0.9999	0.9997
	15	1.0000	1.0000	1.0000	1.0000	1.0000	1.0000	1.0000	1.0000	1.0000
	16	1.0000	1.0000	1.0000	1.0000	1.0000	1.0000	1.0000	1.0000	1.0000
	17	1.0000	1.0000	1.0000	1.0000	1.0000	1.0000	1.0000	1.0000	1.0000
18	0	0.3972	0.1501	0.0536	0.0180	0.0056	0.0016	0.0004	0.0001	0.0000
	1	0.7735	0.4503	0.2241	0.0991	0.0395	0.0142	0.0046	0.0013	0.0003
	2	0.9419	0.7338	0.4797	0.2713	0.1353	0.0600	0.0236	0.0082	0.0025
	3	0.9891	0.9018	0.7202	0.5010	0.3057	0.1646	0.0783	0.0328	0.0120
	4	0.9985	0.9718	0.8794	0.7164	0.5187	0.3327	0.1886	0.0942	0.0411
	5	0.9998	0.9936	0.9581	0.8671	0.7175	0.5344	0.3550	0.2088	0.1077
	6	1.0000	0.9988	0.9882	0.9487	0.8610	0.7217	0.5491	0.3743	0.2258
	7	1.0000	0.9998	0.9973	0.9837	0.9431	0.8593	0.7283	0.5634	0.3915
	8	1.0000	1.0000	0.9995	0.9957	0.9807	0.9404	0.8609	0.7368	0.5778
	9	1.0000	1.0000	0.9999	0.9991	0.9946	0.9790	0.9403	0.8653	0.7473
	10	1.0000	1.0000	1.0000	0.9998	0.9988	0.9939	0.9788	0.9424	0.8720
	11	1.0000	1.0000	1.0000	1.0000	0.9998	0.9986	0.9938	0.9797	0.9463
	12	1.0000	1.0000	1.0000	1.0000	1.0000	0.9997	0.9986	0.9942	0.9817
	13	1.0000	1.0000	1.0000	1.0000	1.0000	1.0000	0.9997	0.9987	0.9951
	14	1.0000	1.0000	1.0000	1.0000	1.0000	1.0000	1.0000	0.9998	0.9990
	15	1.0000	1.0000	1.0000	1.0000	1.0000	1.0000	1.0000	1.0000	0.9999
	16	1.0000	1.0000	1.0000	1.0000	1.0000	1.0000	1.0000	1.0000	1.0000
	17	1.0000	1.0000	1.0000	1.0000	1.0000	1.0000	1.0000	1.0000	1.0000
	18	1.0000	1.0000	1.0000	1.0000	1.0000	1.0000	1.0000	1.0000	1.0000

TABLE A3 (Continued)

n	y	p = 0.50	0.55	0.60	0.65	0.70	0.75	0.80	0.85	0.90	0.95
17	0	0.0000	0.0000	0.0000	0.0000	0.0000	0.0000	0.0000	0.0000	0.0000	0.0000
	1	0.0001	0.0000	0.0000	0.0000	0.0000	0.0000	0.0000	0.0000	0.0000	0.0000
	2	0.0012	0.0003	0.0001	0.0000	0.0000	0.0000	0.0000	0.0000	0.0000	0.0000
	3	0.0064	0.0019	0.0005	0.0001	0.0000	0.0000	0.0000	0.0000	0.0000	0.0000
	4	0.0245	0.0086	0.0025	0.0006	0.0001	0.0000	0.0000	0.0000	0.0000	0.0000
	5	0.0717	0.0301	0.0106	0.0030	0.0007	0.0001	0.0000	0.0000	0.0000	0.0000
	6	0.1662	0.0826	0.0348	0.0120	0.0032	0.0006	0.0001	0.0000	0.0000	0.0000
	7	0.3145	0.1834	0.0919	0.0383	0.0127	0.0031	0.0005	0.0000	0.0000	0.0000
	8	0.5000	0.3374	0.1989	0.0994	0.0403	0.0124	0.0026	0.0003	0.0000	0.0000
	9	0.6855	0.5257	0.3595	0.2128	0.1046	0.0402	0.0109	0.0017	0.0001	0.0000
	10	0.8338	0.7098	0.5522	0.3812	0.2248	0.1071	0.0377	0.0083	0.0008	0.0000
	11	0.9283	0.8529	0.7361	0.5803	0.4032	0.2347	0.1057	0.0319	0.0047	0.0001
	12	0.9755	0.9404	0.8740	0.7652	0.6113	0.4261	0.2418	0.0987	0.0221	0.0012
	13	0.9936	0.9816	0.9536	0.8972	0.7981	0.6470	0.4511	0.2444	0.0826	0.0088
	14	0.9988	0.9959	0.9877	0.9673	0.9226	0.8363	0.6904	0.4802	0.2382	0.0503
	15	0.9999	0.9994	0.9979	0.9933	0.9807	0.9499	0.8818	0.7475	0.5182	0.2078
	16	1.0000	1.0000	0.9998	0.9993	0.9977	0.9925	0.9775	0.9369	0.8332	0.5819
	17	1.0000	1.0000	1.0000	1.0000	1.0000	1.0000	1.0000	1.0000	1.0000	1.0000
18	0	0.0000	0.0000	0.0000	0.0000	0.0000	0.0000	0.0000	0.0000	0.0000	0.0000
	1	0.0001	0.0000	0.0000	0.0000	0.0000	0.0000	0.0000	0.0000	0.0000	0.0000
	2	0.0007	0.0001	0.0000	0.0000	0.0000	0.0000	0.0000	0.0000	0.0000	0.0000
	3	0.0038	0.0010	0.0002	0.0000	0.0000	0.0000	0.0000	0.0000	0.0000	0.0000
	4	0.0154	0.0049	0.0013	0.0003	0.0000	0.0000	0.0000	0.0000	0.0000	0.0000
	5	0.0481	0.0183	0.0058	0.0014	0.0003	0.0000	0.0000	0.0000	0.0000	0.0000
	6	0.1189	0.0537	0.0203	0.0062	0.0014	0.0002	0.0000	0.0000	0.0000	0.0000
	7	0.2403	0.1280	0.0576	0.0212	0.0061	0.0012	0.0002	0.0000	0.0000	0.0000
	8	0.4073	0.2527	0.1347	0.0597	0.0210	0.0054	0.0009	0.0001	0.0000	0.0000
	9	0.5927	0.4222	0.2632	0.1391	0.0596	0.0193	0.0043	0.0005	0.0000	0.0000
	10	0.7597	0.6085	0.4366	0.2717	0.1407	0.0569	0.0163	0.0027	0.0002	0.0000
	11	0.8811	0.7742	0.6257	0.4509	0.2783	0.1390	0.0513	0.0118	0.0012	0.0000
	12	0.9519	0.8923	0.7912	0.6450	0.4656	0.2825	0.1329	0.0419	0.0064	0.0002
	13	0.9846	0.9589	0.9058	0.8114	0.6673	0.4813	0.2836	0.1206	0.0282	0.0015
	14	0.9962	0.9880	0.9672	0.9217	0.8354	0.6943	0.4990	0.2798	0.0982	0.0109
	15	0.9993	0.9975	0.9918	0.9764	0.9400	0.8647	0.7287	0.5203	0.2662	0.0581
	16	0.9999	0.9997	0.9987	0.9954	0.9858	0.9605	0.9009	0.7759	0.5497	0.2265
	17	1.0000	1.0000	0.9999	0.9996	0.9984	0.9944	0.9820	0.9464	0.8499	0.6028
	18	1.0000	1.0000	1.0000	1.0000	1.0000	1.0000	1.0000	1.0000	1.0000	1.0000

TABLE A3 (Continued)

n	y	$p = 0.05$	0.10	0.15	0.20	0.25	0.30	0.35	0.40	0.45
19	0	0.3774	0.1351	0.0456	0.0144	0.0042	0.0011	0.0003	0.0001	0.0000
	1	0.7547	0.4203	0.1985	0.0829	0.0310	0.0104	0.0031	0.0008	0.0002
	2	0.9335	0.7054	0.4413	0.2369	0.1113	0.0462	0.0170	0.0055	0.0015
	3	0.9868	0.8850	0.6841	0.4551	0.2631	0.1332	0.0591	0.0230	0.0077
	4	0.9980	0.9648	0.8556	0.6733	0.4654	0.2822	0.1500	0.0696	0.0280
	5	0.9998	0.9914	0.9463	0.8369	0.6678	0.4739	0.2968	0.1629	0.0777
	6	1.0000	0.9983	0.9837	0.9324	0.8251	0.6655	0.4812	0.3081	0.1727
	7	1.0000	0.9997	0.9959	0.9767	0.9225	0.8180	0.6656	0.4878	0.3169
	8	1.0000	1.0000	0.9992	0.9933	0.9713	0.9161	0.8145	0.6675	0.4940
	9	1.0000	1.0000	0.9999	0.9984	0.9911	0.9674	0.9125	0.8139	0.6710
	10	1.0000	1.0000	1.0000	0.9997	0.9977	0.9895	0.9653	0.9115	0.8159
	11	1.0000	1.0000	1.0000	1.0000	0.9995	0.9972	0.9886	0.9648	0.9129
	12	1.0000	1.0000	1.0000	1.0000	0.9999	0.9994	0.9969	0.9884	0.9658
	13	1.0000	1.0000	1.0000	1.0000	1.0000	0.9999	0.9993	0.9969	0.9891
	14	1.0000	1.0000	1.0000	1.0000	1.0000	1.0000	0.9999	0.9994	0.9972
	15	1.0000	1.0000	1.0000	1.0000	1.0000	1.0000	1.0000	0.9999	0.9995
	16	1.0000	1.0000	1.0000	1.0000	1.0000	1.0000	1.0000	1.0000	0.9999
	17	1.0000	1.0000	1.0000	1.0000	1.0000	1.0000	1.0000	1.0000	1.0000
	18	1.0000	1.0000	1.0000	1.0000	1.0000	1.0000	1.0000	1.0000	1.0000
	19	1.0000	1.0000	1.0000	1.0000	1.0000	1.0000	1.0000	1.0000	1.0000
20	0	0.3585	0.1216	0.0388	0.0115	0.0032	0.0008	0.0002	0.0000	0.0000
	1	0.7358	0.3917	0.1756	0.0692	0.0243	0.0076	0.0021	0.0005	0.0001
	2	0.9245	0.6769	0.4049	0.2061	0.0913	0.0355	0.0121	0.0036	0.0009
	3	0.9841	0.8670	0.6477	0.4114	0.2252	0.1071	0.0444	0.0160	0.0049
	4	0.9974	0.9568	0.8298	0.6296	0.4148	0.2375	0.1182	0.0510	0.0189
	5	0.9997	0.9887	0.9327	0.8042	0.6172	0.4164	0.2454	0.1256	0.0553
	6	1.0000	0.9976	0.9781	0.9133	0.7858	0.6080	0.4166	0.2500	0.1299
	7	1.0000	0.9996	0.9941	0.9679	0.8982	0.7723	0.6010	0.4159	0.2520
	8	1.0000	0.9999	0.9987	0.9900	0.9591	0.8867	0.7624	0.5956	0.4143
	9	1.0000	1.0000	0.9998	0.9974	0.9861	0.9520	0.8782	0.7553	0.5914
	10	1.0000	1.0000	1.0000	0.9994	0.9961	0.9829	0.9468	0.8725	0.7507
	11	1.0000	1.0000	1.0000	0.9999	0.9991	0.9949	0.9804	0.9435	0.8692
	12	1.0000	1.0000	1.0000	1.0000	0.9998	0.9987	0.9940	0.9790	0.9420
	13	1.0000	1.0000	1.0000	1.0000	1.0000	0.9997	0.9985	0.9935	0.9786
	14	1.0000	1.0000	1.0000	1.0000	1.0000	1.0000	0.9997	0.9984	0.9936
	15	1.0000	1.0000	1.0000	1.0000	1.0000	1.0000	1.0000	0.9997	0.9985
	16	1.0000	1.0000	1.0000	1.0000	1.0000	1.0000	1.0000	1.0000	0.9997
	17	1.0000	1.0000	1.0000	1.0000	1.0000	1.0000	1.0000	1.0000	1.0000
	18	1.0000	1.0000	1.0000	1.0000	1.0000	1.0000	1.0000	1.0000	1.0000
	19	1.0000	1.0000	1.0000	1.0000	1.0000	1.0000	1.0000	1.0000	1.0000
	20	1.0000	1.0000	1.0000	1.0000	1.0000	1.0000	1.0000	1.0000	1.0000

TABLE A3 (Continued)

n	y	p = 0.50	0.55	0.60	0.65	0.70	0.75	0.80	0.85	0.90	0.95
19	0	0.0000	0.0000	0.0000	0.0000	0.0000	0.0000	0.0000	0.0000	0.0000	0.0000
	1	0.0000	0.0000	0.0000	0.0000	0.0000	0.0000	0.0000	0.0000	0.0000	0.0000
	2	0.0004	0.0001	0.0000	0.0000	0.0000	0.0000	0.0000	0.0000	0.0000	0.0000
	3	0.0022	0.0005	0.0001	0.0000	0.0000	0.0000	0.0000	0.0000	0.0000	0.0000
	4	0.0096	0.0028	0.0006	0.0001	0.0000	0.0000	0.0000	0.0000	0.0000	0.0000
	5	0.0318	0.0109	0.0031	0.0007	0.0001	0.0000	0.0000	0.0000	0.0000	0.0000
	6	0.0835	0.0342	0.0116	0.0031	0.0006	0.0001	0.0000	0.0000	0.0000	0.0000
	7	0.1796	0.0871	0.0352	0.0114	0.0028	0.0005	0.0000	0.0000	0.0000	0.0000
	8	0.3238	0.1841	0.0885	0.0347	0.0105	0.0023	0.0003	0.0000	0.0000	0.0000
	9	0.5000	0.3290	0.1861	0.0875	0.0326	0.0089	0.0016	0.0001	0.0000	0.0000
	10	0.6762	0.5060	0.3325	0.1855	0.0839	0.0287	0.0067	0.0008	0.0000	0.0000
	11	0.8204	0.6831	0.5122	0.3344	0.1820	0.0775	0.0233	0.0041	0.0003	0.0000
	12	0.9165	0.8273	0.6919	0.5188	0.3345	0.1749	0.0676	0.0163	0.0017	0.0000
	13	0.9682	0.9223	0.8371	0.7032	0.5261	0.3322	0.1631	0.0537	0.0086	0.0002
	14	0.9904	0.9720	0.9304	0.8500	0.7178	0.5346	0.3267	0.1444	0.0352	0.0020
	15	0.9978	0.9923	0.9770	0.9409	0.8668	0.7369	0.5449	0.3159	0.1150	0.0132
	16	0.9996	0.9985	0.9945	0.9830	0.9538	0.8887	0.7631	0.5587	0.2946	0.0665
	17	1.0000	0.9998	0.9992	0.9969	0.9896	0.9690	0.9171	0.8015	0.5797	0.2453
	18	1.0000	1.0000	0.9999	0.9997	0.9989	0.9958	0.9856	0.9544	0.8649	0.6226
	19	1.0000	1.0000	1.0000	1.0000	1.0000	1.0000	1.0000	1.0000	1.0000	1.0000
20	0	0.0000	0.0000	0.0000	0.0000	0.0000	0.0000	0.0000	0.0000	0.0000	0.0000
	1	0.0000	0.0000	0.0000	0.0000	0.0000	0.0000	0.0000	0.0000	0.0000	0.0000
	2	0.0002	0.0000	0.0000	0.0000	0.0000	0.0000	0.0000	0.0000	0.0000	0.0000
	3	0.0013	0.0003	0.0000	0.0000	0.0000	0.0000	0.0000	0.0000	0.0000	0.0000
	4	0.0059	0.0015	0.0003	0.0000	0.0000	0.0000	0.0000	0.0000	0.0000	0.0000
	5	0.0207	0.0064	0.0016	0.0003	0.0000	0.0000	0.0000	0.0000	0.0000	0.0000
	6	0.0577	0.0214	0.0065	0.0015	0.0003	0.0000	0.0000	0.0000	0.0000	0.0000
	7	0.1316	0.0580	0.0210	0.0060	0.0013	0.0002	0.0000	0.0000	0.0000	0.0000
	8	0.2517	0.1308	0.0565	0.0196	0.0051	0.0009	0.0001	0.0000	0.0000	0.0000
	9	0.4119	0.2493	0.1275	0.0532	0.0171	0.0039	0.0006	0.0000	0.0000	0.0000
	10	0.5881	0.4086	0.2447	0.1218	0.0480	0.0139	0.0026	0.0002	0.0000	0.0000
	11	0.7483	0.5857	0.4044	0.2376	0.1133	0.0409	0.0100	0.0013	0.0001	0.0000
	12	0.8684	0.7480	0.5841	0.3990	0.2277	0.1018	0.0321	0.0059	0.0004	0.0000
	13	0.9423	0.8701	0.7500	0.5834	0.3920	0.2142	0.0867	0.0219	0.0024	0.0000
	14	0.9793	0.9447	0.8744	0.7546	0.5836	0.3828	0.1958	0.0673	0.0113	0.0003
	15	0.9941	0.9811	0.9490	0.8818	0.7625	0.5852	0.3704	0.1702	0.0432	0.0026
	16	0.9987	0.9951	0.9840	0.9556	0.8929	0.7748	0.5886	0.3523	0.1330	0.0159
	17	0.9998	0.9991	0.9964	0.9879	0.9645	0.9087	0.7939	0.5951	0.3231	0.0755
	18	1.0000	0.9999	0.9995	0.9979	0.9924	0.9757	0.9308	0.8244	0.6083	0.2642
	19	1.0000	1.0000	1.0000	0.9998	0.9992	0.9968	0.9885	0.9612	0.8784	0.6415
	20	1.0000	1.0000	1.0000	1.0000	1.0000	1.0000	1.0000	1.0000	1.0000	1.0000

[a] Y has the binomial distribution with parameters n and p. The entries are the values of $P(Y \le y) = \sum_{i=0}^{y} \binom{n}{i} p^i (1 - p)^{n-i}$, for p ranging from 0.05 to 0.95.

For n larger than 20, the rth quantile y_r of a binomial random variable may be approximated using $y_r = np + z_r \sqrt{np(1 - p)}$, where z_r is the rth quantile of a standard normal random variable, obtained from Table A1.

TABLE A4 Exact Confidence Intervals for the Binomial Parameter p

		90%		95%		99%	
n	Y	Lower	Upper	Lower	Upper	Lower	Upper
1	0	0.000	0.950	0.000	0.975	0.000	0.995
	1	0.050	1.000	0.025	1.000	0.005	1.000
2	0	0.000	0.776	0.000	0.842	0.000	0.929
	1	0.025	0.975	0.013	0.987	0.003	0.997
	2	0.224	1.000	0.158	1.000	0.071	1.000
3	0	0.000	0.632	0.000	0.708	0.000	0.829
	1	0.017	0.865	0.008	0.906	0.002	0.959
	2	0.135	0.983	0.094	0.992	0.041	0.998
	3	0.368	1.000	0.292	1.000	0.171	1.000
4	0	0.000	0.527	0.000	0.602	0.000	0.734
	1	0.013	0.751	0.006	0.806	0.001	0.889
	2	0.098	0.902	0.068	0.932	0.029	0.971
	3	0.249	0.987	0.194	0.994	0.111	0.999
	4	0.473	1.000	0.398	1.000	0.266	1.000
5	0	0.000	0.451	0.000	0.522	0.000	0.653
	1	0.010	0.657	0.005	0.716	0.001	0.815
	2	0.076	0.811	0.053	0.853	0.023	0.917
	3	0.189	0.924	0.147	0.947	0.083	0.977
	4	0.343	0.990	0.284	0.995	0.185	0.999
	5	0.549	1.000	0.478	1.000	0.347	1.000
6	0	0.000	0.393	0.000	0.459	0.000	0.586
	1	0.009	0.582	0.004	0.641	0.001	0.746
	2	0.063	0.729	0.043	0.777	0.019	0.856
	3	0.153	0.847	0.118	0.882	0.066	0.934
	4	0.271	0.937	0.223	0.957	0.144	0.981
	5	0.418	0.991	0.359	0.996	0.254	0.999
	6	0.607	1.000	0.541	1.000	0.414	1.000
7	0	0.000	0.348	0.000	0.410	0.000	0.531
	1	0.007	0.521	0.004	0.579	0.001	0.685
	2	0.053	0.659	0.037	0.710	0.016	0.797
	3	0.129	0.775	0.099	0.816	0.055	0.882
	4	0.225	0.871	0.184	0.901	0.118	0.945
	5	0.341	0.947	0.290	0.963	0.203	0.984
	6	0.479	0.993	0.421	0.996	0.315	0.999
	7	0.652	1.000	0.590	1.000	0.469	1.000

TABLE A4 (Continued)

n	Y	90%		95%		99%	
		Lower	Upper	Lower	Upper	Lower	Upper
8	0	0.000	0.312	0.000	0.369	0.000	0.484
	1	0.006	0.471	0.003	0.526	0.001	0.632
	2	0.046	0.600	0.032	0.651	0.014	0.742
	3	0.111	0.711	0.085	0.755	0.047	0.830
	4	0.193	0.807	0.157	0.843	0.100	0.900
	5	0.289	0.889	0.245	0.915	0.170	0.953
	6	0.400	0.954	0.349	0.968	0.258	0.986
	7	0.529	0.994	0.474	0.997	0.368	0.999
	8	0.688	1.000	0.631	1.000	0.516	1.000
9	0	0.000	0.283	0.000	0.336	0.000	0.445
	1	0.006	0.429	0.003	0.482	0.001	0.585
	2	0.041	0.550	0.028	0.600	0.012	0.693
	3	0.098	0.655	0.075	0.701	0.042	0.781
	4	0.169	0.749	0.137	0.788	0.087	0.854
	5	0.251	0.831	0.212	0.863	0.146	0.913
	6	0.345	0.902	0.299	0.925	0.219	0.958
	7	0.450	0.959	0.400	0.972	0.307	0.988
	8	0.571	0.994	0.518	0.997	0.415	0.999
	9	0.717	1.000	0.664	1.000	0.555	1.000
10	0	0.000	0.259	0.000	0.308	0.000	0.411
	1	0.005	0.394	0.003	0.445	0.001	0.544
	2	0.037	0.507	0.025	0.556	0.011	0.648
	3	0.087	0.607	0.067	0.652	0.037	0.735
	4	0.150	0.696	0.122	0.738	0.077	0.809
	5	0.222	0.778	0.187	0.813	0.128	0.872
	6	0.304	0.850	0.262	0.878	0.191	0.923
	7	0.393	0.913	0.348	0.933	0.265	0.963
	8	0.493	0.963	0.444	0.975	0.352	0.989
	9	0.606	0.995	0.555	0.997	0.456	0.999
	10	0.741	1.000	0.692	1.000	0.589	1.000
11	0	0.000	0.238	0.000	0.285	0.000	0.382
	1	0.005	0.364	0.002	0.413	0.000	0.509
	2	0.033	0.470	0.023	0.518	0.010	0.608
	3	0.079	0.564	0.060	0.610	0.033	0.693
	4	0.135	0.650	0.109	0.692	0.069	0.767
	5	0.200	0.729	0.167	0.766	0.115	0.831
	6	0.271	0.800	0.234	0.833	0.169	0.885
	7	0.350	0.865	0.308	0.891	0.233	0.931
	8	0.436	0.921	0.390	0.940	0.307	0.967
	9	0.530	0.967	0.482	0.977	0.392	0.990
	10	0.636	0.995	0.587	0.998	0.491	1.000
	11	0.762	1.000	0.715	1.000	0.618	1.000

TABLE A4   (Continued)

n	Y	90%		95%		99%	
		Lower	Upper	Lower	Upper	Lower	Upper
12	0	0.000	0.221	0.000	0.265	0.000	0.357
	1	0.004	0.339	0.002	0.385	0.000	0.477
	2	0.030	0.438	0.021	0.484	0.009	0.573
	3	0.072	0.527	0.055	0.572	0.030	0.655
	4	0.123	0.609	0.099	0.651	0.062	0.728
	5	0.181	0.685	0.152	0.723	0.103	0.792
	6	0.245	0.755	0.211	0.789	0.152	0.848
	7	0.315	0.819	0.277	0.848	0.208	0.897
	8	0.391	0.877	0.349	0.901	0.272	0.938
	9	0.473	0.928	0.428	0.945	0.345	0.970
	10	0.562	0.970	0.516	0.979	0.427	0.991
	11	0.661	0.996	0.615	0.998	0.523	1.000
	12	0.779	1.000	0.735	1.000	0.643	1.000
13	0	0.000	0.206	0.000	0.247	0.000	0.335
	1	0.004	0.316	0.002	0.360	0.000	0.449
	2	0.028	0.410	0.019	0.454	0.008	0.541
	3	0.066	0.495	0.050	0.538	0.028	0.621
	4	0.113	0.573	0.091	0.614	0.057	0.691
	5	0.166	0.645	0.139	0.684	0.094	0.755
	6	0.224	0.713	0.192	0.749	0.138	0.811
	7	0.287	0.776	0.251	0.808	0.189	0.862
	8	0.355	0.834	0.316	0.861	0.245	0.906
	9	0.427	0.887	0.386	0.909	0.309	0.943
	10	0.505	0.934	0.462	0.950	0.379	0.972
	11	0.590	0.972	0.546	0.981	0.459	0.992
	12	0.684	0.996	0.640	0.998	0.551	1.000
	13	0.794	1.000	0.753	1.000	0.665	1.000
14	0	0.000	0.193	0.000	0.232	0.000	0.315
	1	0.004	0.297	0.002	0.339	0.000	0.424
	2	0.026	0.385	0.018	0.428	0.008	0.512
	3	0.061	0.466	0.047	0.508	0.026	0.589
	4	0.104	0.540	0.084	0.581	0.053	0.658
	5	0.153	0.610	0.128	0.649	0.087	0.720
	6	0.206	0.675	0.177	0.711	0.127	0.777
	7	0.264	0.736	0.230	0.770	0.172	0.828
	8	0.325	0.794	0.289	0.823	0.223	0.873
	9	0.390	0.847	0.351	0.872	0.280	0.913
	10	0.460	0.896	0.419	0.916	0.342	0.947
	11	0.534	0.939	0.492	0.953	0.411	0.974
	12	0.615	0.974	0.572	0.982	0.488	0.992
	13	0.703	0.996	0.661	0.998	0.576	1.000
	14	0.807	1.000	0.768	1.000	0.685	1.000

TABLE A4   (Continued)

n	Y	90%		95%		99%	
		Lower	Upper	Lower	Upper	Lower	Upper
15	0	0.000	0.181	0.000	0.218	0.000	0.298
	1	0.003	0.279	0.002	0.319	0.000	0.402
	2	0.024	0.363	0.017	0.405	0.007	0.486
	3	0.057	0.440	0.043	0.481	0.024	0.561
	4	0.097	0.511	0.078	0.551	0.049	0.627
	5	0.142	0.577	0.118	0.616	0.080	0.688
	6	0.191	0.640	0.163	0.677	0.117	0.744
	7	0.244	0.700	0.213	0.734	0.159	0.795
	8	0.300	0.756	0.266	0.787	0.205	0.841
	9	0.360	0.809	0.323	0.837	0.256	0.883
	10	0.423	0.858	0.384	0.882	0.312	0.920
	11	0.489	0.903	0.449	0.922	0.373	0.951
	12	0.560	0.943	0.519	0.957	0.439	0.976
	13	0.637	0.976	0.595	0.983	0.514	0.993
	14	0.721	0.997	0.681	0.998	0.598	1.000
	15	0.819	1.000	0.782	1.000	0.702	1.000
16	0	0.000	0.171	0.000	0.206	0.000	0.282
	1	0.003	0.264	0.002	0.302	0.000	0.381
	2	0.023	0.344	0.016	0.383	0.007	0.463
	3	0.053	0.417	0.040	0.456	0.022	0.534
	4	0.090	0.484	0.073	0.524	0.045	0.599
	5	0.132	0.548	0.110	0.587	0.075	0.658
	6	0.178	0.609	0.152	0.646	0.109	0.713
	7	0.227	0.667	0.198	0.701	0.147	0.764
	8	0.279	0.721	0.247	0.753	0.190	0.810
	9	0.333	0.773	0.299	0.802	0.236	0.853
	10	0.391	0.822	0.354	0.848	0.287	0.891
	11	0.452	0.868	0.413	0.890	0.342	0.925
	12	0.516	0.910	0.476	0.927	0.401	0.955
	13	0.583	0.947	0.544	0.960	0.466	0.978
	14	0.656	0.977	0.617	0.984	0.537	0.993
	15	0.736	0.997	0.698	0.998	0.619	1.000
	16	0.829	1.000	0.794	1.000	0.718	1.000

TABLE A4   (Continued)

n	Y	90% Lower	90% Upper	95% Lower	95% Upper	99% Lower	99% Upper
17	0	0.000	0.162	0.000	0.195	0.000	0.268
	1	0.003	0.250	0.001	0.287	0.000	0.363
	2	0.021	0.326	0.015	0.364	0.006	0.441
	3	0.050	0.396	0.038	0.434	0.021	0.510
	4	0.085	0.461	0.068	0.499	0.043	0.573
	5	0.124	0.522	0.103	0.560	0.070	0.631
	6	0.166	0.580	0.142	0.617	0.101	0.685
	7	0.212	0.636	0.184	0.671	0.137	0.734
	8	0.260	0.689	0.230	0.722	0.176	0.781
	9	0.311	0.740	0.278	0.770	0.219	0.824
	10	0.364	0.788	0.329	0.816	0.266	0.863
	11	0.420	0.834	0.383	0.858	0.315	0.899
	12	0.478	0.876	0.440	0.897	0.369	0.930
	13	0.539	0.915	0.501	0.932	0.427	0.957
	14	0.604	0.950	0.566	0.962	0.490	0.979
	15	0.674	0.979	0.636	0.985	0.559	0.994
	16	0.750	0.997	0.713	0.999	0.637	1.000
	17	0.838	1.000	0.805	1.000	0.732	1.000
18	0	0.000	0.153	0.000	0.185	0.000	0.255
	1	0.003	0.238	0.001	0.273	0.000	0.346
	2	0.020	0.310	0.014	0.347	0.006	0.422
	3	0.047	0.377	0.036	0.414	0.020	0.488
	4	0.080	0.439	0.064	0.476	0.040	0.549
	5	0.116	0.498	0.097	0.535	0.065	0.605
	6	0.156	0.554	0.133	0.590	0.095	0.658
	7	0.199	0.608	0.173	0.643	0.128	0.707
	8	0.244	0.659	0.215	0.692	0.165	0.753
	9	0.291	0.709	0.260	0.740	0.205	0.795
	10	0.341	0.756	0.308	0.785	0.247	0.835
	11	0.392	0.801	0.357	0.827	0.293	0.872
	12	0.446	0.844	0.410	0.867	0.342	0.905
	13	0.502	0.884	0.465	0.903	0.395	0.935
	14	0.561	0.920	0.524	0.936	0.451	0.960
	15	0.623	0.953	0.586	0.964	0.512	0.980
	16	0.690	0.980	0.653	0.986	0.578	0.994
	17	0.762	0.997	0.727	0.999	0.654	1.000
	18	0.847	1.000	0.815	1.000	0.745	1.000

TABLE A4   (Continued)

n	Y	90% Lower	90% Upper	95% Lower	95% Upper	99% Lower	99% Upper
19	0	0.000	0.146	0.000	0.176	0.000	0.243
	1	0.003	0.226	0.001	0.260	0.000	0.331
	2	0.019	0.296	0.013	0.331	0.006	0.404
	3	0.044	0.359	0.034	0.396	0.019	0.468
	4	0.075	0.419	0.061	0.456	0.038	0.527
	5	0.110	0.476	0.091	0.512	0.062	0.582
	6	0.147	0.530	0.126	0.565	0.089	0.633
	7	0.188	0.582	0.163	0.616	0.121	0.681
	8	0.230	0.632	0.203	0.665	0.155	0.726
	9	0.274	0.680	0.244	0.711	0.192	0.768
	10	0.320	0.726	0.289	0.756	0.232	0.808
	11	0.368	0.770	0.335	0.797	0.274	0.845
	12	0.418	0.813	0.384	0.837	0.319	0.879
	13	0.470	0.853	0.435	0.874	0.367	0.911
	14	0.524	0.890	0.488	0.909	0.418	0.938
	15	0.581	0.925	0.544	0.939	0.473	0.962
	16	0.641	0.956	0.604	0.966	0.532	0.981
	17	0.704	0.981	0.669	0.987	0.596	0.994
	18	0.774	0.997	0.740	0.999	0.669	1.000
	19	0.854	1.000	0.824	1.000	0.757	1.000
20	0	0.000	0.139	0.000	0.168	0.000	0.233
	1	0.003	0.216	0.001	0.249	0.000	0.317
	2	0.018	0.283	0.012	0.317	0.005	0.387
	3	0.042	0.344	0.032	0.379	0.018	0.449
	4	0.071	0.401	0.057	0.437	0.036	0.507
	5	0.104	0.456	0.087	0.491	0.058	0.560
	6	0.140	0.508	0.119	0.543	0.085	0.610
	7	0.177	0.558	0.154	0.592	0.114	0.657
	8	0.217	0.606	0.191	0.639	0.146	0.701
	9	0.259	0.653	0.231	0.685	0.181	0.743
	10	0.302	0.698	0.272	0.728	0.218	0.782
	11	0.347	0.741	0.315	0.769	0.257	0.819
	12	0.394	0.783	0.361	0.809	0.299	0.854
	13	0.442	0.823	0.408	0.846	0.343	0.886
	14	0.492	0.860	0.457	0.881	0.390	0.915
	15	0.544	0.896	0.509	0.913	0.440	0.942
	16	0.599	0.929	0.563	0.943	0.493	0.964
	17	0.656	0.958	0.621	0.968	0.551	0.982
	18	0.717	0.982	0.683	0.988	0.613	0.995
	19	0.784	0.997	0.751	0.999	0.683	1.000
	20	0.861	1.000	0.832	1.000	0.767	1.000

TABLE A4   (Continued)

n	Y	90%		95%		99%	
		Lower	Upper	Lower	Upper	Lower	Upper
21	0	0.000	0.133	0.000	0.161	0.000	0.223
	1	0.002	0.207	0.001	0.238	0.000	0.304
	2	0.017	0.271	0.012	0.304	0.005	0.372
	3	0.040	0.329	0.030	0.363	0.017	0.432
	4	0.068	0.384	0.054	0.419	0.034	0.488
	5	0.099	0.437	0.082	0.472	0.055	0.539
	6	0.132	0.487	0.113	0.522	0.080	0.588
	7	0.168	0.536	0.146	0.570	0.108	0.634
	8	0.206	0.583	0.181	0.616	0.138	0.677
	9	0.245	0.628	0.218	0.660	0.171	0.719
	10	0.286	0.672	0.257	0.702	0.205	0.758
	11	0.328	0.714	0.298	0.743	0.242	0.795
	12	0.372	0.755	0.340	0.782	0.281	0.829
	13	0.417	0.794	0.384	0.819	0.323	0.862
	14	0.464	0.832	0.430	0.854	0.366	0.892
	15	0.513	0.868	0.478	0.887	0.412	0.920
	16	0.563	0.901	0.528	0.918	0.461	0.945
	17	0.616	0.932	0.581	0.946	0.512	0.966
	18	0.671	0.960	0.637	0.970	0.568	0.983
	19	0.729	0.983	0.696	0.988	0.628	0.995
	20	0.793	0.998	0.762	0.999	0.696	1.000
	21	0.867	1.000	0.839	1.000	0.777	1.000
22	0	0.000	0.127	0.000	0.154	0.000	0.214
	1	0.002	0.198	0.001	0.228	0.000	0.292
	2	0.016	0.259	0.011	0.292	0.005	0.358
	3	0.038	0.316	0.029	0.349	0.016	0.416
	4	0.065	0.369	0.052	0.403	0.032	0.470
	5	0.094	0.420	0.078	0.454	0.053	0.520
	6	0.126	0.468	0.107	0.502	0.076	0.567
	7	0.160	0.515	0.139	0.549	0.102	0.612
	8	0.196	0.561	0.172	0.593	0.131	0.655
	9	0.233	0.605	0.207	0.636	0.162	0.695
	10	0.271	0.647	0.244	0.678	0.195	0.734
	11	0.311	0.689	0.282	0.718	0.229	0.771
	12	0.353	0.729	0.322	0.756	0.266	0.805
	13	0.395	0.767	0.364	0.793	0.305	0.838
	14	0.439	0.804	0.407	0.828	0.345	0.869
	15	0.485	0.840	0.451	0.861	0.388	0.898
	16	0.532	0.874	0.498	0.893	0.433	0.924
	17	0.580	0.906	0.546	0.922	0.480	0.947
	18	0.631	0.935	0.597	0.948	0.530	0.968
	19	0.684	0.962	0.651	0.971	0.584	0.984
	20	0.741	0.984	0.708	0.989	0.642	0.995

TABLE A4 (Continued)

n	Y	90% Lower	90% Upper	95% Lower	95% Upper	99% Lower	99% Upper
	21	0.802	0.998	0.772	0.999	0.708	1.000
	22	0.873	1.000	0.846	1.000	0.786	1.000
23	0	0.000	0.122	0.000	0.148	0.000	0.206
	1	0.002	0.190	0.001	0.219	0.000	0.281
	2	0.016	0.249	0.011	0.280	0.005	0.345
	3	0.037	0.304	0.028	0.336	0.015	0.401
	4	0.062	0.355	0.050	0.388	0.031	0.453
	5	0.090	0.404	0.075	0.437	0.050	0.502
	6	0.120	0.451	0.102	0.484	0.073	0.548
	7	0.152	0.496	0.132	0.529	0.097	0.592
	8	0.186	0.540	0.164	0.573	0.125	0.634
	9	0.222	0.583	0.197	0.615	0.154	0.674
	10	0.258	0.625	0.232	0.655	0.185	0.712
	11	0.296	0.665	0.268	0.694	0.218	0.748
	12	0.335	0.704	0.306	0.732	0.252	0.782
	13	0.375	0.742	0.345	0.768	0.288	0.815
	14	0.417	0.778	0.385	0.803	0.326	0.846
	15	0.460	0.814	0.427	0.836	0.366	0.875
	16	0.504	0.848	0.471	0.868	0.408	0.903
	17	0.549	0.880	0.516	0.898	0.452	0.927
	18	0.596	0.910	0.563	0.925	0.498	0.950
	19	0.645	0.938	0.612	0.950	0.547	0.969
	20	0.696	0.963	0.664	0.972	0.599	0.985
	21	0.751	0.984	0.720	0.989	0.655	0.995
	22	0.810	0.998	0.781	0.999	0.719	1.000
	23	0.878	1.000	0.852	1.000	0.794	1.000
24	0	0.000	0.117	0.000	0.142	0.000	0.198
	1	0.002	0.183	0.001	0.211	0.000	0.271
	2	0.015	0.240	0.010	0.270	0.004	0.332
	3	0.035	0.292	0.027	0.324	0.015	0.387
	4	0.059	0.342	0.047	0.374	0.029	0.438
	5	0.086	0.389	0.071	0.422	0.048	0.485
	6	0.115	0.435	0.098	0.467	0.069	0.530
	7	0.146	0.479	0.126	0.511	0.093	0.573
	8	0.178	0.521	0.156	0.553	0.119	0.614
	9	0.212	0.563	0.188	0.594	0.146	0.653
	10	0.246	0.603	0.221	0.634	0.176	0.690
	11	0.282	0.642	0.256	0.672	0.207	0.726
	12	0.319	0.681	0.291	0.709	0.240	0.760
	13	0.358	0.718	0.328	0.744	0.274	0.793
	14	0.397	0.754	0.366	0.779	0.310	0.824
	15	0.437	0.788	0.406	0.812	0.347	0.854
	16	0.479	0.822	0.447	0.844	0.386	0.881
	17	0.521	0.854	0.489	0.874	0.427	0.907
	18	0.565	0.885	0.533	0.902	0.470	0.931
	19	0.611	0.914	0.578	0.929	0.515	0.952

TABLE A4   (Continued)

		90%		95%		99%	
n	Y	Lower	Upper	Lower	Upper	Lower	Upper
	20	0.658	0.941	0.626	0.953	0.562	0.971
	21	0.708	0.965	0.676	0.973	0.613	0.985
	22	0.760	0.985	0.730	0.990	0.668	0.996
	23	0.817	0.998	0.789	0.999	0.729	1.000
	24	0.883	1.000	0.858	1.000	0.802	1.000
25	0	0.000	0.113	0.000	0.137	0.000	0.191
	1	0.002	0.176	0.001	0.204	0.000	0.262
	2	0.014	0.231	0.010	0.260	0.004	0.321
	3	0.034	0.282	0.025	0.312	0.014	0.374
	4	0.057	0.330	0.045	0.361	0.028	0.424
	5	0.082	0.375	0.068	0.407	0.046	0.470
	6	0.110	0.420	0.094	0.451	0.066	0.514
	7	0.139	0.462	0.121	0.494	0.089	0.555
	8	0.170	0.504	0.150	0.535	0.114	0.595
	9	0.202	0.544	0.180	0.575	0.140	0.634
	10	0.236	0.583	0.211	0.613	0.168	0.670
	11	0.270	0.621	0.244	0.651	0.197	0.705
	12	0.305	0.659	0.278	0.687	0.228	0.739
	13	0.341	0.695	0.313	0.722	0.261	0.772
	14	0.379	0.730	0.349	0.756	0.295	0.803
	15	0.417	0.764	0.387	0.789	0.330	0.832
	16	0.456	0.798	0.425	0.820	0.366	0.860
	17	0.496	0.830	0.465	0.850	0.405	0.886
	18	0.538	0.861	0.506	0.879	0.445	0.911
	19	0.580	0.890	0.549	0.906	0.486	0.934
	20	0.625	0.918	0.593	0.932	0.530	0.954
	21	0.670	0.943	0.639	0.955	0.576	0.972
	22	0.718	0.966	0.688	0.975	0.626	0.986
	23	0.769	0.986	0.740	0.990	0.679	0.996
	24	0.824	0.998	0.796	0.999	0.738	1.000
	25	0.887	1.000	0.863	1.000	0.809	1.000
26	0	0.000	0.109	0.000	0.132	0.000	0.184
	1	0.002	0.170	0.001	0.196	0.000	0.253
	2	0.014	0.223	0.009	0.251	0.004	0.310
	3	0.032	0.272	0.024	0.302	0.013	0.362
	4	0.054	0.318	0.044	0.349	0.027	0.410
	5	0.079	0.363	0.066	0.393	0.044	0.455
	6	0.106	0.405	0.090	0.436	0.064	0.498
	7	0.134	0.447	0.116	0.478	0.085	0.538
	8	0.163	0.487	0.143	0.518	0.109	0.578
	9	0.194	0.526	0.172	0.557	0.134	0.615
	10	0.226	0.564	0.202	0.594	0.161	0.651
	11	0.258	0.602	0.234	0.631	0.189	0.686
	12	0.292	0.638	0.266	0.666	0.218	0.719
	13	0.327	0.673	0.299	0.701	0.249	0.751
	14	0.362	0.708	0.334	0.734	0.281	0.782

TABLE A4 (Continued)

n	Y	90%		95%		99%	
		Lower	Upper	Lower	Upper	Lower	Upper
	15	0.398	0.742	0.369	0.766	0.314	0.811
	16	0.436	0.774	0.406	0.798	0.349	0.839
	17	0.474	0.806	0.443	0.828	0.385	0.866
	18	0.513	0.837	0.482	0.857	0.422	0.891
	19	0.553	0.866	0.522	0.884	0.462	0.915
	20	0.595	0.894	0.564	0.910	0.502	0.936
	21	0.637	0.921	0.607	0.934	0.545	0.956
	22	0.682	0.946	0.651	0.956	0.590	0.973
	23	0.728	0.968	0.698	0.976	0.638	0.987
	24	0.777	0.986	0.749	0.991	0.690	0.996
	25	0.830	0.998	0.804	0.999	0.747	1.000
	26	0.891	1.000	0.868	1.000	0.816	1.000
27	0	0.000	0.105	0.000	0.128	0.000	0.178
	1	0.002	0.164	0.001	0.190	0.000	0.245
	2	0.013	0.215	0.009	0.243	0.004	0.300
	3	0.031	0.263	0.024	0.292	0.013	0.351
	4	0.052	0.308	0.042	0.337	0.026	0.397
	5	0.076	0.351	0.063	0.381	0.042	0.441
	6	0.101	0.392	0.086	0.423	0.061	0.483
	7	0.129	0.432	0.111	0.463	0.082	0.523
	8	0.157	0.471	0.138	0.502	0.104	0.561
	9	0.186	0.509	0.165	0.540	0.128	0.597
	10	0.217	0.547	0.194	0.576	0.154	0.633
	11	0.248	0.583	0.224	0.612	0.181	0.667
	12	0.280	0.618	0.255	0.647	0.209	0.700
	13	0.313	0.653	0.287	0.681	0.238	0.731
	14	0.347	0.687	0.319	0.713	0.269	0.762
	15	0.382	0.720	0.353	0.745	0.300	0.791
	16	0.417	0.752	0.388	0.776	0.333	0.819
	17	0.453	0.783	0.424	0.806	0.367	0.846
	18	0.491	0.814	0.460	0.835	0.403	0.872
	19	0.529	0.843	0.498	0.862	0.439	0.896
	20	0.568	0.871	0.537	0.889	0.477	0.918
	21	0.608	0.899	0.577	0.914	0.517	0.939
	22	0.649	0.924	0.619	0.937	0.559	0.958
	23	0.692	0.948	0.663	0.958	0.603	0.974
	24	0.737	0.969	0.708	0.976	0.649	0.987
	25	0.785	0.987	0.757	0.991	0.700	0.996
	26	0.836	0.998	0.810	0.999	0.755	1.000
	27	0.895	1.000	0.872	1.000	0.822	1.000
28	0	0.000	0.101	0.000	0.123	0.000	0.172
	1	0.002	0.159	0.001	0.183	0.000	0.237
	2	0.013	0.208	0.009	0.235	0.004	0.291
	3	0.030	0.254	0.023	0.282	0.012	0.340
	4	0.050	0.298	0.040	0.327	0.025	0.385

TABLE A4 (Continued)

n	Y	90% Lower	90% Upper	95% Lower	95% Upper	99% Lower	99% Upper
	5	0.073	0.339	0.061	0.369	0.041	0.428
	6	0.098	0.380	0.083	0.410	0.059	0.469
	7	0.124	0.419	0.107	0.449	0.079	0.508
	8	0.151	0.457	0.132	0.487	0.100	0.545
	9	0.179	0.494	0.159	0.524	0.123	0.581
	10	0.208	0.530	0.186	0.559	0.148	0.615
	11	0.238	0.565	0.215	0.594	0.173	0.649
	12	0.269	0.600	0.245	0.628	0.200	0.681
	13	0.301	0.634	0.275	0.661	0.228	0.713
	14	0.333	0.667	0.306	0.694	0.257	0.743
	15	0.366	0.699	0.339	0.725	0.287	0.772
	16	0.400	0.731	0.372	0.755	0.319	0.800
	17	0.435	0.762	0.406	0.785	0.351	0.827
	18	0.470	0.792	0.441	0.814	0.385	0.852
	19	0.506	0.821	0.476	0.841	0.419	0.877
	20	0.543	0.849	0.513	0.868	0.455	0.900
	21	0.581	0.876	0.551	0.893	0.492	0.921
	22	0.620	0.902	0.590	0.917	0.531	0.941
	23	0.661	0.927	0.631	0.939	0.572	0.959
	24	0.702	0.950	0.673	0.960	0.615	0.975
	25	0.746	0.970	0.718	0.977	0.660	0.988
	26	0.792	0.987	0.765	0.991	0.709	0.996
	27	0.841	0.998	0.817	0.999	0.763	1.000
	28	0.899	1.000	0.877	1.000	0.828	1.000
29	0	0.000	0.098	0.000	0.119	0.000	0.167
	1	0.002	0.153	0.001	0.178	0.000	0.230
	2	0.012	0.202	0.008	0.228	0.004	0.282
	3	0.029	0.246	0.022	0.274	0.012	0.330
	4	0.049	0.288	0.039	0.317	0.024	0.374
	5	0.070	0.329	0.058	0.358	0.039	0.416
	6	0.094	0.368	0.080	0.397	0.056	0.455
	7	0.119	0.406	0.103	0.435	0.076	0.493
	8	0.145	0.443	0.127	0.472	0.096	0.530
	9	0.172	0.479	0.153	0.508	0.119	0.565
	10	0.201	0.514	0.179	0.543	0.142	0.599
	11	0.229	0.549	0.207	0.577	0.167	0.632
	12	0.259	0.583	0.235	0.611	0.192	0.664
	13	0.289	0.616	0.264	0.643	0.219	0.695
	14	0.320	0.648	0.294	0.675	0.247	0.724
	15	0.352	0.680	0.325	0.706	0.276	0.753
	16	0.384	0.711	0.357	0.736	0.305	0.781
	17	0.417	0.741	0.389	0.765	0.336	0.808
	18	0.451	0.771	0.423	0.793	0.368	0.833
	19	0.486	0.799	0.457	0.821	0.401	0.858
	20	0.521	0.828	0.492	0.847	0.435	0.881
	21	0.557	0.855	0.528	0.873	0.470	0.904

TABLE A4 (Continued)

n	Y	90% Lower	90% Upper	95% Lower	95% Upper	99% Lower	99% Upper
	22	0.594	0.881	0.565	0.897	0.507	0.924
	23	0.632	0.906	0.603	0.920	0.545	0.944
	24	0.671	0.930	0.642	0.942	0.584	0.961
	25	0.712	0.951	0.683	0.961	0.626	0.976
	26	0.754	0.971	0.726	0.978	0.670	0.988
	27	0.798	0.988	0.772	0.992	0.718	0.996
	28	0.847	0.998	0.822	0.999	0.770	1.000
	29	0.902	1.000	0.881	1.000	0.833	1.000
30	0	0.000	0.095	0.000	0.116	0.000	0.162
	1	0.002	0.149	0.001	0.172	0.000	0.223
	2	0.012	0.195	0.008	0.221	0.004	0.274
	3	0.028	0.239	0.021	0.265	0.012	0.320
	4	0.047	0.280	0.038	0.307	0.023	0.363
	5	0.068	0.319	0.056	0.347	0.038	0.404
	6	0.091	0.357	0.077	0.386	0.054	0.443
	7	0.115	0.394	0.099	0.423	0.073	0.480
	8	0.140	0.430	0.123	0.459	0.093	0.516
	9	0.166	0.465	0.147	0.494	0.114	0.550
	10	0.193	0.499	0.173	0.528	0.137	0.583
	11	0.221	0.533	0.199	0.561	0.160	0.616
	12	0.250	0.566	0.227	0.594	0.185	0.647
	13	0.279	0.598	0.255	0.626	0.211	0.677
	14	0.308	0.630	0.283	0.657	0.237	0.707
	15	0.339	0.661	0.313	0.687	0.265	0.735
	16	0.370	0.692	0.343	0.717	0.293	0.763
	17	0.402	0.721	0.374	0.745	0.323	0.789
	18	0.434	0.750	0.406	0.773	0.353	0.815
	19	0.467	0.779	0.439	0.801	0.384	0.840
	20	0.501	0.807	0.472	0.827	0.417	0.863
	21	0.535	0.834	0.506	0.853	0.450	0.886
	22	0.570	0.860	0.541	0.877	0.484	0.907
	23	0.606	0.885	0.577	0.901	0.520	0.927
	24	0.643	0.909	0.614	0.923	0.557	0.946
	25	0.681	0.932	0.653	0.944	0.596	0.962
	26	0.720	0.953	0.693	0.962	0.637	0.977
	27	0.761	0.972	0.735	0.979	0.680	0.988
	28	0.805	0.988	0.779	0.992	0.726	0.996
	29	0.851	0.998	0.828	0.999	0.777	1.000
	30	0.905	1.000	0.884	1.000	0.838	1.000

SOURCE: Generated by R. L. Iman. Used with permission.

TABLE A5 Sample Sizes for Nonparametric Tolerance Limits When $r + m = 1$[a]

$1 - \alpha$	$q = 0.500$	0.700	0.750	0.800	0.850	0.900	0.950	0.975	0.980	0.990
0.500	1	2	3	4	5	7	14	28	35	69
0.700	2	4	5	6	8	12	24	48	60	120
0.750	2	4	5	7	9	14	28	55	69	138
0.800	3	5	6	8	10	16	32	64	80	161
0.850	3	6	7	9	12	19	37	75	94	189
0.900	4	7	9	11	15	22	45	91	144	230
0.950	5	9	11	14	19	29	59	119	149	299
0.975	6	11	13	17	23	36	72	146	183	368
0.980	6	11	14	18	25	38	77	155	194	390
0.990	7	13	17	21	29	44	90	182	228	459
0.995	8	15	19	24	33	51	104	210	263	528
0.999	10	20	25	31	43	66	135	273	342	688

[a] The quantity tabled is the sample size n such that $q^n \le \alpha$, for use in finding the tolerance limits

$$P(X^{(1)} \le p \text{ of the population}) \ge 1 - \alpha$$

or

$$P(q \text{ of the population} \le X^{(n)}) \ge 1 - \alpha$$

as described in Section 3.3.

TABLE A6 Sample Sizes for Nonparametric Tolerance Limits When $r + m = 2$[a]

$1 - \alpha$	$q = 0.500$	0.700	0.750	0.800	0.850	0.900	0.950	0.975	0.980	0.990
0.500	3	6	7	9	11	17	34	67	84	168
0.700	5	8	10	12	16	24	49	97	122	244
0.750	5	9	10	13	18	27	53	107	134	269
0.800	5	9	11	14	19	29	59	119	149	299
0.850	6	10	13	16	22	33	67	134	168	337
0.900	7	12	15	18	25	38	77	155	194	388
0.950	8	14	18	22	30	46	93	188	236	473
0.975	9	17	20	26	35	54	110	221	277	555
0.980	9	17	21	27	37	56	115	231	290	581
0.990	11	20	24	31	42	64	130	263	330	662
0.995	12	22	27	34	47	72	146	294	369	740
0.999	14	27	33	42	58	89	181	366	458	920

[a] The quantity tabled is the sample size n such that $q^n + nq^{n-1}(1 - q) \le \alpha$ for use in finding the tolerance limits

$$P(X^{(r)} \le q \text{ of the population} \le X^{(n+1-m)}) \ge 1 - \alpha$$

when $r + m = 2$.

TABLE A7 Quantiles of the Mann–Whitney Test Statistic[a]

n	p	m = 2	3	4	5	6	7	8	9	10	11	12	13	14	15	16	17	18	19	20
2	0.001	3	3	3	3	3	3	3	3	3	3	3	3	3	3	3	3	3	3	3
	0.005	3	3	3	3	3	3	3	3	3	3	3	3	3	3	3	3	3	4	4
	0.01	3	3	3	3	3	3	3	3	3	3	4	4	4	4	4	4	4	5	5
	0.025	3	3	3	3	3	3	4	4	4	5	5	5	5	5	5	6	6	6	6
	0.05	3	3	3	4	4	4	5	5	5	5	6	6	7	7	7	7	8	8	8
	0.10	3	4	4	5	5	5	6	6	7	7	8	8	8	9	9	10	10	11	11
3	0.001	6	6	6	6	6	6	6	6	6	6	6	6	6	6	7	7	7	7	7
	0.005	6	6	6	6	6	6	6	7	7	7	8	8	8	9	9	9	9	10	10
	0.01	6	6	6	6	6	7	7	8	8	8	9	9	9	10	10	10	11	11	12
	0.025	6	6	6	7	8	8	9	9	10	10	11	11	12	12	13	13	14	14	15
	0.05	6	7	7	8	9	9	10	11	11	12	12	13	14	14	15	16	16	17	18
	0.10	7	8	8	9	10	11	12	12	13	14	15	16	17	17	18	19	20	21	22
4	0.001	10	10	10	10	10	10	10	10	11	11	11	12	12	12	13	13	14	14	14
	0.005	10	10	10	10	11	11	12	12	13	13	14	14	15	16	16	17	17	18	19
	0.01	10	10	10	11	12	12	13	14	14	15	16	16	17	18	18	19	20	20	21
	0.025	10	10	11	12	13	14	15	15	16	17	18	19	20	21	22	22	23	24	25
	0.05	10	11	12	13	14	15	16	17	18	19	20	21	22	23	25	26	27	28	29
	0.10	11	12	14	15	16	17	18	20	21	22	23	24	26	27	28	29	31	32	33
5	0.001	15	15	15	15	15	15	16	17	17	18	18	19	19	20	21	21	22	23	23
	0.005	15	15	15	16	17	17	18	19	20	21	22	23	23	24	25	26	27	28	29
	0.01	15	15	16	17	18	19	20	21	22	23	24	25	26	27	28	29	30	31	32
	0.025	15	16	17	18	19	21	22	23	24	25	27	28	29	30	31	33	34	35	36
	0.05	16	17	18	20	21	22	24	25	27	28	29	31	32	34	35	36	38	39	41
	0.10	17	18	20	21	23	24	26	28	29	31	33	34	36	38	39	41	43	44	46
6	0.001	21	21	21	21	21	21	23	24	25	26	26	27	28	29	30	31	32	33	34
	0.005	21	21	22	23	24	25	26	27	28	29	31	32	33	34	35	37	38	39	40
	0.01	21	21	23	24	25	26	28	29	30	31	33	34	35	37	38	40	41	42	44
	0.025	21	23	24	25	27	28	30	32	33	35	36	38	39	41	43	44	46	47	49
	0.05	22	24	25	27	29	30	32	34	36	38	39	41	43	45	47	48	50	52	54
	0.10	23	25	27	29	31	33	35	37	39	41	43	45	47	49	51	53	56	58	60
7	0.001	28	28	28	28	29	30	31	32	34	35	36	37	38	39	40	42	43	44	45
	0.005	28	28	29	30	32	33	35	36	38	39	41	42	44	45	47	48	50	51	53
	0.01	28	29	30	32	33	35	36	38	40	41	43	45	46	48	50	52	53	55	57
	0.025	28	30	32	34	35	37	39	41	43	45	47	49	51	53	55	57	59	61	63
	0.05	29	31	33	35	37	40	42	44	46	48	50	53	55	57	59	62	64	66	68
	0.10	30	33	35	37	40	42	45	47	50	52	55	57	60	62	65	67	70	72	75
8	0.001	36	36	36	37	38	39	41	42	43	45	46	48	49	51	52	54	55	57	58
	0.005	36	36	38	39	41	43	44	46	48	50	52	54	55	57	59	61	63	65	67
	0.01	36	37	39	41	43	44	46	48	50	52	54	56	59	61	63	65	67	69	71
	0.025	37	39	41	43	45	47	50	52	54	56	59	61	63	66	68	71	73	75	78
	0.05	38	40	42	45	47	50	52	55	57	60	63	65	68	70	73	76	78	81	84
	0.10	39	42	44	47	50	53	56	59	61	64	67	70	73	76	79	82	85	88	91

TABLE A7 (Continued)

n	p	m=2	3	4	5	6	7	8	9	10	11	12	13	14	15	16	17	18	19	20
9	0.001	45	45	45	47	48	49	51	53	54	56	58	60	61	63	65	67	69	71	72
	0.005	45	46	47	49	51	53	55	57	59	62	64	66	68	70	73	75	77	79	82
	0.01	45	47	49	51	53	55	57	60	62	64	67	69	72	74	77	79	82	84	86
	0.025	46	48	50	53	56	58	61	63	66	69	72	74	77	80	83	85	88	91	94
	0.05	47	50	52	55	58	61	64	67	70	73	76	79	82	85	88	91	94	97	100
	0.10	48	51	55	58	61	64	68	71	74	77	81	84	87	91	94	98	101	104	108
10	0.001	55	55	55	57	59	61	62	64	66	68	70	73	75	77	79	81	83	85	88
	0.005	55	56	57	60	62	65	67	69	72	74	77	80	82	85	87	90	93	95	98
	0.01	55	57	59	62	64	67	69	72	75	78	80	83	86	89	92	94	97	100	103
	0.025	56	59	61	64	67	70	73	76	79	82	85	89	92	95	98	101	104	108	111
	0.05	57	60	63	67	70	73	76	80	83	87	90	93	97	100	104	107	111	114	118
	0.10	59	62	66	69	73	77	80	84	88	92	95	99	103	107	110	114	118	122	126
11	0.001	66	66	67	69	71	73	75	77	79	82	84	87	89	91	94	96	99	101	104
	0.005	66	67	69	72	74	77	80	83	85	88	91	94	97	100	103	106	109	112	115
	0.01	66	68	71	74	76	79	82	85	89	92	95	98	101	104	108	111	114	117	120
	0.025	67	70	73	76	80	83	86	90	93	97	100	104	107	111	114	118	122	125	129
	0.05	68	72	75	79	83	86	90	94	98	101	105	109	113	117	121	124	128	132	136
	0.10	70	74	78	82	86	90	94	98	103	107	111	115	119	124	128	132	136	140	145
12	0.001	78	78	79	81	83	86	88	91	93	96	98	102	104	106	110	113	116	118	121
	0.005	78	80	82	85	88	91	94	97	100	103	106	110	113	116	120	123	126	130	133
	0.01	78	81	84	87	90	93	96	100	103	107	110	114	117	121	125	128	132	135	139
	0.025	80	83	86	90	93	97	101	105	108	112	116	120	124	128	132	136	140	144	148
	0.05	81	84	88	92	96	100	105	109	111	117	121	126	130	134	139	143	147	151	156
	0.10	83	87	91	96	100	105	109	114	118	123	128	132	137	142	146	151	156	160	165
13	0.001	91	91	93	95	97	100	103	106	109	112	115	118	121	124	127	130	134	137	140
	0.005	91	93	95	99	102	105	109	112	116	119	123	126	130	134	137	141	145	149	152
	0.01	92	94	97	101	104	108	112	115	119	123	127	131	135	139	143	147	151	155	159
	0.025	93	96	100	104	108	112	116	120	125	129	133	137	142	146	151	155	159	164	168
	0.05	94	98	102	107	111	116	120	125	129	134	139	143	148	153	157	162	167	172	176
	0.10	96	101	105	110	115	120	125	130	135	140	145	150	155	160	166	171	176	181	186
14	0.001	105	105	107	109	112	115	118	121	125	128	131	135	138	142	145	149	152	156	160
	0.005	105	107	110	113	117	121	124	128	132	136	140	144	148	152	156	160	164	169	173
	0.01	106	108	112	116	119	123	128	132	136	140	144	149	153	157	162	166	171	175	179
	0.025	107	111	115	119	123	128	132	137	142	146	151	156	161	165	170	175	180	184	189
	0.05	109	113	117	122	127	132	137	142	147	152	157	162	167	172	177	183	188	193	198
	0.10	110	116	121	126	131	137	142	147	153	158	164	169	175	180	186	191	197	203	208
15	0.001	120	120	121	125	128	133	135	138	142	145	149	153	157	161	164	168	172	176	180
	0.005	120	121	124	129	133	137	141	145	150	154	158	163	167	172	176	181	185	190	194
	0.01	121	124	126	132	136	140	145	149	154	158	163	168	172	177	182	187	191	196	201
	0.025	122	126	131	135	140	145	150	155	160	165	170	175	180	185	191	196	201	206	211
	0.05	124	128	133	139	144	150	154	160	165	171	177	182	187	193	198	204	209	215	221
	0.10	126	131	137	143	148	154	160	166	172	178	184	189	195	201	207	213	219	225	231
16	0.001	136	136	136	142	145	148	152	156	160	164	168	172	176	180	185	189	193	197	202
	0.005	136	136	139	146	150	155	159	164	168	173	178	182	187	192	197	202	207	211	216
	0.01	137	140	140	149	153	158	163	168	173	178	183	188	193	198	203	208	213	219	224
	0.025	138	143	143	152	158	163	168	174	179	184	190	196	201	207	212	218	223	229	235
	0.05	140	145	151	156	162	167	173	179	185	191	197	202	208	214	220	226	232	238	244
	0.10	142	148	154	160	166	173	179	185	191	198	204	211	217	223	230	236	243	249	256

TABLE A7 (Continued)

n	p	m = 2	3	4	5	6	7	8	9	10	11	12	13	14	15	16	17	18	19	20
17	0.001	153	154	156	159	163	167	171	175	179	183	188	192	197	201	206	211	215	220	224
	0.005	153	156	160	164	169	173	178	183	188	193	198	203	208	214	219	224	229	235	240
	0.01	154	158	162	167	172	177	182	187	192	198	203	209	214	220	225	231	236	242	247
	0.025	156	160	165	171	176	182	188	193	199	205	211	217	223	229	235	241	247	253	259
	0.05	157	163	169	174	180	187	193	199	205	211	218	224	231	237	243	250	256	263	269
	0.10	160	166	172	179	185	192	199	206	212	219	226	233	239	246	253	260	267	274	281
18	0.001	171	172	175	178	182	186	190	195	199	204	209	214	218	223	228	233	238	243	248
	0.005	171	174	178	183	188	193	198	203	209	214	219	225	230	236	242	247	253	259	264
	0.01	172	176	181	186	191	196	202	208	213	219	225	231	237	242	248	254	260	266	272
	0.025	174	179	184	190	196	202	208	214	220	227	233	239	246	252	258	265	271	278	284
	0.05	176	181	188	194	200	207	213	220	227	233	240	247	254	260	267	274	281	288	295
	0.10	178	185	192	199	206	213	220	227	234	241	249	256	263	270	278	285	292	300	307
19	0.001	190	191	194	198	202	206	211	216	220	225	231	236	241	246	251	257	262	268	273
	0.005	191	194	198	203	208	213	219	224	230	236	242	248	254	260	265	272	278	284	290
	0.01	192	195	200	206	211	217	223	229	235	241	247	254	260	266	273	279	285	292	298
	0.025	193	198	204	210	216	223	229	236	243	249	256	263	269	276	283	290	297	304	310
	0.05	195	201	208	214	221	228	235	242	249	256	263	271	278	285	292	300	307	314	321
	0.10	198	205	212	219	227	234	242	249	257	264	272	280	288	295	303	311	319	326	334
20	0.001	210	211	214	218	223	227	232	237	243	248	253	259	265	270	276	281	287	293	299
	0.005	211	214	219	224	229	235	241	247	253	259	265	271	278	284	290	297	303	310	316
	0.01	212	216	221	227	233	239	245	251	258	264	271	278	284	291	298	304	311	318	325
	0.025	213	219	225	231	238	245	251	259	266	273	280	287	294	301	309	316	323	330	338
	0.05	215	222	229	236	243	250	258	265	273	280	288	295	303	311	318	326	334	341	349
	0.10	218	226	233	241	249	257	265	273	281	289	297	305	313	321	330	338	346	354	362

For n or m greater than 20, the pth quantile w_p of the Mann–Whitney test statistic may be approximated by

$$w_p = n(N + 1)/2 + z_p \sqrt{nm(N + 1)/12}$$

where z_p is the pth quantile of a standard normal random variable, obtained from Table A1, and where $N = m + n$.
[a] The entries in this table are quantiles w_p of the Mann–Whitney test statistic T, given by Equation 5.1.1, for selected values of p. Note that $P(T < w_p) \leq p$. Upper quantiles may be found from the equation

$$w_p = n(n + m + 1) - w_{1-p}$$

Critical regions correspond to values of T less than (or greater than) but not equal to the appropriate quantile.

TABLE A8 Quantiles of the Kruskal-Wallis Test Statistic for Small Sample Sizes[a]

Sample Sizes	$W_{0.90}$	$W_{0.95}$	$W_{0.99}$
2, 2, 2	3.7143	4.5714	4.5714
3, 2, 1	3.8571	4.2857	4.2857
3, 2, 2	4.4643	4.5000	5.3571
3, 3, 1	4.0000	4.5714	5.1429
3, 3, 2	4.2500	5.1389	6.2500
3, 3, 3	4.6000	5.0667	6.4889
4, 2, 1	4.0179	4.8214	4.8214
4, 2, 2	4.1667	5.1250	6.0000
4, 3, 1	3.8889	5.0000	5.8333
4, 3, 2	4.4444	5.4000	6.3000
4, 3, 3	4.7000	5.7273	6.7091
4, 4, 1	4.0667	4.8667	6.1667
4, 4, 2	4.4455	5.2364	6.8727
4, 4, 3	4.7730	5.5758	7.1364
4, 4, 4	4.5000	5.6538	7.5385
5, 2, 1	4.0500	4.4500	5.2500
5, 2, 2	4.2933	5.0400	6.1333
5, 3, 1	3.8400	4.8711	6.4000
5, 3, 2	4.4946	5.1055	6.8218
5, 3, 3	4.4121	5.5152	6.9818
5, 4, 1	3.9600	4.8600	6.8400
5, 4, 2	4.5182	5.2682	7.1182
5, 4, 3	4.5231	5.6308	7.3949
5, 4, 4	4.6187	5.6176	7.7440
5, 5, 1	4.0364	4.9091	6.8364
5, 5, 2	4.5077	5.2462	7.2692
5, 5, 3	4.5363	5.6264	7.5429
5, 5, 4	4.5200	5.6429	7.7914
5, 5, 5	4.5000	5.6600	7.9800

SOURCE: Adapted from Iman, Quade, and Alexander (1975), with permission from the American Mathematical Society.

[a] The null hypothesis may be rejected at the level α if the Kruskal-Wallis test statistic, given by Equation 5.2.5, exceeds the $1 - \alpha$ quantile given in the table.

TABLE A9   Quantiles of the Squared Ranks Test Statistic[a]

n	p	m = 3	4	5	6	7	8	9	10
	0.005	14	14	14	14	14	14	21	21
	0.01	14	14	14	14	21	21	26	26
	0.025	14	14	21	26	29	30	35	41
	0.05	21	21	26	30	38	42	49	54
3	0.10	26	29	35	42	50	59	69	77
	0.90	65	90	117	149	182	221	260	305
	0.95	70	101	129	161	197	238	285	333
	0.975	77	110	138	170	213	257	308	362
	0.99	77	110	149	194	230	285	329	394
	0.995	77	110	149	194	245	302	346	413
	0.005	30	30	30	39	39	46	50	54
	0.01	30	30	39	46	50	51	62	66
	0.025	30	39	50	54	63	71	78	90
	0.05	39	50	57	66	78	90	102	114
4	0.10	50	62	71	85	99	114	130	149
	0.90	111	142	182	222	270	321	375	435
	0.95	119	154	197	246	294	350	413	476
	0.975	126	165	206	255	311	374	439	510
	0.99	126	174	219	270	334	401	470	545
	0.995	126	174	230	281	351	414	494	567
	0.005	55	55	66	75	79	88	99	110
	0.01	55	66	75	82	90	103	115	127
	0.025	66	79	88	100	114	130	145	162
	0.05	75	88	103	120	135	155	175	195
5	0.10	87	103	121	142	163	187	212	239
	0.90	169	214	264	319	379	445	514	591
	0.95	178	228	282	342	410	479	558	639
	0.975	183	235	297	363	433	508	592	680
	0.99	190	246	310	382	459	543	631	727
	0.995	190	255	319	391	478	559	654	754
	0.005	91	104	115	124	136	152	167	182
	0.01	91	115	124	139	155	175	191	210
	0.025	115	130	143	164	184	208	231	255
	0.05	124	139	164	187	211	239	268	299
6	0.10	136	163	187	215	247	280	315	352
	0.90	243	300	364	435	511	592	679	772
	0.95	255	319	386	463	545	634	730	831
	0.975	259	331	406	486	574	670	771	880
	0.99	271	339	424	511	607	706	817	935
	0.995	271	346	431	526	624	731	847	970

SOURCE. Adapted from tables generated by R.L. Iman. Used with permission.

[a] The entries in this table are selected quantiles w_p of the squared ranks test statistic T, given by Equation 5.3.3. Note that $P(T < w_p) \leq p$ and $P(T > w_p) \leq 1 - p$. Critical regions correspond to values less than (or greater than) but not including the appropriate quantile.

TABLE A9   (Continued)

n	p	m = 3	4	5	6	7	8	9	10
7	0.005	140	155	172	195	212	235	257	280
	0.01	155	172	191	212	236	260	287	315
	0.025	172	195	217	245	274	305	338	372
	0.05	188	212	240	274	308	344	384	425
	0.10	203	236	271	308	350	394	440	489
	0.90	335	407	487	572	665	764	871	984
	0.95	347	428	515	608	707	814	929	1051
	0.975	356	443	536	635	741	856	979	1108
	0.99	364	456	560	664	779	900	1032	1172
	0.995	371	467	571	683	803	929	1067	1212
8	0.005	204	236	260	284	311	340	368	401
	0.01	221	249	276	309	340	372	408	445
	0.025	249	276	311	345	384	425	468	513
	0.05	268	300	340	381	426	473	524	576
	0.10	285	329	374	423	476	531	590	652
	0.90	447	536	632	735	846	965	1091	1224
	0.95	464	560	664	776	896	1023	1159	1303
	0.975	476	579	689	807	935	1071	1215	1368
	0.99	485	599	716	840	980	1124	1277	1442
	0.995	492	604	731	863	1005	1156	1319	1489
9	0.005	304	325	361	393	429	466	508	549
	0.01	321	349	384	423	464	508	553	601
	0.025	342	380	423	469	517	570	624	682
	0.05	365	406	457	510	567	626	689	755
	0.10	390	444	501	561	625	694	766	843
	0.90	581	689	803	925	1056	1195	1343	1498
	0.95	601	717	840	972	1112	1261	1420	1587
	0.975	615	741	870	1009	1158	1317	1485	1662
	0.99	624	757	900	1049	1209	1377	1556	1745
	0.995	629	769	916	1073	1239	1417	1601	1798
10	0.005	406	448	486	526	573	620	672	725
	0.01	425	470	513	561	613	667	725	785
	0.025	457	505	560	616	677	741	808	879
	0.05	486	539	601	665	734	806	883	963
	0.10	514	580	649	724	801	885	972	1064
	0.90	742	866	1001	1144	1296	1457	1627	1806
	0.95	765	901	1045	1197	1360	1533	1715	1907
	0.975	778	925	1078	1241	1413	1596	1788	1991
	0.99	793	949	1113	1286	1470	1664	1869	2085
	0.995	798	961	1130	1314	1505	1708	1921	2145

For n or m greater than 10, the pth quantile w_p of the squared ranks test statistic may be approximated by

$$w_p = \frac{n(N + 1)(2N + 1)}{6} + z_p \sqrt{\frac{mn(N + 1)(2N + 1)(8N + 11)}{180}}$$

where $N = n + m$, and where z_p is the pth quantile of a standard normal random variable, obtained from Table A1.

TABLE A10   Quantiles of Spearman's ρ^a

n	$p = 0.900$	0.950	0.975	0.990	0.995	0.999
4	0.8000	0.8000				
5	0.7000	0.8000	0.9000	0.9000		
6	0.6000	0.7714	0.8286	0.8857	0.9429	
7	0.5357	0.6786	0.7500	0.8571	0.8929	0.9643
8	0.5000	0.6190	0.7143	0.8095	0.8571	0.9286
9	0.4667	0.5833	0.6833	0.7667	0.8167	0.9000
10	0.4424	0.5515	0.6364	0.7333	0.7818	0.8667
11	0.4182	0.5273	0.6091	0.7000	0.7455	0.8364
12	0.3986	0.4965	0.5804	0.6713	0.7203	0.8112
13	0.3791	0.4780	0.5549	0.6429	0.6978	0.7857
14	0.3626	0.4593	0.5341	0.6220	0.6747	0.7670
15	0.3500	0.4429	0.5179	0.6000	0.6500	0.7464
16	0.3382	0.4265	0.5000	0.5794	0.6324	0.7265
17	0.3260	0.4118	0.4853	0.5637	0.6152	0.7083
18	0.3148	0.3994	0.4696	0.5480	0.5975	0.6904
19	0.3070	0.3895	0.4579	0.5333	0.5825	0.6737
20	0.2977	0.3789	0.4451	0.5203	0.5684	0.6586
21	0.2909	0.3688	0.4351	0.5078	0.5545	0.6455
22	0.2829	0.3597	0.4241	0.4963	0.5426	0.6318
23	0.2767	0.3518	0.4150	0.4852	0.5306	0.6186
24	0.2704	0.3435	0.4061	0.4748	0.5200	0.6070
25	0.2646	0.3362	0.3977	0.4654	0.5100	0.5962
26	0.2588	0.3299	0.3894	0.4564	0.5002	0.5856
27	0.2540	0.3236	0.3822	0.4481	0.4915	0.5757
28	0.2490	0.3175	0.3749	0.4401	0.4828	0.5660
29	0.2443	0.3113	0.3685	0.4320	0.4744	0.5567
30	0.2400	0.3059	0.3620	0.4251	0.4665	0.5479

For n greater than 30 the approximate quantiles of ρ may be obtained from

$$w_p \cong \frac{z_p}{\sqrt{n-1}}$$

where z_p is the pth quantile of a standard normal random variable obtained from Table A1.

SOURCE: Adapted from Glasser and Winter (1961), with corrections, with permission from the *Biometrika* Trustees.

^{a} The entries in this table are selected quantiles w_p of the Spearman rank correlation coefficient ρ when used as a test statistic. The lower quantiles may be obtained from the equation

$$w_p = -w_{1-p}$$

The critical region corresponds to values of ρ smaller than (or greater than) but not including the appropriate quantile. Note that the median of ρ is 0.

TABLE A11   Quantiles of the Kendall test statistic $T = N_c - N_d$. Quantiles of Kendall's τ are given in parentheses. Lower quantiles are the negative of the upper quantiles, $w_p = -w_{1-p}$.

n	p = 0.900	0.950	0.975	0.990	0.995
4	4 (0.6667)	4 (0.6667)	6 (1.0000)	6 (1.0000)	6 (1.0000)
5	6 (0.6000)	6 (0.6000)	8 (0.8000)	8 (0.8000)	10 (1.0000)
6	7 (0.4667)	9 (0.6000)	11 (0.7333)	11 (0.7333)	13 (0.8667)
7	9 (0.4286)	11 (0.5238)	13 (0.6190)	15 (0.7143)	17 (0.8095)
8	10 (0.3571)	14 (0.5000)	16 (0.5714)	18 (0.6429)	20 (0.7143)
9	12 (0.3333)	16 (0.4444)	18 (0.5000)	22 (0.6111)	24 (0.6667)
10	15 (0.3333)	19 (0.4222)	21 (0.4667)	25 (0.5556)	27 (0.6000)
11	17 (0.3091)	21 (0.3818)	25 (0.4545)	29 (0.5273)	31 (0.5636)
12	18 (0.2727)	24 (0.3636)	28 (0.4242)	34 (0.5152)	36 (0.5455)
13	22 (0.2821)	26 (0.3333)	32 (0.4103)	38 (0.4872)	42 (0.5285)
14	23 (0.2527)	31 (0.3407)	35 (0.3846)	41 (0.4505)	45 (0.4945)
15	27 (0.2571)	33 (0.3143)	39 (0.3714)	47 (0.4476)	51 (0.4857)
16	28 (0.2333)	36 (0.3000)	44 (0.3667)	50 (0.4167)	56 (0.4667)
17	32 (0.2353)	40 (0.2941)	48 (0.3529)	56 (0.4118)	62 (0.4559)
18	35 (0.2288)	43 (0.2810)	51 (0.3333)	61 (0.3987)	67 (0.4379)
19	37 (0.2164)	47 (0.2749)	55 (0.3216)	65 (0.3801)	73 (0.4269)
20	40 (0.2105)	50 (0.2632)	60 (0.3158)	70 (0.3684)	78 (0.4105)
21	42 (0.2000)	54 (0.2571)	64 (0.3048)	76 (0.3619)	84 (0.4000)
22	45 (0.1948)	59 (0.2554)	69 (0.2987)	81 (0.3506)	89 (0.3853)
23	49 (0.1937)	63 (0.2490)	73 (0.2885)	87 (0.3439)	97 (0.3834)
24	52 (0.1884)	66 (0.2391)	78 (0.2826)	92 (0.3333)	102 (0.3696)
25	56 (0.1867)	70 (0.2333)	84 (0.2800)	98 (0.3267)	108 (0.3600)
26	59 (0.1815)	75 (0.2308)	89 (0.2738)	105 (0.3231)	115 (0.3538)
27	61 (0.1738)	79 (0.2251)	93 (0.2650)	111 (0.3162)	123 (0.3504)
28	66 (0.1746)	84 (0.2222)	98 (0.2593)	116 (0.3069)	128 (0.3386)
29	68 (0.1675)	88 (0.2167)	104 (0.2562)	124 (0.3054)	136 (0.3350)
30	73 (0.1678)	93 (0.2138)	109 (0.2506)	129 (0.2966)	143 (0.3287)
31	75 (0.1613)	97 (0.2086)	115 (0.2473)	135 (0.2903)	149 (0.3204)
32	80 (0.1613)	102 (0.2056)	120 (0.2419)	142 (0.2863)	158 (0.3185)
33	84 (0.1591)	106 (0.2008)	126 (0.2386)	150 (0.2841)	164 (0.3106)
34	87 (0.1551)	111 (0.1979)	131 (0.2335)	155 (0.2763)	173 (0.3084)
35	91 (0.1529)	115 (0.1933)	137 (0.2303)	163 (0.2739)	179 (0.3008)
36	94 (0.1492)	120 (0.1905)	144 (0.2286)	170 (0.2698)	188 (0.2984)
37	98 (0.1471)	126 (0.1892)	150 (0.2252)	176 (0.2643)	198 (0.2943)

TABLE A11 (Continued)

n	p = 0.900	0.950	0.975	0.990	0.995
38	103 (0.1465)	131 (0.1863)	155 (0.2205)	183 (0.2603)	203 (0.2888)
39	107 (0.1444)	137 (0.1849)	161 (0.2173)	191 (0.2578)	211 (0.2848)
40	110 (0.1372)	142 (0.1821)	168 (0.2154)	198 (0.2538)	220 (0.2821)
41	114 (0.1390)	146 (0.1780)	174 (0.2122)	206 (0.2512)	228 (0.2780)
42	119 (0.1382)	151 (0.1754)	181 (0.2102)	213 (0.2474)	235 (0.2729)
43	123 (0.1362)	157 (0.1739)	187 (0.2071)	221 (0.2447)	245 (0.2713)
44	128 (0.1353)	162 (0.1712)	194 (0.2051)	228 (0.2410)	252 (0.2664)
45	132 (0.1333)	168 (0.1697)	200 (0.2020)	236 (0.2383)	262 (0.2646)
46	135 (0.1304)	173 (0.1671)	207 (0.2000)	245 (0.2367)	271 (0.2618)
47	141 (0.1304)	179 (0.1656)	213 (0.1970)	253 (0.2340)	279 (0.2581)
48	144 (0.1277)	186 (0.1649)	220 (0.1950)	260 (0.2305)	288 (0.2553)
49	150 (0.1276)	190 (0.1616)	228 (0.1939)	268 (0.2279)	296 (0.2517)
50	153 (0.1249)	197 (0.1608)	233 (0.1902)	277 (0.2261)	305 (0.2490)
51	159 (0.1247)	203 (0.1592)	241 (0.1890)	285 (0.2235)	315 (0.2471)
52	162 (0.1222)	208 (0.1569)	248 (0.1870)	294 (0.2217)	324 (0.2443)
53	168 (0.1219)	214 (0.1553)	256 (0.1858)	302 (0.2192)	334 (0.2424)
54	173 (0.1209)	221 (0.1544)	263 (0.1838)	311 (0.2173)	343 (0.2397)
55	177 (0.1192)	227 (0.1529)	269 (0.1811)	319 (0.2148)	353 (0.2377)
56	182 (0.1182)	232 (0.1506)	276 (0.1792)	328 (0.2130)	362 (0.2351)
57	186 (0.1165)	240 (0.1504)	284 (0.1779)	336 (0.2105)	372 (0.2331)
58	191 (0.1155)	245 (0.1482)	291 (0.1760)	345 (0.2087)	381 (0.2305)
59	197 (0.1151)	251 (0.1467)	299 (0.1748)	355 (0.2075)	391 (0.2285)
60	202 (0.1141)	258 (0.1458)	306 (0.1729)	364 (0.2056)	402 (0.2271)

For n greater than 60, approximate quantiles of T may be obtained from

$$w_p \cong z_p \sqrt{\frac{n(n-1)(2n+5)}{18}}$$

where z_p is from the standard normal distribution given by Table A1. Approximate quantiles of τ may be obtained from

$$w_p \cong z_p \frac{\sqrt{2(2n+5)}}{3\sqrt{n(n-1)}}$$

Critical regions correspond to values of T greater than (or less than) but not including the appropriate quantile. Note that the median of T is 0. Quantiles for τ are obtained by dividing the quantiles of T by $n(n-1)/2$.

SOURCE. Adapted from Table 1, Best (1974), with permission from the author.

TABLE A12 Quantiles of the Wilcoxon Signed Ranks Test Statistic

	$W_{0.005}$	$W_{0.01}$	$W_{0.025}$	$W_{0.05}$	$W_{0.10}$	$W_{0.20}$	$W_{0.30}$	$W_{0.40}$	$W_{0.50}$	$\frac{n(n+1)}{2}$
$n = 4$	0	0	0	0	1	3	3	4	5	10
5	0	0	0	1	3	4	5	6	7.5	15
6	0	0	1	3	4	6	8	9	10.5	21
7	0	1	3	4	6	9	11	12	14	28
8	1	2	4	6	9	12	14	16	18	36
9	2	4	6	9	11	15	18	20	22.5	45
10	4	6	9	11	15	19	22	25	27.5	55
11	6	8	11	14	18	23	27	30	33	66
12	8	10	14	18	22	28	32	36	39	78
13	10	13	18	22	27	33	38	42	45.5	91
14	13	16	22	26	32	39	44	48	52.5	105
15	16	20	26	31	37	45	51	55	60	120
16	20	24	30	36	43	51	58	63	68	136
17	24	28	35	42	49	58	65	71	76.5	153
18	28	33	41	48	56	66	73	80	85.5	171
19	33	38	47	54	63	74	82	89	95	190
20	38	44	53	61	70	83	91	98	105	210
21	44	50	59	68	78	91	100	108	115.5	231
22	49	56	67	76	87	100	110	119	126.5	253
23	55	63	74	84	95	110	120	130	138	276
24	62	70	82	92	105	120	131	141	150	300
25	69	77	90	101	114	131	143	153	162.5	325
26	76	85	99	111	125	142	155	165	175.5	351
27	84	94	108	120	135	154	167	178	189	378
28	92	102	117	131	146	166	180	192	203	406
29	101	111	127	141	158	178	193	206	217.5	435
30	110	121	138	152	170	191	207	220	232.5	465
31	119	131	148	164	182	205	221	235	248	496
32	129	141	160	176	195	219	236	250	264	528
33	139	152	171	188	208	233	251	266	280.5	561
34	149	163	183	201	222	248	266	282	297.5	595
35	160	175	196	214	236	263	283	299	315	630
36	172	187	209	228	251	279	299	317	333	666
37	184	199	222	242	266	295	316	335	351.5	703
38	196	212	236	257	282	312	334	353	370.5	741
39	208	225	250	272	298	329	352	372	390	780
40	221	239	265	287	314	347	371	391	410	820
41	235	253	280	303	331	365	390	411	430.5	861
42	248	267	295	320	349	384	409	431	451.5	903

TABLE A12   **(Continued)**

	$w_{0.005}$	$w_{0.01}$	$w_{0.025}$	$w_{0.05}$	$w_{0.10}$	$w_{0.20}$	$w_{0.30}$	$w_{0.40}$	$w_{0.50}$	$\dfrac{n(n+1)}{2}$
43	263	282	311	337	366	403	429	452	473	946
44	277	297	328	354	385	422	450	473	495	990
45	292	313	344	372	403	442	471	495	517.5	1035
46	308	329	362	390	423	463	492	517	540.5	1081
47	324	346	379	408	442	484	514	540	564	1128
48	340	363	397	428	463	505	536	563	588	1176
49	357	381	416	447	483	527	559	587	612.5	1225
50	374	398	435	467	504	550	583	611	637.5	1275

For n larger than 50, the pth quantile w_p of the Wilcoxon signed ranks test statistic may be approximated by $w_p = [n(n+1)/4] + z_p\sqrt{n(n+1)(2n+1)/24}$, where z_p is the pth quantile of a standard normal random variable, obtained from Table A1.

Source. Adapted from Harter and Owen (1970), with permission from the American Mathematical Society.

[a] The entries in this table are quantiles w_p of the Wilcoxon signed ranks test statistic T^+, given by Equation 5.7.3, for selected values of $p \leq 0.50$. Quantiles w_p for $p > 0.50$ may be computed from the equation

$$w_p = n(n+1)/2 - w_{1-p}$$

where $n(n+1)/2$ is given in the right hand column in the table. Note that $P(T^+ < w_p) \leq p$ and $P(T^+ > w_p) \leq 1 - p$ if H_0 is true. Critical regions correspond to values of T^+ less than (or greater than) but not including the appropriate quantile.

TABLE A13   Quantiles of the Kolmogorov Test Statistic[a]

One-Sided Test $p = 0.90$	0.95	0.975	0.99	0.995		One-Sided Test $p = 0.90$	0.95	0.975	0.99	0.995
Two-Sided Test $p = 0.80$	0.90	0.95	0.98	0.99		Two-Sided Test $p = 0.80$	0.90	0.95	0.98	0.99
$n = 1$ 0.900	0.950	0.975	0.990	0.995	$n = 21$	0.226	0.259	0.287	0.321	0.344
2 0.684	0.776	0.842	0.900	0.929	22	0.221	0.253	0.281	0.314	0.337
3 0.565	0.636	0.708	0.785	0.829	23	0.216	0.247	0.275	0.307	0.330
4 0.493	0.565	0.624	0.689	0.734	24	0.212	0.242	0.269	0.301	0.323
5 0.447	0.509	0.563	0.627	0.669	25	0.208	0.238	0.264	0.295	0.317
6 0.410	0.468	0.519	0.577	0.617	26	0.204	0.233	0.259	0.290	0.311
7 0.381	0.436	0.483	0.538	0.576	27	0.200	0.229	0.254	0.284	0.305
8 0.358	0.410	0.454	0.507	0.542	28	0.197	0.225	0.250	0.279	0.300
9 0.339	0.387	0.430	0.480	0.513	29	0.193	0.221	0.246	0.275	0.295
10 0.323	0.369	0.409	0.457	0.489	30	0.190	0.218	0.242	0.270	0.290
11 0.308	0.352	0.391	0.437	0.468	31	0.187	0.214	0.238	0.266	0.285
12 0.296	0.338	0.375	0.419	0.449	32	0.184	0.211	0.234	0.262	0.281
13 0.285	0.325	0.361	0.404	0.432	33	0.182	0.208	0.231	0.258	0.277
14 0.275	0.314	0.349	0.390	0.418	34	0.179	0.205	0.227	0.254	0.273
15 0.266	0.304	0.338	0.377	0.404	35	0.177	0.202	0.224	0.251	0.269
16 0.258	0.295	0.327	0.366	0.392	36	0.174	0.199	0.221	0.247	0.265
17 0.250	0.286	0.318	0.355	0.381	37	0.172	0.196	0.218	0.244	0.262
18 0.244	0.279	0.309	0.346	0.371	38	0.170	0.194	0.215	0.241	0.258
19 0.237	0.271	0.301	0.337	0.361	39	0.168	0.191	0.213	0.238	0.255
20 0.232	0.265	0.294	0.329	0.352	40	0.165	0.189	0.210	0.235	0.252
			Approximation for $n > 40$			$\dfrac{1.07}{\sqrt{n}}$	$\dfrac{1.22}{\sqrt{n}}$	$\dfrac{1.36}{\sqrt{n}}$	$\dfrac{1.52}{\sqrt{n}}$	$\dfrac{1.63}{\sqrt{n}}$

SOURCE. Adapted from Table 1 of Miller (1956). Used with permission of the American Statistical Association.

[a] The entries in this table are selected quantiles w_p of the Kolmogorov test statistics T, T^+, and T^- as defined by Equation 6.1.1 for two-sided tests and by Equations 6.1.2 and 6.1.3 for one-sided tests. Reject H_0 at the level α if T exceeds the $1 - \alpha$ quantile given in this table. These quantiles are exact for $n \leq 40$ in the two-tailed test. The other quantiles are approximations that are equal to the exact quantiles in most cases. A better approximation for $n > 40$ results if $(n + \sqrt{n}/10)^{1/2}$ is used instead of $\sqrt{n}$ in the denominator.

TABLE A14 Quantiles of the Lilliefors Test Statistic for Normality[a]

	$p = 0.80$	0.85	0.90	0.95	0.99
Sample size $n = 4$	0.303	0.320	0.344	0.374	0.414
5	0.290	0.302	0.319	0.344	0.398
6	0.268	0.280	0.295	0.321	0.371
7	0.252	0.264	0.280	0.304	0.353
8	0.239	0.251	0.266	0.290	0.333
9	0.227	0.239	0.253	0.275	0.319
10	0.217	0.228	0.241	0.262	0.303
11	0.209	0.219	0.232	0.252	0.291
12	0.201	0.210	0.223	0.243	0.281
13	0.193	0.203	0.215	0.233	0.270
14	0.187	0.196	0.209	0.227	0.264
15	0.181	0.190	0.202	0.219	0.256
16	0.176	0.184	0.195	0.212	0.248
17	0.170	0.179	0.190	0.207	0.241
18	0.166	0.174	0.185	0.201	0.234
19	0.162	0.171	0.181	0.197	0.230
20	0.159	0.167	0.177	0.192	0.223
21	0.155	0.163	0.173	0.188	0.219
22	0.152	0.160	0.170	0.185	0.214
23	0.149	0.156	0.165	0.181	0.210
24	0.145	0.153	0.162	0.177	0.205
25	0.144	0.151	0.159	0.173	0.202
26	0.141	0.147	0.156	0.170	0.198
27	0.138	0.145	0.153	0.166	0.193
28	0.136	0.142	0.151	0.165	0.191
29	0.134	0.140	0.149	0.162	0.188
30	0.132	0.138	0.146	0.159	0.183
≥ 31	$\dfrac{0.741}{d_n}$	$\dfrac{0.775}{d_n}$	$\dfrac{0.819}{d_n}$	$\dfrac{0.895}{d_n}$	$\dfrac{1.035}{d_n}$

$$d_n = (\sqrt{n} - 0.01 + 0.83/\sqrt{n})$$

SOURCE. Table L.5, Mason and Bell (1986). Used with permission from Marcel Dekker, Inc.
[a] The entries in this table are the approximate quantiles w_p of the Lilliefors test statistic T_1 as defined by Equation 6.2.4. Reject H_0 at the level α if T_1 exceeds $w_{1-\alpha}$ for the particular sample size n.

TABLE A15 Quantiles of the Lilliefors Test Statistic for the Exponential Distribution[a]

	p = 0.05	0.10	0.20	0.30	0.50	0.70	0.80	0.90	0.95	0.99	0.999
n = 2	0.3127	0.3200	0.3337	0.3617	0.4337	0.5034	0.5507	0.5934	0.6133	0.6284	0.6317
3	0.2299	0.2544	0.2899	0.3166	0.3645	0.4122	0.4508	0.5111	0.5508	0.6003	0.6296
4	0.2072	0.2281	0.2545	0.2766	0.3163	0.3685	0.4007	0.4442	0.4844	0.5574	0.6215
5	0.1884	0.2052	0.2290	0.2483	0.2877	0.3317	0.3603	0.4045	0.4420	0.5127	0.5814
6	0.1726	0.1882	0.2102	0.2290	0.2645	0.3045	0.3320	0.3732	0.4085	0.4748	0.5497
7	0.1604	0.1750	0.1961	0.2136	0.2458	0.2838	0.3098	0.3481	0.3811	0.4459	0.5181
8	0.1506	0.1646	0.1845	0.2006	0.2309	0.2671	0.2914	0.3274	0.3590	0.4208	0.4913
9	0.1426	0.1561	0.1746	0.1897	0.2186	0.2529	0.2758	0.3101	0.3404	0.3995	0.4679
10	0.1359	0.1486	0.1661	0.1805	0.2082	0.2407	0.2626	0.2955	0.3244	0.3813	0.4473
12	0.1249	0.1364	0.1524	0.1657	0.1912	0.2209	0.2411	0.2714	0.2981	0.3511	0.4132
14	0.1162	0.1268	0.1418	0.1542	0.1778	0.2054	0.2242	0.2525	0.2774	0.3272	0.3858
16	0.1091	0.1191	0.1332	0.1448	0.1669	0.1929	0.2105	0.2371	0.2606	0.3076	0.3632
18	0.1032	0.1127	0.1260	0.1369	0.1578	0.1824	0.1990	0.2242	0.2465	0.2911	0.3441
20	0.0982	0.1073	0.1199	0.1303	0.1501	0.1735	0.1893	0.2132	0.2345	0.2771	0.3277
22	0.0939	0.1025	0.1146	0.1245	0.1434	0.1657	0.1809	0.2038	0.2241	0.2649	0.3135
24	0.0901	0.0984	0.1099	0.1195	0.1376	0.1590	0.1735	0.1954	0.2150	0.2542	0.3010
26	0.0868	0.0947	0.1058	0.1150	0.1324	0.1530	0.1670	0.1881	0.2069	0.2447	0.2899
28	0.0838	0.0914	0.1021	0.1110	0.1278	0.1477	0.1611	0.1815	0.1997	0.2362	0.2799
30	0.0811	0.0885	0.0988	0.1074	0.1236	0.1428	0.1559	0.1756	0.1932	0.2286	0.2709
35	0.0754	0.0822	0.0918	0.0997	0.1148	0.1326	0.1447	0.1630	0.1793	0.2123	0.2517
40	0.0707	0.0771	0.0861	0.0935	0.1077	0.1243	0.1356	0.1528	0.1681	0.1990	0.2361
45	0.0668	0.0729	0.0814	0.0884	0.1017	0.1174	0.1281	0.1443	0.1588	0.1880	0.2231
50	0.0636	0.0693	0.0774	0.0840	0.0966	0.1116	0.1217	0.1371	0.1509	0.1787	0.2121
60	0.0582	0.0635	0.0708	0.0769	0.0885	0.1021	0.1114	0.1255	0.1381	0.1635	0.1943
70	0.0541	0.0589	0.0658	0.0714	0.0821	0.0946	0.1033	0.1164	0.1281	0.1517	b
80	0.0507	0.0553	0.0616	0.0669	0.0769	0.0887	0.0968	0.1090	0.1200	0.1421	b
90	0.0479	0.0522	0.0582	0.0632	0.0726	0.0838	0.0914	0.1029	0.1132	0.1341	b
n = 100	0.0455	0.0496	0.0553	0.0600	0.0690	0.0796	0.0868	0.0977	0.1075	0.1274	b
Approximation for n > 100	$\frac{0.4550}{\sqrt{n}}$	$\frac{0.4959}{\sqrt{n}}$	$\frac{0.5530}{\sqrt{n}}$	$\frac{0.6000}{\sqrt{n}}$	$\frac{0.6898}{\sqrt{n}}$	$\frac{0.7957}{\sqrt{n}}$	$\frac{0.8678}{\sqrt{n}}$	$\frac{0.9773}{\sqrt{n}}$	$\frac{1.0753}{\sqrt{n}}$	$\frac{1.2743}{\sqrt{n}}$	b

SOURCE. Adapted from Durbin (1975), with permission from the *Biometrika* Trustees.

[a] The entries in this table are selected quantiles w_p of the Lilliefors test statistic T_2 as given by Equation 6.2.6. Reject at the level of significance α if T_2 is greater than the $1 - \alpha$ quantile given in the table. The approximation for $n > 100$ is merely the exact value for $n = 100$. More accurate approximations for $n > 100$ may be obtained from Table 54 of Pearson and Hartley (1972).

[b] These quantiles are not presently available.

TABLE A16 Coefficients for the Shapiro-Wilk Test[a]

i \ n	2	3	4	5	6	7	8	9	10
1	0.7071	0.7071	0.6872	0.6646	0.6431	0.6233	0.6052	0.0588	0.5739
2	—	0.0000	0.1667	0.2413	0.2806	0.3031	0.3164	0.3244	0.3291
3	—	—	—	0.0000	0.0875	0.1401	0.1743	0.1976	0.2141
4	—	—	—	—	—	0.0000	0.0561	0.0947	0.1224
5	—	—	—	—	—	—	—	0.0000	0.0399

i \ n	11	12	13	14	15	16	17	18	19	20
1	0.5601	0.5475	0.5359	0.5251	0.5150	0.5056	0.4968	0.4886	0.4808	0.4734
2	0.3315	0.3325	0.3325	0.3318	0.3306	0.3290	0.3273	0.3253	0.3232	0.3211
3	0.2260	0.2347	0.2412	0.2460	0.2495	0.2521	0.2540	0.2553	0.2561	0.2565
4	0.1429	0.1586	0.1707	0.1802	0.1878	0.1939	0.1988	0.2027	0.2059	0.2085
5	0.0695	0.0922	0.1099	0.1240	0.1353	0.1447	0.1524	0.1587	0.1641	0.1686
6	0.0000	0.0303	0.0539	0.0727	0.0880	0.1005	0.1109	0.1197	0.1271	0.1334
7	—	—	0.0000	0.0240	0.0433	0.0593	0.0725	0.0837	0.0932	0.1013
8	—	—	—	—	0.0000	0.0196	0.0359	0.0496	0.0612	0.0711
9	—	—	—	—	—	—	0.0000	0.0163	0.0303	0.0422
10	—	—	—	—	—	—	—	—	0.0000	0.0140

i \ n	21	22	23	24	25	26	27	28	29	30
1	0.4643	0.4590	0.4542	0.4493	0.4450	0.4407	0.4366	0.4328	0.4291	0.4254
2	0.3185	0.3156	0.3126	0.3098	0.3069	0.3043	0.3018	0.2992	0.2968	0.2944
3	0.2578	0.2571	0.2563	0.2554	0.2543	0.2533	0.2522	0.2510	0.2499	0.2487
4	0.2119	0.2131	0.2139	0.2145	0.2148	0.2151	0.2152	0.2151	0.2150	0.2148
5	0.1736	0.1764	0.1787	0.1807	0.1822	0.1836	0.1848	0.1857	0.1864	0.1870
6	0.1399	0.1443	0.1480	0.1512	0.1539	0.1563	0.1584	0.1601	0.1616	0.1630
7	0.1092	0.1150	0.1201	0.1245	0.1283	0.1316	0.1346	0.1372	0.1395	0.1415
8	0.0804	0.0878	0.0941	0.0997	0.1046	0.1089	0.1128	0.1162	0.1192	0.1219
9	0.0530	0.0618	0.0696	0.0764	0.0823	0.0876	0.0923	0.0965	0.1002	0.1036
10	0.0263	0.0368	0.0459	0.0539	0.0610	0.0672	0.0728	0.0778	0.0822	0.0862
11	0.0000	0.0122	0.0228	0.0321	0.0403	0.0476	0.0540	0.0598	0.0650	0.0697
12	—	—	0.0000	0.0107	0.0200	0.0284	0.0358	0.0424	0.0483	0.0537
13	—	—	—	—	0.0000	0.0094	0.0178	0.0253	0.0320	0.0381
14	—	—	—	—	—	—	0.0000	0.0084	0.0159	0.0227
15	—	—	—	—	—	—	—	—	0.0000	0.0076

TABLE A16   (Continued)

i\n	31	32	33	34	35	36	37	38	39	40
1	0.4220	0.4188	0.4156	0.4127	0.4096	0.4068	0.4040	0.4015	0.3989	0.3964
2	0.2921	0.2898	0.2876	0.2854	0.2834	0.2813	0.2794	0.2774	0.2755	0.2737
3	0.2475	0.2462	0.2451	0.2439	0.2427	0.2415	0.2403	0.2391	0.2380	0.2368
4	0.2145	0.2141	0.2137	0.2132	0.2127	0.2121	0.2116	0.2110	0.2104	0.2098
5	0.1874	0.1878	0.1880	0.1882	0.1883	0.1883	0.1883	0.1881	0.1880	0.1878
6	0.1641	0.1651	0.1660	0.1667	0.1673	0.1678	0.1683	0.1686	0.1689	0.1691
7	0.1433	0.1449	0.1463	0.1475	0.1487	0.1496	0.1505	0.1513	0.1520	0.1526
8	0.1243	0.1265	0.1284	0.1301	0.1317	0.1331	0.1344	0.1356	0.1366	0.1376
9	0.1066	0.1093	0.1118	0.1140	0.1160	0.1179	0.1196	0.1211	0.1225	0.1237
10	0.0899	0.0931	0.0961	0.0988	0.1013	0.1036	0.1056	0.1075	0.1092	0.1108
11	0.0739	0.0777	0.0812	0.0844	0.0873	0.0900	0.0924	0.0947	0.0967	0.0986
12	0.0585	0.0629	0.0669	0.0706	0.0739	0.0770	0.0798	0.0824	0.0848	0.0870
13	0.0435	0.0485	0.0530	0.0572	0.0610	0.0645	0.0677	0.0706	0.0733	0.0759
14	0.0289	0.0344	0.0395	0.0441	0.0484	0.0523	0.0559	0.0592	0.0622	0.0651
15	0.0144	0.0206	0.0262	0.0314	0.0361	0.0404	0.0444	0.0481	0.0515	0.0546
16	0.0000	0.0068	0.0131	0.0187	0.0239	0.0287	0.0331	0.0372	0.0409	0.0444
17	—	—	0.0000	0.0062	0.0119	0.0172	0.0220	0.0264	0.0305	0.0343
18	—	—	—	—	0.0000	0.0057	0.0110	0.0158	0.0203	0.0244
19	—	—	—	—	—	—	0.0000	0.0053	0.0101	0.0146
20	—	—	—	—	—	—	—	—	0.0000	0.0049

i\n	41	42	43	44	45	46	47	48	49	50
1	0.3940	0.3917	0.3894	0.3872	0.3850	0.3830	0.3808	0.3789	0.3770	0.3751
2	0.2719	0.2701	0.2684	0.2667	0.2651	0.2635	0.2620	0.2604	0.2589	0.2574
3	0.2357	0.2345	0.2334	0.2323	0.2313	0.2302	0.2291	0.2281	0.2271	0.2260
4	0.2091	0.2085	0.2078	0.2072	0.2065	0.2058	0.2052	0.2045	0.2038	0.2032
5	0.1876	0.1874	0.1871	0.1868	0.1865	0.1862	0.1859	0.1855	0.1851	0.1847
6	0.1693	0.1694	0.1695	0.1695	0.1695	0.1695	0.1695	0.1693	0.1692	0.1691
7	0.1531	0.1535	0.1539	0.1542	0.1545	0.1548	0.1550	0.1551	0.1553	0.1554
8	0.1384	0.1392	0.1398	0.1405	0.1410	0.1415	0.1420	0.1423	0.1427	0.1430
9	0.1249	0.1259	0.1269	0.1278	0.1286	0.1293	0.1300	0.1306	0.1312	0.1317
10	0.1123	0.1136	0.1149	0.1160	0.1170	0.1180	0.1189	0.1197	0.1205	0.1212
11	0.1004	0.1020	0.1035	0.1049	0.1062	0.1073	0.1085	0.1095	0.1105	0.1113
12	0.0891	0.0909	0.0927	0.0943	0.0959	0.0972	0.0986	0.0998	0.1010	0.1020
13	0.0782	0.0804	0.0824	0.0842	0.0860	0.0876	0.0892	0.0906	0.0919	0.0932
14	0.0677	0.0701	0.0724	0.0745	0.0765	0.0783	0.0801	0.0817	0.0832	0.0846
15	0.0575	0.0602	0.0628	0.0651	0.0673	0.0694	0.0713	0.0731	0.0748	0.0764

TABLE A16   (Continued)

i \ n	41	42	43	44	45	46	47	48	49	50
16	0.0476	0.0506	0.0534	0.0560	0.0584	0.0607	0.0628	0.0648	0.0667	0.0685
17	0.0379	0.0411	0.0442	0.0471	0.0497	0.0522	0.0546	0.0568	0.0588	0.0608
18	0.0283	0.0318	0.0352	0.0383	0.0412	0.0439	0.0465	0.0489	0.0511	0.0532
19	0.0188	0.0227	0.0263	0.0296	0.0328	0.0357	0.0385	0.0411	0.0436	0.0459
20	0.0094	0.0136	0.0175	0.0211	0.0245	0.0277	0.0307	0.0335	0.0361	0.0386
21	0.0000	0.0045	0.0087	0.0126	0.0163	0.0197	0.0229	0.0259	0.0288	0.0314
22	—	—	0.0000	0.0042	0.0081	0.0118	0.0153	0.0185	0.0215	0.0244
23	—	—	—	—	0.0000	0.0039	0.0076	0.0111	0.0143	0.0174
24	—	—	—	—	—	—	0.0000	0.0037	0.0071	0.0104
25	—	—	—	—	—	—	—	—	0.0000	0.0035

SOURCE. Reprinted from Vol. 2 of Pearson and Hartley (1976), with permission from the *Biometrika* Trustees.

[a] The entries in this table are the coefficients a_i for use in the Shapiro-Wilk test statistic for normality given by Equation 6.2.9.

TABLE A17   Quantiles of the Shapiro-Wilk Test Statistic[a]

n	0.01	0.02	0.05	0.10	0.50	0.90	0.95	0.98	0.99
3	0.753	0.756	0.767	0.789	0.959	0.998	0.999	1.000	1.000
4	0.687	0.707	0.748	0.792	0.935	0.987	0.992	0.996	0.997
5	0.686	0.715	0.762	0.806	0.927	0.979	0.986	0.991	0.993
6	0.713	0.743	0.788	0.826	0.927	0.974	0.981	0.986	0.989
7	0.730	0.760	0.803	0.838	0.928	0.972	0.979	0.985	0.988
8	0.749	0.778	0.818	0.851	0.932	0.972	0.978	0.984	0.987
9	0.764	0.791	0.829	0.859	0.935	0.972	0.978	0.984	0.986
10	0.781	0.806	0.842	0.869	0.938	0.972	0.978	0.983	0.986
11	0.792	0.817	0.850	0.876	0.940	0.973	0.979	0.984	0.986
12	0.805	0.828	0.859	0.883	0.943	0.973	0.979	0.984	0.986
13	0.814	0.837	0.866	0.889	0.945	0.974	0.979	0.984	0.986
14	0.825	0.846	0.874	0.895	0.947	0.975	0.980	0.984	0.986
15	0.835	0.855	0.881	0.901	0.950	0.975	0.980	0.984	0.987
16	0.844	0.863	0.887	0.906	0.952	0.976	0.981	0.985	0.987
17	0.851	0.869	0.892	0.910	0.954	0.977	0.981	0.985	0.987
18	0.858	0.874	0.897	0.914	0.956	0.978	0.982	0.986	0.988
19	0.863	0.879	0.901	0.917	0.957	0.978	0.982	0.986	0.988
20	0.868	0.884	0.905	0.920	0.959	0.979	0.983	0.986	0.988
21	0.873	0.888	0.908	0.923	0.960	0.980	0.983	0.987	0.989
22	0.878	0.892	0.911	0.926	0.961	0.980	0.984	0.987	0.989
23	0.881	0.895	0.914	0.928	0.962	0.981	0.984	0.987	0.989
24	0.884	0.898	0.916	0.930	0.963	0.981	0.984	0.987	0.989

TABLE A17   (Continued)

n	0.01	0.02	0.05	0.10	0.50	0.90	0.95	0.98	0.99
25	0.888	0.901	0.918	0.931	0.964	0.981	0.985	0.988	0.989
26	0.891	0.904	0.920	0.933	0.965	0.982	0.985	0.988	0.989
27	0.894	0.906	0.923	0.935	0.965	0.982	0.985	0.988	0.990
28	0.896	0.908	0.924	0.936	0.966	0.982	0.985	0.988	0.990
29	0.898	0.910	0.926	0.937	0.966	0.982	0.985	0.988	0.990
30	0.900	0.912	0.927	0.939	0.967	0.983	0.985	0.988	0.990
31	0.902	0.914	0.929	0.940	0.967	0.983	0.986	0.988	0.990
32	0.904	0.915	0.930	0.941	0.968	0.983	0.986	0.988	0.990
33	0.906	0.917	0.931	0.942	0.968	0.983	0.986	0.989	0.990
34	0.908	0.919	0.933	0.943	0.969	0.983	0.986	0.989	0.990
35	0.910	0.920	0.934	0.944	0.969	0.984	0.986	0.989	0.990
36	0.912	0.922	0.935	0.945	0.970	0.984	0.986	0.989	0.990
37	0.914	0.924	0.936	0.946	0.970	0.984	0.987	0.989	0.990
38	0.916	0.925	0.938	0.947	0.971	0.984	0.987	0.989	0.990
39	0.917	0.927	0.939	0.948	0.971	0.984	0.987	0.989	0.991
40	0.919	0.928	0.940	0.949	0.972	0.985	0.987	0.989	0.991
41	0.920	0.929	0.941	0.950	0.972	0.985	0.987	0.989	0.991
42	0.922	0.930	0.942	0.951	0.972	0.985	0.987	0.989	0.991
43	0.923	0.932	0.943	0.951	0.973	0.985	0.987	0.990	0.991
44	0.924	0.933	0.944	0.952	0.973	0.985	0.987	0.990	0.991
45	0.926	0.934	0.945	0.953	0.973	0.985	0.988	0.990	0.991
46	0.927	0.935	0.945	0.953	0.974	0.985	0.988	0.990	0.991
47	0.928	0.936	0.946	0.954	0.974	0.985	0.988	0.990	0.991
48	0.929	0.937	0.947	0.954	0.974	0.985	0.988	0.990	0.991
49	0.929	0.937	0.947	0.955	0.974	0.985	0.988	0.990	0.991
50	0.930	0.938	0.947	0.955	0.974	0.985	0.988	0.990	0.991

[a] The entries in this table are quantiles w_p of the Shapiro-Wilk test statistic given by Equation 6.2.9. Reject H_0 at the level p if $T_3 < w_p$.

TABLE A18   A Method for Converting the Shapiro-Wilk Statistic to Approximate Normality

$v \backslash (d_n)$	3 (0.7500)	4 (0.6297)	5 (0.5521)	6 (0.4963)	$v (d_n)$	3 (0.7500)	4 (0.6297)	5 (0.5521)	6 (0.4963)
−7.0	−3.29	—	—	—	2.2	0.52	0.74	0.75	0.64
−5.4	−2.81	—	—	—	2.6	0.67	1.00	1.09	1.06
−5.0	−2.68	—	—	—	3.0	0.81	1.23	1.40	1.45
−4.6	−2.54	—	—	—	3.4	0.95	1.44	1.67	1.83
−4.2	−2.40	—	—	—	3.8	1.07	1.65	1.91	2.17
−3.8	−2.25	−3.50	—	—	4.2	1.19	1.85	2.15	2.50
−3.4	−2.10	−3.27	—	—	4.6	1.31	2.03	2.47	2.77
−3.0	−1.94	−3.05	−4.01	—	5.0	1.42	2.19	2.85	3.09
−2.6	−1.77	−2.84	−3.70	—	5.4	1.52	2.34	3.24	3.54
−2.2	−1.59	−2.64	−3.38	—	5.8	1.62	2.48	3.64	—
−1.8	−1.40	−2.44	−3.11	—	6.2	1.72	2.62	—	—
−1.4	−1.21	−2.22	−2.87	—	6.6	1.81	2.75	—	—
−1.0	−1.01	−1.96	−2.56	−3.72	7.0	1.90	2.87	—	—
−0.6	−0.80	−1.66	−2.20	−2.88	7.4	1.98	2.97	—	—
−0.2	−0.60	−1.31	−1.81	−2.27	7.8	2.07	3.08	—	—
0.2	−0.39	−0.94	−1.41	−1.85	8.2	2.15	3.22	—	—
0.6	−0.19	−0.57	−0.97	−1.38	8.6	2.23	3.36	—	—
1.0	0.00	−0.19	−0.51	−0.84	9.0	2.31	—	—	—
1.4	0.18	0.15	−0.06	−0.33	9.4	2.38	—	—	—
1.8	0.35	0.45	0.37	0.18	9.8	2.45	—	—	—

For $3 \leq n \leq 6$, first compute $v = \ln [(T - d_n)/(1 - T)]$ where d_n is given at the top of the table and T is the Shapiro-Wilk statistic. Then enter the table with v and n to find G, which is approximately normal.

TABLE A18   (Continued)

n	b_n	c_n	d_n	n	b_n	c_n	d_n
7	−2.356	1.245	0.4533	29	−6.074	1.934	0.1907
8	−2.696	1.333	0.4186	30	−6.150	1.949	0.1872
9	−2.968	1.400	0.3900				
10	−3.262	1.471	0.3600	31	−6.248	1.965	0.1840
				32	−6.324	1.976	0.1811
11	−3.485	1.515	0.3451	33	−6.402	1.988	0.1781
12	−3.731	1.571	0.3270	34	−6.480	2.000	0.1755
13	−3.936	1.613	0.3111	35	−6.559	2.012	0.1727
14	−4.155	1.655	0.2969				
15	−4.373	1.695	0.2842	36	−6.640	2.024	0.1702
				37	−6.721	2.037	0.1677
16	−4.567	1.724	0.2727	38	−6.803	2.049	0.1656
17	−4.713	1.739	0.2622	39	−6.887	2.062	0.1633
18	−4.885	1.770	0.2528	40	−6.961	2.075	0.1612
19	−5.018	1.786	0.2440				
20	−5.153	1.802	0.2359	41	−7.035	2.088	0.1591
				42	−7.111	2.101	0.1572
21	−5.291	1.818	0.2264	43	−7.188	2.114	0.1552
22	−5.413	1.835	0.2207	44	−7.266	2.128	0.1534
23	−5.508	1.848	0.2157	45	−7.345	2.141	0.1516
24	−5.605	1.862	0.2106				
25	−5.704	1.876	0.2063	46	−7.414	2.155	0.1499
				47	−7.484	2.169	0.1482
26	−5.803	1.890	0.2020	48	−7.555	2.183	0.1466
27	−5.905	1.905	0.1980	49	−7.615	2.198	0.1451
28	−5.988	1.919	0.1943	50	−7.677	2.212	0.1436

SOURCE. Reprinted from Vol. 2 of Pearson and Hartley (1976), with permission from the *Biometrika* Trustees.

For $7 \leq n \leq 50$, enter the table above with n to find the coefficients b_n, c_n, and d_n. Then compute

$$G = b_n + c_n \ln \{(T - d_n)/(1 - T)\}$$

which is approximately standard normal.

TABLE A19   Quantiles of the Smirnov Test Statistic for Two Samples of Equal Size n^a

One-Sided Test: $p = 0.90$	0.95	0.975	0.99	0.995	One-Sided Test: $p = 0.90$	0.95	0.975	0.99	0.995
Two-Sided Test: $p = 0.80$	0.90	0.95	0.98	0.99	Two-Sided Test: $p = 0.80$	0.90	0.95	0.98	0.99
$n = 3$ 2/3	2/3				$n = 22$ 7/22	8/22	8/22	10/22	10/22
4 3/4	3/4	3/4			23 7/23	8/23	9/23	10/23	10/23
5 3/5	3/5	4/5	4/5	4/5	24 7/24	8/24	9/24	10/24	11/24
6 3/6	4/6	4/6	5/6	5/6	25 7/25	8/25	9/25	10/25	11/25
7 4/7	4/7	5/7	5/7	5/7	26 7/26	8/26	9/26	10/26	11/26
8 4/8	4/8	5/8	5/8	6/8	27 7/27	8/27	9/27	11/27	11/27
9 4/9	5/9	5/9	6/9	6/9	28 8/28	9/28	10/28	11/28	12/28
10 4/10	5/10	6/10	6/10	7/10	29 8/29	9/29	10/29	11/29	12/29
11 5/11	5/11	6/11	7/11	7/11	30 8/30	9/30	10/30	11/30	12/30
12 5/12	5/12	6/12	7/12	7/12	31 8/31	9/31	10/31	11/31	12/31
13 5/13	6/13	6/13	7/13	8/13	32 8/32	9/32	10/32	12/32	12/32
14 5/14	6/14	7/14	7/14	8/14	33 8/33	9/33	11/33	12/33	13/33
15 5/15	6/15	7/15	8/15	8/15	34 8/34	10/34	11/34	12/34	13/34
16 6/16	6/16	7/16	8/16	9/16	35 8/35	10/35	11/35	12/35	13/35
17 6/17	7/17	7/17	8/17	9/17	36 9/36	10/36	11/36	12/36	13/36
18 6/18	7/18	8/18	9/18	9/18	37 9/37	10/37	11/37	13/37	13/37
19 6/19	7/19	8/19	9/19	9/19	38 9/38	10/38	11/38	13/38	14/38
20 6/20	7/20	8/20	9/20	10/20	39 9/39	10/39	11/39	13/39	14/39
21 6/21	7/21	8/21	9/21	10/21	40 9/40	10/40	12/40	13/40	14/40
					Approximation for $n > 40$:	$\dfrac{1.52}{\sqrt{n}}$	$\dfrac{1.73}{\sqrt{n}}$	$\dfrac{1.92}{\sqrt{n}}$	$\dfrac{2.15}{\sqrt{n}}$	$\dfrac{2.30}{\sqrt{n}}$

SOURCE. Adapted from Birnbaum and Hall (1960), with permission from the Institute of Mathematical Statistics.

^{a} The entries in this table are selected quantiles w_p of the Smirnov two-sample test statistic T defined by Equations 6.3.2 and 6.3.3 for the one-tailed test and defined by Equation 6.3.1 for the two-tailed test. Reject H_0 at the level α if T exceeds the $1 - \alpha$ quantile of T as given in this table. The test statistic is a discrete random variable, so the exact level of significance may be less than the apparent α used in this table.

TABLE A20 Quantiles of the Smirnov Test Statistic for Two Samples of Different Size n and m^a

One-Sided Test:			$p = 0.90$	0.95	0.975	0.99	0.995
Two-Sided Test:			$p = 0.80$	0.90	0.95	0.99	0.99
$N_1 = 1$	$N_2 =$	9	17/18				
		10	9/10				
$N_1 = 2$	$N_2 =$	3	5/6				
		4	3/4				
		5	4/5	4/5			
		6	5/6	5/6			
		7	5/7	6/7			
		8	3/4	7/8	7/8		
		9	7/9	8/9	8/9		
		10	7/10	4/5	9/10		
$N_1 = 3$	$N_2 =$	4	3/4	3/4			
		5	2/3	4/5	4/5		
		6	2/3	2/3	5/6		
		7	2/3	5/7	6/7	6/7	
		8	5/8	3/4	3/4	7/8	
		9	2/3	2/3	7/9	8/9	8/9
		10	3/5	7/10	4/5	9/10	9/10
		12	7/12	2/3	3/4	5/6	11/12
$N_1 = 4$	$N_2 =$	5	3/5	3/4	4/5	4/5	
		6	7/12	2/3	3/4	5/6	5/6
		7	17/28	5/7	3/4	6/7	6/7
		8	5/8	5/8	3/4	7/8	7/8
		9	5/9	2/3	3/4	7/9	8/9
		10	11/20	13/20	7/10	4/5	4/5
		12	7/12	2/3	2/3	3/4	5/6
		16	9/16	5/8	11/16	3/4	13/16
$N_1 = 5$	$N_2 =$	6	3/5	2/3	2/3	5/6	5/6
		7	4/7	23/35	5/7	29/35	6/7
		8	11/20	5/8	27/40	4/5	4/5
		9	5/9	3/5	31/45	7/9	4/5
		10	1/2	3/5	7/10	7/10	4/5
		15	8/15	3/5	2/3	11/15	11/15
		20	1/2	11/20	3/5	7/10	3/4
$N_1 = 6$	$N_2 =$	7	23/42	4/7	29/42	5/7	5/6
		8	1/2	7/12	2/3	3/4	3/4
		9	1/2	5/9	2/3	13/18	7/9
		10	1/2	17/30	19/30	7/10	11/15
		12	1/2	7/12	7/12	2/3	3/4
		18	4/9	5/9	11/18	2/3	13/18
		24	11/24	1/2	7/12	5/8	2/3

TABLE A20   (Continued)

One-Sided Test:		$p = 0.90$	0.95	0.975	0.99	0.995
Two-Sided Test:		$p = 0.80$	0.90	0.95	0.99	0.99
$N_1 = 7$	$N_2 = 8$	27/56	33/56	5/8	41/56	3/4
	9	31/63	5/9	40/63	5/7	47/63
	10	33/70	39/70	43/70	7/10	5/7
	14	3/7	1/2	4/7	9/14	5/7
	28	3/7	13/28	15/28	17/28	9/14
$N_1 = 8$	$N_2 = 9$	4/9	13/24	5/8	2/3	3/4
	10	19/40	21/40	23/40	27/40	7/10
	12	11/24	1/2	7/12	5/8	2/3
	16	7/16	1/2	9/16	5/8	5/8
	32	13/32	7/16	1/2	9/16	19/32
$N_1 = 9$	$N_2 = 10$	7/15	1/2	26/45	2/3	31/45
	12	4/9	1/2	5/9	11/18	2/3
	15	19/45	22/45	8/15	3/5	29/45
	18	7/18	4/9	1/2	5/9	11/18
	36	13/36	5/12	17/36	19/36	5/9
$N_1 = 10$	$N_2 = 15$	2/5	7/15	1/2	17/30	19/30
	20	2/5	9/20	1/2	11/20	3/5
	40	7/20	2/5	9/20	1/2	—
$N_1 = 12$	$N_2 = 15$	23/60	9/20	1/2	11/20	7/12
	16	3/8	7/16	23/48	13/24	7/12
	18	13/36	5/12	17/36	19/36	5/9
	20	11/30	5/12	7/15	31/60	17/30
$N_1 = 15$	$N_2 = 20$	7/20	2/5	13/30	29/60	31/60
$N_1 = 16$	$N_2 = 20$	27/80	31/80	17/40	19/40	41/80
Large sample approximation		$1.07\sqrt{\dfrac{m+n}{mn}}$	$1.22\sqrt{\dfrac{m+n}{mn}}$	$1.36\sqrt{\dfrac{m+n}{mn}}$	$1.52\sqrt{\dfrac{m+n}{mn}}$	$1.63\sqrt{\dfrac{m+n}{mn}}$

SOURCE. Adapted from Massey (1952), with permission from the Institute of Mathematical Statistics.

[a] The entries in this table are selected quantiles w_p of the Smirnov test statistic T for two samples, defined by Equations 6.3.1, 6.3.2, and 6.3.3. To enter the table let N_1 be the smaller sample size and let N_2 be the larger sample size. Reject H_0 at the level α if T exceeds $w_{1-\alpha}$ as given in this table. If n and m are not covered by this table, use the large sample approximation given at the end of the table, or consult exact tables by Kim and Jennrich, which appear in Harter and Owen (1970) for $n, m \leq 100$.

TABLE A2I   The t Distribution[a]

Degrees of Freedom	$p = 0.6$	0.75	0.9	0.95	0.975	0.99	0.995	0.9975	0.999	0.9995
1	0.325	1.000	3.078	6.314	12.706	31.821	63.657	127.32	318.31	636.62
2	0.289	0.816	1.886	2.920	4.303	6.965	9.925	14.089	22.327	31.598
3	0.277	0.765	1.638	2.353	3.182	4.541	5.841	7.453	10.214	12.924
4	0.271	0.741	1.533	2.132	2.776	3.747	4.604	5.598	7.173	8.610
5	0.267	0.727	1.476	2.015	2.571	3.365	4.032	4.773	5.893	6.869
6	0.265	0.718	1.440	1.943	2.447	3.143	3.707	4.317	5.208	5.959
7	0.263	0.711	1.415	1.895	2.365	2.998	3.499	4.029	4.785	5.408
8	0.262	0.706	1.397	1.860	2.306	2.896	3.355	3.833	4.501	5.041
9	0.261	0.703	1.383	1.833	2.262	2.821	3.250	3.690	4.297	4.781
10	0.260	0.700	1.372	1.812	2.228	2.764	3.169	3.581	4.144	4.587
11	0.260	0.697	1.363	1.796	2.201	2.718	3.106	3.497	4.025	4.437
12	0.259	0.695	1.356	1.782	2.179	2.681	3.055	3.428	3.930	4.318
13	0.259	0.694	1.350	1.771	2.160	2.650	3.012	3.372	3.852	4.221
14	0.258	0.692	1.345	1.761	2.145	2.624	2.977	3.326	3.787	4.140
15	0.258	0.691	1.341	1.753	2.131	2.602	2.947	3.286	3.733	4.073
16	0.258	0.690	1.377	1.746	2.120	2.583	2.921	3.252	3.686	4.015
17	0.257	0.689	1.333	1.740	2.110	2.567	2.898	3.222	3.646	3.965
18	0.257	0.688	1.330	1.734	2.101	2.552	2.878	3.197	3.610	3.922
19	0.257	0.688	1.328	1.729	2.093	2.539	2.861	3.174	3.579	3.883
20	0.257	0.687	1.325	1.725	2.086	2.528	2.845	3.153	3.552	3.850
21	0.257	0.686	1.323	1.721	2.080	2.518	2.831	3.135	3.527	3.819
22	0.256	0.686	1.321	1.717	2.074	2.508	2.819	3.119	3.505	3.792
23	0.256	0.685	1.319	1.714	2.069	2.500	2.807	3.104	3.485	3.767
24	0.256	0.685	1.318	1.711	2.064	2.492	2.797	3.091	3.467	3.745
25	0.256	0.684	1.316	1.708	2.060	2.485	2.787	3.078	3.450	3.725
26	0.256	0.684	1.315	1.706	2.056	2.479	2.779	3.067	3.435	3.707
27	0.256	0.684	1.314	1.703	2.052	2.473	2.771	3.057	3.421	3.690
28	0.256	0.683	1.313	1.701	2.048	2.467	2.763	3.047	3.408	3.674
29	0.256	0.683	1.311	1.699	2.045	2.462	2.756	3.038	3.396	3.659
30	0.256	0.683	1.310	1.697	2.042	2.457	2.750	3.030	3.385	3.646
40	0.255	0.681	1.303	1.684	2.021	2.423	2.704	2.971	3.307	3.551
60	0.254	0.679	1.296	1.671	2.000	2.390	2.660	2.915	3.232	3.460
120	0.254	0.677	1.289	1.658	1.980	2.358	2.617	2.860	3.160	3.373
∞	0.253	0.674	1.282	1.645	1.960	2.326	2.576	2.807	3.090	3.291

SOURCE. Reprinted from Vol. I of Pearson and Hartley (1976), with permission from the *Biometrika* Trustees.

[a] The entries in this table are quantiles w_p of the t distribution for various degrees of freedom. Quantiles w_p for $p < 0.5$ may be computed from the equation

$$w_p = -w_{1-p}$$

Note that $w_{0.50} = 0$ for all degrees of freedom.

TABLE A22   The F Distribution with k_1 and k_2 Degrees of Freedom (0.75 Quantiles)

k_2 \ k_1	1	2	3	4	5	6	7	8	9
1	5.83	7.50	8.20	8.58	8.82	8.98	9.10	9.19	9.26
2	2.57	3.00	3.15	3.23	3.28	3.31	3.34	3.35	3.37
3	2.02	2.28	2.36	2.39	2.41	2.42	2.43	2.44	2.44
4	1.81	2.00	2.05	2.06	2.07	2.08	2.08	2.08	2.08
5	1.69	1.85	1.88	1.89	1.89	1.89	1.89	1.89	1.89
6	1.62	1.76	1.78	1.79	1.79	1.78	1.78	1.78	1.77
7	1.57	1.70	1.72	1.72	1.71	1.71	1.70	1.70	1.69
8	1.54	1.66	1.67	1.66	1.66	1.65	1.64	1.64	1.63
9	1.51	1.62	1.63	1.63	1.62	1.61	1.60	1.60	1.59
10	1.49	1.60	1.60	1.59	1.59	1.58	1.57	1.56	1.56
11	1.47	1.58	1.58	1.57	1.56	1.55	1.54	1.53	1.53
12	1.46	1.56	1.56	1.55	1.54	1.53	1.52	1.51	1.51
13	1.45	1.55	1.55	1.53	1.52	1.51	1.50	1.49	1.49
14	1.44	1.53	1.53	1.52	1.51	1.50	1.49	1.48	1.47
15	1.43	1.52	1.52	1.51	1.49	1.48	1.47	1.46	1.46
16	1.42	1.51	1.51	1.50	1.48	1.47	1.46	1.45	1.44
17	1.42	1.51	1.50	1.49	1.47	1.46	1.45	1.44	1.43
18	1.41	1.50	1.49	1.48	1.46	1.45	1.44	1.43	1.42
19	1.41	1.49	1.49	1.47	1.46	1.44	1.43	1.42	1.41
20	1.40	1.49	1.48	1.47	1.45	1.44	1.43	1.42	1.41
21	1.40	1.48	1.48	1.46	1.44	1.43	1.42	1.41	1.40
22	1.40	1.48	1.47	1.45	1.44	1.42	1.41	1.40	1.39
23	1.39	1.47	1.47	1.45	1.43	1.42	1.41	1.40	1.39
24	1.39	1.47	1.46	1.44	1.43	1.41	1.40	1.39	1.38
25	1.39	1.47	1.46	1.44	1.42	1.41	1.40	1.39	1.38
26	1.38	1.46	1.45	1.44	1.42	1.41	1.39	1.38	1.37
27	1.38	1.46	1.45	1.43	1.42	1.40	1.39	1.38	1.37
28	1.38	1.46	1.45	1.43	1.41	1.40	1.39	1.38	1.37
29	1.38	1.45	1.45	1.43	1.41	1.40	1.38	1.37	1.36
30	1.38	1.45	1.44	1.42	1.41	1.39	1.38	1.37	1.36
40	1.36	1.44	1.42	1.40	1.39	1.37	1.36	1.35	1.34
60	1.35	1.42	1.41	1.38	1.37	1.35	1.33	1.32	1.31
120	1.34	1.40	1.39	1.37	1.35	1.33	1.31	1.30	1.29
∞	1.32	1.39	1.37	1.35	1.33	1.31	1.29	1.28	1.27

TABLE A22 (Continued)

10	12	15	20	24	30	40	60	120	∞
9.32	9.41	9.49	9.58	9.63	9.67	9.71	9.76	9.80	9.85
3.38	3.39	3.41	3.43	3.43	3.44	3.45	3.46	3.47	3.48
2.44	2.45	2.46	2.46	2.46	2.47	2.47	2.47	2.47	2.47
2.08	2.08	2.08	2.08	2.08	2.08	2.08	2.08	2.08	2.08
1.89	1.89	1.89	1.88	1.88	1.88	1.88	1.87	1.87	1.87
1.77	1.77	1.76	1.76	1.75	1.75	1.75	1.74	1.74	1.74
1.69	1.68	1.68	1.67	1.67	1.66	1.66	1.65	1.65	1.65
1.63	1.62	1.62	1.61	1.60	1.60	1.59	1.59	1.58	1.58
1.59	1.58	1.57	1.56	1.56	1.55	1.54	1.54	1.53	1.53
1.55	1.54	1.53	1.52	1.52	1.51	1.51	1.50	1.49	1.48
1.52	1.51	1.50	1.49	1.49	1.48	1.47	1.47	1.46	1.45
1.50	1.49	1.48	1.47	1.46	1.45	1.45	1.44	1.43	1.42
1.48	1.47	1.46	1.45	1.44	1.43	1.42	1.42	1.41	1.40
1.46	1.45	1.44	1.43	1.42	1.41	1.41	1.40	1.39	1.38
1.45	1.44	1.43	1.41	1.41	1.40	1.39	1.38	1.37	1.36
1.44	1.43	1.41	1.40	1.39	1.38	1.37	1.36	1.35	1.34
1.43	1.41	1.40	1.39	1.38	1.37	1.36	1.35	1.34	1.33
1.42	1.40	1.39	1.38	1.37	1.36	1.35	1.34	1.33	1.32
1.41	1.40	1.38	1.37	1.36	1.35	1.34	1.33	1.32	1.30
1.40	1.39	1.37	1.36	1.35	1.34	1.33	1.32	1.31	1.29
1.39	1.38	1.37	1.35	1.34	1.33	1.32	1.31	1.30	1.28
1.39	1.37	1.36	1.34	1.33	1.32	1.31	1.30	1.29	1.28
1.38	1.37	1.35	1.34	1.33	1.32	1.31	1.30	1.28	1.27
1.38	1.36	1.35	1.33	1.32	1.31	1.30	1.29	1.28	1.26
1.37	1.36	1.34	1.33	1.32	1.31	1.29	1.28	1.27	1.25
1.37	1.35	1.34	132	1.31	1.30	1.29	1.28	1.26	1.25
1.36	1.35	1.33	1.32	1.31	1.30	1.28	1.27	1.26	1.24
1.36	1.34	1.33	1.31	1.30	1.29	1.28	1.27	1.25	1.24
1.35	1.34	1.32	1.31	1.30	1.29	1.27	1.26	1.25	1.23
1.35	1.34	1.32	1.30	1.29	1.28	1.27	1.26	1.24	1.23
1.33	1.31	1.30	1.28	1.26	1.25	1.24	1.22	1.21	1.19
1.30	1.29	1.27	1.25	1.24	1.22	1.21	1.19	1.17	1.15
1.28	1.26	1.24	1.22	1.21	1.19	1.18	1.16	1.13	1.10
1.25	1.24	1.22	1.19	1.18	1.16	1.14	1.12	1.08	1.00

TABLE A22   (Continued) (0.90 Quantiles)

k_2 \ k_1	1	2	3	4	5	6	7	8	9
1	39.86	49.50	53.59	55.83	57.24	58.20	58.91	59.44	59.86
2	8.53	9.00	9.16	9.24	9.29	9.33	9.35	9.37	9.38
3	5.54	5.46	5.39	5.34	5.31	5.28	5.27	5.25	5.24
4	4.54	4.32	4.19	4.11	4.05	4.01	3.98	3.95	3.94
5	4.06	3.78	3.62	3.52	3.45	3.40	3.37	3.34	3.32
6	3.78	3.46	3.29	3.18	3.11	3.05	3.01	2.98	2.96
7	3.59	3.26	3.07	2.96	2.88	2.83	2.78	2.75	2.72
8	3.46	3.11	2.92	2.81	2.73	2.67	2.62	2.59	2.56
9	3.36	3.01	2.81	2.69	2.61	2.55	2.51	2.47	2.44
10	3.29	2.92	2.73	2.61	2.52	2.46	2.41	2.38	2.35
11	3.23	2.86	2.66	2.54	2.45	2.39	2.34	2.30	2.27
12	3.18	2.81	2.61	2.48	2.39	2.33	2.28	2.24	2.21
13	3.14	2.76	2.56	2.43	2.35	2.28	2.23	2.20	2.16
14	3.10	2.73	2.52	2.39	2.31	2.24	2.19	2.15	2.12
15	3.07	2.70	2.49	2.36	2.27	2.21	2.16	2.12	2.09
16	3.05	2.67	2.46	2.33	2.24	2.18	2.13	2.09	2.06
17	3.03	2.64	2.44	2.31	2.22	2.15	2.10	2.06	2.03
18	3.01	2.62	2.42	2.29	2.20	2.13	2.08	2.04	2.00
19	2.99	2.61	2.40	2.27	2.18	2.11	2.06	2.02	1.98
20	2.97	2.59	2.38	2.25	2.16	2.09	2.04	2.00	1.96
21	2.96	2.57	2.36	2.23	2.14	2.08	2.02	1.98	1.95
22	2.95	2.56	2.35	2.22	2.13	2.06	2.01	1.97	1.93
23	2.94	2.55	2.34	2.21	2.11	2.05	1.99	1.95	1.92
24	2.93	2.54	2.33	2.19	2.10	2.04	1.98	1.94	1.91
25	2.92	2.53	2.32	2.18	2.09	2.02	1.97	1.93	1.89
26	2.91	2.52	2.31	2.17	2.08	2.01	1.96	1.92	1.88
27	2.90	2.51	2.30	2.17	2.07	2.00	1.95	1.91	1.87
28	2.89	2.50	2.29	2.16	2.06	2.00	1.94	1.90	1.87
29	2.89	2.50	2.28	2.15	2.06	1.99	1.93	1.89	1.86
30	2.88	2.49	2.28	2.14	2.05	1.98	1.93	1.88	1.85
40	2.84	2.44	2.23	2.09	2.00	1.93	1.87	1.83	1.79
60	2.79	2.39	2.18	2.04	1.95	1.87	1.82	1.77	1.74
120	2.75	2.35	2.13	1.99	1.90	1.82	1.77	1.72	1.68
∞	2.71	2.30	2.08	1.94	1.85	1.77	1.72	1.67	1.63

TABLE A22   (Continued)

10	12	15	20	24	30	40	60	120	∞
60.19	60.71	61.22	61.74	62.00	62.26	62.53	62.79	63.06	63.33
9.39	9.41	9.42	9.44	9.45	9.46	9.47	9.47	9.48	9.49
5.23	5.22	5.20	5.18	5.18	5.17	5.16	5.15	5.14	5.13
3.92	3.90	3.87	3.84	3.83	3.82	3.80	3.79	3.78	3.76
3.30	3.27	3.24	3.21	3.19	3.17	3.16	3.14	3.12	3.10
2.94	2.90	2.87	2.84	2.82	2.80	2.78	2.76	2.74	2.72
2.70	2.67	2.63	2.59	2.58	2.56	2.54	2.51	2.49	2.47
2.54	2.50	2.46	2.42	2.40	2.38	2.36	2.34	2.32	2.29
2.42	2.38	2.34	2.30	2.28	2.25	2.23	2.21	2.18	2.16
2.32	2.28	2.24	2.20	2.18	2.16	2.13	2.11	2.08	2.06
2.25	2.21	2.17	2.12	2.10	2.08	2.05	2.03	2.00	1.97
2.19	2.15	2.10	2.06	2.04	2.01	1.99	1.96	1.93	1.90
2.14	2.10	2.05	2.01	1.98	1.96	1.93	1.90	1.88	1.85
2.10	2.05	2.01	1.96	1.94	1.91	1.89	1.86	1.83	1.80
2.06	2.02	1.97	1.92	1.90	1.87	1.85	1.82	1.79	1.76
2.03	1.99	1.94	1.89	1.87	1.84	1.81	1.78	1.75	1.72
2.00	1.96	1.91	1.86	1.84	1.81	1.78	1.75	1.72	1.69
1.98	1.93	1.89	1.84	1.81	1.78	1.75	1.72	1.69	1.68
1.96	1.91	1.86	1.81	1.79	1.76	1.73	1.70	1.67	1.63
1.94	1.89	1.84	1.79	1.77	1.74	1.71	1.68	1.64	1.61
1.92	1.87	1.83	1.78	1.75	1.72	1.69	1.66	1.62	1.59
1.90	1.86	1.81	1.76	1.73	1.70	1.67	1.64	1.60	1.57
1.89	1.84	1.80	1.74	1.72	1.69	1.66	1.62	1.59	1.55
1.88	1.83	1.78	1.73	1.70	1.67	1.64	1.61	1.57	1.53
1.87	1.82	1.77	1.72	1.69	1.66	1.63	1.59	1.56	1.52
1.86	1.81	1.76	1.71	1.68	1.65	1.61	1.58	1.54	1.50
1.85	1.80	1.75	1.70	1.67	1.64	1.60	1.57	1.53	1.49
1.84	1.79	1.74	1.69	1.66	1.63	1.59	1.56	1.52	1.48
1.83	1.78	1.73	1.68	1.65	1.62	1.58	1.55	1.51	1.47
1.82	1.77	1.72	1.67	1.64	1.61	1.57	1.54	1.50	1.46
1.76	1.71	1.66	1.61	1.57	1.54	1.51	1.47	1.42	1.38
1.71	1.66	1.60	1.54	1.51	1.48	1.44	1.40	1.35	1.29
1.65	1.60	1.55	1.48	1.45	1.41	1.37	1.32	1.26	1.19
1.60	1.55	1.49	1.42	1.38	1.34	1.30	1.24	1.17	1.00

TABLE A22 (Continued) (0.95 Quantiles)

k_2 \ k_1	1	2	3	4	5	6	7	8	9
1	161.4	199.5	215.7	224.6	230.2	234.0	236.8	238.9	240.5
2	18.51	19.00	19.16	19.25	19.30	19.33	19.35	19.37	19.38
3	10.13	9.55	9.28	9.12	9.01	8.94	8.89	8.85	8.81
4	7.71	6.94	6.59	6.39	6.26	6.16	6.09	6.04	6.00
5	6.61	5.79	5.41	5.19	5.05	4.95	4.88	4.82	4.77
6	5.99	5.14	4.76	4.53	4.39	4.28	4.21	4.15	4.10
7	5.59	4.74	4.35	4.12	3.97	3.87	3.79	3.73	3.68
8	5.32	4.46	4.07	3.84	3.69	3.58	3.50	3.44	3.39
9	5.12	4.26	3.86	3.63	3.48	3.37	3.29	3.23	3.18
10	4.96	4.10	3.71	3.48	3.33	3.22	3.14	3.07	3.02
11	4.84	3.98	3.59	3.36	3.20	3.09	3.01	2.95	2.90
12	4.75	3.89	3.49	3.26	3.11	3.00	2.91	2.85	2.80
13	4.67	3.81	3.41	3.18	3.03	2.92	2.83	2.77	2.71
14	4.60	3.74	3.34	3.11	2.96	2.85	2.76	2.70	2.65
15	4.54	3.68	3.29	3.06	2.90	2.79	2.71	2.64	2.59
16	4.49	3.63	3.24	3.01	2.85	2.74	2.66	2.59	2.54
17	4.45	3.59	3.20	2.96	2.81	2.70	2.61	2.55	2.49
18	4.41	3.55	3.16	2.93	2.77	2.66	2.58	2.51	2.46
19	4.38	3.52	3.13	2.90	2.74	2.63	2.54	2.48	2.42
20	4.35	3.49	3.10	2.87	2.71	2.60	2.51	2.45	2.30
21	4.32	3.47	3.07	2.84	2.68	2.57	2.49	2.42	2.37
22	4.30	3.44	3.05	2.82	2.66	2.55	2.46	2.40	2.34
23	4.28	3.42	3.03	2.80	2.64	2.53	2.44	2.37	2.32
24	4.26	3.40	3.01	2.78	2.62	2.51	2.42	2.36	2.30
25	4.24	3.39	2.99	2.76	2.60	2.49	2.40	2.34	2.28
26	4.23	3.37	2.98	2.74	2.59	2.47	2.39	2.32	2.27
27	4.21	3.35	2.96	2.73	2.57	2.46	2.37	2.31	2.25
28	4.20	3.34	2.95	2.71	2.56	2.45	2.36	2.29	2.24
29	4.18	3.33	2.93	2.70	2.55	2.43	2.35	2.28	2.22
30	4.17	3.32	2.92	2.69	2.53	2.42	2.33	2.27	2.21
40	4.08	3.23	2.84	2.61	2.45	2.34	2.25	2.18	2.12
60	4.00	3.15	2.76	2.53	2.37	2.25	2.17	2.10	2.04
120	3.92	3.07	2.68	2.45	2.29	2.17	2.09	2.02	1.96
∞	3.84	3.00	2.60	2.37	2.21	2.10	2.01	1.94	1.88

TABLE A22 (Continued)

10	12	15	20	24	30	40	60	120	∞
241.9	243.9	245.9	248.0	249.1	250.1	251.1	252.2	253.3	254.3
19.40	19.41	19.43	19.45	19.45	19.46	19.47	19.48	19.49	19.50
8.79	8.74	8.70	8.66	8.64	8.62	8.59	8.57	8.55	8.53
5.96	5.91	5.86	5.80	5.77	5.75	5.72	5.69	5.66	5.63
4.74	4.68	4.62	4.56	4.53	4.50	4.46	4.43	4.40	4.36
4.06	4.00	3.94	3.87	3.84	3.81	3.77	3.74	3.70	3.67
3.64	3.57	3.51	3.44	3.41	3.38	3.34	3.30	3.27	3.23
3.35	3.28	3.22	3.15	3.12	3.08	3.04	3.01	2.97	2.93
3.14	3.07	3.01	2.94	2.90	2.86	2.83	2.79	2.75	2.71
2.98	2.91	2.85	2.77	2.74	2.70	2.66	2.62	2.58	2.54
2.85	2.79	2.72	2.65	2.61	2.57	2.53	2.49	2.45	2.40
2.75	2.69	2.62	2.54	2.51	2.47	2.43	2.38	2.34	2.30
2.67	2.60	2.53	2.46	2.42	2.38	2.34	2.30	2.25	2.21
2.60	2.53	2.46	2.39	2.35	2.31	2.27	2.22	2.18	2.13
2.54	2.48	2.40	2.33	2.29	2.25	2.20	2.16	2.11	2.07
2.49	2.42	2.35	2.28	2.24	2.19	2.15	2.11	2.06	2.01
2.45	2.38	2.31	2.23	2.19	2.15	2.10	2.06	2.01	1.96
2.41	2.34	2.27	2.19	2.15	2.11	2.06	2.02	1.97	1.92
2.38	2.31	2.23	2.16	2.11	2.07	2.03	1.98	1.93	1.88
2.35	2.28	2.20	2.12	2.08	2.04	1.99	1.95	1.90	1.84
2.32	2.25	2.18	2.10	2.05	2.01	1.96	1.92	1.87	1.81
2.30	2.23	2.15	2.07	2.03	1.98	1.94	1.89	1.84	1.78
2.27	2.20	2.13	2.05	2.01	1.96	1.91	1.86	1.81	1.76
2.25	2.18	2.11	2.03	1.98	1.94	1.89	1.84	1.79	1.73
2.24	2.16	2.09	2.01	1.96	1.92	1.87	1.82	1.77	1.71
2.22	2.15	2.07	1.99	1.95	1.90	1.85	1.80	1.75	1.69
2.20	2.13	2.06	1.97	1.93	1.88	1.84	1.79	1.73	1.67
2.19	2.12	2.04	1.96	1.91	1.87	1.82	1.77	1.71	1.65
2.18	2.10	2.03	1.94	1.90	1.85	1.81	1.75	1.70	1.64
2.16	2.09	2.01	1.93	1.89	1.84	1.79	1.74	1.68	1.62
2.08	2.00	1.92	1.84	1.79	1.74	1.69	1.64	1.58	1.51
1.99	1.92	1.84	1.75	1.70	1.65	1.59	1.53	1.47	1.39
1.91	1.83	1.75	1.66	1.61	1.55	1.50	1.43	1.35	1.25
1.83	1.75	1.67	1.57	1.52	1.46	1.39	1.32	1.22	1.00

TABLE A22 (Continued) (0.975 Quantiles)

k_2 \ k_1	1	2	3	4	5	6	7	8	9
1	647.8	799.5	864.2	899.6	921.8	937.1	948.2	956.7	963.3
2	38.51	39.00	39.17	39.25	39.30	39.33	39.36	39.37	39.39
3	17.44	16.04	15.44	15.10	14.88	14.73	14.62	14.54	14.47
4	12.22	10.65	9.98	9.60	9.36	9.20	9.07	8.98	8.90
5	10.01	8.43	7.76	7.39	7.15	6.98	6.85	6.76	6.68
6	8.81	7.26	6.60	6.23	5.99	5.82	5.70	5.60	5.52
7	8.07	6.54	5.89	5.52	5.29	5.12	4.99	4.90	4.82
8	7.57	6.06	5.42	5.05	4.82	4.65	4.53	4.43	4.36
9	7.21	5.71	5.08	4.72	4.48	4.32	4.20	4.10	4.03
10	6.94	5.46	4.83	4.47	4.24	4.07	3.95	3.85	3.78
11	6.72	5.26	4.63	4.28	4.04	3.88	3.76	3.66	3.59
12	6.55	5.10	4.47	4.12	3.89	3.73	3.61	3.51	3.44
13	6.41	4.97	4.35	4.00	3.77	3.60	3.48	3.39	3.31
14	6.30	4.86	4.24	3.89	3.66	3.50	3.38	3.29	3.21
15	6.20	4.77	4.15	3.80	3.58	3.41	3.29	3.20	3.12
16	6.12	4.69	4.08	3.73	3.50	3.34	3.22	3.12	3.05
17	6.04	4.62	4.01	3.66	3.44	3.28	3.16	3.06	2.98
18	5.98	4.56	3.95	3.61	3.38	3.22	3.10	3.01	2.93
19	5.92	4.51	3.90	3.56	3.33	3.17	3.05	2.96	2.88
20	5.87	4.46	3.86	3.51	3.29	3.13	3.01	2.91	2.84
21	5.83	4.42	3.82	3.48	3.25	3.09	2.97	2.87	2.80
22	5.79	4.38	3.78	3.44	3.22	3.05	2.93	2.84	2.76
23	5.75	4.35	3.75	3.41	3.18	3.02	2.90	2.81	2.73
24	5.72	4.32	3.72	3.38	3.15	2.99	2.87	2.78	2.70
25	5.69	4.29	3.69	3.35	3.13	2.97	2.85	2.75	2.68
26	5.66	4.27	3.67	3.33	3.10	2.94	2.82	2.73	2.65
27	5.63	4.24	3.65	3.31	3.08	2.92	2.80	2.71	2.63
28	5.61	4.22	3.63	3.29	3.06	2.90	2.78	2.69	2.61
29	5.59	4.20	3.61	3.27	3.04	2.88	2.76	2.67	2.59
30	5.57	4.18	3.59	3.25	3.03	2.87	2.75	2.65	2.57
40	5.42	4.05	3.46	3.13	2.90	2.74	2.62	2.53	2.45
60	5.29	3.93	3.34	3.01	2.79	2.63	2.51	2.41	2.33
120	5.15	3.80	3.23	2.89	2.67	2.52	2.39	2.30	2.22
∞	5.02	3.69	3.12	2.79	2.57	2.41	2.29	2.19	2.11

TABLE A22   (Continued)

10	12	15	20	24	30	40	60	120	∞
968.6	976.7	984.9	993.1	997.2	1001	1006	1010	1014	1018
39.40	39.41	39.43	39.45	39.46	39.46	39.47	39.48	39.49	39.50
14.42	14.34	14.25	14.17	14.12	14.08	14.04	13.99	13.95	13.90
8.84	8.75	8.66	8.56	8.51	8.46	8.41	8.36	8.31	8.26
6.62	6.52	6.43	6.33	6.28	6.23	6.18	6.12	6.07	6.02
5.46	5.37	5.27	5.17	5.12	5.07	5.01	4.96	4.90	4.85
4.76	4.67	4.57	4.47	4.42	4.36	4.31	4.25	4.20	4.14
4.30	4.20	4.10	4.00	3.95	3.89	3.84	3.78	3.73	3.67
3.96	3.87	3.77	3.67	3.61	3.56	3.51	3.45	3.39	3.33
3.72	3.62	3.52	3.42	3.37	3.31	3.26	3.20	3.14	3.08
3.53	3.43	3.33	3.23	3.17	3.12	3.06	3.00	2.94	2.88
3.37	3.28	3.18	3.07	3.02	2.96	2.91	2.85	2.79	2.72
3.25	3.15	3.05	2.95	2.89	2.84	2.78	2.72	2.66	2.60
3.15	3.05	2.95	2.84	2.79	2.73	2.67	2.61	2.55	2.49
3.06	2.96	2.80	2.76	2.70	2.64	2.59	2.52	2.46	2.40
2.99	2.89	2.79	2.68	2.63	2.57	2.51	2.45	2.38	2.32
2.92	2.82	2.72	2.62	2.56	2.50	2.44	2.38	2.32	2.25
2.87	2.77	2.67	2.56	2.50	2.44	2.38	2.32	2.26	2.19
2.82	2.72	2.62	2.51	2.45	2.39	2.33	2.27	2.20	2.13
2.77	2.68	2.57	2.46	2.41	2.35	2.29	2.22	2.16	2.09
2.73	2.64	2.53	2.42	2.37	2.31	2.25	2.18	2.11	2.04
2.70	2.60	2.50	2.39	2.33	2.27	2.21	2.14	2.08	2.00
2.67	2.57	2.47	2.36	2.30	2.24	2.18	2.11	2.04	1.97
2.64	2.54	2.44	2.33	2.27	2.21	2.15	2.08	2.01	1.94
2.61	2.51	2.41	2.30	2.24	2.18	2.12	2.05	1.98	1.91
2.59	2.49	2.39	2.28	2.22	2.16	2.09	2.03	1.95	1.88
2.57	2.47	2.36	2.25	2.19	2.13	2.07	2.00	1.93	1.85
2.55	2.45	2.34	2.23	2.17	2.11	2.05	1.98	1.91	1.83
2.53	2.43	2.32	2.21	2.15	2.09	2.03	1.96	1.89	1.81
2.51	2.41	2.31	2.20	2.14	2.07	2.01	1.94	1.87	1.79
2.39	2.29	2.18	2.07	2.01	1.94	1.88	1.80	1.72	1.64
2.27	2.17	2.06	1.94	1.88	1.82	1.74	1.67	1.58	1.48
2.16	2.05	1.94	1.82	1.76	1.69	1.61	1.53	1.43	1.31
2.05	1.94	1.83	1.71	1.64	1.57	1.48	1.39	1.27	1.00

TABLE A22   (Continued) (0.99 Quantiles)

k_1 / k_2	1	2	3	4	5	6	7	8	9
1	4052	4999.5	5403	5625	5764	5859	5928	5981	6022
2	98.50	99.00	99.17	99.25	99.30	99.33	99.36	99.37	99.39
3	34.12	30.82	29.46	28.71	28.24	27.91	27.67	27.49	27.35
4	21.20	18.00	16.69	15.98	15.52	15.21	14.98	14.80	14.66
5	16.26	13.27	12.06	11.39	10.97	10.67	10.46	10.29	10.16
6	13.75	10.92	9.78	9.15	8.75	8.47	8.26	8.10	7.98
7	12.25	9.55	8.45	7.85	7.46	7.19	6.99	6.84	6.72
8	11.26	8.65	7.59	7.01	6.63	6.37	6.18	6.03	5.91
9	10.56	8.02	6.99	6.42	6.06	5.80	5.61	5.47	5.35
10	10.04	7.56	6.55	5.99	5.64	5.39	5.20	5.06	4.94
11	9.65	7.21	6.22	5.67	5.32	5.07	4.89	4.74	4.63
12	9.33	6.93	5.95	5.41	5.06	4.82	4.64	4.50	4.39
13	9.07	6.70	5.74	5.21	4.86	4.62	4.44	4.30	4.19
14	8.86	6.51	5.56	5.04	4.69	4.46	4.28	4.14	4.03
15	8.68	6.36	5.42	4.89	4.56	4.32	4.14	4.00	3.89
16	8.53	6.23	5.29	4.77	4.44	4.20	4.03	3.89	3.78
17	8.40	6.11	5.18	4.67	4.34	4.10	3.93	3.79	3.68
18	8.29	6.01	5.09	4.58	4.25	4.01	3.84	3.71	3.60
19	8.18	5.93	5.01	4.50	4.17	3.94	3.77	3.63	3.52
20	8.10	5.85	4.94	4.43	4.10	3.87	3.70	3.56	3.46
21	8.02	5.78	4.87	4.37	4.04	3.81	3.64	3.51	3.40
22	7.95	5.72	4.82	4.31	3.99	3.76	3.59	3.45	3.35
23	7.88	5.66	4.76	4.26	3.94	3.71	3.54	3.41	3.30
24	7.82	5.61	4.72	4.22	3.90	3.67	3.50	3.36	3.26
25	7.77	5.57	4.68	4.18	3.85	3.63	3.46	3.32	3.22
26	7.72	5.53	4.64	4.14	3.82	3.59	3.42	3.29	3.18
27	7.68	5.49	4.60	4.11	3.78	3.56	3.39	3.26	3.15
28	7.64	5.45	4.57	4.07	3.75	3.53	3.36	3.23	3.12
29	7.60	5.42	4.54	4.04	3.73	3.50	3.33	3.20	3.09
30	7.56	5.39	4.51	4.02	3.70	3.47	3.30	3.17	3.07
40	7.31	5.18	4.31	3.83	3.51	3.29	3.12	2.99	2.89
60	7.08	4.98	4.13	3.65	3.34	3.12	2.95	2.82	2.72
120	6.85	4.79	3.95	3.48	3.17	2.96	2.79	2.66	2.56
∞	6.63	4.61	3.78	3.32	3.02	2.80	2.64	2.51	2.41

TABLE A22 (Continued)

10	12	15	20	24	30	40	60	120	∞
6056	6106	6157	6209	6235	6261	6287	6313	6339	6366
99.40	99.42	99.43	99.45	99.46	99.47	99.47	99.48	99.49	99.50
27.23	27.05	26.87	26.69	26.60	26.50	26.41	26.32	26.22	26.13
14.55	14.37	14.20	14.02	13.93	13.84	13.75	13.65	13.56	13.46
10.05	9.89	9.72	9.55	9.47	9.38	9.29	9.20	9.11	9.02
7.87	7.72	7.56	7.40	7.31	7.23	7.14	7.06	6.97	6.88
6.62	6.47	6.31	6.16	6.07	5.99	5.91	5.82	5.74	5.65
5.81	5.67	5.52	5.36	5.28	5.20	5.12	5.03	4.95	4.86
5.26	5.11	4.96	4.81	4.73	4.65	4.57	4.48	4.40	4.31
4.85	4.71	4.56	4.41	4.33	4.25	4.17	4.08	4.00	3.91
4.54	4.40	4.25	4.10	4.02	3.94	3.86	3.78	3.69	3.60
4.30	4.16	4.01	3.86	3.78	3.70	3.62	3.54	3.45	3.36
4.10	3.96	3.82	3.66	3.59	3.51	3.43	3.34	3.25	3.17
3.94	3.80	3.66	3.51	3.43	3.35	3.27	3.18	3.09	3.00
3.80	3.67	3.52	3.37	3.29	3.21	3.13	3.05	2.96	2.87
3.69	3.55	3.41	3.26	3.18	3.10	3.02	2.93	2.84	2.75
3.59	3.46	3.31	3.16	3.08	3.00	2.92	2.83	2.75	2.65
3.51	3.37	3.23	3.08	3.00	2.92	2.84	2.75	2.66	2.57
3.43	3.30	3.15	3.00	2.92	2.84	2.76	2.67	2.58	2.49
3.37	3.23	3.09	2.94	2.86	2.78	2.69	2.61	2.52	2.42
3.31	3.17	3.03	2.88	2.80	2.72	2.64	2.55	2.46	2.36
3.26	3.12	2.98	2.83	2.75	2.67	2.58	2.50	2.40	2.31
3.21	3.07	2.93	2.78	2.70	2.62	2.54	2.45	2.35	2.26
3.17	3.03	2.89	2.74	2.66	2.58	2.49	2.40	2.31	2.21
3.13	2.99	2.85	2.70	2.62	2.54	2.45	2.36	2.27	2.17
3.09	2.96	2.81	2.66	2.58	2.50	2.42	2.33	2.23	2.13
3.06	2.93	2.78	2.63	2.55	2.47	2.38	2.29	2.20	2.10
3.03	2.90	2.75	2.60	2.52	2.44	2.35	2.26	2.17	2.06
3.00	2.87	2.73	2.57	2.49	2.41	2.33	2.23	2.14	2.03
2.98	2.84	2.70	2.55	2.47	2.39	2.30	2.21	2.11	2.01
2.80	2.66	2.52	2.37	2.29	2.20	2.11	2.02	1.92	1.80
2.63	2.50	2.35	2.20	2.12	2.03	1.94	1.84	1.73	1.60
2.47	2.34	2.19	2.03	1.95	1.86	1.76	1.66	1.53	1.38
2.32	2.18	2.04	1.88	1.79	1.70	1.59	1.47	1.32	1.00

ANSWERS TO
ODD-NUMBERED EXERCISES

CHAPTER 1

Sec. 1.1: 1. 10,000; 3. 24; 5. 220; 7. 105; 9. 14; 11. 232/729.

Sec. 1.2: 1. *HHH, HHT, HTH, THH, TTH, THT, HTT, TTT*; 3. 0.85; 5. $9/2^8$; 7. 0.007125; 9. 0.06; 11. 1/7; 13. 1/4; 15. 4/15; 17. (a) 6/1000 for 3 different digits, 3/1000 if one digit appears twice, or 1/1000 if one digit appears 3 times, (b) no.

Sec. 1.3: 1. (a) 1/729, (b) 64/729, (c) 0, (d) 496/729, (e) 0, (f) 1; 3. (a) 1/128, (b) 4/128, (c) 3/128, (d) 1/128, (e) 0, (f) 1/128, (g) 12/128, (h) 1; 5. (a) *A, B, C, D, E, F*, (b) $P(A) = 1/6$, $P(B) = 1/6, \ldots, P(F) = 1/6$, (c) $X(A) = 1, X(B) = 2$, etc.

Sec. 1.4: 1. (a) 7/6, (b) 41/36, (c) 29/6, (d) 1, (e) 0.5, (f) 1; 3. (a) 1/2, (b) 1/2, (c) 1/4, (d) 1, (e) 0, (f) 1/2, (g) 1/2, (h) yes; 5. 2211; 7. (a) 7/2, (b) 35/12, (c) 56/3; 9. $\mu = 11$, $\sigma^2 = 44/3$, range = 16; 11. (a) $P(X = 0) = 2/3, P(X = 1) = 1/3$, (b) 0, (c) $-1/9$, (d) $-1/2$; 13. $\mu = 13.5$, $\sigma^2 = 11.25$.

Sec. 1.5: 1. (a) 0.5, (b) 0.975, (c) 0.159, (d) 0.682, (e) 0.5, (f) 0.6745; 3. 0.001; 5. 207; 7. (a) 9.488, (b) 15.51, (c) 233.7; 9. 0.266; 11. <0.0005.

CHAPTER 2

Sec. 2.1: 1. (a) all high schools in the U.S.A., (b) all high schools in the Washington D.C. area, (c) 75,287,520, (d) 1/75,287,520; 3. (a) ordinal, (b) the sum of the points, for awarding the trophy; 5. all nominal; 7. no.

Sec. 2.2: 1. (b) $4300, (c) $8600, (d) $97,296,000, (e) $9863.87; 3. 12.6, 12.1, 0.5; 5. (number of sample values $\leq c)/n$, 1/5; 7. $1050 to $1262; 9. 0.05 to 0.55; 11. $\hat{P}(x) = 1$ to $x = 187$, 6/7 to $x = 196$, 9/14 to $x = 206$, 3/7 to $x = 210$, 3/14 to $x = 273$, then = 0.

Sec. 2.3: 1. (a) H_0: The new method is no better than the existing one, H_1: The new method is better than the existing one, (b) the probability of deciding the new method is better, when it is not, (c) the probability of deciding the new method is better, when it is better; 3. (a) Fertilizer B is not as good as Fertilizer A, (b) My opponent is cheating, (c) The occurrence of sun spots does affect the economic cycle; 5. $\alpha = 0.1817$, power $= 0.33696$; 7. 0.10, 0.91.

Sec. 2.4: 1. (a) $1/32$, (b) p^5, $p > 0.5$, (d) yes; 3. yes; 5. (a) 0.57, (b) 1.75.

CHAPTER 3

Sec. 3.1: 1. no, $T = 0$, $p = 0.036$; 3. (a) maybe, $p = 0.055$, (b) (0.15, 0.27); 5. (0.057, 0.437); 7. (a) (0.087, 0.491), (b) (0.060, 0.440); 9. $p = 0.991$; 11. (a) (0.544, 0.896), (b) (0.591, 0.909).

Sec. 3.2: 1. $T_1 = 6$, $T_2 = 4$, $p = 0.1154$; 3. $T_1 = 4$, $p = 0.2375$; 5. no, $T_1 = 90$, $T_2 = 90$, $p = 0.022$; 7. $T_1 = 10$, $p < 0.00001$.

Sec. 3.3: 1. (a) 77, (b) 76.35, so use 77; 3. (a) 15, (b) 14.198, so use 15; 5. 26; 7. (a) 94.6%, 95.6%, (b) 91.2%, 92.4%; 9. 210.

Sec. 3.4: 1. $T = 1$, $p = 0.1094$; 3. $T = 23$, $p < 0.0002$; 5. $T = 60$, $p = 0.046$.

Sec. 3.5: 1. $T_1 = 0.5455$, $p > 0.25$; 3. $T = 7$, $p = 0.0078$; 5. $T = 4$, $p = 0.006$; 7. $T = 6$, $p = 0.0312$; 9. $T_2 = 3$, $n = 17$, $p = 0.0128$.

CHAPTER 4

Sec. 4.1: 1. $T_1 = 0.800$, $p = 0.424$; 3. $T_1 = 0.935$, $p = 0.35$; 5. $T_2 = 1$, $p = 0.005$; 7. (a) $T_4 = 0.8482$ with cont. corr., $p = 0.198$, (b) $T_5 = 1.0599$, $p = 0.145$, (c) T_5.

Sec. 4.2: 1. 2×3 with equal column totals ($R_1 = 8$, $R_2 = 10$), $T = 6.3$, $p < 0.05$; 3. $T = 6.81$, $0.05 < p < 0.10$, fixed marginal totals; 5. Combine C with B, $T = 2.496$, $0.10 < p < 0.25$; 7. $T = 6.605$, $p > 0.25$.

Sec. 4.3: 1. exact $T = 1.10$, approximate $T = 1.20$, $p > 0.25$; 3. $T = 34.875$, $p < 0.001$; 5. $T = 5.067$, $0.05 < p < 0.10$.

Sec. 4.4: 1. (a) 6.34, (b) 0.178, (c) 0.0317, (d) 0.244, (e) 0.0634, (f) 0.1780; 3. (a) $R_5 = 0.149$, $p = 0.117$, (b) 0.3636, (c) 0.

Sec. 4.5: 1. The answers depend on the choice of intervals; 3. from Sec. 4.2, $T = 19.03$, $p > 0.25$; 5. $T = 4.247$, $0.10 < p < 0.25$.

Sec. 4.6: 1. (a) $T = 4$, $0.025 < p < 0.05$, (b) $T = 4$, $0.025 < p < 0.05$, (c) $T = 2.67$, $0.10 < p < 0.25$; 3. $T = 10.4$, $0.001 < p < 0.005$.

CHAPTER 5

Sec. 5.1:   1. $T = 117$, $T' = 45$, $p < 0.001$; 3. $(0, 14)$; 5. $T = 24$, $T' = 12$, $0.05 < p < 0.10$, exact $p = 4/70 = 0.057$.

Sec. 5.2:   1. $T = 8.40$, $p < 0.01$ from Table A8, differences between A and B, A and C at $\alpha = 0.05$; 3. $T = 40.83$, $p < 0.001$, differences between minimum tillage and terrace, minimum tillage and other, contour and terrace, contour and other at $\alpha = 0.05$; 5. $T = 27.99$, $p < 0.001$, differences between all pairs except B and E at $\alpha = 0.05$.

Sec. 5.3:   1. $T = 1107.5$, $0.05 < p < 0.10$; 3. $T_2 = 5.15$, $0.05 < p < 0.10$.

Sec. 5.4:   1. (a) -0.603, (b) -0.511, (c) $0.05 < p < 0.10$, (d) $0.02 < p < 0.05$; 3. (a) $\rho = -0.9025$, $p < 0.001$, (b) $T = -57$, $p < 0.005$; 5. (a) 0.8721, $0.025 < p < 0.05$, (b) 0.7778, $0.025 < p < 0.05$; 7. $N_c = 98$, $N_d = 27$, $\tau = 0.5680$, $p = 0.014$.

Sec. 5.5:   1. (b) $a = 8.69$, $b = 13.6$, (d) $\rho = -0.6606$, $0.02 < p < 0.05$, (e) $(S^{(12)}, S^{(34)}) = (10.7, 17.4)$; 3. $\rho = -0.8373$, $p < 0.001$.

Sec. 5.6:   1. (a) It might be linear in the middle; it certainly looks monotonic, (b) 46.14, (c) 53.51, (d) $R(y) = 0.4333 + 0.9212 R(x)$, line segments connecting $(X, Y) = (0.50, 0)$ to $(0.85, 0)$ to $(1, 2.76)$ to $(1.5, 11.97)$ to $(1.94, 20)$ to $(2, 21.58)$ to $(2.5, 33.86)$ to $(2.75, 40)$ to $(3, 46.14)$ to $(3.5, 58.42)$ to $(3.56, 60)$ to $(4, 76.06)$ to $(4.11, 80)$ to $(4.5, 89.66)$ to $(4.92, 100)$ to $(5.00, 100)$; 3. $R(y) = 3.9375 + 0.5625R(x)$, line segments connecting $(X, Y) = (4.0, 0)$ to $(4.36, 0)$ to $(4.4, 0.01)$ to $(4.7, 0.07)$ to $(5.1, 0.14)$ to $(9.3, 0.21)$ to $(10.7, 0.27)$ to $(11.2, 0.34)$ to $(13.7, 0.35)$ to $(13.9, 0.47)$ to $(15.0, 0.54)$ to $(15.8, 0.60)$ to $(17.0, 0.67)$ to $(18.1, 0.74)$ to $(19.7, 0.80)$ to $(19.9, 0.87)$ to $(20.7, 0.93)$ to $(21.6, 1.0)$.

Sec. 5.7:   1. $T = 2.988$, $p = 0.003$; 3. $T^+ = 39$, $p \approx 0.50$; 5. 5.5 to 7.45, 6.5; 7. (a) -18 to 23, (b) $61/64 = 95.3\%$.

Sec. 5.8:   1. (a) $T_3 = 4.43$, $0.01 < p < 0.025$, differences between spring and the other three seasons, (b) $T_2 = 2.946$, $0.05 < p < 0.10$, (c) the larger hospitals received more weight in the first test; 3. $T_2 = 8.328$, $0.05 < p < 0.10$; 5. $T_5 = 4.736$, $p < 0.0001$.

Sec. 5.9:   1. $T_2 = 7.54$, $p < 0.01$, tires 3 and 6 are better than the others, and tires 5 and 7 are better than tire 2; 3. $T_2 = 10.30$, $p < 0.01$, team 5 is better than teams 1, 2, and 4, team 3 is better than teams 1 and 2, and team 4 is better than team 2.

Sec. 5.10:   1. $T_1 = 7.97$, $p = 0.020$, same differences as with K-W test; 3. $T_2 = -2.938$, $p = 0.004$; 5. $T_3 = 1.7140$, $p = 0.043$; 7. $\rho = -0.6925$, larger in absolute value, $p = 0.038$, slightly smaller.

Sec. 5.11:   1. $T_1 = 5$, $p = 6/252 = 0.0238$; 3. $T_2 = 7$, $p = 8/256 = 0.03125$.

CHAPTER 6

Sec. 6.1:   1. $S_n(x) \pm 0.509$; 3. $T = 0.425$, $p > 0.20$; 5. $T = 0.492$, $p < 0.01$.

Sec. 6.2:   1. $T_1 = 0.103$, $p > 0.20$; 3. $T_3 = 0.9569$, $p > 0.50$; 5. $T_2 = 0.2155$, $p \approx 0.20$; 7. $Z = -1.9083$, $p = 0.028$.

Sec. 6.3:   1. $T_1^+ \doteq 0.8$, $p = 0.40$; 3. $T_2 = 0.556$, $0.01 < p < 0.05$; 5. $T_1 = \frac{6}{16}$, $p = 0.052$.

WILEY SERIES IN PROBABILITY
AND MATHEMATICAL STATISTICS

ESTABLISHED BY WALTER A. SHEWHART AND SAMUEL S. WILKS

Editors

Ralph A. Bradley   David G. Kendall
J. Stuart Hunter   Geoffrey S. Watson

Probability and Mathematical Statistics

573

PURI and SEN · Nonparametric Methods in Multivariate Analysis

RAGHAVARAO · Constructions and Combinatorial Problems in Design of Experiments

RANDLES and WOLFE · Introduction to the Theory of Nonparametric Statistics

RAO · Linear Statistical Inference and Its Applications, *Second Edition*

ROHATGI · An Introduction to Probability Theory and Mathematical Statistics

SCHEFFE · The Analysis of Variance

SEBER · Linear Regression Analysis

SERFLING · Approximation Theorems of Mathematical Statistics

WILKS · Mathematical Statistics

WILLIAMS · Diffusions, Markov Processes, and Martingales, Volume I: Foundations

ZACKS · The Theory of Statistical Inference

Applied Probability and Statistics

ANDERSON, AUQUIER, HAUCK, OAKES, VANDAELE, and WEISBERG ·-Statistical Methods in Comparative Studies

BAILEY · The Elements of Stochastic Processes with Applications to the Nature Sciences

BAILEY · Mathematics, Statistics and Systems for Health

BARNETT and LEWIS · Outliers in Statistical Data

BARTHOLOMEW · Stochastic Models for Social Processes, *Second Edition*

BARTHOLOMEW and FORBES · Statistical Techniques for Manpower Planning

BECK and ARNOLD · Parameter Estimation in Engineering and Science

BELSLEY, KUH, and WELSCH · Regression Diagnostics: Identifying Influential Data and Sources of Collinearity

BENNETT and FRANKLIN · Statistical Analysis in Chemistry and the Chemical Industry

BHAT · Elements of Applied Stochastic Processes

BLOOMFIELD · Fourier Analysis of Time Series: An Introduction

BOX · R. A. Fisher, The Life of a Scientist

BOX and DRAPER · Evolutionary Operation: A Statistical Method for Process Improvement

BOX, HUNTER, and HUNTER · Statistics for Experimenters: An Introduction to Design, Data Analysis, and Model Building

BROWN and HOLLANDER · Statistics: A Biomedical Introduction

BROWNLEE · Statistical Theory and Methodology in Science and Engineering, *Second Edition*

BURY · Statistical Models in Applied Science

CHAMBERS · Computational Methods for Data Analysis

CHATTERJEE and PRICE · Regression Analysis by Example

HOEL · Elementary Statistics, *Fourth Edition*
HOLLANDER and WOLFE · Nonparametric Statistical Methods
HUANG · Regression and Econometric Methods
JAGERS · Branching Processes with Biological Applications
JESSEN · Statistical Survey Techniques
JOHNSON and KOTZ · Distributions in Statistics
    Discrete Distributions
    Continuous Univariate Distributions—1
    Continuous Univariate Distributions—2
    Continuous Multivariate Distributions
JOHNSON and KOTZ · Urn Models and Their Application: An Approach
    to Modern Discrete Probability Theory
JOHNSON and LEONE · Statistics and Experimental Design in Engineering
    and the Physical Sciences, Volumes I and II, *Second Edition*
JUDGE, GRIFFTHS, HILL · The Theory and Practice of Econometrics
KALBFLEISCH and PRENTICE · The Statistical Analysis of Failure Time
    Data
KEENEY and RAIFFA · Decisions with Multiple Objectives
LANCASTER · An Introduction to Medical Statistics
LEAMER · Specification Searches: Ad Hoc Inference with Nonexperimen-
    tal Data
McNEIL · Interactive Data Analysis
MANN, SCHAFER and SINGPURWALLA · Methods for Statistical Analysis
    of Reliability and Life Data
MEYER · Data Analysis for Scientists and Engineers
MILLER, EFRON, BROWN, and MOSES · Biostatistics Casebook
OTNES and ENOCHSON · Applied Time Series Analysis: Volume I, Basic
    Techniques
OTNES and ENOCHSON · Digital Time Series Analysis
POLLOCK · The Algebra of Econometrics
PRENTER · Splines and Variational Methods
RAO and MITRA · Generalized Inverse of Matrices and Its Applications
SARD and WEINTRAUB · A Book of Splines
SCHUSS · Theory and Applications of Stochastic Differential Equations
SEAL · Survival Probabilities: The Goal of Risk Theory
SEARLE · Linear Models
SPRINGER · The Algebra of Random Variables
THOMAS · An Introduction to Applied Probability and Random Processes
UPTON · The Analysis of Cross-Tabulated Data
WHITTLE · Optimization Under Constraints
WILLIAMS · A Sampler on Sampling
WONNACOTT and WONNACOTT · Econometrics, *Second Edition*
WONNACOTT and WONNACOTT · Introductory Statistics, *Third Edition*

WONNACOTT and WONNACOTT · Introductory Statistics for Business and Economics, *Second Edition*

ZELLNER · An Introduction to Bayesian Inference in Econometrics

Tracts on Probability and Statistics

BARNDORFF-NEILSEN · Information and Exponential Families in Statistical Theory

BHATTACHARYYA and RAO · Normal Approximation and Asymptotic Expansions

BIBBY and TOUTENBURG · Prediction and Improved Estimation in Linear Models

BILLINGSLEY · Convergence of Probability Measures

JARDINE and SIBSON · Mathematical Taxonomy

KELLY · Reversibility and Stochastic Networks

KINGMAN · Regenerative Phenomena

INDEX

Printed in the United States
121535LV00002B/37-40/A

9 780471 160687